Biological Activities of Alkaloids

Biological Activities of Alkaloids

From Toxicology to Pharmacology

Special Issue Editors

Sabino Aurelio Bufo
Linda L. Blythe
Zbigniew Adamski
Luigi Milella

MDPI • Basel • Beijing • Wuhan • Barcelona • Belgrade • Manchester • Tokyo • Cluj • Tianjin

Special Issue Editors
Sabino Aurelio Bufo
University of Basilicata
Italy

Linda L. Blythe
Oregon State University
USA

Zbigniew Adamski
Adam Mickiewicz University
Poland

Luigi Milella
University of Basilicata
Italy

Editorial Office
MDPI
St. Alban-Anlage 66
4052 Basel, Switzerland

This is a reprint of articles from the Special Issue published online in the open access journal *Toxins* (ISSN 2072-6651) (available at: https://www.mdpi.com/journal/toxins/special_issues/alkaloids_toxicology_pharmacology).

For citation purposes, cite each article independently as indicated on the article page online and as indicated below:

LastName, A.A.; LastName, B.B.; LastName, C.C. Article Title. *Journal Name* **Year**, *Article Number*, Page Range.

ISBN 978-3-03928-927-1 (Pbk)
ISBN 978-3-03928-928-8 (PDF)

Contents

About the Special Issue Editors vii

Preface to "Biological Activities of Alkaloids" ix

Zbigniew Adamski, Linda L. Blythe, Luigi Milella and Sabino A. Bufo
Biological Activities of Alkaloids: From Toxicology to Pharmacology
Reprinted from: *Toxins* **2020**, *12*, 210, doi:10.3390/toxins12040210 1

Filomena Lelario, Susanna De Maria, Anna Rita Rivelli, Daniela Russo, Luigi Milella, Sabino Aurelio Bufo and Laura Scrano
A Complete Survey of Glycoalkaloids Using LC-FTICR-MS and IRMPD in a Commercial Variety and a Local Landrace of Eggplant (*Solanum melongena* L.) and their Anticholinesterase and Antioxidant Activities
Reprinted from: *Toxins* **2019**, *11*, 230, doi:10.3390/toxins11040230 5

Anna Petruczynik, Tomasz Plech, Tomasz Tuzimski, Justyna Misiurek, Barbara Kaproń, Dorota Misiurek, Małgorzata Szultka-Młyńska, Bogusław Buszewski and Monika Waksmundzka-Hajnos
Determination of Selected Isoquinoline Alkaloids from *Mahonia aquifolia; Meconopsis cambrica; Corydalis lutea; Dicentra spectabilis; Fumaria officinalis; Macleaya cordata* Extracts by HPLC-DAD and Comparison of Their Cytotoxic Activity
Reprinted from: *Toxins* **2019**, *11*, 575, doi:10.3390/toxins11100575 23

Okiemute Rosa Johnson-Ajinwo, Alan Richardson and Wen-Wu Li
Palmatine from Unexplored *Rutidea parviflora* Showed Cytotoxicity and Induction of Apoptosis in Human Ovarian Cancer Cells
Reprinted from: *Toxins* **2019**, *11*, 237, doi:10.3390/toxins11040237 41

Anna Och, Daniel Zalewski, Łukasz Komsta, Przemysław Kołodziej, Janusz Kocki and Anna Bogucka-Kocka
Cytotoxic and Proapoptotic Activity of Sanguinarine, Berberine, and Extracts of *Chelidonium majus* L. and *Berberis thunbergii* DC. toward Hematopoietic Cancer Cell Lines
Reprinted from: *Toxins* **2019**, *11*, 485, doi:10.3390/toxins11090485 53

Jaime Ribeiro-Filho, Fagner Carvalho Leite, Andrea Surrage Calheiros, Alan de Brito Carneiro, Juliana Alves Azeredo, Edson Fernandes de Assis, Celidarque da Silva Dias, Márcia Regina Piuvezam and Patrícia T. Bozza
Curine Inhibits Macrophage Activation and Neutrophil Recruitment in a Mouse Model of Lipopolysaccharide-Induced Inflammation
Reprinted from: *Toxins* **2019**, *11*, 705, doi:10.3390/toxins11120705 73

Dmitry I. Osmakov, Sergey G. Koshelev, Victor A. Palikov, Yulia A. Palikova, Elvira R. Shaykhutdinova, Igor A. Dyachenko, Yaroslav A. Andreev and Sergey A. Kozlov
Alkaloid Lindoldhamine Inhibits Acid-Sensing Ion Channel 1a and Reveals Anti-Inflammatory Properties
Reprinted from: *Toxins* **2019**, *11*, 542, doi:10.3390/toxins11090542 85

Fang Zhao, Qinglian Tang, Jian Xu, Shuangyan Wang, Shaoheng Li, Xiaohan Zou and Zhengyu Cao
Dehydrocrenatidine Inhibits Voltage-Gated Sodium Channels and Ameliorates Mechanic Allodia in a Rat Model of Neuropathic Pain
Reprinted from: *Toxins* **2019**, *11*, 229, doi:10.3390/toxins11040229 97

Chih-Hsiang Chang, Mei-Chih Chen, Te-Huan Chiu, Yu-Hsuan Li, Wan-Chen Yu, Wan-Ling Liao, Muhammet Oner, Chang-Tze Ricky Yu, Chun-Chi Wu, Tsung-Ying Yang, Chieh-Lin Jerry Teng, Kun-Yuan Chiu, Kun-Chien Chen, Hsin-Yi Wang, Chia-Herng Yue, Chih-Ho Lai, Jer-Tsong Hsieh and Ho Lin
Arecoline Promotes Migration of A549 Lung Cancer Cells through Activating the EGFR/Src/FAK Pathway
Reprinted from: *Toxins* **2019**, *11*, 185, doi:10.3390/toxins11040185 . . . . **111**

Bruno Casciaro, Andrea Calcaterra, Floriana Cappiello, Mattia Mori, Maria Rosa Loffredo, Francesca Ghirga, Maria Luisa Mangoni, Bruno Botta and Deborah Quaglio
Nigritanine as a New Potential Antimicrobial Alkaloid for the Treatment of *Staphylococcus aureus*-Induced Infections
Reprinted from: *Toxins* **2019**, *11*, 511, doi:10.3390/toxins11090511 . . . . **127**

Sylwia Zielińska, Magdalena Wójciak-Kosior, Magdalena Dziągwa-Becker, Michał Gleńsk, Ireneusz Sowa, Karol Fijałkowski, Danuta Rurańska-Smutnicka, Adam Matkowski and Adam Junka
The Activity of Isoquinoline Alkaloids and Extracts from *Chelidonium majus* against Pathogenic Bacteria and *Candida* sp.
Reprinted from: *Toxins* **2019**, *11*, 406, doi:10.3390/toxins11070406 . . . . **153**

Amin Thawabteh, Salma Juma, Mariam Bader, Donia Karaman, Laura Scrano, Sabino A. Bufo and Rafik Karaman
The Biological Activity of Natural Alkaloids against Herbivores, Cancerous Cells and Pathogens
Reprinted from: *Toxins* **2019**, *11*, 656, doi:10.3390/toxins11110656 . . . . **167**

Paweł Marciniak, Angelika Kolińska, Marta Spochacz, Szymon Chowański, Zbigniew Adamski, Laura Scrano, Patrizia Falabella, Sabino A. Bufo and Grzegorz Rosiński
Differentiated Effects of Secondary Metabolites from *Solanaceae* and *Brassicaceae* Plant Families on the Heartbeat of *Tenebrio molitor* Pupae
Reprinted from: *Toxins* **2019**, *11*, 287, doi:10.3390/toxins11050287 . . . . **195**

Xiaoyu Ji, Mengbi Yang, Ka Hang Or, Wan Sze Yim and Zhong Zuo
Tissue Accumulations of Toxic Aconitum Alkaloids after Short-Term and Long-Term Oral Administrations of Clinically Used *Radix Aconiti Lateralis* Preparations in Rats
Reprinted from: *Toxins* **2019**, *11*, 353, doi:10.3390/toxins11060353 . . . . **209**

Sarah C. Finch, John S. Munday, Jan M. Sprosen and Sweta Bhattarai
Toxicity Studies of Chanoclavine in Mice
Reprinted from: *Toxins* , *11*, 249, doi:10.3390/toxins11050249 . . . . **231**

Rebecca K. Poole and Daniel H. Poole
Impact of Ergot Alkaloids on Female Reproduction in Domestic Livestock Species
Reprinted from: *Toxins* **2019**, *11*, 364, doi:10.3390/toxins11060364 . . . . **243**

About the Special Issue Editors

Sabino Aurelio Bufo (Prof.) is a Full Professor of Soil and Agricultural Chemistry at the University of Basilicata, Department of Sciences; coordinator of the International Ph.D. Program Applied Biology & Environmental Safeguard. His current research topics include pesticide chemistry and biochemistry, chemistry of natural substances in plants and soil, pharmaceuticals from plants and microorganisms, and the fate of xenobiotics in the environment.

Linda L. Blythe received her Doctor of Veterinary Medicine in 1974 and Ph.D. in 1978 from UC Davis, USA. After graduation, she was recruited to the newly funded College of Veterinary Medicine at Oregon State University, Corvallis, Oregon, USA. She has taught veterinary neurology, toxicology, and sports medicine for 39 years to veterinary students and has received a number of awards for her teaching.

Zbigniew Adamski, Ph.D. works at the Faculty of Biology, Adam Mickiewicz University in Poznań, Poland. He shares his time between the Electron and Confocal Microscope Laboratory, and the Department of Animal Physiology and Development, implementing electron microscopy in research on the activity of natural and synthetic pesticides against insect pests and nematodes.

Luigi Milella is aggregate Professor in Pharmaceutical Biology SSD BIO15, Faculty of Pharmacy, University of Basilicata (Italy), where he carries out research focused on the study of the pharmacological activity, chemical structure, and biosynthesis of natural compounds coming from different sources, either foods, marine sources, or medicinal plant species.

Preface to "Biological Activities of Alkaloids"

Plants produce many substances, including the secondary metabolites that have biological activity. They are distinct from the components of the primary metabolism as they are generally not essential for the basic metabolic processes of plants. However, they are often physiologically active compounds that are applicable in different fields, for instance medicine or agriculture. A broad spectrum of physiological activity is demonstrated by alkaloids. Their rich diversity results in part from an evolutionary process driven by selection for the acquisition of an improved defense against microbial attacks or the predation of herbivores. Their main role in plants is to protect them from diseases caused by pests. However, some alkaloids are of concern to veterinary toxicology due to their occurrence in plant species involved in animal poisoning, which usually occurs when plants contaminate hay or silage or when forage alternatives are not available. At times, some toxicity effects have been highlighted in human nutrition. Other components of this class of compounds exhibit antioxidant, anti-inflammatory, anti-aggregation, hypo-cholesteric, immunostimulant, or anticancer properties. The effects of toxicity can be both harmful and beneficial depending on the ecological or pharmacological context, and, as often reported, are dose-dependent. Researchers remain keenly interested in the study of the bioactivities of plant alkaloids.

In this Special Issue, the ecological, biological, pharmacological, and toxicological effects, as well as structural and analytical aspects of plant alkaloids, are collected. Reviews, original research articles on alkaloid biosynthesis and action mechanisms, metabolism and accumulation of alkaloids, studies describing the biological activity of natural alkaloids against pathogens, herbivores and cancerous cells, as well as research describing molecular and cytological mode of action of alkaloids, are presented in this Issue.

Sabino Aurelio Bufo, Linda L. Blythe, Zbigniew Adamski, Luigi Milella
Special Issue Editors

Editorial

Biological Activities of Alkaloids: From Toxicology to Pharmacology

Zbigniew Adamski [1,*], Linda L. Blythe [2], Luigi Milella [3] and Sabino A. Bufo [3,*]

[1] Department of Animal Physiology and Development/Electron and Confocal Microscope Laboratory, Faculty of Biology, Adam Mickiewicz University, 61-614 Poznań, Poland

[2] Department of Veterinary Medicine, Oregon State University, Corvallis, 97331 OR, USA; Linda.Blythe@oregonstate.edu

[3] Department of Science, University of Basilicata, 85100 Potenza, Italy; luigi.milella@unibas.it

* Correspondence: zbigniew.adamski@amu.edu.pl (Z.A.); sabino.bufo@unibas.it (S.A.B.)

Received: 24 February 2020; Accepted: 25 March 2020; Published: 26 March 2020

Plants produce many secondary metabolites, which reveal biological activity. Among them, alkaloids demonstrate a broad spectrum of activities. In nature, they not only are produced against herbivores but also reduce bacterial or fungal infestation. Therefore, they are substances that possess high potential in medicine, plant protection, veterinary, or toxicology. Hence, the research on these substances and their properties develops intensively in many areas. The studies describing the physiological, pharmacological, and toxicological activity of alkaloids for different organisms belonging to every kingdom are of very wide interest. Both pure alkaloids and extracts are studied, and their activities are compared. In the Special Issue "Biological Activities of Alkaloids: From Toxicology to Pharmacology", 15 manuscripts describing ecological, biological, pharmacological, and toxicological effects as well as structural and analytical aspects of plant alkaloids, their mode of action, and possible application in veterinary, medicine, and plan protection were collected. The subjects focused on two main areas of interest, the structure/activity nexus and the application of alkaloids against pathogens.

Although the number of research articles on alkaloids increases, our knowledge of them is still far from completeness. This is due to the very high number of alkaloids produced by many different organisms, mostly plants, diffused all over the world. Therefore, the identification, characterization, and quantification of alkaloids present in plant species and their parts is very important and brings interesting data [1,2]. The spectrum of alkaloids' activity is also very wide. Among them, there are substances showing antiviral, antibacterial, anti-inflammatory, and anticancer properties. Thus, many studies deal with curative aspects of alkaloids and their mode of action. *Mahonia aquifolia*, *Meconopsis cambrica*, *Corydalis lutea*, *Dicentra spectabilis*, *Fumaria officinalis*, and *Macleaya cordata* plant extracts showed cytotoxic activity against the tested human squamous carcinoma and adenocarcinoma cells [1]. The extracts obtained from the stem bark of *Rutidea parviflora* (*R. parviflora*) revealed significant cytotoxic activity against ovarian cancer. In this study, palmatine from the stem bark of *R. parviflora* was more toxic for human ovarian cancer cells than for human ovarian noncancerous cells [3]. Such basic studies are necessary and determine a very important point for the development of new anticancer drugs and therapies. In addition, sanguinarine and berberine, the isoquinoline alkaloids, revealed cytotoxic activity against hematopoietic cancer cell lines and induced apoptosis in the tested cell lines [4]. Curine—a bisbenzylisoquinoline alkaloid—was proven to modulate inflammatory effects in mice, due to the inhibition of macrophage activation and neutrophil recruitment, the inhibition of the production of cytokines and the decreased level of nitric oxide. The effects may be probably linked to the decreased level of nitric oxide and induced possibly by negatively modulating a Ca^{2+} influx [5]. The regulatory mode of the action of alkaloids refers also to other mechanisms within cellular membranes. Lindoldhamine (a bisbenzylisoquinoline alkaloid) was shown as a novel antagonist of acid-sensing ion channels (ASICs). Lindoldhamine significantly inhibited the ASIC1a channel's response to physiologically relevant stimuli [6]. This observation is especially important, since only some molecules were described as modulators of ASIC1. That opens a new research area about bisbenzylisoquinoline alkaloids as important molecules in neurobiology. On the other hand, dehydrocrenatidine, a β-carboline alkaloid, suppresses voltage-gated sodium channels and leads to decreased allodynia. The alkaloid is the main component of *Picrasma quassioides*—a plant used in medicine, since it reveals antiviral activity, which is also known as an anti-inflammatory and analgesic agent. The research of Zhao and co-workers [7] brought important data on the mode of the action of this alkaloid.

Unfortunately, not all gold glitters: the consumption of some alkaloids may lead to toxic effects. Among them, there is arecoline, an alkaloid found for example in betel nuts. Overconsumption may lead to cancerogenesis and tumor formation. The mechanism of this effect is not fully known. Chang and co-workers described important aspects of the cancerogenic activity of arecoline [8]. The authors postulated that the mechanism uses a muscarinic acetylcholine receptor and the pathway that is triggered by the activation of this receptor. The authors described the effects of arecoline on cell migration and actin organization. The studies of that type may appear to be very important from the cytotoxicological, pharmacological, and clinical points of view.

Not only are cancer cells susceptible to alkaloids. The antiviral and antibacterial activity of alkaloids has already been described. This area of research appears to be important especially in the light of increasing the resistance of pathogenic bacteria to antibiotics. Casciaro and his co-workers presented an interesting study showing that nigritanine, an alkaloid obtained from *Strychnos nigritana*—a flowering plant that belongs to the family of Loganiaceae - possess high antibacterial activity against *Staphylococcus aureus* (*S. aureus*), which is recognised to be one of the most important pathogenic bacteria diffused worldwide [9]. What appeared extremely important is the tested alkaloid did not reveal significant toxicity for mammalian red blood cells and human keratinocytes. The authors compared also the monomer/dimer structure–antibacterial activity relationship, which brought important information on the mechanism of activity against *S. aureus*. The research presented by Zielińska and her colleagues [10] included them in the same area of research. The authors showed a range of research on the presence of alkaloids in organs of *Chelidonium majus* and combined these observations with the activity of extracts and single metabolites against certain microorganisms: *S. aureus*, *Pseudomonas aeruginosa*, *Klebsiella pneumonia*, *Escherichia coli*, and *Candida albicans*. The results are in tune with the abovementioned research of Casciaro et al. [9] due to the described overall lower toxicity against eukaryotic cells (fibroblasts) than against microorganisms.

However, there are alkaloids that reveal toxic activity against animals. This seems obvious, since one of their main roles is to deter herbivory. Therefore, the wide range of alkaloids is described not only as substances with antimicrobial or anticancer agents but also as substances revealing insecticidal activity [11]. However, the nature of the toxic action of alkaloids on insects is still insufficiently described. In this issue, the effects of the activity of crude extracts obtained from *Solanum tuberosum*, *Solanum lycopersicum*, *Solanum nigrum* (Solanaceae), and *Armoracia rusticana* (Brassicaceae), as well as purified alkaloids, on the heart contractility of *Tenebrio molitor*—a pest of stored products—have been described [12]. In this research, chaconine was stated to be the most cardioactive substance among those tested. Apart from the information on the activity of alkaloids in insect science, the investigation methods issued in this kind of research can be of interest in medical research. Due to economical and ethical reasons, invertebrates, including insects, became important models in the first stage of drug designing.

The pharmacological ranges of concentrations and toxic levels are often close. Therefore, emphasis must be put on concentrations and doses, which may cause lethal and sublethal effects in mammals. This is important in the case of substances that are used in plant protection, food preservation, and hygiene of storage chambers and containers. From the human point of view, the toxic activity of substances, which are used as medicines, is equally, if not more important. Aconitum alkaloids are used in ethnomedicine and modern medicine, and their toxicity may be lethal for mammals. The data on the distribution of toxic alkaloids within the organs of the exposed individual is crucial for clinical toxicology [13]. In addition, some endophytes, like *Epichloe*, produce secondary metabolites that are toxic to insects. Therefore, they are potential sources of insecticides. Chanoclavine, an ergot alkaloid, was tested by Finch and co-workers against mice, to estimate their toxicity for a mammal model organism [14]. Although the mice revealed some neurotoxic symptoms, they were not permanent, and the median lethal dose was higher than 2000 mg per kg body weight. That suggested that the substance is relatively safe for mammals. However, further research is necessary, due to the reported toxicity of ergot alkaloids to mammals, including human. Additionally, the livestock that consumes ergot alkaloids shows various toxic symptoms, including endocrine disruption, reproductive and developmental malfunctions, and blood circulation [15]. The two review manuscripts present in this Special Issue proved the need for further extensive studies on the activity of alkaloids [11,15].

All the abovementioned studies proved the enormous potential of alkaloids in veterinary, pharmacology, medicine, and plant protection. Additionally, they showed multifold aspects of alkaloids and alkaloid-containing extracts toxicity from cytotoxicity through the malfunctions of organs and systems to lethal effects. Due to the increasing resistance of bacteria to antibiotics, they may become crucial for fighting microbial diseases. The description of postulated metabolic pathways influenced by the tested substances appeared to be very important for the planning of possible drugs in veterinary and medicine, as well as for basic science, like neurobiology or cell physiology. Similarly to bacteria developing resistance to antibiotics, insects develop resistance to insecticides. Hence, there is a need for new formulas, which may fight herbivore insects, with high selectivity against pests. Alkaloids are among the substances that are postulated as such novel insecticides. To sum up, the scientific and applicatory potential of alkaloids is immense. The research on their structure and activity develops intensively in various fields of science, which was proven by the variety of research topics present in this Special Issue. For sure, the number of research papers showing interesting and applicable pharmacological and toxicological aspects of alkaloids' activity will be increasing.

References

1. Lelario, F.; De Maria, S.; Rivelli, A.R.; Russo, D.; Milella, L.; Bufo, S.A.; Scrano, L. A Complete Survey of Glycoalkaloids Using LC-FTICR-MS and IRMPD in a Commercial Variety and a Local Landrace of Eggplant (*Solanum melongena* L.) and their Anticholinesterase and Antioxidant Activities. *Toxins* **2019**, *11*, 230. [CrossRef]
2. Petruczynik, A.; Plech, T.; Tuzimski, T.; Misiurek, J.; Kaproń, B.; Misiurek, D.; Szultka-Młyńska, M.; Buszewski, B.; Waksmundzka-Hajnos, M. Determination of Selected Isoquinoline Alkaloids from *Mahonia aquifolia; Meconopsis cambrica; Corydalis lutea; Dicentra spectabilis; Fumaria officinalis; Macleaya cordata* Extracts by HPLC-DAD and Comparison of Their Cytotoxic Activity. *Toxins* **2019**, *11*, 575. [CrossRef]
3. Johnson-Ajinwo, O.R.; Richardson, A.; Li, W.-W. Palmatine from Unexplored *Rutidea parviflora* Showed Cytotoxicity and Induction of Apoptosis in Human Ovarian Cancer Cells. *Toxins* **2019**, *11*, 237. [CrossRef]
4. Och, A.; Zalewski, D.; Komsta, Ł.; Kołodziej, P.; Kocki, J.; Bogucka-Kocka, A. Cytotoxic and Proapoptotic Activity of Sanguinarine, Berberine, and Extracts of *Chelidonium majus* L. and *Berberis thunbergii* DC. toward Hematopoietic Cancer Cell Lines. *Toxins* **2019**, *11*, 485. [CrossRef]
5. Ribeiro-Filho, J.; Carvalho Leite, F.; Surrage Calheiros, A.; de Brito Carneiro, A.; Alves Azeredo, J.; Fernandes de Assis, E.; da Silva Dias, C.; Regina Piuvezam, M.T.; Bozza, P. Curine Inhibits Macrophage Activation and Neutrophil Recruitment in a Mouse Model of Lipopolysaccharide-Induced Inflammation. *Toxins* **2019**, *11*, 705. [CrossRef]
6. Osmakov, D.I.; Koshelev, S.G.; Palikov, V.A.; Palikova, Y.A.; Shaykhutdinova, E.R.; Dyachenko, I.A.; Andreev, Y.A.; Kozlov, S.A. Alkaloid Lindoldhamine Inhibits Acid-Sensing Ion Channel 1a and Reveals Anti-Inflammatory Properties. *Toxins* **2019**, *11*, 542. [CrossRef] [PubMed]
7. Zhao, F.; Tang, Q.; Xu, J.; Wang, S.; Li, S.; Zou, X.; Cao, Z. Dehydrocrenatidine Inhibits Voltage-Gated Sodium Channels and Ameliorates Mechanic Allodia in a Rat Model of Neuropathic Pain. *Toxins* **2019**, *11*, 229. [CrossRef] [PubMed]

8. Chang, C.-H.; Chen, M.-C.; Chiu, T.-H.; Li, Y.-H.; Yu, W.-C.; Liao, W.-L.; Oner, M.; Yu, C.-T.R.; Wu, C.-C.; Yang, T.-Y.; et al. Arecoline Promotes Migration of A549 Lung Cancer Cells through Activating the EGFR/Src/FAK Pathway. *Toxins* **2019**, *11*, 185. [CrossRef] [PubMed]
9. Casciaro, B.; Calcaterra, A.; Cappiello, F.; Mori, M.; Loffredo, M.R.; Ghirga, F.; Mangoni, M.L.; Botta, B.; Quaglio, D. Nigritanine as a New Potential Antimicrobial Alkaloid for the Treatment of *Staphylococcus aureus*-Induced Infections. *Toxins* **2019**, *11*, 511. [CrossRef] [PubMed]
10. Zielińska, S.; Wójciak-Kosior, M.; Dziągwa-Becker, M.; Gleńsk, M.; Sowa, I.; Fijałkowski, K.; Rurańska-Smutnicka, D.; Matkowski, A.; Junka, A. The Activity of Isoquinoline Alkaloids and Extracts from *Chelidonium majus* against Pathogenic Bacteria and *Candida* sp. *Toxins* **2019**, *11*, 406. [CrossRef]
11. Thawabteh, A.; Juma, S.; Bader, M.; Karaman, D.; Scrano, L.; Bufo, S.A.; Karaman, R. The Biological Activity of Natural Alkaloids against Herbivores, Cancerous Cells and Pathogens. *Toxins* **2019**, *11*, 656. [CrossRef] [PubMed]
12. Marciniak, P.; Kolińska, A.; Spochacz, M.; Chowański, S.; Adamski, Z.; Scrano, L.; Falabella, P.; Bufo, S.A.; Rosiński, G. Differentiated Effects of Secondary Metabolites from *Solanaceae* and *Brassicaceae* Plant Families on the Heartbeat of *Tenebrio molitor* Pupae. *Toxins* **2019**, *11*, 287. [CrossRef] [PubMed]
13. Ji, X.; Yang, M.; Or, K.H.; Yim, W.S.; Zuo, Z. Tissue Accumulations of Toxic Aconitum Alkaloids after Short-Term and Long-Term Oral Administrations of Clinically Used *Radix Aconiti Lateralis* Preparations in Rats. *Toxins* **2019**, *11*, 353. [CrossRef] [PubMed]
14. Finch, S.C.; Munday, J.S.; Sprosen, J.M.; Bhattarai, S. Toxicity Studies of Chanoclavine in Mice. *Toxins* **2019**, *11*, 249. [CrossRef] [PubMed]
15. Poole, R.K.; Poole, D.H. Impact of Ergot Alkaloids on Female Reproduction in Domestic Livestock Species. *Toxins* **2019**, *11*, 364. [CrossRef] [PubMed]

Article

A Complete Survey of Glycoalkaloids Using LC-FTICR-MS and IRMPD in a Commercial Variety and a Local Landrace of Eggplant (*Solanum melongena* L.) and their Anticholinesterase and Antioxidant Activities

Filomena Lelario [1], Susanna De Maria [2], Anna Rita Rivelli [2], Daniela Russo [1,3,*], Luigi Milella [1,3,*], Sabino Aurelio Bufo [1,4] and Laura Scrano [5]

1 Department of Sciences, University of Basilicata, Via dell'Ateneo Lucano, 85100 Potenza (PZ), Italy; filomena.lelario@unibas.it (F.L.); sabino.bufo@unibas.it (S.A.B.)
2 School of Agricultural, Forest, Food and Environmental Sciences, University of Basilicata, Via dell'Ateneo Lucano, 85100 Potenza (PZ), Italy; demariasusanna@libero.it (S.D.M.); annarita.rivelli@unibas.it (A.R.R.)
3 Spinoff Accademico BioActiPlant, Via dell'Ateneo Lucano, 85100 Potenza (PZ), Italy
4 Department of Geography, Environmental Management & Energy Studies, University of Johannesburg, Auckland Park Kingsway Campus, Johannesburg 2092, South Africa
5 Department of European and Mediterranean Cultures, University of Basilicata, Via San Rocco, 75100 Matera, Italy; laura.scrano@unibas.it
* Correspondence: daniela.russo@unibas.it (D.R.); luigi.milella@unibas.it (L.M.); Tel.: +39-0971-205-525 (L.M.)

Received: 15 March 2019; Accepted: 15 April 2019; Published: 19 April 2019

Abstract: Eggplant contains glycoalkaloids (GAs), a class of nitrogen-containing secondary metabolites of great structural variety that may have both adverse and beneficial biological effects. In this study, we performed a complete survey of GAs and their malonylated form, in two genotypes of eggplants: A commercial cultivated type, Mirabella (Mir), with purple peel and bitter taste and a local landrace, named *Melanzana Bianca di Senise* (Sen), characterized by white peel with purple strip and a typical sweet aroma. Besides the analysis of their morphological traits, nineteen glycoalkaloids were tentatively identified in eggplant berry extracts based upon LC-ESI-FTICR-MS analysis using retention times, elution orders, high-resolution mass spectra, as well as high-resolution fragmentation by IRMPD. The relative signal intensities (i.e., ion counts) of the GAs identified in Mir and Sen pulp extracts showed as solamargine, and its isomers are the most abundant. In addition, anticholinesterase and antioxidant activities of the extracts were evaluated. Pulp tissue was found to be more active in inhibiting acetylcholinesterase enzyme than peel showing an inhibitory effect higher than 20% for Mir pulp. The identification of new malonylated GAs in eggplant is proposed.

Keywords: *Solanum melongena* L.; malonylated form; glycoalkaloids; secondary metabolites; solasonine; solamargine; malonyl-solamargine; acetylcholinesterase; antioxidant

Key Contribution: Structural characterization of nineteen glycoalkaloids (GAs) and malonyl-GAs in eggplant berries was performed successfully by LC-ESI-FTICR-MS and infrared multiphoton dissociation. Anticholinesterase and antioxidant activities of plant extracts were evaluated.

---

1. Introduction

Solanum melongena L., commonly known as eggplant or aubergine, is an economically important vegetable crop, belonging to the Solanaceae family, growing in tropical and temperate areas. It is the most widely consumed vegetable together with tomatoes and potatoes. The global production of eggplant has largely increased, reaching 52.3 million tons in 2017 [1]. This species includes a large number of commercial cultivars or varieties and local landraces that produce fruits differing in shape (ovoid, oblong, cylindrical, club-shaped), colour (purple, green, purple with white stripes) and size [2]. However, the elongated ovoid fruit with a dark purple/black peel is today the most-marketed worldwide, obtaining a general acceptance of its high nutritional value [3]. Eggplant is an inexpensive low-fat food source, providing energy, protein, fibre and vitamins, but it is actually studied as a source of health-promoting metabolites, including antioxidant and nutraceutical compounds, mainly anthocyanins and chlorogenic acid [4].

Moreover, eggplant also contains glycoalkaloids and saponins, which are responsible for the typical bitter taste of the pulp and are usually considered as anti-nutritional compounds, and are potentially toxic for humans as they can affect the absorption of nutrients [5]. Glycoalkaloids (GAs), a class of nitrogen-containing secondary metabolites, are commonly found in the Solanaceae family and play an important role in the defence of the plant against pests [6]. Although toxic for human health at certain levels, GAs also exhibit a wide range of pharmacological properties, including anticancer activity [7–10]. Many GAs exhibit acetylcholinesterase (AChE) inhibitory activity, which is associated with the treatment of several diseases such as Alzheimer's disease (AD), *Myasthenia gravis*, and glaucoma as well as the mechanisms of insecticidal activity and anthelmintic drugs [11]. Furthermore, it has been demonstrated that the progression of neurodegenerative diseases is also related to the oxidative stress mechanism [12]. Several natural compounds have been shown to be useful tools for preventing oxidative stress and its damages and many plants have been studied, by different approaches, for the identification of new acetylcholinesterase inhibitors (AChE-Is). Both non-alkaloids and alkaloid-derivative compounds have been demonstrated to be new potential lead compounds for AD treatment [13].

Thus, in recent years, medicinal uses of GAs have been the focus of scientific and pharmacological attention and their identification in plants has become a topic of increasing interest. However, nowadays most of the studies focus only on the presence of major GAs in each species, although it has been reported that accessions of the same species can have different GAs patterns (Figure 1) [14].

For common eggplant, only the two major GAs are usually reported and analysed: The spirosolane-type GAs solasonine and solamargine. Such GAs are structurally similar compounds and share the same aglycone, the solasodine, but differ in their carbohydrate component. They contain either glucose (solamargine) or galactose (solasonine) as the primary glycosylic residue. Both GAs, reported also in pepper, possess anti-proliferative activity on many human tumour cells [4].

Recently, some authors showed the occurrence of minor GAs and malonyl-GAs in Solanaceae plants. In particular, Wu et al. [15] showed the presence in *Solanum melongena* of minor GAs solanandaine, robeneoside B and 3′ or 6′ malonyl-solamargine by using reversed-phase LC-TOF-MS methods. The occurrence of a malonylated form of GAs in eggplant was also reported by Docimo et al. [16], which tentatively identified this compound as malonyl-solamargine based on the retention time data and by matching the MS/MS spectra with those reported in the *S. melongena* secondary metabolite database.

Nevertheless, a complete survey of glycoalkaloids and their malonylated form in the eggplant still needs to be recognized.

Several methods have been proposed for the identification and quantification of GAs in different species [17–21]. The presence of only small structural differences among various GAs requires the use of accurate and reproducible methods to identify and efficiently characterize them [22,23]. Electrospray ionization Fourier transform ion cyclotron resonance mass spectrometry (ESI-FTICR-MS) provides a highly selective tool for the unambiguous identification of molecules, which can be extended to minor components without significant interferences from other compounds in plant extracts [24–26]. In order to obtain a high degree of GAs structural information, infrared multiphoton dissociation (IRMPD) is widely used as the method of excellence, since it produces a larger number of fragments [27].

The hypothesis of this study was to verify the differences, by using a high-resolution LC-ESI-FTICR-MS method and IRMPD ion fragmentations, of the entire family of GAs and their malonylated forms in extracts from two types of eggplant grown in Mediterranean area beside a comparison of their Acetylcholinesterase (AChE) inhibitory and antioxidant activities. A commercial variety characterized by purple peel and bitter taste, and a local landrace, the *Melanzana Bianca di Senise*, recently included in the Traditional National Food Products (by the Italian Ministry of Agriculture Decree No. 168 issued on 17 June 2015), locally consumed and appreciated for the intense and fruity aroma of the berry, the sweetness and delayed turning brown of the pulp after cutting, have been used for this study.

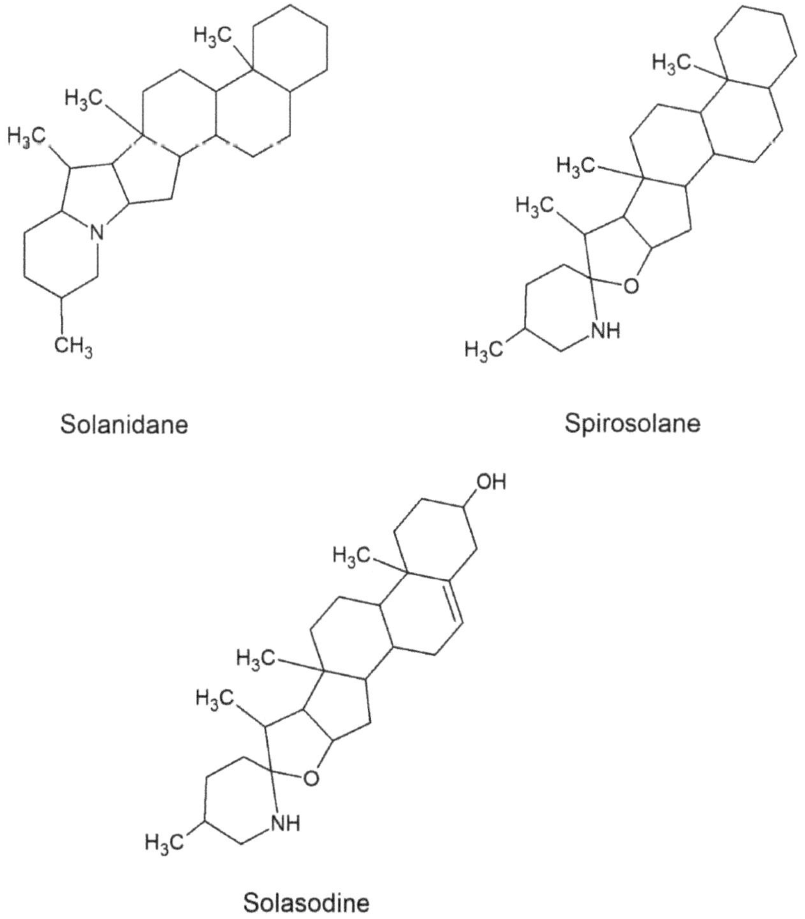

Figure 1. Basic structural formulas of the most common glycoalkaloid aglycons of *S. melongena*.

2. Results and Discussion

2.1. Acetylcholinesterase Inhibition of Eggplant Extracts

The Solanaceae family contains members relevant to human nutrition and health. These include peppers, eggplant, tomato and potato as well as black nightshade and jimson weed seeds and tobacco. These plants produce different classes of compounds including alkaloids and glycoalkaloids (GAs). GAs exhibit also a wide range of pharmacological properties, including anticancer activity or acetylcholinesterase (AChE) inhibitory activity [7,8]. This latter is associated with the treatment of several diseases such as Alzheimer's (AD) or Parkinson diseases.

In this study, freeze-dried pulp and peel of two varieties of eggplant genotypes of *Solanum melongena* (Mirabella and Melanzana Bianca di Senise) were extracted by 1% (*v/v*) aqueous acetic acid solution. This extraction solution was used to recovery mainly glycoalkaloid compounds from plant tissues as reported by other studies [28]. Many species belonging to the Solanaceae family reported AChE inhibitory properties, but the importance of the chemical structure and the heterocyclic nitrogen of steroidal alkaloids play an important role in AChE inhibition [12]. Results of our study reported that peel and pulp extracts have a mild acetylcholinesterase inhibitory activity (Figure 2); no butyrylcholinesterase inhibition was shown at tested concentrations. The inhibition activity of the extracts was expressed as the % of inhibition at 5 mg/mL. Galanthamine was used as the reference standard and at the same concentration of extracts; it showed 100% AChE inhibition. Pulp tissue was found to be more active in inhibiting acetylcholinesterase enzyme than peel in Mirabella (Mir) sample ($p < 0.05$), so this part has been used for further analysis.

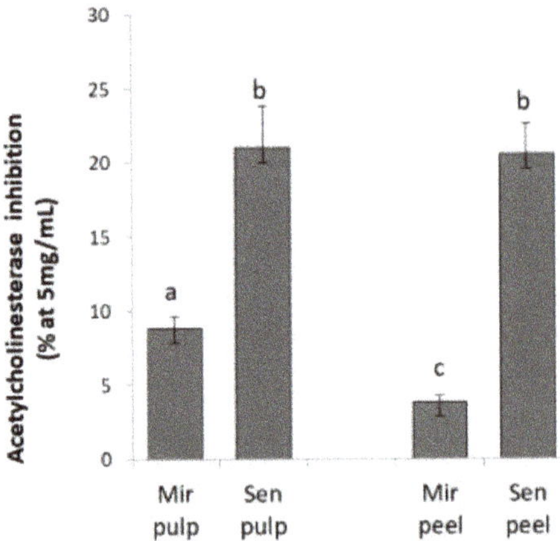

Figure 2. Inhibition of acetylcholinesterase (AChE) enzyme by pulp and peel aqueous extracts (1% acetic acid) from Senise (Sen) and common eggplant (Mir). Different letters indicate significant differences between mean values of a particular index of the given species $p < 0.05$ (according to Tukey's test).

2.2. GAs Profile of S. melongena var. Mirabella Pulp Extracts

Due to the diversity of GAs in plants, which is considerably greater than previously thought, there is a demand to improve GAs identification methods. The direct analysis of secondary metabolites in plant extracts by reverse-phase liquid chromatography (LC) with electrospray ionization (ESI) and FTICR-MS has shown to be feasible in conjunction with IRMPD as a structural elucidation and/or confirmation tool [24–26,29].

The present work extends the previous efforts of investigators to elucidate the GAs profile of *Solanum melongena* L., which takes advantage of an optimized LC-ESI-FTICR-MS method. Separation and subsequent identification of GAs and their malonylated form were achieved upon direct extraction, using an aqueous acidified solution, and high-resolution mass spectral analysis of putative compounds [15,27]. Firstly, most naturally occurring GAs and malonyl-GAs of eggplant extracts were examined and characterized by MS and IRMPD MS^2, then a comparison between the two genotypes was accomplished.

Several minor GAs could be displayed by LC-ESI-FTICR-MS in positive ion mode through the narrow window extracted ion chromatograms (XICs) of each compound (± 1 mDa) from the complex matrix of berry pulp. This strategy decreased the background or co-eluted interferences in the chromatographic peaks. Surprisingly, XICs data analysis of peel extracts did not show the presence of a detectable level of GAs, so our study focused on pulp extracts.

A representative XICs obtained for a Mirabella pulp extract is reported in Figure 3A. As shown in Figure 3A, sixteen main peaks corresponding to eleven GAs and five malonyl-GAs have been identified with accurate monoisotopic values.

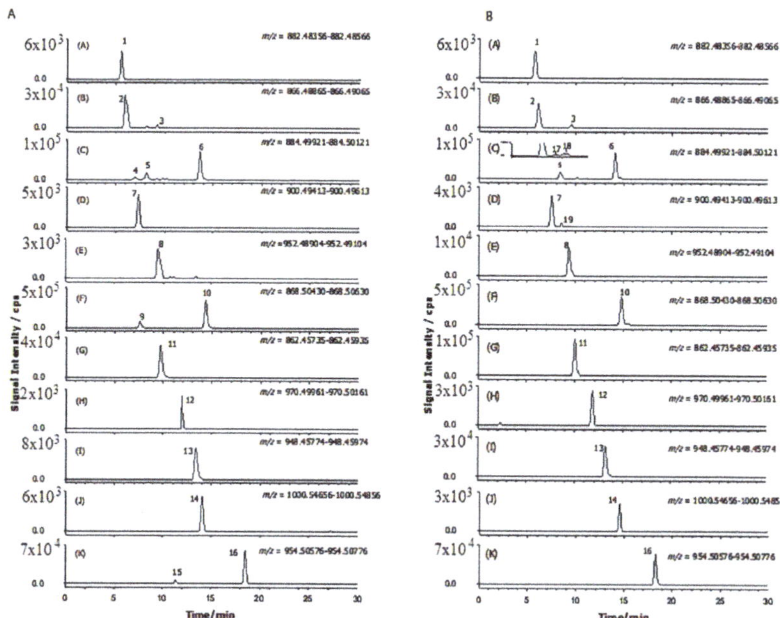

Figure 3. Extracted ion chromatograms (LC-ESI-FTICR-MS, high resolution) to the respective molecular ion peak $[M+H]^+$ acquired in positive mode of "Mirabella" (**A**) and "Melanzana di Senise" (**B**) pulp extracts. The ion monitored are displayed in each trace (plots A–K) and peak numbers. Peak numbers are (1) solanidenetriol chacotriose, (2) solanidenediol chacotriose, (3) dehydrosolamargine, (4) solanandaine isomer I, (5) solanandaine, (6) solasonine (7) robenoside B, (8) malonyl-solanidenediol chacotriose, (9) solamargine isomer, (10) solamargine, (11) solanidatetraenol chacotriose, (12) malonyl-solanandaine, (13) malonyl-solanidatetraenol chacotriose, (14) arudonine, (15) malonyl-solamargine isomer, (16) malonyl-solamargine, (17) solanandaine isomer II, (18) solanandaine Isomer III, (19) robenoside B isomer.

In Table 1, the common name, molecular formula of protonated compounds, monoisotopic exact value as $[M+H]^+$ ion and retention time of all detected GAs and malonylated GAs (listed with the same peak number used to identify each compound in the XICs) are reported; the examined *Solanum melongena* genotype is included as well.

Interestingly, accurate mass data of GAs and malonylated-GAs were found as protonated molecules, with a mass error lower than 1.3 ppm, suggesting a very good mass accuracy.

Among all types of GAs and malonyl-GAs found in pulp extracts, only solasonine and solamargine were identified using commercial standard compounds. Identification of the other compounds was based on chromatographic behaviour, accurate mass measurements, IRMPD MS^2 analyses and comparison with data from literature.

As frequently encountered for most plant secondary metabolites, rich fragmentation patterns were produced during ionization of GAs and malonyl-GAs, which provide sufficient resolution for a priori structure elucidation. By analysing the chromatographic behaviour and the MS and MS^2

spectra of compounds 8, 12, 13, 15 and 16 according to related literature [15,27,30], we can assume that these compounds are malonylated form of compounds 2, 5, 11, 9, 10, respectively. Compared with non-malonylated GAs, their chromatographic behaviour is characterized by higher retention time (Table 1).

Table 1. Peak number, common name, molecular formulae, monoisotopic exact value and retention time of the glycoalkaloids detected in the two eggplant genotypes analysed.

Peak Number	Common Name	Molecular Formulae	Monoisotopic Exact Value [M+H]+ (*m/z*) (Δm) [a]	Retention Time (Rt, min)	Genotype
1	Solanidenetriol chacotriose	$C_{45}H_{71}NO_{16}$	882.48456 (−0.4)	5.6	Mir, Sen
2	Solanidenediol chacotriose	$C_{45}H_{71}NO_{15}$	866.48965 (0.5)	6.0	Mir, Sen
3	Dehydrosolamargine	$C_{45}H_{71}NO_{15}$	866.48965 (−0.2)	9.2	Mir, Sen
4	Solanandaine isomer I	$C_{45}H_{73}NO_{16}$	884.50021 (0.4)	7.0	Mir, /
5	Solanandaine	$C_{45}H_{73}NO_{16}$	884.50021 (−0.4)	8.2	Mir, Sen
6	Solasonine (spirosolenol solatriose)	$C_{45}H_{73}NO_{16}$	884.50021 (−0.2)	13.6	Mir, Sen
7	Robenoside B (solanidenediol chacotriose)	$C_{45}H_{73}NO_{17}$	900.49513 (0.6)	7.3	Mir, Sen
8	Malonyl-solanidenediol chacotriose	$C_{48}H_{73}NO_{18}$	952.49004 (−0.5)	9.5	Mir, Sen
9	Solamargine Isomer (spirosolenol chacotriose)	$C_{45}H_{73}NO_{15}$	868.50530 (1.2)	7.6	Mir, /
10	Solamargine (spirosolenol chacotriose)	$C_{45}H_{73}NO_{15}$	868.50530 (1.3)	14.3	Mir, Sen
11	Solanidatetraenol chacotriose	$C_{45}H_{67}NO_{15}$	862.45835 (0.3)	9.7	Mir, Sen
12	Malonyl-solanandaine	$C_{48}H_{76}NO_{19}$	970.50061 (0.7)	11.9	Mir, Sen
13	Malonyl-solanidatetraenol chacotriose	$C_{48}H_{69}NO_{18}$	948.45874 (−0.9)	13.4	Mir, Sen
14	Arudonine	$C_{50}H_{82}NO_{19}$	1000.54756 (0.2)	14.1	Mir, Sen
15	Malonyl-solamargine isomer	$C_{48}H_{76}NO_{18}$	954.50569 (1.2)	11.3	Mir, /
16	Malonyl-solamargine	$C_{48}H_{76}NO_{18}$	954.50569 (−0.5)	18.5	Mir, Sen
17	Solanandaine Isomer II (spirosolendiol chacotriose)	$C_{45}H_{73}NO_{16}$	884.50021 (−0.9)	9.5	Mir, Sen
18	Solanandaine Isomer III (spirosolendiol chacotriose)	$C_{45}H_{73}NO_{16}$	884.50021 (−0.3)	10.1	Mir, Sen
19	Robenoside B Isomer (solanidenediol chacotriose)	$C_{45}H_{73}NO_{17}$	900.49513 (1.3)	8.6	/Sen

[a] Mass error in part per million (ppm) $=10^6 \times$ (accurate mass-exact mass)/exact mass.

On the basis of the above considerations, peak numbers were tentatively attributed to (**1**) solanidenetriol chacotriose, (2) solanidenediol chacotriose, (3) dehydrosolamargine, (4) solanandaine isomer I, (5) solanandaine, (6) solasonine, (7) robenoside B, (8) malonyl-solanidenediol chacotriose, (9) solamargine isomer, (10) solamargine, (11) solanidatetraenol chacotriose, (12) malonyl-solanandaine, (13) malonyl-solanidatetraenol chacotriose, (14) arudonine, (15) malonyl-solamargine isomer, (16) malonyl-solamargine, (17) solanandaine isomer II, (18) solanandaine isomer III. According to our knowledge, compounds 1, 2, 8, 11, 12, 13, 14 and 15 have not been previously described in eggplant.

It should be noted that there is a direct relationship between the level of malonyl-GAs and the level of correspondent-free GAs; thus, the relatively high intensity of solamargine also accounts for the high signal intensity of its malonyl conjugated compound.

Noticeably, only the compounds that contain chacotriose as a carbohydrate component showed the formation of malonylated forms. This is probably due to their presence only in chacotriose of a glucose moiety as the directly joined to the aglycone (vide infra). We suppose that malonylation can occur to the primary OH-group of the glucose more easily than to the OH- located in other positions. Unsurprisingly, the peak signals of solasonine 6 and solamargine 10 were the most intense (Figure 3A). As previously demonstrated [27], ESI-IRMPD in positive ion mode of solasonine and solamargine led to the formation of numerous diagnostic signals (Supplementary Table S1). In particular, solasonine showed signals at accurate *m/z* values of 720.43324, 558.37856 and 396.32658 due to the sugar residue losses from dehydrated solasonine: The ion at accurate *m/z* 720.43324 corresponding to the loss of a rhamnose moiety (Rha, 146 Da), one of the three sugars composing the aglycone-linked solatriose, through glycosidic cleavage. The ion at *m/z* 558.37856 resulted from the loss of a glucose moiety (Glc, 162 Da) and the ion at *m/z* 396.32658 from the further loss of galactose (Gal, 162 Da).

Solamargine underwent similar fragmentation pathways as 6, generating characteristic fragments at accurate *m/z* values of 704.44910 ($[M+H-H_2O-Rha]^+$), 558.37844 ($[M+H-H_2O-2Rha]^+$) and 396.32617($[M+H-H_2O-2Rha-Glc]^+$) due to the sequential loss of the sugar residues composing chacotriose sugar chain. As expected, for both compounds 6 and 10, the ion at nominal *m/z* 414 (accurate *m/z* 414.33592 and 414.33688 for solasonine and solamargine, respectively) corresponding to protonated solasodine aglycone was detected.

The unknown peak with accurate *m/z* 866.48965 (3) exhibited a fragmentation pathway similar to that of solamargine. As reported in Supplementary Table S1, accurate mass measurements of the main diagnostic ions under laser irradiation dissociation of compound 3 led to molecular formula fragments with two hydrogen atoms less than those detected for solamargine (10). Therefore, this GA is probably a dehydrogenated form of solamargine, where a chacotriose sugar chain is linked to an aglycone of nominal *m/z* 412 (accurate *m/z* 412.32129), which is likely a solasodine derivative with an additional double bond (molecular formula $C_{27}H_{41}NO_2$). This hypothesis was confirmed by the elution order (Table 1) of compound 3 (Rt = 9.2 min), which eluted earlier than compound 10 (Rt = 14.3 min). In fact, it is known that an increase of double bonds in a given GA causes it to elute earlier. Based on the reported considerations, compound 3 is tentatively determined as dehydrosolamargine ($C_{45}H_{72}NO_{15}^+$, $\Delta m = -0.2$ ppm).

An isomer of compound **3** eluting at Rt of 6.0 min was also displayed in the XICs trace of the protonated precursor ion at *m/z* 866.48965 (compound 2, $C_{45}H_{72}NO_{15}$, $\Delta m = 0.5$ ppm).

The characteristic IRMPD MS^2 product ions reported for this compound (2) in Supplementary Table 1 at *m/z* 720.43188 ($[M+H-Rha]^+$), 574.37402 ($[M+H-2Rha]^+$) and 412.31111 ($[M+H-2Rha-Glc]^+$) suggest the presence of the chacotriose moiety linked to the skeleton of solanidenediol aglycone ($C_{27}H_{42}NO^+$, $\Delta m = 0.2$ ppm). Generally, solanidane GAs elute earlier than spirosolanes (Figure 1) and addition of hydroxy groups makes their structure even more polar, thus, compound 2, which should have two hydroxy groups on the aglycone, was characterized as solanidadienol chacotriose (Table 1).

Similarly, the IRMPD MS^2 mass spectrum of compound 1 at *m/z* 882.48419 shows the neutral losses of Rha and Glc sugars from chacotriose moiety yielding the aglycone ion at *m/z* 428.31628 ($C_{27}H_{42}NO_3$). Subsequently, the aglycone undergoes sequential loss of three water molecule giving rise to the *m/z* 374.28439 ion. This, together with the product ion at *m/z* 267.17441 ($C_{19}H_{23}O^+$, $\Delta m = 0.2$), infers that in compound 1 an additional oxygen atom is localized to solanidadienol aglycone. Therefore, compound 1, which elutes before compound 2, was identified as a solanidenetriol chacotriose.

Through the XICs of the protonated precursor ion at *m/z* 862.45835 in the full scan (±1 mDa), peak 11 was obviously observed at Rt of 9.7 min, providing an intensive $[M+H]^+$ ion at *m/z* 862.45864 ($C_{45}H_{68}NO_{15}$, Δm = 0.3 ppm). The fragmentation (Supplementary Table S1) was consistent with that of a structure composed of a chacotriose sugar chain and an aglycone of nominal *m/z* 408 ($C_{27}H_{38}NO_2$, Δm= 0.1 ppm) which have not been previously reported in eggplant. The occurrence of *m/z* 716.39879 $[M+H\text{–}Rha]^+$, *m/z* 570.34136 $[M+H\text{–}2Rha]^+$ and *m/z* 408.28930 $[M+H\text{–}2Rha\text{–}Glc]^+$ product ions correspond to the typical sequential loss of Rha-Rha-Glc already displayed for solamargine.

One possibility is that this compound corresponds to solanidatetraenol chacotriose, which would account for a *m/z* 408.28930 ($C_{27}H_{38}NO_2$, $\Delta m = -1.0$ ppm) aglycone. To confirm this hypothesis, the elution time of compound 11 was compared to that of solamargine and the elution order was consistent with what would be the expected for a GA with additional double bonds [20].

Similar to the peak 11, the XICs of the protonated precursor ion at *m/z* 900.49513 in the MS full scan (±1 mDa) for *Mirabella* pulp extract showed the presence of a single chromatographic peak (7) at a Rt of 7.3 min (Figure 3A).

In Supplementary Figure S1, the IRMPD-FTICR mass spectrum of the precursor ion from peak 7 ($[C_{45}H_{73}NO_{17}]^+$, Δm = +0.6 ppm) is reported. It exhibits a remarkable fragmentation upon 290 ms of laser irradiation at the highest energy, leading to the formation of many diagnostic fragments. This compound eluting at 7.8 min and consisting of solatriose and an aglycone at nominal *m/z* 430 seems to be a hydroxyl-solasonine.

A typical set of signals already described for solasonine and due to the sequential neutral losses of Rha, Glc and Gal was detected. Product ions of *m/z* 430.33142 $[M+H\text{–}solatriose]^+$, 412.32111 $[M+H\text{–}H_2O\text{–}solatriose]^+$, 394.31055 $[M+H\text{–}2H_2O\text{–}solatriose]^+$ and 376.26370 $[M+H\text{–}3H_2O\text{–}solatriose]^+$, corresponded to the sequential loss of the hydroxy groups. The order of elution of compound 5 in respect to solasonine (6) is also consistent with those expected for a GAs with an additional hydroxy group.

Based on the suggested fragmentation pattern, literature data [15] supported by the derived elemental compositions from the accurate mass measurements of all the product ions (Supplementary Table S1) and taking into account that two GAs, namely, robeneoside A and robeneoside B have been isolated and identified from the fruits of *Solanum lycocarpum* [31], peak 5 was tentatively identified as robeneoside B.

Another compound with *m/z* 900.49628 was detected at 8.6 min only in the genotype *Melanzana Bianca di Senise* (Compound 19, Table 1). It underwent similar fragmentation pathways as 5, so it was tentatively identified as a 12-hydroxysolasonine. These results are in agreement with those reported in *Solanum lycocarpum* [31].

When subjected to IRMPD MS^2 analysis, the protonated molecules of all malonyl-GAs share a characteristic set of neutral losses (Supplementary Table S1); these correspond to the loss of H_2O (18.0106 Da) from the aglycone and chacotriose, CO_2 (43.9892 Da) and C_2H_2O (42.01001 Da) from the malonyl group, one or two rhamnose (146.0579 Da = $C_6H_{10}O_4$) and glucose (162.0528 Da = $C_6H_{10}O_5$) from chacotriose. The neutral loss of the whole group $C_3H_2O_3$ (85.99984 Da) was also observed in their MS^2 mass spectra, which confirmed the presence of malonic acid (Supplementary Table S1). Thus, the spectra of malonylated-GAs were similar to those previously reported for the protonated molecules of the corresponding non-malonylated forms [27].

For example, the IRMPD MS^2 spectrum of compound 16 (molecular formula of protonated molecule $C_{45}H_{73}O_{16}N$, calculated for $C_{45}H_{74}O_{16}N$, *m/z* 954.50569; Δm = 0.5 ppm), is similar to mass spectra of solamargine [27]. As summarized in Figure 4, in addition to the characteristic fragments already described for solamargine at *m/z* 576.38941 $[M+H\text{–}2Rha]^+$, 558.37898 $[M+H\text{–}2Rha\text{–}H_2O]^+$, 414.33672 $[M+H\text{–}2Rha\text{–}Glc]^+$, 396.32690 $[M+H\text{–}2Rha\text{–}Glc\text{–}H_2O]^+$, 378.31576 $M+H\text{–}2Rha\text{–}Glc\text{–}2H_2O]^+$, $[C_{19}H_{27}O+H]^+$ 271.20587 and 253.19502 $[C_{19}H_{25}+H]^+$, due to the fragmentation of solasodine aglycone and chacotriose sugar residue, the isolated protonated compound $[M+H]^+$ at *m/z* 954.50525 under laser irradiation led to the formation of other diagnostic fragment ions. The occurrence of

$[M+H-CO_2]^+$ at *m/z* 910.51751, $[M+H-CO_2-H_2O]^+$ at *m/z* 892.50543, $[M+H-CO_2-Rha]^+$ at *m/z* 764.45806, $[M+H-CO_2-H_2O-Rha]^+$ at *m/z* 746.44784 and $[M+H-C_3H_2O_3-Rha]^+$ at *m/z* 722.44865, suggests the presence of a malonate ester in the sugar group (Figure 4), most probably in glucose residue in position 3′ or 6′ [15]. This compound with molecular formula $C_{48}H_{75}O_{18}N$ and retention time of 18.5 min was tentatively assigned the identity of malonyl-solamargine.

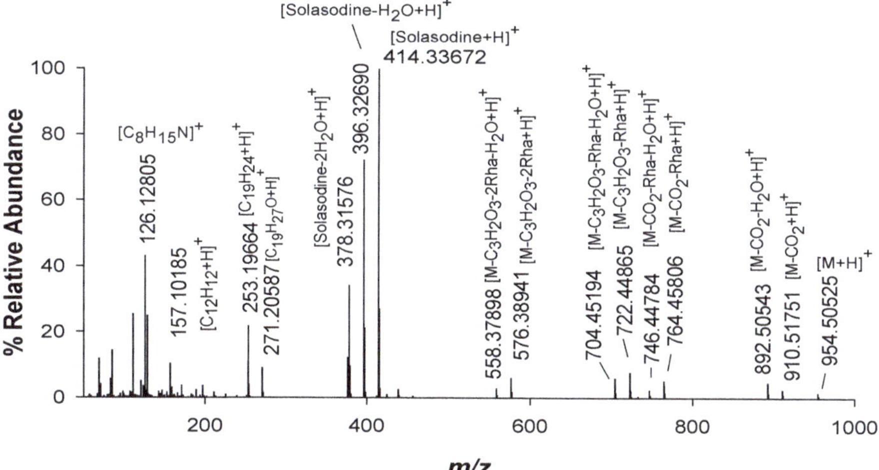

Figure 4. High-resolution IRMPD FTICR mass spectrum of the protonated malonyl-solamargine at *m/z* 954. Following transfer to the ICR cell, precursor ion populations were photon-irradiated for 290 ms at 100% laser power.

The location of the malonyl group in the sugar is not clear, but no neutral loss of terminal rhamnose with malonic ester was detected under fragmentation of malonyl-solamargine (Figure 4), consistent with the possibility that the malonyl groups are connected to glucose moieties in position 3′ or 6′ [15]. Since the occurrence of a malonyl ester in the glucosyl moiety causes a negative charge and could facilitate the efficient transport of these compounds into the vacuole by anionic transporters, malonylation of secondary metabolites is considered an important step in plant defence process [31].

Taking into account that in many secondary metabolites produced in Solanaceae plants and involved in plant defences against insect herbivory, malonyl moiety is typically connected to the C′-6 hydroxyl groups of the glucose [32–35] we can suppose that this is the preferential position of attachment of malonyl group also for GAs.

As shown in Figure 3A and Table 1, four structural isomers of solasonine (compounds 4, 5, 17 and 18) and one isomer of solamargine (compound 9, named solamargine isomer) have been displayed through the XICs of *m/z* 884.50021 (±1 mDa) and *m/z* 868.50530 (±1 mDa) in full-scan MS data obtained for the commercial variety *Mirabella* berry pulp extract analysis.

Compounds 4 and 9 were detected only in *Mirabella* while all the other compounds occurred in both *Mirabella* and *Melanzana Bianca di Senise* extracts.

Supplementary Figure S2 shows the IRMPD spectra of the charged precursor ions $[M+H]^+$ at *m/z* 884.49982 (A) and $[M+H]^+$ at *m/z* 970.50159 (B), tentatively assigned to solanandaine (5) and malonyl-solanandaine. The positive IRMPD ion spectrum of solanandaine showed a peak at accurate *m/z* 884.49982 $[M+H]^+$ corresponding to the molecular formula $C_{45}H_{73}O_{16}N$ (calculated for $C_{45}H_{74}O_{16}N$, *m/z* 884.50021; $\Delta m = -0.4$ ppm). In addition, significant peaks at *m/z* 720.43724 ($[M+H–H_2O–Rha]^+$), 574.37463 ($[M+H–H_2O–2Rha]^+$), 430.33215 ($[M+H–2Rha–Glc]^+$), 412.32132 ($[M+H–H_2O–2Rha–Glc]^+$) and 142.12265 ($[C_8H_{15}NO+H]^+$) were detected. This GA, therefore, contains three hexose units. The sequential loss of Rha, Rha and Glc indicates that solanandaine has a chacotriose side chain attached to an aglycone moiety.

The aglycone *m/z* 430.33215 $[Agly+H]^+$ is likely a solaparnaine, a solasodine derivative with an additional hydroxyl group in ring F, as confirmed by the occurrence of fragment at *m/z* 142.12265 ($[C_8H_{15}NO+H]^+$) corresponding to the characteristic spirosolane fragment originated from ring E breaking at *m/z* 126.12769 ($[C_8H_{15}N+H]^+$) plus an OH group.

As expected, compound 12 showed the same fragmentation behaviour of solanandaine (5) due to the facile breaking of ester link, by the losses of CO_2 (44 Da) and of an additional 42 Da-moiety (C_2H_2O) (Supplementary Figure S3). Additional neutral loss of 146 Da, 146 Da and 162 Da and accurate mass measurements of the remaining aglycone protonated at *m/z* 430.33176 (calculated *m/z* for $C_{27}H_{44}NO_3^+ = 430.33157$; $\Delta m = 0.5$ ppm) suggest the presence of a malonyl-chacotriose linked to solaparnaine aglycone.

The same rationale used for malonyl-solamargine and malonyl-solanandaine was applied to explain the occurrence of malonyl-solanidenediol chacotriose (8), malonyl-solanandaine (12), malonyl-solanidatetraenol chacotriose (13) and malonyl-solamargine isomer (15), presumably formed because of the relatively high content of solanidenediol chacotriose, solanandaine, solanidatetraenol chacotriose and solamargine isomer, peaks 2, 5, 11 and 9, respectively. To the author's knowledge, the only detailed data on malonylated GAs compounds reported in eggplant, thus far, in the literature are those on malonyl-solamargine [15].

The more informative fragments in the IRMPD MS^2 spectrum from the protonated ion $[M+H]^+$ of compound 4 (which was of low intensity) were $[M+H–H_2O]^+$ at *m/z* 866.49023 and $[M+H–H_2O–chacotriose]^+$ at *m/z* 412.32104 in addition to the characteristic fragments of spirosolane type aglycone. The peak at nominal *m/z* 430 corresponding to aglycone solasodine, or its isomer, with an additional hydroxyl group, was not observed in the spectrum 4 but can be attributed to facile water loss. Therefore, we can suppose that compound 4 was an isomer of solanandaine.

The same considerations could be done for compounds 17 and 18, named solanandaine isomer II and solanandaine isomer III, because their fragmentations (Supplementary Table S1) were consistent with a structure composed of chacotriose linked to an hydroxysolasodine or its isomer detected at *m/z* 430.33118 for compound 17 and at *m/z* 430.33173 for compound 18.

Compound 14 was eluted at Rt = 14.1 min and the ESI-MS spectrum showed an $[M+H]^+$ ion at accurate *m/z* 1000.54779 with an elemental composition of non-protonated compound $C_{50}H_{81}NO_{19}$.

Although the mass spectrum was less rich in diagnostic fragments than other GAs shown above, the presence of a solasodine core linked to a tetrasaccharide moiety could be established.

The IRMPD MS^2 mass spectrum (Supplementary Figure S2) gave fragments at *m/z* 982.53912 ($[M+H–18]^+$) and *m/z* 836.47998 ($[M+H–18–146]^+$), which correspond to the loss of a molecule of water and an additional rhamnose. The peaks at *m/z* 704.43732 ($[M+H–18–132–146]^+$) and *m/z* 558.37921 ($[M+H–18–132–146–146]^+$) corresponded to the loss from $[M+H–18]^+$ of a xylose and a rhamnose, and a xylose and two rhamnose moieties, respectively. The fragment at *m/z* 414.33667 corresponding to the protonated aglycone ([solasodine + H]$^+$) could be originated from the loss of the entire sugar moiety composed of a xylose, two rhamnose and a glucose.

Based on the mass difference between this compound and solasonine (132 Da), IRMPD data and according to a previous report [36], compound 14 was deduced as arudonine, an allelopathic steroidal glycoalkaloid found in the root bark of *Solanum arundo* Mattei.

Interestingly, no malonylated form of this GAs was detected, probably due to the low level of arudonine in the sample.

2.3. GAs Profile of *Melanzana Bianca di Senise* Pulp Extracts

In view of the worldwide interest in local germplasm conservation and considering the risks resulting from genetic erosion of agricultural plant resources, several research projects have addressed the safeguarding and conservation of agro-biodiversity [37,38], including endemic or local eggplant genotypes [39].

Local landraces, unlike the commercial varieties, are often poorly studied, and little scientific information is available on their characteristics and distinctive features, despite being generally associated with better flavour, local tradition and environmentally friendly production [40]. This is the case of *Melanzana Bianca di Senise*, an eggplant locally appreciated for the intense and fruity aroma of the berry, its sweetness and particularly for delaying of turning brown of the pulp after cutting, of which there are no studies in the literature.

For this reason, the optimized, positive-ion LC-FTICR-MS method used to characterize the commercial variety *Mirabella* was then used to profile and compare GAs from *Melanzana Bianca di Senise* pulp extracts. The resultant XICs are presented in Figure 3B. Identification of individual compounds, provided in Table 1 and Supplementary Table S1, revealed a profile very similar to that of Mirabella variety, lacking only compounds 9 and its malonylated form 15 and showing, as mentioned above, the occurrence of three additional compound in the XICs trace, one at *m/z* 900.49513 (±1 mDa), tentatively assigned to a robenoside B isomer, probably compound 12-hydroxysolasonine [31] and two at *m/z* 884.50021 (±1 mDa), tentatively assigned to Solanandaine Isomer II (spirosolendiol chacotriose) and Solanandaine Isomer III (spirosolendiol chacotriose).

On the basis of the GAs profile of the local eggplant landrace that resulted similar to Mirabella and considering the excellent recognized organoleptic characteristics (sweetness and delaying of turning brown of the pulp after cutting), the nutritional value of the *Melanzana Bianca di Senise* should be further analysed aiming at the protection and preservation of agro-biodiversity.

2.4. Composition Profile of GAs in *Solanum Melongena*

As far as quantification of GAs in sample extracts of *Solanum melongena*, the use of solasonine and solamargine as calibration standards established that the LC-FTICR-MS method was linear over at least two orders of magnitude, 0.5–50 ppm (data not shown). Due to the lack of commercially available standards, only these two compounds were quantified. Solamargine was found the most representative with relatively high content, i.e., 37.3 ± 1.8 mg/100g (dry weight (dw), three samples); the level of solasonine was considerably lower than those of solamargine, i.e., 6.2 ± 0.5 mg/100g (dry weight (dw), three samples).

Figure 5 depicts the relative proportions in terms of ion counts of each GAs occurring in the Mirabella and Melanzana Bianca di Senise extracts. To the best of our knowledge, this is the broadest spectrum of GAs known to be identified in *Solanum melongena*. As expected, solamargine (peak 10, nominal m/z 868), solasonine (peak 6, nominal m/z 884) and malonyl-solamargine (peak 11, nominal m/z 954) are the most representative GAs in both pulp extracts, amounting to >70% of the total GAs present in samples. Interestingly, only in Mirabella extract it was possible to detect high level of a solamargine isomer (peak 9, nominal m/z 868), i.e 11.06%, and of the its manolynated form at nominal 954 (peak 15), i.e 1.07%; on the contrary, in the Melanzana Bianca di Senise it was possible to highlight the occurrence of a high level of solanidatetraenol chacotriose i.e 14.35%, (peak 11, nominal m/z 862) and its malonylated form (peak 13, nominal m/z 948), which are at low levels in the Mirabella variety.

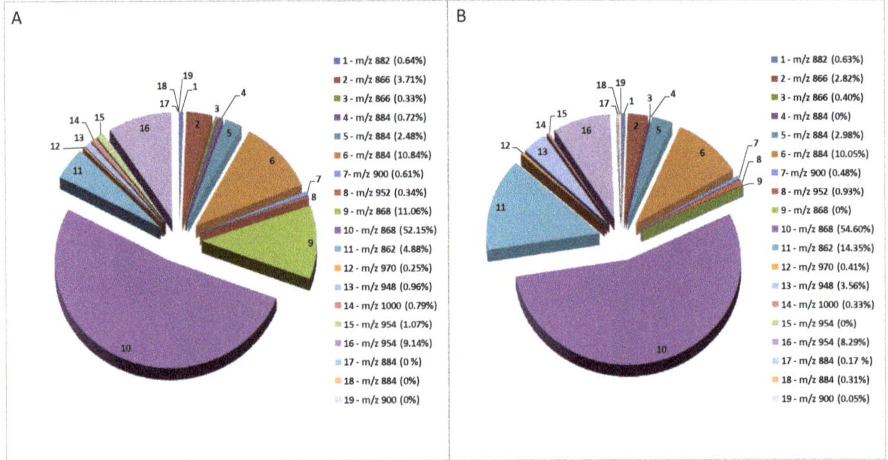

Figure 5. Pie charts illustrating the relative signal intensities (i.e. ion counts) of the GAs identified in "Mirabella" (**A**) and "Melanzana di Senise" (**B**) pulp extracts. (1) solanidenetriol chacotriose, (2) solanidenediol chacotriose, (3) dehydrosolamargine, (4) solanandaine isomer I, (5) solanandaine, (6) solasonine (7) robenoside B, (8) malonyl-solanidenediol chacotriose, (9) solamargine isomer, (10) solamargine, (11) solanidatetraenol chacotriose, (12) malonyl-solanandaine, (13) malonyl-solanidatetraenol chacotriose, (14) arudonine, (15) malonyl-solamargine isomer, (16) malonyl-solamargine, (17) solanandaine isomer II, (18) solanandaine Isomer III, (19) robenoside B isomer.

As presented, the method is basically semi-quantitative. However, by the use of appropriate calibration standards, it would be straightforward to convert the method into a fully quantitative procedure and, thus, provide precise concentrations of individual GAs in pulp extracts of *Solanum melongena*.

2.5. Acetylcholinesterase Inhibition Activity of Eggplant GAs

Solamargine was tested to evaluate the cholinesterase inhibition (Supplementary Figure S4) and then compared to galanthamine to estimate its contribution to extract activity. Solamargine showed an IC_{25} of 327.88 μM = 0.28 mg/mL (IC_{50} was not reached at the tested concentrations), whereas galanthamine showed a higher value of inhibition (IC_{25} of 5.31 μM). A previous study reported that the pure steroidal glycoalkaloids α-solanine and α-chaconine significantly inhibited acetylcholinesterase enzyme, whereas solasonine and solamargine showed a very reduced activity [41]. Nevertheless, it was found that combinations of solanine and chaconine, and also of solasonine and solamargine, produced effects which were slightly antagonistic or non-interactive. This could explain why the extracts showed lower AChE inhibition than pure compounds. Referring to literature results, other GAs could be more active than solamargine in eggplant extracts. Our further studies will be directed to the isolation, and activity determination of other GAs identified in the extracts.

2.6. Antioxidant Activity of Eggplant Extracts

The antioxidant activity was also evaluated by different assays able to measure the radical scavenging activity of extracts and the possible inhibition of lipid peroxidation. The antioxidant activity was determined by widely used antioxidant methods as 2,2-diphenyl-1-picrylhydrazyl radical (DPPH) and beta carotene bleaching assay (BCB).

The results of the antioxidant activity reported in Figure 6 did not demonstrate significant differences in DPPH assay ($p > 0.05$), while Melanzana di Senise was shown to be two folds more active than the Mirabella sample when the antioxidant activity was measured using the BCB test (45% *vs.* 23% respectively; $p < 0.05$). It was demonstrated that the antioxidant activity of Solanaceae species is not directly correlated to the amount of glycoalkaloids but, rather, to phenolic compounds [6]. This means that the antioxidant activity showed by the extract was preferably due to the presence of phenolic compounds solubilised in the extraction solvent (acidified water) together with GAs.

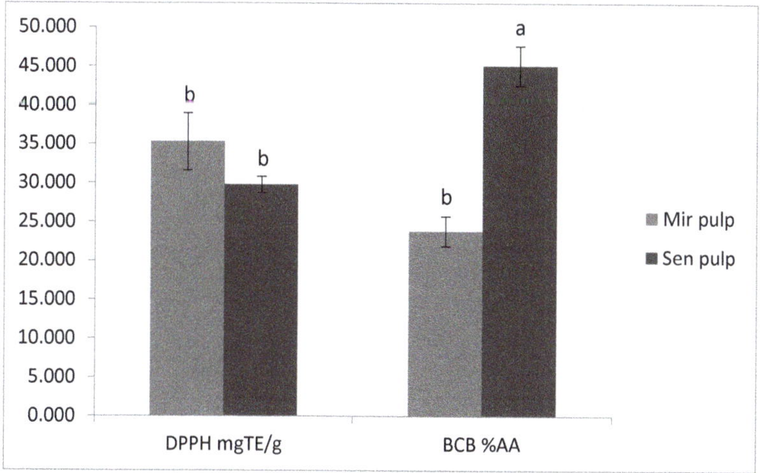

Figure 6. Antioxidant activity of the pulp extract from "Mirabella" and "Melanzana di Senise" samples. DPPH (2,2-diphenyl-1-picrylhydrazyl) results were expressed as mg of Trolox® equivalent (TE) per mg of dried extract; BCB (beta Carotene Bleaching) results were expressed as % antioxidant activity (% A.A.). Different letters indicate significant differences between mean values of a particular index of the given species $p < 0.05$ (according to Tukey's test).

3. Conclusions

In conclusion, the investigation and the structural characterization of the GAs and malonyl-GAs in the two considered eggplant berries were performed successfully by LC-ESI-FTICR-MS. Assisted by the high mass accuracy, high-resolution and flexible MS/MS capability of a Fourier transform ion cyclotron resonance (FTICR) mass spectrometer, infrared multiphoton dissociation (IRMPD) fragmentation of GAs has been accomplished allowing the tentative identification of all compounds under investigation and showing the presence of some GAs and malonyl-GAs not previously reported for this vegetable. The study based on MS methods is a valuable source of information for further isolation and structural investigations (e.g., by NMR) aimed at elucidating new GA structures. Secondary metabolites identified in the plant extracts demonstrated to inhibit a key enzyme involved in several neurological disorders. Even that some GAs were demonstrated to be toxic in human and animal models, in other experiments a wide range of biological activities have been evidenced [7–10,42,43]. Our results reinforced the hypothesis that GAs could serve as lead compounds for the design of new drugs active vs. cholinesterase effective in neurological disorders.

4. Materials and Methods

4.1. Chemicals

Acetonitrile (ACN; LC-MS grade) and HCOOH (99%) were from Carlo Erba (Milan, Italy). Ultrapure H_2O was produced using a Milli-Q system (Millipore, Billerica, MA, USA). Pure solamargine (97.5%) and solasonine (98.3%) were purchased from Glycomix (Reading, UK).

Acetylthiocholine iodide, 5,5′-Dithiobis(2-nitrobenzoic acid) (DNTB), Tris-HCl, bovine serum albumin (BSA), acetylcholinesterase (AChE), butyrylcholinesterase (BChE), butyrylthiocholine chloride, galanthamine (99.0% purity), sodium carbonate, 2,2-diphenyl-1-picrylhydrazyl radical (DPPH), beta-carotene, linoleic acid, Trolox® and Tween 20 were purchased by Sigma-Aldrich (Milan, Italy).

4.2. Plant Material and Sample Preparation

The experiment was carried out on the berries of two eggplant genotypes of *Solanum melongena* L.: The commercial variety *Mirabella* (Mir) and the local landrace *Melanzana Bianca di Senise* (Sen), grown in an experimental field located in Rotonda (PZ, Italy) at the Experimental Agricultural Farm Station sited in the Pollino National Park. The plants were grown and samples were collected under special permission from the Farm Director to conduct the study on this area. This study did not involve endangered or protected species. For both genotypes, at the commercial fruit ripening stage, ten fruits from ten plants were harvested and cleaned, then morphological traits were recorded (peel colour, length, diameter and weight of the berries are shown in Supplementary Table S2). Subsequently, the fruits were longitudinally divided into four equal parts, peeled to separate the peel and pulp, quickly frozen in liquid N_2, and lyophilized. The freeze-dried tissue samples were ground to a fine powder and stored at −80 °C until analysis for glycoalkaloids. The extraction procedure was based on the method of Cataldi, Lelario and Bufo [28]: A 1.5 g powdered sample was extracted with 20 mL of 1% (*v/v*) aqueous acetic acid solution. The suspension was mixed for about two hours to facilitate the passage of soluble substances in the extraction solvent and subsequently centrifuged at 6000 rpm for 30 min. The supernatant was collected using a syringe and filtered through a 0.22 μm nylon filter (Whatman, Maidstone, UK). Afterwards, the pellet was re-suspended in 5 mL of 1% (*v/v*) aqueous acetic acid solution, stirred, centrifuged, collected and then filtered. The two extracts were joined in a unique sample.

4.3. Analysis of Glycoalkaloids (GAs)

GAs analyses were performed using a Surveyor autosampling LC system, interfaced to a LIT-FTICR-mass spectrometer (Thermo Fisher Scientific, Bremen, Germany) via electrospray source, equipped with a 20 W CO_2-laser (Synrad, Mukilteo, WA, USA; 10.6 μm) described elsewhere [27]. Source settings used for the ionization of GAs were ESI needle voltage, +4.5 kV; capillary voltage, +35 V; temperature of the heated capillary, 300 °C; and sheath gas (N_2) flow rate of 80 arbitrary units (a.u.). The instrument was externally calibrated with appropriate standards.

Chromatographic separation of GAs was performed at ambient temperature on a Supelcosil LC-ABZ, amide-C_{16} (5 μm, 250 × 4.6 mm), equipped with a guard column of the same material (Supelco Inc., Bellefonte, PA, USA), and a mobile phase consisting of 0.1% HCOOH in H_2O (solvent A) and MeOH (solvent B). The following gradient at 0.8 mL/min was applied: 30–43% B in 0–8 min; 43–60% B in 8–20 min and 60% B in 20–24 min. Prior to the next injection, the column was equilibrated for 6 min. The injection volume was 20 μL and the flow to the source was reduced to 200 μL/min by a post-column splitter. Mass spectrometric data were acquired in positive ion mode while scanning *m/z* 50–1300 at a rate of 2 scans/s.

For the high-resolution MS/MS experiments by infrared multiphoton dissociation (IRMPD), precursor ions were isolated in the linear ion trap (LIT) and transferred into the ICR cell. The duration of laser irradiation was adapted to generate optimal fragmentation and varied between 100 and 350 ms. Typically, a medium IRMPD pulse, of 290 ms at 100% energy was applied as photon irradiation for

GAs. Data were collected in full MS scan mode and processed post-acquisition to reconstruct the elution profile for the ions of interest with a given *m/z* value and tolerance. The simplest method to identify analytes by eXtracted Ion Chromatograms (XICs) in LC–FTICR-MS was used. The reduction of the interferences in the XICs significantly facilitates the identification of potential GAs, including the minor or uncommon compounds, which otherwise could have been missed. Data were acquired and processed by the Xcalibur software package (version 2.0 SR1; Thermo Fisher Scientific). The chromatographic raw data were imported, elaborated, and plotted using SigmaPlot 10.0 software (Systat Software, Inc., London, UK). Precursor and product ion structures were drawn by ChemDraw Pro 14.0 (CambridgeSoft Corporation, Cambridge, MA, USA).

4.4. Acetylcholinesterase (AChE) Inhibitory Activity

The inhibition of AChE activity was determined based on Ellman's method, as previously reported by [43]. For this assay, 25 µL of acetylthiocholine iodide (15 mM), 125 µL of DTNB (3 mM), 25 µL of buffer B (50mM Tris-HCl, pH 8 containing 0.1% BSA) and 50 µL of each test extract solution at the different concentrations were mixed. The mixture was incubated for 10 min at 37 °C. The reaction was started by adding 25 µL of 0.44 U/mL AChE. The absorbance was measured at 405 nm for 10 min and the rates of reactions were calculated by SpectroStar Nano.

4.5. Antioxidant Activity

4.5.1. Radical Scavenging Activity

The radical scavenging activity was measured using 2,2-diphenyl-1-picrylhydrazyl radical (DPPH) [13]. Several concentrations (from 5.0 mg mL^{-1} to 0.3 mg mL^{-1}) of extracts (50 µL) were added to 200 µL of methanol DPPH solution (100 µM). Absorbance was read at 515 nm after 30 min of incubation in the darkness. Results were expressed as mg of Trolox equivalent (TE) per mg of dried extract.

4.5.2. Inhibition of Lipid Peroxidation

The inhibition of lipid peroxidation was tested by beta carotene bleaching assay (BCB). For the analysis, 950 µL of beta carotene emulsion (0.4% *w/w*) was added to 50 µL of extract (0.25 mg mL^{-1}). the beta carotene solution was prepared as previously reported [44]. BHT was used as a positive control and results were expressed as a percentage of antioxidant activity (% A.A.) [45].

4.6. Statistical Analysis

The experimental results were expressed as mean ± standard deviation (SD) [46] of three independent replicates (n = 3). Data were analysed by Graph Pad version 5 and they were subjected to one-way analysis of variance (ANOVA) and differences between samples were determined by Tukey test, p values less than 0.05 were considered statistically significant.

Supplementary Materials: The following are available online at http://www.mdpi.com/2072-6651/11/4/230/s1, Figure S1: High-resolution IRMPD FTICR mass spectrum of the protonated Robenoside B at *m/z* 900, Figure S2: High-resolution IRMPD FTICR mass spectrum of the protonated arudonine at *m/z* 1000, Figure S3: High-resolution IRMPD FTICR mass spectra of the protonated solanandaine (A) at *m/z* 884 and malonyl-solanandaine (B) at *m/z* 970, Figure S4: Inhibition of Acetylcholinesterase (AChE) enzyme by Galantamine and Solamargine, Table S1: IRMPD FTICR MS product ions obtained from glycoalkaloids and malonyl glycoalkaloid and identified in S. melongena pulp by high-resolution LC-ESI-IRMPD-FTICR MS, Table S2: Fruit morphological traits of Mirabella and Bianca di Senise eggplants.

Author Contributions: Conceptualization, A.R.R., L.M., S.A.B. and L.S.; Data curation, F.L. and L.S.; Formal analysis, F.L., S.D.M. and D.R.; Funding acquisition, L.M. and S.A.B.; Investigation, F.L., A.R.R. and D.R.; Methodology, S.D.M., D.R., L.M., S.A.B. and L.S.; Project administration, L.S.; Supervision, L.M. and S.A.B.; Validation, S.A.B.; Writing – original draft, F.L., A.R.R., S.A.B. and L.S.; Writing – review & editing, D.R. and L.M.

Funding: This work was performed by using the instrumental facilities of CIGAS Center founded by EU (Project No. 2915/12), Regione Basilicata and Università degli Studi della Basilicata. This work received financial support from DGR n. 1490 del 4/12/2014, "Monitoriaggio delle acque Marine e Costiere della Basilicata, vs rep. n. 163 n8, from the Regione Basilicata and Fondazione Enrico Mattei and Regional Project ALIMINTEGRA, GO NUTRIBAS financed on 16.1 PSR Basilicata founding ex D.G.R. n° 312/17 CUP: C31G18000210002.

Conflicts of Interest: The authors declare no conflict of interest.

References

1. Food and Agriculture Organization of United Nations (FAO). Available online: http://www.fao.org/faostat/en/#data/QC (accessed on 9 March 2019).
2. Marsic, N.K.; Mikulic-Petkovsek, M.; Stampar, F. Grafting Influences Phenolic Profile and Carpometric Traits of Fruits of Greenhouse-Grown Eggplant (*Solanum melongena* L.). *J. Agric. Food Chem.* **2014**, *62*, 10504–10514. [CrossRef] [PubMed]
3. Zhang, Y.J.; Hu, Z.L.; Chu, G.H.; Huang, C.; Tian, S.B.; Zhao, Z.P.; Chen, G.P. Anthocyanin Accumulation and Molecular Analysis of Anthocyanin Biosynthesis-Associated Genes in Eggplant (*Solanum melongena* L.). *J. Agric. Food Chem.* **2014**, *62*, 2906–2912. [CrossRef] [PubMed]
4. Friedman, M. Chemistry and anticarcinogenic mechanisms of glycoalkaloids produced by eggplants, potatoes, and tomatoes. *J. Agric. Food Chem.* **2015**, *63*, 3323–3337. [CrossRef] [PubMed]
5. Aubert, S.; Daunay, M.; Pochard, E. Saponosides stéroïdiques de l'aubergine (*Solanum melongena* L.) I. Intérêt alimentaire, méthodologie d'analyse, localisation dans le fruit. *Agronomie* **1989**, *9*, 641–651. [CrossRef]
6. Chowański, S.; Adamski, Z.; Marciniak, P.; Rosiński, G.; Büyükgüzel, E.; Büyükgüzel, K.; Falabella, P.; Scrano, L.; Ventrella, E.; Lelario, F. A review of bioinsecticidal activity of Solanaceae alkaloids. *Toxins* **2016**, *8*, 60. [CrossRef]
7. Hameed, A.; Ijaz, S.; Mohammad, I.S.; Muhammad, K.S.; Akhtar, N.; Khan, H.M.S. Aglycone solanidine and solasodine derivatives: A natural approach towards cancer. *Biomed. Pharmacother.* **2017**, *94*, 446–457. [CrossRef] [PubMed]
8. Milner, S.E.; Brunton, N.P.; Jones, P.W.; O'Brien, N.M.; Collins, S.G.; Maguire, A.R. Bioactivities of Glycoalkaloids and Their Aglycones from Solanum Species. *J. Agric. Food Chem.* **2011**, *59*, 3454–3484. [CrossRef] [PubMed]
9. Zhong, Y.; Li, S.; Chen, L.; Liu, Z.; Luo, X.; Xu, P.; Chen, L. In Vivo Toxicity of Solasonine and Its Effects on cyp450 Family Gene Expression in the Livers of Male Mice from Four Strains. *Toxins* **2018**, *10*, 487. [CrossRef]
10. Fekry, M.I.; Ezzat, S.M.; Salama, M.M.; Alshehri, O.Y.; Al-Abd, A.M. Bioactive glycoalkaloides isolated from *Solanum melongena* fruit peels with potential anticancer properties against hepatocellular carcinoma cells. *Sci. Rep.* **2019**, *9*, 1746. [CrossRef]
11. Spochacz, M.; Chowański, S.; Szymczak, M.; Lelario, F.; Bufo, S.; Adamski, Z. Sublethal Effects of *Solanum nigrum* Fruit Extract and Its Pure Glycoalkaloids on the Physiology of Tenebrio molitor (Mealworm). *Toxins* **2018**, *10*, 504. [CrossRef]
12. Fidelis, Q.C.; Faraone, I.; Russo, D.; Aragao Catunda, F.E., Jr.; Vignola, L.; de Carvalho, M.G.; de Tommasi, N.; Milella, L. Chemical and Biological insights of *Ouratea hexasperma* (A. St.-Hil.) Baill.: A source of bioactive compounds with multifunctional properties. *Nat. Prod. Res.* **2018**, 1–4. [CrossRef] [PubMed]
13. Russo, D.; Malafronte, N.; Frescura, D.; Imbrenda, G.; Faraone, I.; Milella, L.; Fernandez, E.; De Tommasi, N. Antioxidant activities and quali-quantitative analysis of different *Smallanthus sonchifolius* [(Poepp. and Endl.) H. Robinson] landrace extracts. *Nat. Prod. Res.* **2015**, *29*, 1673–1677. [CrossRef]
14. Van Gelder, W.; Vinke, J.; Scheffer, J. Steroidal glycoalkaloids in tubers and leaves of Solanum species used in potato breeding. *Euphytica* **1988**, *39*, 147–158. [CrossRef]
15. Wu, S.B.; Meyer, R.S.; Whitaker, B.D.; Litt, A.; Kennelly, E.J. A new liquid chromatography-mass spectrometry-based strategy to integrate chemistry, morphology, and evolution of eggplant (*Solanum*) species. *J. Chromatogr. A* **2013**, *1314*, 154–172. [CrossRef] [PubMed]
16. Docimo, T.; Francese, G.; De Palma, M.; Mennella, D.; Toppino, L.; Lo Scalzo, R.; Mennella, G.; Tucci, M. Insights in the fruit flesh browning mechanisms in *Solanum melongena* genetic lines with opposite postcut behavior. *J. Agric. Food Chem.* **2016**, *64*, 4675–4685. [CrossRef]
17. Caprioli, G.; Cahill, M.G.; Vittori, S.; James, K.J. Liquid Chromatography–Hybrid Linear Ion Trap–High-Resolution Mass Spectrometry (LTQ-Orbitrap) Method for the Determination of Glycoalkaloids and Their Aglycons in Potato Samples. *Food Anal. Methods* **2014**, *7*, 1367–1372. [CrossRef]

18. Laurila, J.; Laakso, I.; Vaananen, T.; Kuronen, P.; Huopalahti, R.; Pehu, E. Determination of solanidine- and tomatidine-type glycoalkaloid aglycons by gas chromatography/mass spectrometry. *J. Agric. Food Chem.* **1999**, *47*, 2738–2742. [CrossRef] [PubMed]
19. Sanchez-Mata, M.C.; Yokoyama, W.E.; Hong, Y.J.; Prohens, J. Alpha-solasonine and alpha-solamargine contents of gboma (*Solanum macrocarpon* L.) and scarlet (*Solanum aethiopicum* L.) eggplants. *J. Agric. Food Chem.* **2010**, *58*, 5502–5508. [CrossRef]
20. Shakya, R.; Navarre, D.A. LC-MS analysis of solanidane glycoalkaloid diversity among tubers of four wild potato species and three cultivars (*Solanum tuberosum*). *J. Agric. Food Chem.* **2008**, *56*, 6949–6958. [CrossRef]
21. Sotelo, A.; Serrano, B. High-performance liquid chromatographic determination of the glycoalkaloids alpha-solanine and alpha-chaconine in 12 commercial varieties of Mexican potato. *J. Agric. Food Chem.* **2000**, *48*, 2472–2475. [CrossRef]
22. Claeys, M.; Van den Heuvel, H.; Chen, S.; Derrick, P.J.; Mellon, F.A.; Price, K.R. Comparison of high-and low-energy collision-induced dissociation tandem mass spectrometry in the analysis of glycoalkaloids and their aglycons. *J. Am. Soc. Mass Spectrom.* **1996**, *7*, 173–181. [CrossRef]
23. Zywicki, B.; Catchpole, G.; Draper, J.; Fiehn, O. Comparison of rapid liquid chromatography-electrospray ionization-tandem mass spectrometry methods for determination of glycoalkaloids in transgenic field-grown potatoes. *Anal. Biochem.* **2005**, *336*, 178–186. [CrossRef]
24. Agneta, R.; Rivelli, A.R.; Ventrella, E.; Lelario, F.; Sarli, G.; Bufo, S.A. Investigation of glucosinolate profile and qualitative aspects in sprouts and roots of horseradish (*Armoracia rusticana*) using LC-ESI–hybrid linear ion trap with Fourier transform ion cyclotron resonance mass spectrometry and infrared multiphoton dissociation. *J. Agric. Food Chem.* **2012**, *60*, 7474–7482.
25. Bianco, G.; Lelario, F.; Battista, F.G.; Bufo, S.A.; Cataldi, T.R. Identification of glucosinolates in capers by LC-ESI-hybrid linear ion trap with Fourier transform ion cyclotron resonance mass spectrometry (LC-ESI-LTQ-FTICR MS) and infrared multiphoton dissociation. *J. Mass Spectrom.* **2012**, *47*, 1160–1169. [CrossRef]
26. Lelario, F.; Bianco, G.; Bufo, S.A.; Cataldi, T.R.I. Establishing the occurrence of major and minor glucosinolates in Brassicaceae by LC-ESI-hybrid linear ion-trap and Fourier-transform ion cyclotron resonance mass spectrometry. *Phytochemistry* **2012**, *73*, 74–83. [CrossRef]
27. Lelario, F.; Labella, C.; Napolitano, G.; Scrano, L.; Bufo, S.A. Fragmentation study of major spirosolane-type glycoalkaloids by collision-induced dissociation linear ion trap and infrared multiphoton dissociation Fourier transform ion cyclotron resonance mass spectrometry. *Rapid Commun. Mass Spectrom.* **2016**, *30*, 2395–2406. [CrossRef]
28. Cataldi, T.R.; Lelario, F.; Bufo, S.A. Analysis of tomato glycoalkaloids by liquid chromatography coupled with electrospray ionization tandem mass spectrometry. *Rapid Commun. Mass Spectrom.* **2005**, *19*, 3103–3110. [CrossRef]
29. Agneta, R.; Lelario, F.; De Maria, S.; Möllers, C.; Bufo, S.A.; Rivelli, A.R. Glucosinolate profile and distribution among plant tissues and phenological stages of field-grown horseradish. *Phytochemistry* **2014**, *106*, 178–187. [CrossRef]
30. Sun, T.T.; Liang, X.L.; Zhu, H.Y.; Peng, X.L.; Guo, X.J.; Zhao, L.S. Rapid separation and identification of 31 major saponins in Shizhu ginseng by ultra-high performance liquid chromatography-electron spray ionization-MS/MS. *J. Ginseng Res.* **2016**, *40*, 220–228. [CrossRef] [PubMed]
31. Yoshikawa, M.; Nakamura, S.; Ozaki, K.; Kumahara, A.; Morikawa, T.; Matsuda, H. Structures of steroidal alkaloid oligoglycosides, robeneosides A and B, and antidiabetogenic constituents from the Brazilian medicinal plant Solanum lycocarpum. *J. Nat. Prod.* **2007**, *70*, 210–214. [CrossRef]
32. Yu, X.H.; Chen, M.H.; Liu, C.J. Nucleocytoplasmic-localized acyltransferases catalyze the malonylation of 7-O-glycosidic (iso)flavones in *Medicago truncatula*. *Plant J.* **2008**, *55*, 382–396. [CrossRef] [PubMed]
33. Heiling, S.; Schuman, M.C.; Schoettner, M.; Mukerjee, P.; Berger, B.; Schneider, B.; Jassbi, A.R.; Baldwin, I.T. Jasmonate and ppHsystemin Regulate Key Malonylation Steps in the Biosynthesis of 17-Hydroxygeranyllinalool Diterpene Glycosides, an Abundant and Effective Direct Defense against Herbivores in *Nicotiana attenuata*. *Plant Cell* **2010**, *22*, 273–292. [CrossRef] [PubMed]
34. Jassbi, A.R.; Zamanizadehnajari, S.; Baldwin, I.T. 17-Hydroxygeranyllinalool glycosides are major resistance traits of *Nicotiana obtusifolia* against attack from tobacco hornworm larvae. *Phytochemistry* **2010**, *71*, 1115–1121. [CrossRef] [PubMed]

35. Taguchi, G.; Ubukata, T.; Nozue, H.; Kobayashi, Y.; Takahi, M.; Yamamoto, H.; Hayashida, N. Malonylation is a key reaction in the metabolism of xenobiotic phenolic glucosides in Arabidopsis and tobacco. *Plant J.* **2010**, *63*, 1031–1041. [CrossRef]
36. Fukuhara, K.; Shimizu, K.; Kubo, I. Arudonine, an allelopathic steroidal glycoalkaloid from the root bark of Solanum arundo Mattei. *Phytochemistry* **2004**, *65*, 1283–1286. [CrossRef]
37. Elia, A.; Santamaria, P. Biodiversity in vegetable crops, a heritage to save: The case of Puglia region. *Ital. J. Agron.* **2013**, *8*, e4. [CrossRef]
38. Hammer, K.; Laghetti, G. Genetic erosion—Examples from Italy. *Genet. Resour. Crop Evol.* **2005**, *52*, 629–634. [CrossRef]
39. Hurtado, M.; Vilanova, S.; Plazas, M.; Gramazio, P.; Andujar, I.; Herraiz, F.J.; Castro, A.; Prohens, J. Enhancing conservation and use of local vegetable landraces: The Almagro eggplant (*Solanum melongena* L.) case study. *Genet. Resour. Crop Evol.* **2014**, *61*, 787–795. [CrossRef]
40. Trichopoulou, A.; Soukara, S.; Vasilopoulou, E. Traditional foods: A science and society perspective. *Trends Food Sci. Technol.* **2007**, *18*, 420–427. [CrossRef]
41. Roddick, J.G. The acetylcholinesterase-inhibitory activity of steroidal glycoalkaloids and their aglycones. *Phytochemistry* **1989**, *28*, 2631–2634. [CrossRef]
42. Sucha, L.; Tomsik, P. The Steroidal Glycoalkaloids from Solanaceae: Toxic Effect, Antitumour Activity and Mechanism of Action. *Planta Med.* **2016**, *82*, 379–387. [CrossRef] [PubMed]
43. Mensinga, T.T.; Sips, A.J.A.M.; Rompelberg, C.J.M.; Van Twillert, K.; Meulenbelt, J.; Van Den Top, H.J.; Van Egmond, H.P. Potato glycoalkaloids and adverse effects in humans: An ascending dose study. *Regul. Toxicol. Pharmacol.* **2005**, *41*, 66–72. [CrossRef] [PubMed]
44. Todaro, L.; Russo, D.; Cetera, P.; Milella, L. Effects of thermo-vacuum treatment on secondary metabolite content and antioxidant activity of poplar (*Populus nigra* L.) wood extracts. *Ind. Crops Prod.* **2017**, *109*, 384–390. [CrossRef]
45. Russo, D.; Bonomo, M.; Salzano, G.; Martelli, G.; Milella, L. Nutraceutical properties of Citrus clementina juices. *Pharmacologyonline* **2012**, *1*, 84–93.
46. Saltos, M.B.V.; Puente, B.F.N.; Milella, L.; De Tommasi, N.; Dal Piaz, F.; Braca, A. Antioxidant and Free Radical Scavenging Activity of Phenolics from Bidens humilis. *Planta Med.* **2015**, *81*, 1056–1064. [CrossRef] [PubMed]

Article

Determination of Selected Isoquinoline Alkaloids from *Mahonia aquifolia; Meconopsis cambrica; Corydalis lutea; Dicentra spectabilis; Fumaria officinalis; Macleaya cordata* Extracts by HPLC-DAD and Comparison of Their Cytotoxic Activity

Anna Petruczynik [1,*], Tomasz Plech [2], Tomasz Tuzimski [3], Justyna Misiurek [1], Barbara Kaproń [4], Dorota Misiurek [5], Małgorzata Szultka-Młyńska [6], Bogusław Buszewski [6] and Monika Waksmundzka-Hajnos [1,*]

1 Department of Inorganic Chemistry, Medical University of Lublin, Chodźki 4a, 20-093 Lublin, Poland; justyna.misiurek@umlub.pl
2 Department of Pharmacology, Medical University of Lublin, Chodźki 4a, 20-093 Lublin, Poland; tomasz.plech@umlub.pl
3 Department of Physical Chemistry, Medical University of Lublin, Chodźki 4a, 20-093 Lublin, Poland; tomasz.tuzimski@umlub.pl
4 Department of Clinical Genetics, Medical University of Lublin, Radziwiłłowska 11, 20-080 Lublin, Poland; barbara.kapron@umlub.pl
5 Botanical Garden of Maria Curie-Skłodowska University in Lublin, Sławinkowska 3, 20-810 Lublin, Poland; dorota.misiurek@poczta.umcs.lublin.pl
6 Department of Environmental Chemistry and Bioanalytics, Nicolaus Copernicus University, Faculty of Chemistry Gagarina 7, PL-87-100 Torun, Poland; szultka.malgorzata@wp.pl (M.S.-M.); bbusz@chem.umk.pl (B.B.)
* Correspondence: annapetruczynik@poczta.onet.pl (A.P.); monika.hajnos@umlub.pl (M.W.-H.); Tel.: +48-81-448-7169 (A.P.); +48-81-448-7162 (M.W.-H.)

Received: 31 July 2019; Accepted: 28 September 2019; Published: 2 October 2019

Abstract: Alkaloids have protective functions for plants and can play an important role in living organisms. Alkaloids may have a wide range of pharmacological activities. Many of them have cytotoxic activity. Nowadays, cancer has become a serious public health problem. Searching for effective drugs with anticancer activity is one of the most significant challenges of modern scientific research. The aim of this study was the investigation of cytotoxic activity of extracts obtained from *Corydalis lutea* root and herb, *Dicentra spectabilis* root and herb, *Fumaria officinalis, Macleaya cordata* leaves and herb, *Mahonia aquifolia* leaves and cortex, *Meconopsis cambrica* root and herb on FaDu, SCC-25, MCF-7, and MDA-MB-231 cancer cell lines. The cytotoxic activity of these extracts has not been previously tested for these cell lines. The aim was also to quantify selected alkaloids in the investigated extracts by High Performance Liquid Chromatography (HPLC). The analyses of alkaloid content were performed using HPLC in reversed phase (RP) mode using Polar RP column and mobile phase containing acetonitrile, water and ionic liquid (IL). Cytotoxic effect of the tested plant extracts and respective alkaloid standards were examined using human pharyngeal squamous carcinoma cells (FaDu), human tongue squamous carcinoma cells (SCC-25), human breast adenocarcinoma cell line (MCF-7), human triple-negative breast adenocarcinoma cell line (MDA-MB-231). All investigated plant extracts possess cytotoxic activity against tested cancer cell lines: FaDu, SCC-25, MCF-7, and MDA-MB-231. The highest cytotoxic activity against FaDu, SCC-25, and MCF-7 cell lines was estimated for *Macleaya cordata* leaf extract, while the highest cytotoxic activity against MDA-MB-231 cell line was obtained for *Macleaya cordata* herb extract. Differences in cytotoxic activity were observed for extracts obtained from various parts of investigated plants. In almost all cases the cytotoxic activity of investigated plant extracts, especially at the highest concentration against tested cell lines was significantly higher than the activity of anticancer drug etoposide. Our investigations exhibit that these plant extracts can be recommended for further in vivo experiments to confirm their anticancer activity.

Keywords: isoquinoline alkaloids; HPLC-DAD; cytotoxic activity; *Mahonia aquifolia; Meconopsis cambrica; Corydalis lutea; Dicentra spectabilis; Fumaria officinalis; Macleaya cordata*

Key Contribution: Cytotoxic effect of the tested plant extracts: from *Corydalis lutea*, *Dicentra spectabilis*, *Fumaria officinalis*, *Macleaya cordata*, *Mahonia aquifolia*, *Meconopsis cambrica* and respective alkaloid standards were examined using human pharyngeal squamous carcinoma cells (FaDu), human tongue squamous carcinoma cells (SCC-25), human breast adenocarcinoma cell line (MCF-7), human triple-negative breast adenocarcinoma cell line (MDA-MB-231). All investigated plant extracts possess cytotoxic activity against tested cancer cell lines: FaDu, SCC-25, MCF-7, and MDA-MB-231. Higher cytotoxicity was found for extracts containing highly cytotoxic alkaloids: chelerythrine, sanguinarine, and berberine. The highest cytotoxic activity against all tested cancer cell lines was observed after applying the *Macleaya cordata* leaf extract containing high concentrations of chelerythrine, and sanquinarine.

---

1. Introduction

Cancer is one of the most prominent diseases in humans and currently there is considerable scientific interest shown towards the exploration of new anticancer agents from natural sources including plants. Many plants are the source of a variety of substances, including secondary metabolites which exhibit the anticancer activity. Most of the anticancer drugs obtained from plants inhibit the nucleic acid synthesis, but the mechanism of action differs widely.

Isoquinoline alkaloids are a group of natural bioactive products with widespread occurrence in nature. Some isoquinoline alkaloids have antibacterial, antifungal, anti-tumor and other biological activities. For the determination of them in plants, the modern chromatographic methods are most often applied.

Species of *Mahonia* are used in folk medicine worldwide as a cure for tuberculosis, dysentery, various skin disorders and showed antibacterial, antifungal, and anti-inflammatory properties [1]. The stem bark of *Mahonia aquifolium* contains a lot of isoquinoline alkaloids, including berberine, palmatine, jatrorrhizine, magnoflorine and berbamine [2]. Plant extracts obtained from various species of *Mahonia* genus were previously analyzed by HPLC most often on octadecyl stationary phases with mobile phases containing acetonitrile, water with the addition of formic acid [3], trifluoroacetic acid [4], sodium dihydrogen phosphate [5], phosphate buffer at pH 3.0 [6]. Previous investigations showed that representatives of genus *Mahonia*, such as *Mahonia balei*, *Mahonia oiwakensis* and *Mahonia aquifolium* have cytotoxic activity against various human cancer cells, e.g., human colon cancer (HT-29) [5], human lung adenocarcinoma cells (A549 and H1355) [1], large-cell lung carcinoma (H1299) [1], and squamous-cell carcinoma of the lung (H226) [1], human cervical adenocarcinoma cell line (HeLa) [2].

Fumaria species are sometimes used in herbal medicine. *Fumaria* extracts are components of several phytotherapeutic preparations, which are used mostly in cases of minor hepatobiliary dysfunction, gastrointestinal diseases, diuretic agents and for skin disorders [7]. Alkaloids in extracts obtained from various *Fumaria* species were determined usually on C18 columns. Mobile phases contained a mixture of acetonitrile, water with the addition of formic acid [7,8] was most often applied. Rarely, cytotoxic properties of extracts from *Fumaria* species were investigated. Anti-proliferative activity of *Fumaria vaillantii* extract was investigated on malignant melanoma SKMEL-3, human breast adenocarcinoma MCF-7 and human myelogenous leukemia K562 cells [9]. Anticancer activity of *Fumaria indica* was investigated on rat hepatic carcinoma cells [10]. Chemopreventive effect of the plant extract against N-nitrosodiethylamine and CCl_4-induced hepatocellular carcinoma was determined.

Macleaya cordata has antiviral, anti-inflammatory, and insecticidal properties. The plant has been used to cure cervical cancer and thyroid cancer in traditional folk medicine [11]. The alkaloid analyses in extracts obtained from *Macleaya microcarpa, Macleaya cordata* were carried out most often by HPLC on C18 column using mobile phases containing acetonitrile, water, and formic acid [11–14]. Alkaloids in *Macleaya cordata* extract were also analyzed on (C12) column [15]. A mixture of acetonitrile, water, trimethylamine, and 1-heptanesulfonic acid was applied as mobile phase. Anticancer activity of *Macleaya* species was rarely investigated. Cytotoxic effects of *Macleaya cordata* extract was observed against adenocarcinomic lung cells [16]. The extract obtained from roots of *Macleaya microcarpa* exhibit cytotoxicity against Bel-7402, BGC-823, HCT-8, A2780, and A549 human cell-lines [17].

Several species of the genus Corydalis have been used over the past centuries in traditional Asian medicine. Corydalis sp. have antioxidant and anticancer activities, pain relief, and promotion of blood circulation pharmacological effects [18]. The major active constituents of Corydalis species are isoquinoline alkaloids: Berberine, palmatine, coptisine, isocorydine, and glaucine. These have a wide range of pharmacological activities such as: Antioxidant, antibacterial, antiviral, anticancer, analgesic, and promotion of blood circulation. For analysis of alkaloids in extracts obtained from *Corydalis* species, C18 column and mobile phases consisted of acetonitrile, water and formic acid [19–22], ammonium acetate [23], formic acid and ammonium acetate [24], acetate buffer at pH 5.0 [20], ammonium hydroxide [25], acetonitrile, water, acetic acid, and diethylamine [26], or acetic acid with triethylamine [27] rarely methanol, water with the addition of formic acid [28] or formic acid and ammonium acetate [29] were applied. The inhibitory effects on human tumor cell lines: A549, SK-OV-3, SK-MEL-2, and HCT-15 of tubers of *Corydalis ternata* were found [30].

Antiinflammatory, fungitoxic, and apoptosis-inducing activities of compounds from *Dicentra spectabilis* were described [31,32]. The roots of *Dicentra spectabilis* have been used for the treatment of strokes, bruises, improvement of blood circulation. Several alkaloids have been detected from *Dicentra* species: Isocorydine, corydine, dicentrine, protopine, dihydrosanguinarine, sanguinarine, cheilanthifoline, bicuculline, lederine, scoulerine, isoboldine, predicentrine, reticuline, and allocryptopine. The extract from the plant was analysed on C18 column with mobile phase containing acetonitrile, water, ammonium acetate, adjusted to pH 3.0 with acetic acid [33]. The cytotoxicity of compounds from *Dicentra spectabilis* was determined on Raw 264.7 cells [32].

The whole plant of *Meconopsis* species is traditionally used as a Tibetan medicine for the treatment of various diseases, such as inflammation, pain, hepatitis, tuberculosis, headache, and fractures [34,35]. Analysis of alkaloids in extracts obtained from various *Meconopsis* species by HPLC was rarely described and was performed on C18 columns with mobile phases containing acetonitrile, water and ammonia [36] or methanol, water, ammonium acetate, and formic acid [35]. The cytotoxic activity of extracts obtained from *Meconopsis* species was also rarely investigated. Extract from *Meconopsis integrifolia* significantly inhibited human leukemia K562 cell viability, mainly by targeting apoptosis induction and cell cycle arrest in G2/M phase [34]. Extract from *Meconopsis horridula* induced murine leukemia L1210 cell apoptosis and inhibited proliferation through G2/M phase arrest [37].

The aim of this work was to investigate alkaloid compositions by HPLC-DAD and HPLC-MS/MS and anticancer activities of different isoquinoline alkaloids and plant extracts obtained from *Corydalis lutea, Dicentra spectabilis, Fumaria officinalis, Macleaya cordata, Mahonia aquifolia, Meconopsis cambrica* containing these alkaloids against various cancer cell lines. These extracts have not been previously tested against the cancer cell lines.

2. Results and Discussion

2.1. HPLC Analysis of Plant Extracts

Alkaloid standards (see Table 1) were chromatographed on Hydro RP and Polar RP columns in eluent system containing acetonitrile, water and 0.04 ML^{-1} of 1-butyl-3-methylimidazolium tetrafluoroborate in gradient elution mode described in section "Experimental". Because on the Hydro RP column with the octadecyl phase there was a worse shape of the peaks, lower theoretical plates number, and poorer separation selectivity of the investigated alkaloids, the RP Polar column was selected for further investigations (Table 1, Figure 1 and Figure S1). Retention times (t_R), asymmetry factors (As), and theoretical plate number per meter (N/m) for investigated alkaloid standards are presented in Table 1. Application of chromatographic system with double protection against undesirable interactions of analytes with free silanol groups: Phenyl stationary phase with π-π interaction and mobile phase with the addition of ionic liquid as free silanol blocker allow to obtain high system efficiency, symmetrical peaks, and full separation of investigated alkaloids. For all alkaloids, As values between 0.82 and 1.42 and high N/m values (from 33,000 to 700,000) were obtained (Table 1).

Table 1. Values of retention time (t_R), asymmetry factor (A_S), and theoretical plate number per meter (N/m) obtained for alkaloid standards.

Name of Compound	Hydro RP (Octadecyl Stationary Phase)			Polar RP (Phenyl Stationary Phase)		
	t_R	As	N/m	t_R	As	N/m
Berberine	23.57	0.72	33650	34.74	1.42	243280
Chelerythrine	32.29	0.82	190010	40.75	1.37	708700
Magnoflorine	3.12	0.84	8780	3.89	0.82	33190
Palmatine	19.70	*	*	29.78	1.18	126770
Protopine	9.04	0.68	30160	12.54	0.97	49630
Sanquinarine	19.07	*	*	34.87	1.01	413350
Stylopine	13.01	1.28	58300	19.36	0.98	355570

* Very asymmetrical peak.

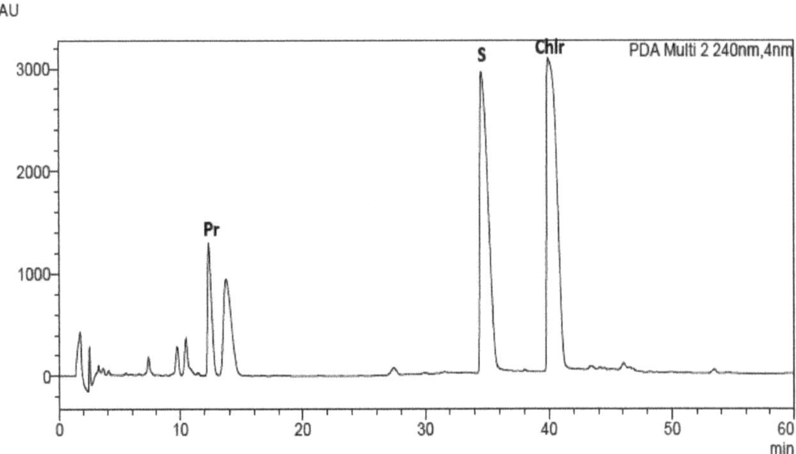

Figure 1. Chromatogram obtained for *Macleaya cordata* leaf extract obtained on Polar RP column with mobile phase containing MeCN, water, and 0.04 ML^{-1} IL. For gradient, see experimental section.

The same chromatographic system was used for the analysis of alkaloids in plant extracts obtained from *Corydalis lutea* root and herb, *Dicentra spectabilis*, *Fumaria officinalis*, *Macleaya cordata* leaves and herb, *Mahonia aquifolia* leaves and cortex, *Meconopsis cambrica* root and herb. The presence of alkaloids in plant extracts was identified by comparison of retention times with retention times of alkaloid standards, UV-Vis spectra and additionally confirmed by MS spectra (Figures S11–S21).

The quantitative analysis was performed by a calibration curve method. The number of replicates was three for all concentrations of all alkaloids. Calibration curve equations, correlation coefficients (r), limit of detection (LOD), and limit of quantification (LOQ) obtained for alkaloids are presented in Table 2.

Table 2. Equation of calibration curve, correlation coefficients (r), limit of detection (LOD) and limit of quantification (LOQ) values.

Alkaloid	Equation of Calibration Curve	r	LOD [mg/mL]	LOQ [mg/mL]
Berberine	$y = 72178227x - 370170$	0.9973	0.0151	0.0457
Chelerythrine	$y = 84228691x + 413980$	0.9998	0.0040	0.0123
Magnoflorine	$y = 23972503x + 263324$	0.9992	0.0094	0.0287
Palmatine	$y = 51166752x + 511129$	0.9991	0.0108	0.0327
Protopine	$y = 7344826x + 64160$	0.9992	0.0095	0.0288
Sanguinarine	$y = 80589787x + 606317$	0.9997	0.0123	0.0371
Stylopine	$y = 879342x - 13994$	0.9996	0.0241	0.0729

The obtained results also show that the content of alkaloids varied greatly not only with the kind of species but also in different parts of plants (Table 3). The identities of the analyte peaks in plant extracts were confirmed by the comparison of their retention times, UV-Vis spectra with the retention times and spectra of alkaloid standards and by MS detection. An example of the obtained chromatogram is presented in Figure 1. Chromatograms obtained for other plant extracts are presented in Supplementary Materials (Figures S2–S9). MS spectra obtained for alkaloid standards and alkaloids from extracts are presented in Figures S11–S21. Berberine was identified in *Mahonia aquifolium* cortex (above 0.13 mg/g of plant material). Chelerytrine was identified in three investigated extracts obtained from: *Fumaria officinalis*, *Macleaya cordata* leaves, and *Macleaya cordata* herb. A high content of this alkaloid was found in extracts obtained from *Macleaya cordata* especially from leaves (above 5.6 mg/g of plant material). Magnoflorine was determined only in extracts from *Mahonia aquifolium*. The cortex of this plant species contained only 0.086 mg/g of plant material, while in the leaves, the content of magnoflorine was higher than 0.32 mg/g of plant material. Palmatine was identified in *Corydalis lutea* root and herb and *Mahonia aquifolium* cortex extracts. The highest content of this alkaloid was found in the root of *Corydalis lutea*, however, its content was only about 0.1 mg/g of plant material. Protopine was identified in the most investigated extracts. The highest amount of these alkaloids was determined in the extracts obtained from the roots of *Corydalis lutea* and *Dicentra spectabilis*. In both roots, near 5.4 mg of protopine per g of plant material was quantified. Sanquinarine was determined in extracts obtained from *Dicentra spectabilis* herb and root, *Fumaria officinalis*, *Macleaya cordata* leaves and herb, *Meconopsis cambrica* root. Especially a lot of sanguinarine content was found in the leaves of *Macleaya cordata*, above 3 mg/g of plant material. Stylopine was identified in *Corydalis lutea* root and herb and *Fumaria officinalis* extracts. In all plant material, the content of stylopine was above 2 mg/g and in *Corydalis lutea* root was higher than 4 mg/g of plant material.

Table 3. Content of alkaloids in plant samples.

Name of Compound	Content of Alkaloids (mg/g of Plant Material)									
	Corydalis lutea Root	*Corydalis lutea* Herb	*Dicentra speclebilis*	*Fumaria officinalis*	*Macleya cordata* Leaves	*Macleya cordata* Herb	*Mahonia aquifalium* Cortex	*Mahonia aquifalium* Leaves	*Meconopsis cambrica* Root	*Meconopsis cambrica* Herb
Berberine	ND	ND	ND	ND	ND	ND	0.1332	ND	ND	ND
Chelerythrine	ND	ND	ND	0.0598	5.6061	1.7654	ND	ND	ND	ND
Magnoflorine	ND	ND	ND	ND	ND	ND	0.0863	0.3251	ND	ND
Palmatine	0.1041	0.03168	ND	ND	ND	ND	0.0360	ND	ND	ND
Protopine	5.4562	0.5526	5.3756	2.7873	1.7621	0.4731	ND	ND	0.0236	0.0787
Sanquinarine	ND	ND	0.0940	0.0097	3.1253	0.7699	ND	ND	0.0504	0.0542
Stylopine	4.0774	2.0725	ND	2.8251	ND	ND	ND	ND	ND	ND

ND, not detected.

2.2. Investigation of In Vitro Anticancer Activity of Alkaloid Standards

The cytotoxic activity of alkaloid standards: Berberine, chelerythrine, magnoflorine, palmatine, protopine, sanquinarine, and stylopine were carried out using the following human cancer cell lines: Human pharyngeal squamous carcinoma cells (FaDu), human tongue squamous carcinoma cells (SCC-25), human breast adenocarcinoma cell line (MCF-7) and human triple-negative breast adenocarcinoma cell line (MDA-MB-231). The results were expressed as IC50 values, which represent the concentrations that inhibited cell growth by 50% (Table 4).

Table 4. Cytotoxic effect of the investigated alkaloids and etoposide against FaDu, SCC-25, MCF-7, and MDA-MB-231 cells.

	IC_{50} [μM] ± SD			
	FaDu	**SCC 25**	**MCF-7**	**MDA-MB-231**
Berberine	27.51± 6.72	84.24 ± 7.75	113.42 ± 14.69	51.05 ± 9.07
Chelerythrine	6.11 ± 0.32	7.49 ± 0.77	9.10 ± 0.55	7.11 ± 0.26
Magnoflorine	>500	>500	>500	>500
Palmatine	94.27 ± 9.39	320.23 ± 46.34	454.77 ± 24.52	423.38 ± 34.22
Protopine	234.95 ± 37.55	298.73 ± 33.42	429.54 ± 34.92	370.13 ± 22.18
Sanguinarine	0.84 ± 0.03	1.41 ± 0.12	0.84 ± 0.06	1.26 ± 0.03
Stylopine	193.26 ± 3.80	340.40 ± 31.21	207.18 ± 16.95	489.51 ± 40.86
Etoposide	38.73 ± 1.56	223.94 ± 24.81	136.48 ± 8.95	219.31 ± 24.47

Varied cytotoxicity of alkaloid standards against the investigated cell lines was determined. The lowest IC_{50} values were obtained for sanquinarine on all tested cell lines (from 0.84 μM against FaDu and MCF-7 to 1.41 μM against SCC-25 cell lines). Very high cytotoxicity against investigated cell lines was also observed for chelerythrine (IC_{50} from 6.11 μM against FaDu to 9.10 μM against MCF-7 cell lines). Low IC_{50} values were found in berberine (IC_{50} from 27.51 against FaDu μM to 113.42 μM against MCF-7 cell lines). Higher IC_{50} values were obtained for the other alkaloids, which indicates their lower cytotoxicity. However, IC_{50} values obtained against all tested cell lines were higher than 500 μM only for magnoflorine.

2.3. Investigation of In Vitro Anticancer Activity of Plant Extracts

In vitro cytotoxic activity of the investigated plant extracts was examined on the same cancer cell lines as previously investigated alkaloids, FaDu and SCC-25 cell lines belong to so-called head and neck squamous cell carcinomas (HNSCC), which account for nearly 90% of head and neck cancers. These types of cancers are often resistant to chemotherapy, including even targeted drug therapy [38,39]. Therefore, HNSCCs are characterized by a high recurrence rate and five-year survival rate in patients with HNSCCs is about 40–60%. They are also the eighth most frequent cause of cancer-related deaths. Both FaDu and SCC-25 cell lines are commonly used for testing small molecules and biological samples during cancer drug development [40].

In turn, breast cancer is the most common cancer in women worldwide. The metastasis of cancer cells is the main reason for deaths of patients suffering from breast cancer. Especially the triple-negative breast cancer (TNBC), which is characterized by the lack of expression of HER-2, progesterone (PR) and estrogen (ER) receptors, exhibits poor prognosis. TNBC accounts for 10% of all cases of breast cancers. The five-year survival rate is estimated to be around 62% in TNBC patients and 75% in non-TNBC patients [41]. Taking into consideration the above-mentioned facts, in our current studies, MDA-MB-231, and MCF-7 cells have been investigated as recognized models of TNBC and non-TNBC, respectively [42].

All human cancer lines were treated by plant extracts in concentrations 10, 25, 50, and 100 μg/mL for preliminary evaluation of cytotoxic properties of these extracts. For comparison of cytotoxic activity, experiments were also performed for anticancer drug, etoposide, on the same cell lines and at the same concentrations as plant extracts. Results were reported as the percentage of relative viability of the treated cells when compared to the untreated control cells (Figures 2–5). All investigated plant extracts exhibit very high cytotoxic activity, especially at higher concentrations. At a concentration of 100 μg/mL, almost all plant extracts showed greater cytotoxic activity against all tested human cancer cell lines compared to cytotoxic activity obtained for etoposide. Lower cytotoxic activity at a concentration of 100 μg/mL was obtained only for the extract from *Fumaria officinalis* against FaDu and MCF-7 cells. Due to the promising results of the preliminary studies, the extract was examined in at least eight different concentrations in order to determine median inhibitory concentration (IC_{50}) values (Table 5).

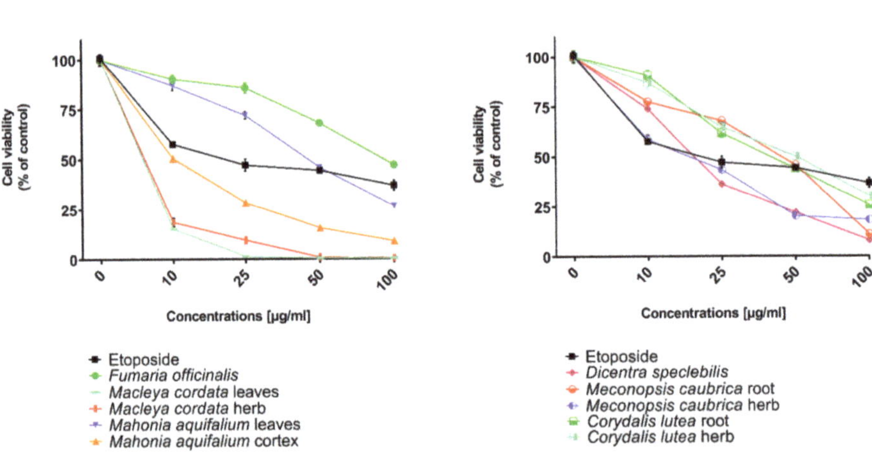

Figure 2. Influence of plant extracts and etoposide concentrations on human pharyngeal squamous carcinoma cells (FaDu) viability.

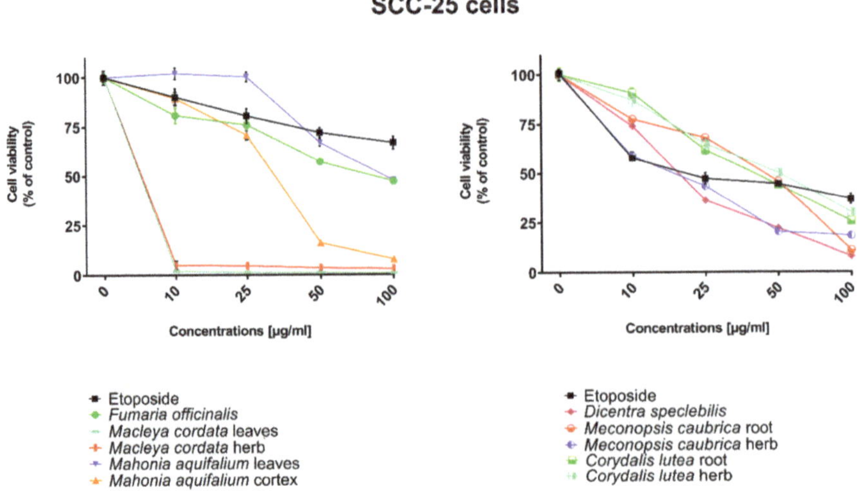

Figure 3. Influence of plant extracts and etoposide concentrations on human tongue squamous carcinoma cells (SCC-25) viability.

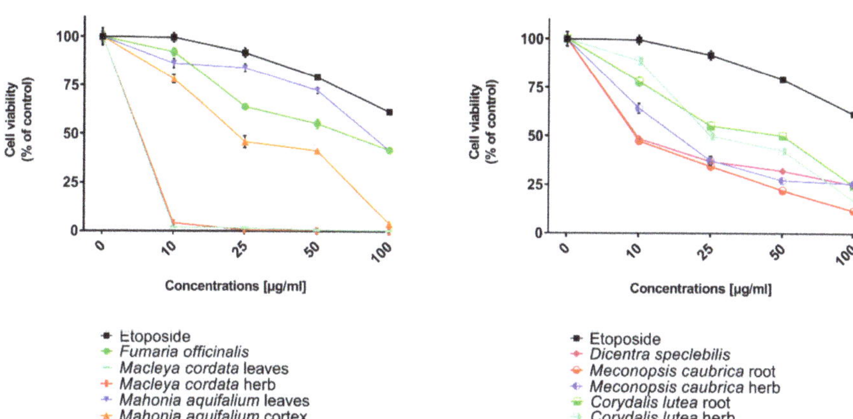

Figure 4. Influence of plant extracts and etoposide concentrations on human triple-negative breast adenocarcinoma cell line (MDA-MB-231) viability.

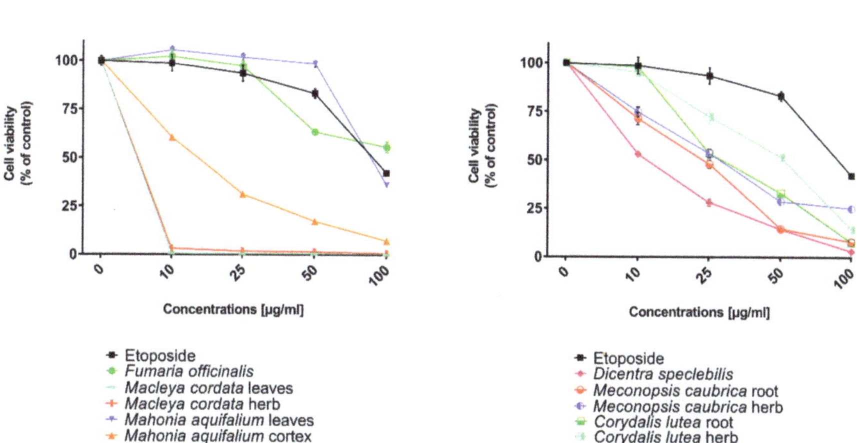

Figure 5. Influence of plant extracts and etoposide concentrations on human breast adenocarcinoma cell line (MCF-7) viability.

The highest cytotoxic effect on all tested cell lines was observed for extracts obtained from the herb, and especially leaves, of *Macleaya cordata*. The extract obtained from leaves exhibit greater cytotoxic effect on MCF-7, FaDu, and SCC-25 cell lines (viability of cells at the concentration of extracts at 100 µg/mL were 0.06%, 0.33%, and 0.90%, respectively), while the extract obtained from the herb of the plant was more potent against MDA-MB-231 cell lines (viability of cells at concentrations of 100 µg/mL was 0.10%). At a concentration of only 10 µg/mL of the extract obtained from *Macleaya cordata* leaves, the viability of SCC-25 and MDA-MB-231 cell lines was less than 2%. Viability of MCF-7 cell line at a concentration of 10 µg/mL was only 0.66%. This indicated very high cytotoxic activity of the extract, repeatedly greater than the cytotoxicity of etoposide in relation to these cell lines. Moreover, the extract obtained from the herb of this plant showed high cytotoxicity against the same cell lines (cell viability was about or below 5%). IC_{50} values obtained from *Macleaya cordata* leaf extract were very low (in the range of 1.86 µg/mL against MCF-7 cell line–2.19 µg/mL against SCC-25 cell line). IC_{50} values obtained for extracts obtained from *Macleaya cordata* herb were only slightly higher (in the range of 2.14–2.57 µg/mL). In these extracts, a very high content of alkaloids with strong cytotoxic properties against the tested cancer cells was detected. The extract obtained from the leaves contained above 5.6 mg/g of plant material of chelerythrine which is 46.6% of the dry mass of the extract and 3.1 mg/g of plant material of sanquinqrine which is 25.98% of the dry mass of the extract (Table 2). Sanguinarine and chelerythrine have very high cytotoxicity against all tested cancer cell lines. The content of these alkaloids is 72.58% in dry mass of the extract. The plant material seems to be a good candidate for the obtaining of these alkaloids as well as for further research on its anticancer activity. Slightly smaller cytotoxic activity was observed for herb extract obtained from the same plant which corresponds to a lower content of these two alkaloids in this extract (29.42% and 12.83% of chelerythrine and sanguinarine in dry mass of the extract, respectively). These results may also indicate the accumulation of these alkaloids in the leaves of this plant.

Table 5. Cytotoxic activity of the investigated plant extracts against different cancer cell lines.

	IC_{50} [µg/mL] ± SD			
Plant Sample	**FaDu**	**SCC-25**	**MCF-7**	**MDA-MB-231**
Fumaria officinalis	102.76 ± 13.03	101.46 ± 5.96	>200	85.60 ± 13.25
Macleaya cordata leaves	1.94 ± 0.27	2.19 ± 0.09	1.86 ± 0.08	2.09 ±0.10
Macleaya cordata herb	2.57 ± 0.24	2.50 ± 0.26	2.14 ± 0.18	2.42 ± 0.21
Mahonia aquifalium leaves	46.77 ± 7.84	97.25 ± 8.07	89.14 ± 2.73	90.71 ± 7.29
Mahonia aquifalium cortex	7.67 ± 0.82	31.37 ± 2.29	15.71 ± 1.92	31.87 ± 4.35
Dicentra speclebilis	19.88 ± 2.26	29.55 ± 4.09	11.66 ± 1.36	9.66 ± 0.42
Meconopsis caubrica root	43.66 ± 4.78	27.96 ± 1.03	23.26 ± 3.69	9.98 ± 1.34
Meconopsis caubrica herb	13.70 ± 1.24	48.02 ± 4.89	31.60 ± 2.42	21.10 ± 1.96
Corydalis lutea root	38.08 ± 3.50	142.14 ± 10.58	29.37 ± 4.01	57.98 ± 10.67
Corydalis lutea herb	47.47 ± 4.50	48.06 ± 0.86	49.34 ± 5.05	31.39 ± 1.82
Etoposide	22.80 ± 0.92	131.80 ± 14.6	80.33 ± 5.27	129.08 ± 14.4

Very high cytotoxic activity was also obtained after treating all investigated cell lines by the extract obtained from *Mahonia aquifolium* cortex. The viability of all tested cancer cell lines treated by the extract at a concentration of 100 µg/mL was below 10%. The extract showed the greatest cytotoxic effect on the MDA-MB-231 cell line, the viability was only 3.88%. Significantly lower cytotoxic activity was observed for the extract obtained from *Mahonia aquifolium* leaves. The viability of cells after treating with the extract at a concentration of 100 µg/mL was from 26.5% for FaDu line to 47.63% for SCC-25 cell line. The content of alkaloids in extracts obtained from leaves and cortex of *Mahonia aquifolium* was significantly different. The IC50 values of *Mahonia aquifolium* cortex extract were very low, especially against FaDu and MCF-7 cell lines, 7.67 and 15.71 µg/mL respectively. Higher IC_{50} values were observed for extract from *Mahonia aquifalium* leaves (46.77–97.25 µg/mL). In *Mahonia aquifolium* leaves, only magnoflorine with low cytotoxic activity was determined, but in cortex, besides magnoflorine and palmatine with low and medium cytotoxic activity, also berberine with high cytotoxicity was found. The extract obtained from *Mahonia aquifalium* cortex contained 0.1332 µg/g of plant material of berberine, which is 2.26% of the dry mass of the extract. The extract exhibits the highest cytotoxicity against FaDu cell line. Berberine has also the highest cytotoxic activity against the FaDu cell line. In the extract obtained from *Mahonia aquifalium* leaves, berberine was not identified and cytotoxicity of the extract was also significantly lower. This may indicate a significant effect of berberine content on the cytotoxicity of extracts. Cytotoxicity of the plant extract can be caused by other components of the extract not identified in our investigations. This requires further investigations.

Potential anticancer activity was found for the extract obtained from *Dicentra spectabilis*. The cytotoxic activity of the plant was different from various types of applied cell lines and also significantly changed with the change of extract concentration. Viability of MDA-MB-231 cells were about 25% at the concentration of extract equaled 100 μg/mL. For other cell lines viability of cells treated by *Dicentra spectabilis* extract was significantly lower: 8.04%, 3.11%, and 3.05% for FaDu, MCF-7 and SCC-25 cell lines, respectively. In the plant, the high content of protopine (above 5 mg/g of plant material) and about 0.1 mg/g of plant material of sanquinarine with high cytotoxicity were determined. Extract obtained from *Dicentra spectabilis* exhibits the lowest IC_{50} values against MDA-MB-231 cell line (only 9.66 μg/mL). In the extract, very high content of protopine was found (95.65% of the dry mass of the extract), but protopine exhibits low cytotoxicity against all cell lines tested by us. High cytotoxic activity of the extract can be caused by the presence of high cytotoxic sanguinarine (1.68% of the dry mass of the extract).

High cytotoxic activity against all tested cell lines was also found for the extract obtained from the root of *Meconopsis cambrica*. Viability of cancer cells treated by the extract significantly decreased with the increase of the extract concentration. At a concentration 100 μg/mL, it was 11.82%, 11.10%, 7.98, and 4.40% for MDA-MB-231, FaDu, MCF-7 and SCC-25 cell lines, respectively. The root extract contains sanquinarine having a very strong cytotoxic effect on all cell lines used in the investigations. Less cytotoxicity was found when the extract obtained from the herb of this plant was applied. The lowest viability was obtained for SCC-25 cell line after treating by the extract at a concentration of 100 μg/mL of *Meconopsis cambrica* root. Great differences were obtained in IC_{50} values for extracts from *Meconopsis caubrica* root and herb. For example, IC_{50} for root extract against FaDu cell line was 43.66 μg/mL, while IC_{50} obtained for extracts from the herb was 13.79 μg/mL against the same cell line. Whereas the extract obtained from *Meconopsis caubrica* root exhibits the lowest IC_{50} values against the other cell lines. The cytotoxic properties of the extract may be caused by the presence of very high cytotoxic sanguinarine which is 1.12% and 0.8% of the dry mass of extracts obtained from herb and root, respectively.

Potential anticancer activity was also determined for *Corydalis lutea* root and herb extracts. Higher cytotoxic activity in relation to MCF-7 and FaDu cell lines was observed for the extract obtained from the root in comparison with the extract obtained from the herb, while higher cytotoxicity in relation to SCC-25 and MDA-MB-231 cell lines was obtained for the herb extract in comparison to the extract from root. The highest cytotoxic effect was found when MCF-7 cells were treated by extract obtained from the root at a concentration of 100 μg/mL (viability of cells was only 7.15%). The lowest cytotoxicity was observed for *Corydalis lutea* root against SCC-25 cell line, but at concentrations of 50 and 100 μg/mL was higher than cytotoxic activity of etoposide at the same concentrations. In extracts obtained from this plant alkaloids having the highest cytotoxic activity were not detected. The IC50 values of *Corydalis lutea* root extract to the investigated cell lines were in the range of 29.37–142.14 μg/mL, while the IC50 values of *Corydalis lutea* herb extract in the range of 31.39–49.34 μg/mL. Lower cytotoxic activity of plant extracts obtained from *Corydalis lutea* against the tested cell lines may be the result of a lack of alkaloids, such as chelerythrine, sanguinarine or berberine with high cytotoxicity in these extracts.

The lowest cytotoxic activity against the tested cell lines was found for the extract obtained from *Fumaria officinalis*. The viability of cells after treatment by the extract at a concentration of 100 μg/mL was from 41.74% for MDA-MB-231 cell line to 55.64% for MCF-7 cell line. The extract at the highest concentration showed greater cytotoxic activity in comparison with the cytotoxicity of etoposide against MDA-MB-231 and SCC-25 cell lines., respectively. The highest IC_{50} values from all investigated plant extracts were also obtained for the extract from *Fumaria officinalis* (from 85.60 μg/mL against MDA-MB-231 to >200 μg/mL against. MCF-7 cell line). In this extract, chelerythrine, protopine, sanquinarine, and stylopine were determined, but alkaloids possessing the highest cytotoxic activity (chelerythrine and sanquinarine) were detected in very small concentrations (0.0598 and 0.0097 mg/g of plant material which is 0.49 and 0.08% of the dry mass of the extract).

3. Experimental

3.1. Chemicals and Plant Material

Acetonitrile (MeCN), methanol (MeOH), 1-butyl-3-methylimidazolium tetrafluoroborate of chromatographic quality were obtained from E. Merck (Darmstadt, Germany), dimethyl sulfoxide (DMSO) was from Sigma-Aldrich (Saint Louis, MO, USA), dimethyl sulfoxide (DMSO) was from Sigma-Aldrich (Saint Louis, MO, USA).

Alkaloid standards (berberine, magnoflorine, palmatine, protopine, sanguinarine, chelerythrine and stylopine) were purchased from Chem Faces Biochemical Co. Ltd. (Wuhan, China). Berberine was purchased from Sigma-Aldrich (St. Louis, MO, USA).

Plant material was collected and identified in the Botanical Garden of Maria Curie-Skłodowska University in Lublin (Poland) in spring and summer 2018.

Plants were divided into roots and aboveground parts. Plants organs were cut into pieces and dried at ambient temperature for one to two weeks.

3.2. Apparatus and HPLC Conditions

HPLC-DAD

The analysis was performed using an LC-20AD Shimadzu (Shimadzu Corporation, Canby, OR, USA) liquid chromatograph equipped with Synergi Hydro RP 80A (150 × 4.6 mm, 5 μm) and Synergi Polar RP 80A (150 × 4.6 mm, 5 μm) columns. The chromatograph was equipped with a Shimadzu SPD-M20A detector (Shimadzu Corporation, Canby, OR, USA). Detection was carried out at a wavelength of 240 nm. All chromatographic measurements were controlled by a CTO-10ASVP thermostat (Shimadzu Corporation, Canby, OR, USA). The eluent flow rate was 1.0 mL/min. Extracts were injected into the columns using the Rheodyne 20 μL injector. The DAD detector was set in the 200–800 nm range. Data acquisition and processing were carried out with LabSolutions software (Shimadzu Corporation, Kyoto, Japan). The mobile phase was composed of 0.04 ML^{-1} 1-butyl-3-methylimidazolium tetrafluoroborate in water (solvent A) and 1-butyl-3-methylimidazolium tetrafluoroborate in acetonitrile (solvent B) in gradient elution: 0–20 min, 25% B; 20–30 min, 25%–32% B; 30–40 min, 32%–40% B, 40–60 min, 40% B. Flow rate was 1 mL/min.

Calibration curves were constructed by analyzing the alkaloid standards at eight concentrations, ranging from 0.001 to 0.2 mg/mL. The calibration curves were obtained by means of the least square method. The limit of detection (LOD) and limit of quantification (LOQ) obtained for alkaloids were calculated according to the formula: LOD = 3.3 (SD/S), and LOQ = 10 (SD/S), where SD is the standard deviation of response (peak area) and S is the slope of the calibration curve.

HPLC analyses of alkaloid standards and plant extracts were repeated three times.

HPLC-MS

Determination of alkaloids was carried out using an HPLC system equipped with the Agilent XDB-C18 1.8 μm 4.6 × 50 mm column. The column was maintained at 20 °C. The injected sample volume was 20 μL, while the mobile phase was composed ACN + 0.1% HCOOOH (30:70) dosed at a flow rate of 0.6 mL/min. The mass spectral analysis was performed on a UHPLC-QTOF/MS model 1260, 6530 Accurate-Mass QTOF LC/MS; Agilent Technologies (Santa Clara, CA, USA) equipped with an ESI interface operating in positive ion mode, with the following set of operation parameters: Capillary voltage, 3500 V; nebulizer pressure, 40 psi; drying gas flow, 7 L/min; drying gas temperature, 295 °C; LC–MS mass spectra were recorded across the range mass range 40–370 m/z; fragmentor 195 V. The HPLC–MS data were acquired and quantified with the use of MassHunter Workstation software. The data were further processed using Microsoft Excel. The instrument was operated in selected ions monitoring mode (SIM) and multiple reaction monitoring (MRM) as well. The monitored pseudomolecular ions $[M^{+}H]^{+}$ are presented in Table 6.

Table 6. The monitored pseudomolecular ions $[M^{+}H]^{+}$ parameters.

	m/z (~)	Q1 (~)	Q3 (~)	Iso. Width	Collison Energy
Berberine	336	320	292	Medium(~4m/z)	35
Chelerythrine	348	332	304	Medium(~4m/z)	35
Magnoflorine	342	296	236	Medium(~4m/z)	30
Palmatine	352	336	308	Medium(~4m/z)	25
Protopine	354	189	149	Medium(~4m/z)	35
Sanguinarine	332	273	316	Medium(~4m/z)	25
Stylopine	324	176	149	Medium(~4m/z)	35

3.3. Extraction Procedure

The previously described procedure of alkaloids extraction from plant material was applied after minor modifications [43,44].

Samples (5 g) of each plant were macerated with 100 mL ethanol for 72 h and continuously extracted in an ultrasonic bath for 5 h. Extracts were filtered, the solvent evaporated under vacuum, and the residues dissolved in 30 mL of 2% sulfuric acid and defatted with diethyl ether (3 × 40 mL). The aqueous layers were subsequently basified with 25% ammonia to a pH of 9.5–10 and the alkaloids extracted with chloroform (3 × 50 mL). After evaporation of the organic solvent, the dried extracts were dissolved in 5 mL MeOH prior to HPLC analysis. Recovery (%) obtained for alkaloids by the extraction procedure is presented in Table S1.

3.4. Investigation of Cytotoxic Activity

Cytotoxic properties of the tested plant extracts and respective secondary metabolites' standards were examined using human pharyngeal squamous carcinoma cells (FaDu), human tongue squamous carcinoma cells (SCC-25), human breast adenocarcinoma cell line (MCF-7), human triple-negative breast adenocarcinoma cell line (MDA-MB-231). Cell lines used in experiments were obtained from the American Type Culture Collection (ATCC, Manassas, VA, USA). All media, buffers, nutrients, and solutions necessary for cell culturing were from Sigma-Aldrich (St. Louis, MO, USA). FaDu cells were cultured using Eagle's minimum essential medium (MEM) supplemented with 10% fetal bovine serum, 100 U/mL of penicillin, and 100 mg/mL of streptomycin. SCC-25 was cultured in Dulbecco's modified Eagle's medium/nutrient mixture F-12 Ham (DMEM/F12) supplemented with 10% fetal bovine serum, 400 ng/mL hydrocortisone, 100 U/mL of penicillin, and 100 mg/mL of streptomycin (all from Sigma-Aldrich). MCF-7 and MDA-MB-231 cells were cultured using Dulbecco's modified Eagle's medium-high glucose (DMEM) containing 10% fetal bovine serum, 100 U/mL of penicillin and 100 mg/mL of streptomycin. Cells were maintained at 37 °C in a 5% CO_2 atmosphere. Both the dried plant extracts and standards were dissolved in DMSO in order to obtain stock solutions at a concentration of 50 mg/mL. On the day of the experiment, suspension of cells (1 × 105 cells/mL) in the respective medium containing 10% FBS was applied to a 96-well plate at 100µL per well. After 24 h of incubation, the medium was removed from wells and replaced by different concentrations (10–100 µg/mL) of plant extracts or standards in a medium containing 2% FBS. Control cells were cultured only with a medium containing 2% FBS. Cytotoxicity of DMSO was also checked at concentrations present in respective dilutions of stock solutions. After 24 h incubation, 15 µL of MTT working solution (5mg/mL in PBS) was added to each well. The plate was incubated for 3 h. Subsequently, 100 uL of 10% SDS solution was added to each well. Cells were incubated overnight at 37 °C to dissolve the precipitated formazan crystals. The concentration of the dissolved formazan was evaluated by measuring the absorbance at λ = 570 nm using a microplate reader (Epoch, BioTek Instruments, Inc., USA). Two independent experiments were performed in triplicate. The viability of cells incubated with the increased concentrations of plant extracts was expressed as % of the viability of control (untreated) cells (Figures 2–5). The investigated standards and extracts were subsequently assayed in at least eight different concentrations in order to calculate the respective IC_{50} from the logarithmic dose-response curve (Tables 4 and 5). The results of the MTT assay were expressed as mean ± SD. DMSO in the concentrations present in the dilutions of stock solutions did not influence the viability of the tested cells.

4. Conclusions

Our studies confirmed a strong ability of the examined alkaloids to inhibit proliferation of cancer cells. The highest cytotoxic activity against all tested cancer cell lines was observed for sanguinarine (IC_{50} < 1.5 μM). Very high cytotoxicity was also obtained for chelerythrine (IC_{50} < 10 μM against all investigated cell lines).

Our data demonstrated that the extracts obtained from *Corydalis lutea* root and herb, *Dicentra spectabilis*, *Fumaria officinalis*, *Macleaya cordata* leaves and herb, *Mahonia aquifolia* leaves and cortex, *Meconopsis cambrica* root and herb have very high cytotoxic activity against MDA-MB-231, FaDu, MCF-7 and SCC-25 cancer cell lines. To the best of our knowledge, the cytotoxic activity of the extracts has not been yet investigated against the cell lines tested by us.

Almost all investigated extracts, especially at higher concentrations, showed cytotoxic activity against tested cell lines significantly higher than cytotoxic activity of an anticancer drug, etoposide.

Higher cytotoxicity was found for extracts containing highly cytotoxic alkaloids: chelerythrine, sanguinarine, and berberine. The highest cytotoxic activity against all tested cancer cell lines (IC_{50} were between 1.86 to 2.19 μg/mL) was observed after applying the *Macleaya cordata* leaf extract containing 46.6 % of the dry mass of the extract of chelerythrine, and 25.98% of sanquinarine.

The differences between the cytotoxic activities of the different parts of investigated plants strongly depend on alkaloids content and a synergic effect of the different alkaloids may influence on extract activities.

The investigated plant material, especially that obtained from *Macleaya cordata*, *Mahonia aquifalium*, *Dicentra speclebilis* and *Meconopsis caubrica*, seems to be promising for further research on its anticancer activity.

Supplementary Materials: The following are available online at http://www.mdpi.com/2072-6651/11/10/575/s1, Figure S1A. Chromatogram obtained for alkaloid standards obtained on Hydro RP column with mobile phase containing MeCN, water and 0.04 ML-1 IL. Gradient see experimental section, Figure S1B. Chromatogram obtained for alkaloid standards obtained on Polar RP column with mobile phase containing MeCN, water and 0.04 ML-1 IL. Gradient see experimental section, Figure S2. Chromatogram obtained for Meconopsis caubrica root extract obtained on Polar RP column with mobile phase containing MeCN, water and 0.04 ML-1 IL. Gradient see experimental section, Figure S3. Chromatogram obtained for Meconopsis caubrica herb extract obtained on Polar RP column with mobile phase containing MeCN, water and 0.04 ML-1 IL. Gradient see experimental section, Figure S4. Chromatogram obtained for *Mahonia aquifalium* leaves extract obtained on Polar RP column with mobile phase containing MeCN, water and 0.04 ML-1 IL. Gradient see experimental section, Figure S5. Chromatogram obtained for *Mahonia aquifalium* cortex extract obtained on Polar RP column with mobile phase containing MeCN, water and 0.04 ML-1 IL. Gradient see experimental section, Figure S6. Chromatogram obtained for *Macleaya cordata* herb extract obtained on Polar RP column with mobile phase containing MeCN, water and 0.04 ML-1 IL. Gradient see experimental section, Figure S7. Chromatogram obtained for Dicentra speclebilis extract obtained on Polar RP column with mobile phase containing MeCN, water and 0.04 ML-1 IL. Gradient see experimental section, Figure S8. Chromatogram obtained for *Fumaria officinalis* extract obtained on Polar RP column with mobile phase containing MeCN, water and 0.04 ML-1 IL. Gradient see experimental section, Figure S9. Chromatogram obtained for *Corydalis lutea* root extract obtained on Polar RP column with mobile phase containing MeCN, water and 0.04 ML-1 IL. Gradient see experimental section, Figure S10. Chromatogram obtained for *Corydalis lutea* herb extract obtained on Polar RP column with mobile phase containing MeCN, water and 0.04 ML-1 IL. Gradient see experimental section, Figure S11A. MS spectrum obtained for standard of berberine, Figure S11B. MS spectrum obtained for standard of chelerythrine, Figure S11C. MS spectrum obtained for standard of magnoflorine, Figure S11D. MS spectrum obtained for standard of palmatine, Figure S11E. MS spectrum obtained for standard of protropine, Figure S11F. MS spectrum obtained for standard of sanguinarine, Figure S11G. MS spectrum obtained for standard of stylopine, Figure S12A. MS spectrum obtained for *Mahonia aquifalium* cortex extract, Figure S12B. MS spectrum obtained for berberine from *Mahonia aquifalium* cortex extract, Figure S12C. MS spectrum obtained for palmatine from *Mahonia aquifalium* cortex extract, Figure S12D. MS spectrum obtained for magnoflorine from *Mahonia aquifalium* cortex extract, Figure S13A. MS spectrum obtained for *Mahonia aquifalium* leaves, Figure S13B.

MS spectrum obtained for magnoflorine from *Mahonia aquifalium* leaves, Figure S14A. MS spectrum obtained for *Fumaria officinalis* extract, Figure S14B. MS spectrum obtained for chelerythrine from *Fumaria officinalis* extract, Figure S14C. MS spectrum obtained for protopine from *Fumaria officinalis* extract, Figure S14D. MS spectrum obtained for sanguinarine from *Fumaria officinalis* extract, Figure S14E. MS spectrum obtained for stylopine from *Fumaria officinalis* extract, Figure S15A. MS spectrum obtained for *Macleaya cordata* leaves extract, Figure S15B. MS spectrum obtained for chelerythrine from *Macleaya cordata* leaves extract, Figure S15C. MS spectrum obtained for protopine from *Macleaya cordata* leaves extract, Figure S15D. MS spectrum obtained for sanguinarine from *Macleaya cordata* leaves extract, Figure S16A. MS spectrum obtained for chelerythrine from *Macleaya cordata* herb extract, Figure S16B. MS spectrum obtained for chelerythrine from *Macleaya cordata* herb extract, Figure S16C. MS spectrum obtained for protopine from *Macleaya cordata* herb extract, Figure S16D. MS spectrum obtained for sanguinarine from *Macleaya cordata* herb extract, Figure S17A. MS spectrum obtained for *Corydalis lutea* root extract, Figure S17B. MS spectrum obtained for palmatine from *Corydalis lutea* root extract, Figure S17C. MS spectrum obtained for protopine from *Corydalis lutea* root extract, Figure S17D. MS spectrum obtained for stylopine from *Corydalis lutea* root extract, Figure S18A. MS spectrum obtained for *Corydalis lutea* herb extract, Figure S18B. MS spectrum obtained for palmatine from *Corydalis lutea* herb extract, Figure S18C. MS spectrum obtained for protopine from *Corydalis lutea* herb extract, Figure S18D. MS spectrum obtained for stylopine from *Corydalis lutea* herb extract, Figure S19A. MS spectrum obtained for *Dicentra spectabilis* herb extract, Figure S19B. MS spectrum obtained for protopine from *Dicentra spectabilis* herb extract, Figure S19C. MS spectrum obtained for sanguinarine from *Dicentra spectabilis* herb extract, Figure S20A. MS spectrum obtained for *Meconopsis cambrica* root extract, Figure S20B. MS spectrum obtained for protopine from *Meconopsis cambrica* root extract, Figure S20C. MS spectrum obtained for sanguinarinee from *Meconopsis cambrica* root extract, Figure S21A. MS spectrum obtained for *Meconopsis cambrica* herb extract, Figure S21B. MS spectrum obtained for protopine from *Meconopsis cambrica* herb extract, Table S1. Exemplary of recoveries of extraction (%) obtained for alkaloids.

Author Contributions: Conceptualization, A.P., T.T., T.P. and M.W.-H.; methodology, A.P., T.T., T.P., M.W.-H., M.S.-M and B.B.; software, J.M. and T.P.; validation, A.P., T.T., T.P., J.M., K.S.; formal analysis, A.P., T.T., T.P., J.M., M.S.-M., K.S.; investigation, A.P., T.T., T.P., M.S.-M, J.M., B.K.; resources, D.M.; data curation, A.P., T.T., T.P., J.M., B.K., M.S.-M.; writing—original draft preparation, A.P., T.T., and T.P. writing—review and editing, A.P., T.T., T.P., and M.W.-H. visualization, J.M. and T.P.; supervision, A.P., T.T., T.P., M.W.-H. project administration, A.P. and M.W.-H., funding acquisition, M.W.-H., A.P., T.T., B.B., M.S.-M. and T.P.

Funding: This research received no external funding.

Conflicts of Interest: The authors declare no conflict of interest.

References

1. Wong, B.-S.; Hsiao, Y.-C.; Lin, T.-W.; Chen, K.-S.; Chen, P.-N.; Kuo, W.-H.; Chue, S.-C.; Hsieh, Y.-S. The in vitro and in vivo apoptotic effects of *Mahonia oiwakensis* on human lung cancer cell. *Chem. Biol. Interact.* **2009**, *180*, 165–174. [CrossRef] [PubMed]
2. Godevac, D.; Damjanovic, A.; Stanojkovic, T.P.; Andelkovic, B.; Zdunic, G. Identification of cytotoxic metabolites from *Mahonia aquifolium* using1H NMR-based metabolomics approach. *J. Pharm. Biomed. Anal.* **2018**, *150*, 9–14. [CrossRef] [PubMed]
3. Zhang, S.-L.; Li, H.; He, X.; Zhang, R.-Q.; Sun, Y.-H.; Zhang, C.-F.; Wang, C.-Z.; Yuan, C.-S. Alkaloids from *Mahonia bealei* posses anti-H+/K+-ATPase andanti-gastrin effects on pyloric ligation-induced gastric ulcer in rats. *Phytomedicine* **2014**, *21*, 1356–1363. [CrossRef] [PubMed]
4. Zhang, L.; Zhu, W.; Zhang, Y.; Yang, B.; Fu, Z.; Li, X.; Tian, J. Proteomics analysis of *Mahonia bealei* leaves with induction of alkaloids via combinatorial peptide ligand libraries. *J. Proteom. S* **2014**, *110*, 59–71. [CrossRef] [PubMed]
5. Hu, W.; Yu, L.; Wang, M.-H. Antioxidant and antiproliferative properties of water extract from *Mahonia bealei* (Fort.) Carr. Leaves. *Food Chem. Toxicol.* **2011**, *49*, 799–806. [CrossRef]
6. Wang, W.; Ma, X.; Guo, X.; Zhao, M.; Tu, P.; Jiang, Y. A series of strategies for solving the shortage of reference standardsfor multi-components determination of traditional Chinese medicine, *Mahoniae Caulis* as a case. *J. Chromatogr. A* **2015**, *1412*, 100–111. [CrossRef] [PubMed]
7. Bribi, N.; Algieri, F.; Rodriguez-Nogales, A.; Vezza, T.; Garrido-Mesa, J.; Utrilla, M.P.; del Mar Contreras, M.; Maiza, F.; Segura-Carretero, A.; Rodriguez-Cabezas, M.E.; et al. Intestinal anti-inflammatory effects of total alkaloid extract from Fumaria capreolata in the DNBS model of mice colitis and intestinal epithelial CMT93 cells. *Phytomedicine* **2016**, *23*, 901–913. [CrossRef]

8. Del Mar Contreras, M.; Bribi, N.; Gómez-Caravaca, A.M.; Gálvez, J.; Segura-Carretero, A. Alkaloids Profiling of Fumaria capreolata by Analytical Platforms Based on the Hyphenation of Gas Chromatography and Liquid Chromatography with Quadrupole-Time-of-Flight Mass Spectrometry. *Int. J. Anal. Chem.* **2017**. [CrossRef]
9. Tabrizi, F.H.A.; Irian, S.; Amanzadeh, A.; Heidarnejad, F.; Gudarzi, H.; Salimi, M. Anti-proliferative activity of *Fumaria vaillantii* extracts on different cancer cell lines. *Res. Pharm. Sci.* **2016**, *11*, 152–159.
10. Hussain, T.; Siddiqui, H.H.; Fareed, S.; Vijayakumar, M.; Rao, C.V. Evaluation of chemopreventive effect of *Fumaria indica* against N-nitrosodiethylamine and CCl_4-induced hepatocellular carcinoma in Wistar rats. *Asian Pac. J. Trop. Med.* **2012**, *5*, 623–629. [CrossRef]
11. Li, L.; Huang, M.; Shao, J.; Lin, B.; Shen, Q. Rapid determination of alkaloids in *Macleaya cordata* using ionic liquid extraction followed by multiple reaction monitoring UPLC–MS/MS analysis. *J. Pharm. Biomed. Anal.* **2017**, *135*, 61–66. [CrossRef] [PubMed]
12. Sun, M.; Zhao, L.; Wang, K.; Han, L.; Shan, J.; Wu, L.; Xue, X. Rapid identification of "mad honey" from Tripterygium wilfordii Hook. f. and *Macleaya cordata* (Willd) R. Br using UHPLC/Q-TOF-MS. *Food Chem.* **2019**, *294*, 67–72. [CrossRef] [PubMed]
13. Qinga, Z.-X.; Cheng, P.; Liu, X.-B.; Liu, Y.-S.; Zeng, J.-G. Systematic identification of alkaloids in *Macleaya microcarpa* fruits by liquid chromatography tandem mass spectrometry combined with the isoquinoline alkaloids biosynthetic pathway. *J. Pharm. Biomed. Anal.* **2015**, *103*, 26–34. [CrossRef] [PubMed]
14. Chena, Y.-Z.; Liu, G.-Z.; Shen, Y.; Chen, B.; Zeng, J.-G. Analysis of alkaloids in *Macleaya cordata* (Willd.) R. Br. using high-performance liquid chromatography with diode array detection and electrospray ionization mass spectrometry. *J. Chromatogr. A* **2009**, *1216*, 2104–2110. [CrossRef] [PubMed]
15. Kosina, P.; Gregorova, J.; Gruz, J.; Vacek, J.; Kolar, M.; Vogel, M.; Roos, W.; Naumann, K.; Simanek, V.; Ulrichova, J. Phytochemical and antimicrobial characterization of *Macleaya cordata* herb. *Fitoterapia* **2010**, *81*, 1006–1012. [CrossRef]
16. Liu, M.; Lin, Y.-l.; Chen, X.-R.; Liao, C.-C.; Poo, W.-K. In vitro assessment of *Macleaya cordata* crude extract bioactivity and anticancer properties in normal and cancerous human lung cells. *Exp. Toxicol. Pathol.* **2013**, *65*, 775–787. [CrossRef] [PubMed]
17. Deng, A.-J.; Qin, H.-L. Cytotoxic disbenzophenanthridine alkaloids from the roots of *Macleaya microcarpa*. *Phytochemistry* **2010**, *71*, 816–822. [CrossRef]
18. Jeong, E.-K.; Lee, S.Y.; Yu, S.M.; Park, N.H.; Lee, H.-S.; Yim, Y.-H.; Hwang, G.-S.; Cheong, C.; Jung, J.H.; Hong, J. Identification of structurally diverse alkaloids in *Corydalis* species by liquid chromatography/electrospray ionization tandem mass spectrometry. *Rapid Commun. Mass Spectrom.* **2012**, *26*, 1661–1674. [CrossRef]
19. Wang, Y.; Li, T.; Meng, X.; Bao, Y.; Wang, S.; Chang, X.; Yang, G.; Bo, T. Metabolomics and genomics: revealing the mechanism of corydalis alkaloid on anti-inflammation in vivo and in vitro. *Med. Chem. Res.* **2018**, *2*, 669–678. [CrossRef]
20. Baia, R.; Yina, X.; Fenga, X.; Caoa, Y.; Wu, Y.; Zhua, Z.; Lia, C.; Tua, P.; Chai, X. Corydalis hendersonii Hemsl. protects against myocardial injury by attenuating inflammation and fibrosis via NF-κB and JAK2-STAT3 signaling pathways. *J. Ethnopharmacol.* **2017**, *207*, 174–183.
21. Mao, Z.; Wang, X.; Liu, Y.; Huang, Y.; Liu, Y.; Di, X. Simultaneous determination of seven alkaloids from Rhizoma *Corydalis Decumbentis* in rabbit aqueous humor by LC–MS/MS: Application to ocular pharmacokinetic studies. *J. Chromatogr. B* **2017**, *1057*, 46–53. [CrossRef] [PubMed]
22. Zhang, J.; Jin, Y.; Dong, J.; Xiao, Y.; Feng, J.; Xue, X.; Zhang, X.; Liang, X. Systematic screening and characterization of tertiary and quaternary alkaloids from *corydalis yanhusuo* W.T. Wang using ultra-performance liquid chromatography–quadrupole-time-of-flight mass spectrometry. *Talanta* **2009**, *78*, 513–522. [CrossRef] [PubMed]
23. Wang, C.; Wang, S.; Fan, G.; Zou, H. Screening of antinociceptive components in Corydalis yanhusuo W.T. Wang by comprehensive two-dimensional liquid chromatography/tandem mass spectrometry. *Anal. Bioanal. Chem.* **2010**, *396*, 1731–1740. [CrossRef] [PubMed]
24. Zheng, X.; Zheng, W.; Zhou, J.; Gao, X.; Liu, Z.; Han, N.; Yin, J. Study on the discrimination between Corydalis Rhizoma and its adulterants based on HPLC-DAD-Q-TOF-MS associated with chemometric analysis. *J. Chromatogr. B* **2018**, *1090*, 110–121. [CrossRef] [PubMed]

25. Wei, X.; Shen, H.; Wang, L.; Meng, Q.; Liu, W. Analyses of Total Alkaloid Extract of Corydalis yanhusuo by Comprehensive RP × RP Liquid Chromatography with pH Difference. *J. Anal. Methods Chem.* **2016**. [CrossRef] [PubMed]
26. Zhang, Q.; Chen, C.; Wang, F.-Q.; Li, C.-H.; Zhang, Q.-H.; Hu, Y.-J.; Xia, Z.-N.; Yang, F.-Q. Simultaneous screening and analysis of antiplatelet aggregation active alkaloids from Rhizoma Corydalis. *Pharm. Biol.* **2016**, *54*, 3113–3120. [CrossRef] [PubMed]
27. Wu, H.; Waldbauer, K.; Tang, L.; Xie, L.; McKinnon, R.; Zehl, M.; Yang, H.; Xu, H.; Kopp, B. Influence of Vinegar and Wine Processing on the Alkaloid Content and Composition of the Traditional Chinese Medicine Corydalis Rhizoma (Yanhusuo). *Molecules* **2014**, *19*, 11487–11504. [CrossRef]
28. Sun, M.; Liun, J.; Lin, C.; Miao, L.; Lin, L. Alkaloid profiling ofthetraditional Chinese medicine Rhizoma corydalis using high performance liquid chromatography-tandem quadrupole time-of-flight mass spectrometry. *Acta Pharm. Sin. B* **2014**, *4*, 208–216. [CrossRef]
29. Shi, J.; Zhang, X.; Ma, Z.; Zhang, M.; Sun, F. Characterization of Aromatase Binding Agents from the Dichloromethane Extract of Corydalis yanhusuo Using Ultrafiltration and Liquid Chromatography Tandem Mass Spectrometry. *Molecules* **2010**, *15*, 3556–3566. [CrossRef]
30. Kim, K.H.; Lee, I.K.; Piao, C.J.; Choi, S.U.; Lee, J.H.; Kim, Y.S.; Lee, K.R. Benzylisoquinoline alkaloids from the tubers of Corydalis ternate and their cytotoxicity. *Bioorg. Med. Chem. Lett.* **2010**, *20*, 4487–4490. [CrossRef]
31. Ma, W.g.; Fukushi, Y.; Tahara, S.; Osawa, T. Fungitoxic alkaloids from Hokkaido Papaveraceae. *Fitoterapia* **2000**, *71*, 527–534. [CrossRef]
32. Kim, A.H.; Jang, J.H.; Woo, K.W.; Park, J.E.; Lee, K.H.; Jung, H.K.; An, B.; Jung, W.S.; Ham, S.H.; Cho, H.W. Chemical constituents of *Dicentra spectabilis* and their anti-inflammation effect. *J. Appl. Biol. Chem.* **2018**, *61*, 39–46. [CrossRef]
33. Och, A.; Szewczyk, K.; Pecio, Ł.; Stochmal, A.; Załuski, D.; Bogucka-Kocka, A. UPLC-MS/MS Profile of Alkaloids with Cytotoxic Properties of Selected Medicinal Plants of the Berberidaceae and Papaveraceae Families. *Oxid. Med. Cell. Longev.* **2017**. [CrossRef]
34. Fan, J.; Wang, P.; Wang, X.; Tang, W.; Liu, C.; Wang, Y.; Yuan, W.; Kong, L.; Liu, Q. Induction of Mitochondrial Dependent Apoptosis in Human Leukemia K562 Cells by Meconopsis integrifolia: A Species from Traditional Tibetan Medicine. *Molecules* **2015**, *20*, 11981–11993. [CrossRef] [PubMed]
35. Liu, J.; Wu, H.; Zheng, F.; Liu, W.; Feng, F.; Xie, N. Chemical constituents of *Meconopsis horridula* and their simultaneous quantification by high-performance liquid chromatography coupled with tandem mass spectrometry. *J. Sep. Sci.* **2014**, *37*, 2513–2522. [CrossRef] [PubMed]
36. Zhou, Y.; Song, J.-Z.; Choi, F.F.-K.; Wu, H.-F.; Qiao, C.-F.; Ding, L.-S.; Gesang, S.-L.; Xu, H.-X. An experimental design approach using response surface techniques to obtain optimal liquid chromatography and mass spectrometry conditions to determine the alkaloids in *Meconopsi* species. *J. Chromatogr. A* **2009**, *1216*, 7013–7023. [CrossRef]
37. Fana, J.; Wanga, Y.; Wanga, X.; Wanga, P.; Tanga, W.; Yuana, W.; Konga, L.; Liu, Q. The Antitumor Activity of Meconopsis Horridula Hook, a Traditional Tibetan Medical Plant, in Murine Leukemia L1210 Cells. *Cell. Physiol. Biochem.* **2015**, *37*, 1055–1065. [CrossRef] [PubMed]
38. Nema, R.; Vishwakarma, S.; Agarwal, R.; Panday, R.K.; Kumar, A. Emerging role of sphingosine-1-phosphate signaling in head and neck squamous cell carcinoma. *Onco. Targets Ther.* **2016**, *9*, 3269–3280.
39. Leemans, C.R.; Braakhuis, B.J.; Brakenhoff, R.H. The molecular biology of head and neck cancer. *Nat. Rev. Cancer* **2011**, *11*, 9–22. [CrossRef]
40. Nichols, A.C.; Yoo, J.; Palma, D.A.; Fung, K.; Franklin, J.H.; Koropatnick, J.; Mymryk, J.S.; Batada, N.N.; Barrett, J.W. Frequent mutations in TP53 and CDKN2A found by next-generation sequencing of head and neck cancer cell lines. *Arch. Otolaryngol. Head Neck Surg.* **2012**, *138*, 732–739. [CrossRef]
41. Gu, J.; Xu, T.; Huang, Q.H.; Zhang, C.M.; Chen, H.Y. HMGB3 silence inhibits breast cancer cell proliferation and tumor growth by interacting with hypoxia-inducible factor 1α. *Cancer Manag. Res.* **2019**, *11*, 5075–5089. [CrossRef] [PubMed]
42. Holliday, D.L.; Speirs, V. Choosing the right cell line for breast cancer research. *Breast Cancer Res.* **2011**, *13*, 215. [CrossRef] [PubMed]

43. Berkov, S.; Bastida, J.; Sidjimova, B.; Viladomata, F.; Codina, C. Phytochemical differentiation of Galanthus nivalisandGalanthus elwesii(Amaryllidaceae): A case study. *Biochem. Syste. Ecol.* **2008**, *36*, 638–645. [CrossRef]
44. Petruczynik, A.; Misiurek, J.; Tuzimski, T.; Uszyński, R.; Szymczak, G.; Chernetskyy, M.; Waksmundzka-Hajnos, M. Comparison of different HPLC systems for analysis of galantamine and lycorine in various species of Amaryllidaceae family. *J. Liq. Chromatogr.* **2016**, *39*, 574–579. [CrossRef]

Article

Palmatine from Unexplored *Rutidea parviflora* Showed Cytotoxicity and Induction of Apoptosis in Human Ovarian Cancer Cells†

Okiemute Rosa Johnson-Ajinwo [1,2], Alan Richardson [1] and Wen-Wu Li [1,*]

1 Guy Hilton Research Centre, School of Pharmacy and Bioengineering, Keele University, Stoke-on-Trent ST4 7QB, UK; okiemute_2002@yahoo.co.uk (O.R.J.-A.); a.richardson1@keele.ac.uk (A.R.)
2 Faculty of Pharmaceutical Sciences, University of Port Harcourt, Port Harcourt, PMB 5323, Nigeria
* Correspondence: w.li@keele.ac.uk; Tel.: +44-(0)1782-674382
† Part of this study was presented at the 5th International Conference on the Mechanism of Action of Nutraceuticals, Aberdeen, UK.

Received: 21 March 2019; Accepted: 18 April 2019; Published: 25 April 2019

Abstract: Ovarian cancer ranks amongst the deadliest cancers in the gynaecological category of cancers. This research work aims to evaluate in vitro anti-ovarian cancer activities and identify phytochemical constituents of a rarely explored plant species—*Rutidea parviflora* DC. The aqueous and organic extracts of the plant were evaluated for cytotoxicity using sulforhodamine B assay in four ovarian cancer cell lines and an immortalized human ovarian epithelial (HOE) cell line. The bioactive compounds were isolated and characterized by gas/liquid chromatography mass spectrometry and nuclear magnetic resonance spectroscopy. Caspase 3/7 activity assay, western blotting and flow cytometry were carried out to assess apoptotic effects of active compounds. The extracts/fractions of *R. parviflora* showed promising anti-ovarian cancer activities in ovarian cancer cell lines. A principal cytotoxic alkaloid was identified as palmatine whose IC_{50} was determined as 5.5–7.9 μM. Palmatine was relatively selective towards cancer cells as it was less cytotoxic toward HOE cells, also demonstrating interestingly absence of cross-resistance in cisplatin-resistant A2780 cells. Palmatine further induced apoptosis by increasing caspase 3/7 activity, poly-ADP-ribose polymerase cleavage, and annexin V and propidium iodide staining in OVCAR-4 cancer cells. Our studies warranted further investigation of palmatine and *R. parviflora* extracts in preclinical models of ovarian cancer.

Keywords: Ovarian cancer; *Rutidea parviflora*; Palmatine; Apoptosis

Key Contribution: Palmatine, a quaternary protoberberine alkaloid, has been isolated and identified from an unexplored Nigerian plant—*Rutidea parviflora* DC. for the first time. It showed potent cytotoxicity against ovarian cancer cells via apoptosis induction.

1. Introduction

Ovarian cancer is a significant and global threat to life in women. American cancer society estimated that 22,240 of new ovarian cancer diagnosed and 14,070 ovarian cancer deaths are projected to occur in the United States in 2018 [1]. It is ranked 5th most common cause of death among women in the UK. There were 7270 new cases of ovarian cancer in 2015 and 4227 cases of ovarian cancer-related deaths in 2016 in the UK. The ten year survival rate (2010–2011) remains just 35% [2]. Epithelial ovarian cancer can be subdivided into at least four major histological subtypes: serous, endometrioid, clear cell and mucinous carcinoma [3]. High-grade serous ovarian cancer (HGSOC), the most aggressive subtype, is responsible for 70–80% of all ovarian cancer deaths; the overall survival rate has not changed significantly for several decades [4,5]. Ovarian cancer is typically diagnosed at a late stage and no effective screening strategy exists. The current standard treatment for ovarian cancer entails surgery aimed at removing most of the cancerous cells, followed by the administration of chemotherapeutics, often resulting in multiyear survival [4]. However, use of chemotherapy introduces drug resistance and consequently subsequent relapse can lead to the death of the cancer patients [6,7]. For example, platinum-based drugs are initially effective against HGSOC, but recurrent tumors resistant to these agents have developed later on [8]. Recently, three novel poly-ADP-ribose polymerase (PARP) inhibitors such as olaparib, rucaparib, and niraparib have been approved by the US Food and Drug Administration and European Medicines Agency for the treatment of ovarian cancer caused by the alteration of DNA damage repair pathways [9]. Despite of these significant achievements, investigation into the discovery of new drugs that may offer wider therapeutic benefits by overcoming resistance mechanisms and drug toxicity in ovarian cancer therapy is still an unmet need in the light of the ovarian cancer menace.

The use of plants in the treatment of cancer is not new because there is documented evidence of the use of medicinal plants for the treatment of cancer [10–12]. A number of plant-derived or semi-synthetic anti-ovarian cancer drugs including paclitaxel, etoposide and topotecan have been approved and widely used in clinic [13,14]. Previously, we identified cytotoxic cyclotides from a Chinese medicinal plant [15], a cytotoxic indolizine alkaloid, securinine [16,17], and three cytotoxic bisbenzylisoquinoline alkaloids, cycleanine [18–20], isochondodendrine and 2′-norcocsuline [21,22] in Nigerian medicinal plants. Semi-synthetic cycleanine [23] and thymoquinone [24] analogues were further prepared and evaluated for their in vitro anti-ovarian cancer activities. In continuation of our search for novel anti-ovarian cancer compounds, we evaluated an unexplored Nigerian medicinal plant—*Rutidea parviflora* DC. (family *Rubiaceae*) [17].

R. parviflora has been used for anti-inflammatory and anti-cancer activities among the indigenous communities in Delta state of Nigeria. The fruits are taken to induce vomiting and for the treatment of convulsions, epilepsy, spasm and paralysis [25]. However, there are no pharmacological and phytochemical studies reported. In this study, we report the extraction, isolation and identification of cytotoxic palmatine from *R. parviflora*, and its induction of apoptosis leading to ovarian cancer cell death.

2. Results

2.1. Bioassay-Guided Isolation and Identification of Palmatine

Both the organic and aqueous extracts of *R. parviflora* inhibited growth of the cultures with IC_{50} values of <10 μg/ml in OVCAR-4, OVCAR-8, A2780, and cisplatin resistant A2780 (A2780cis) ovarian cancer cell lines. Solvent partition of the organic extract yielded *n*-hexane, ethyl acetate, *n*-butanol and aqueous fractions. The *n*-butanol fraction showed the most potent cytotoxic effects followed by the ethyl acetate fraction (Table 1).

Bioassay-guided fractionation and isolation of the bioactive compounds of *R. parviflora* was carried on the most potent *n*-butanol and ethyl acetate fractions. Nine sub-fractions were obtained from the column fractionation of the *n*-butanol fraction. HPLC purification of the most potent sub-fraction 3 was carried out to obtain a yellow powder, which was determined as palmatine (**1**) (Figure 1) by liquid chromatography coupled with mass spectrometry (LC-MS) (Figure S1), NMR spectroscopy and comparing with a standard palmatine.

Table 1. Cytotoxicity (IC_{50}) and selectivity index (SI) of the extracts, fractions, and isolated compounds of *R. parviflora* in ovarian cancer cell lines and an immortalized human ovarian epithelial (HOE) cell line after 72 h treatment. IC_{50} is the half maximal inhibitory concentration of extracts, fractions, or compounds. SI is a ratio of the measured IC_{50} value against HOE to the measured IC_{50} value against each cancer cell line. The results are expressed as mean ±SEM (n = 3). n.d., not determined.

Extract, Fraction and Compounds	OVCAR-4	OVCAR-8	A2780	A2780cis	HOE
			(µg/ml)		
Organic extract	6.6 ± 1.6	8.7 ± 0.5	3.2 ± 0.3	n.d.	n.d.
Aqueous extract	n.d.	5.9 ± 0.03	2.2 ± 0.5	3.7 ± 0.03	n.d.
n-Hexane fraction	23.3 ± 1.0	18.3 ± 0.3	7.3 ± 0.8	n.d.	n.d.
Ethyl acetate fraction	5.4 ± 0.3	5.8 ± 0.4	2.5 ± 0.2	n.d.	n.d.
n-Butanol fraction	2.6 ± 0.1	2.6 ± 0.3	1.7 ± 0.2	n.d.	n.d.
Aqueous fraction	22.9 ± 1.3	22.1 ± 1.1	12.8 ± 1.3	n.d.	n.d.
			(µM)		
Palmatine (**1**)	7.4 ± 0.3	7.9 ± 0.5	6.6 ± 0.5	5.5 ± 0.9	25.1 ± 5.0
SI for **1**	3.4	3.2	3.8	4.6	-
Urs-12-en-24-oic acid, 3-oxo-, methyl ester (**2**)	85.4 ± 2.4	48.9 ± 2.0	31.6 ± 3.3	n.d.	>200
SI for **2**	>2	>4	>6.5	-	-
Carboplatin	11.1 ± 0.4	10.8 ± 1.3	16.0 ± 1.0	>100	15.2 ± 3.0
SI for carboplatin	1.4	1.4	0.95	<0.15	-

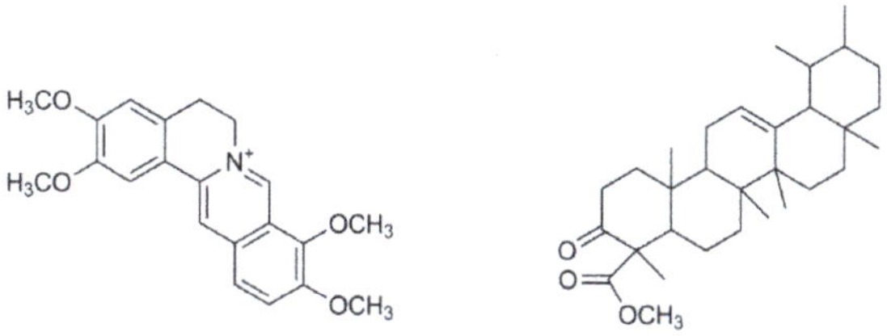

H_3CO H_3CO N^+ OCH_3 OCH_3 **1**

O O OCH_3 **2**

Figure 1. The chemical structure of palmatine (**1**) and urs-12-ene-24-oic acid, 3-oxo, methyl ester (**2**) isolated from *R. parviflora*.

Palmatine (**1**) showed potent cytotoxicity with IC_{50} (5.5–7.9 µM) in four ovarian cancer cell lines, but less cytotoxic to human ovarian epithelial (HOE) cell. The selectivity indexes (SI) of palmatine for cancer cells compared to HOE cells ranged from 3–5, while the SI of the clinically used carboplatin and paclitaxcel [24] only showed around 1–1.5 (Table 1). Further bioassay-guided fractionation of ethyl acetate fraction yielded urs-12-ene-24-oic acid, 3-oxo, methyl ester (**2**), which exhibited moderate inhibition of the growth of ovarian cancer cell cultures and also showed apparently mild effect on HOE cells (Table 1). Compound **2** has been previously identified by gas chromatography coupled with mass spectrometry (GC-MS) analysis from the ethanolic extract of *Canscora perfoliata* used in the treatment of poisonous bites [26].

2.2. Apoptosis Studies

2.2.1. Caspase 3/7 Activity and Western Blotting Analysis

In order to investigate the possible route of cell death caused by the compounds, the effect of palmatine on the activity of caspase 3/7 to evaluate apoptosis was determined in a selected OVCAR-4 cell line. It was derived from HGSOC tumor sample and regarded as one of the most suitable models of ovarian cancer [3]. Figure 2A demonstrated that palmatine as well as positive control-carboplatin significantly increased caspase 3/7 activity in comparison to the vehicle-treated cells for an experimental period of 48 h.

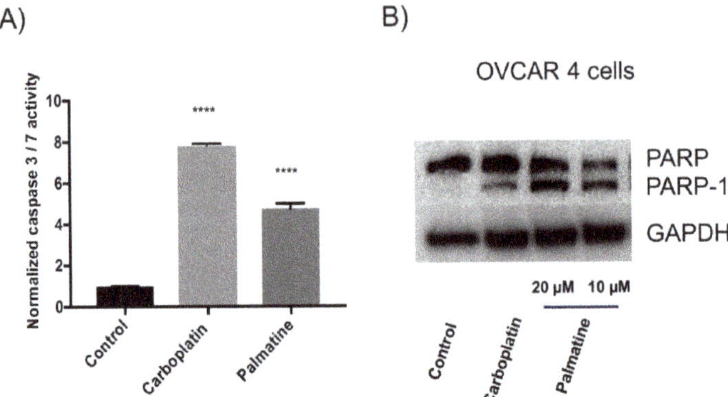

Figure 2. (**A**) The effect of carboplatin and palmatine (each 10 μM) on the caspase 3/7 activity at 48 h in OVCAR-4 cells. The caspase activity was measured and normalized with corresponding sulforhodamine B (SRB)-stained cells to estimate the surviving cell number. **** denotes that the result is significantly different ($p < 0.001$). The results were expressed as mean ±SEM, $n = 3$. (**B**) Detection of poly-ADP-ribose polymerase (PARP) cleavage by immunoblotting. OVCAR-4 cells were treated with palmatine (10 or 20 μM) and carboplatin (40 μM) for 48 h. The vehicle-treated cells served as the control.

To confirm that the compounds induced apoptosis, caspases mediated cleavage of PARP was assessed by immuno-staining. As expected, significant PARP cleavage in OVCAR-4 cells was observed after treatment of palmatine and carboplatin (Figure 2B).

2.2.2. Flow Cytometric Analysis

Further investigation of the apoptosis inducted by palmatine (**1**) was carried out by means of annexin V/propidium iodide (PI) labelling, followed by flow cytometry analysis. Palmatine (**1**) induced concentration–dependent increase in the population of OVCAR-4 cells in the early and late stage of apoptosis compared to the control (Figure 3). The morphological changes of cells treated with palmatine were monitored microscopically at 48–72 h. The characteristic features of apoptosis such as blebbing and shrinkage of cells were also clearly observable by microscopy (Figure S2).

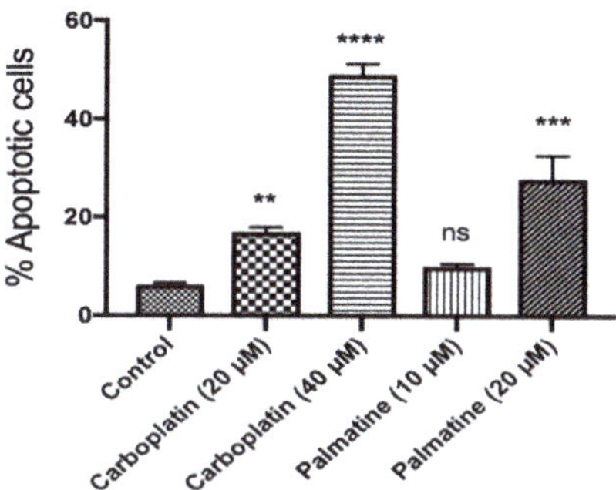

Figure 3. Flow cytometry analysis of the apoptotic effect of palmatine on OVCAR-4 cells after annexin V/PI staining. OVCAR-4 cells were treated with palmatine (10 or 20 µM) and carboplatin (20 or 40 µM) for 48 h. The vehicle-treated cells were used as the control. A representation of the quantification of the combined early and late phase apoptotic cells is shown. **, ***, and **** denote that the results are significantly different from the control with $p < 0.01$, 0.005, and 0.001, respectively. n.s. indicates $p > 0.05$. The results were expressed as mean ±SEM, $n = 3$.

3. Discussion

The phytochemical and pharmacological characterization of *R. parviflora* was carried out in this study. Significant cytotoxic activities of both organic and aqueous extracts were demonstrated in ovarian cancer cells. Here we focused on the organic extract, however, different water-soluble and bioactive compounds could be discovered in the aqueous extract. From the organic extract, two cytotoxic compounds: palmatine (**1**), a quaternary protoberberine alkaloid, and urs-12-ene-24-oic-acid, 3-oxo, methyl ester (**2**), a triterpenoid, were isolated and identified for the first time. Palmatine showed significant inhibitory activity in the cell growth assay. Palmatine did show some preferential selective cytotoxicity for cancer cells, with slightly less potent inhibition of the growth of HOE cells (Table 1). The killing of cancer cells by the compound was likely to occur through induction of apoptosis, evidenced by significant increase in caspase 3/7 activity (Figure 2A), PARP cleavage (Figure 2B), annexin V/propidium iodide labelling of cells (Figure 3). PARP cleavage is a well-established method of demonstrating apoptosis as PARP is a substrate for caspase 3 and 7 and is cleaved in the course of apoptosis into two fragments [27,28]. PARP is a DNA repair nuclear enzyme which detects DNA fragmentation [29]. The cleavage of PARP-1 thus is a confirmation of apoptosis. This enzyme has been a validated drug target of developing successful PARP inhibitors for the treatment of ovarian cancer [10]. It is of interest to note that palmatine showed greater potency and selectivity compared to carboplatin. Importantly, it demonstrated an absence of cross-resistance in cisplatin-resistant A2780 cells (IC_{50} = 6.6 µM for A2780, and 5.5 µM for A2780cis cells). Because drug resistance still remains one of the main causes of failure for ovarian cancer treatment using platinum compounds [8].

Palmatine was previously found to be present in *Rhizoma coptidis*, an important medicinal plant commonly used in the traditional Chinese medicine [30] and the butanol fraction of *Phellodendron amurense* bark extract [31]. Our result is in agreement with the reported cytotoxicity of palmatine in prostate cancer cells [31] and breast cancer MCF-7 cell line [32]. In prostate cancer cells, the ribosomal protein S6, a downstream target of p70S6K and the Akt/mTOR signaling cascade was identified as a potential target of palmatine. Selective cytotoxicity of palmatine against prostate cancer cells was also reported [31]. Palmatine was shown to inhibit growth of pancreatic stellate cells (PSCs) and cancer cells alone or in combination with gemcitabine [33]. Such inhibition of growth and migration of pancreatic cancer cells was due to its suppression of glutamine-mediated changes in glioma-associated oncogene 1 (GL1) signalling, and induction of apoptosis [33]. Palmatine was found to be mainly located in endoplasmic reticulum and mitochondria of MCF-7 cells [33]. Further photodynamic treatment demonstrated photocytotoxicity of the naturally occurring photosensitizer palmatine in breast cancer

MCF-7 [33] and colon adenocarcinoma HT-29 cells [34], causing significant cell apoptosis and increased intracellular reactive oxygen species levels. Palmatine has also been found to bind and stabilize parallel G-quadruplex DNA, indicating that it may be an inhibitor of telomere elongation and oncogene expression in humans [35]. Recently, palmatine from another traditional Chinese medicine *Mahonia bealei* was demonstrated to improve the survival of mice with colorectal cancer via the inhibition of inflammatory cytokines [36]. In the future work, it is necessary to test if palmatine behaves similar mechanism of action in ovarian cancer cells as those found in other cancers, and also to investigate its efficacy and safety in animal models of ovarian cancer.

4. Materials and Methods

4.1. Plant Material and Reagents

The stem bark of *R. parviflora* was collected in Delta state, Nigeria in February 2014. The plant was authenticated by A.O. Oziokowith of University of Benin, Nigeria. A voucher specimen (INTERCEDD/1588) was deposited in the herbarium at the International Centre for Ethnomedicine and Drug Development (INTERCEDD), Enugu state, Nigeria. Trichloroacetic acid (TCA) was purchased from Fisher Scientific (Loughborough, United Kingdom) Carboplatin, glacial acetic acid, N,O-bis(trimethylsily) trifluoroacetamide (BSTFA) with 1% chlorotrimethylsilane (TMCS), palmatine chloride, pyridine, sulforhodamine B (SRB) sodium salt, Trypsin-EDTA solution, and Trizma base were purchased from Sigma Aldrich (Gillingham, United Kingdom). Fetal bovine serum (FBS), penicillin-streptomycin, and RPMI 1640 medium were purchased from Lonza (Visp, Switzerland).

4.2. Extraction of Plant Materials

The plant materials were extracted according to the reported method [37]. The stem bark powder (1 kg) was macerated in a mixture of dichloromethane and methanol (1:1) for three times. The obtained residue was further macerated in methanol to yield the methanol extract. Both extracts were combined and evaporated to yield the total organic extract (7.2 g). Water was then added to the plant residue to obtain the aqueous extract (1.5 g) after freeze drying. The organic extract of *R. parviflora* was further partitioned in water using different organic solvents to obtain the *n*-hexane fraction (2.0 g), the ethyl acetate fraction (1.3 g), the *n*-butanol fraction (0.9 g) and the aqueous fraction (0.5 g).

4.3. Analysis of the Bioactive Fraction of R. parviflora by Gas Chromatography–Mass Spectrometry (GC–MS)

The bioactive fraction (1.0 mg) and isolated compound (**2**) of *R. parviflora* was incubated with 10 µl of pyridine and 50 µl of BSTFA (with 1% TMCS) in the oven at 37 °C for 2 h. The resulting trimethylsilyl (TMSi) derivatives were submitted to GC–MS analysis [16,38,39].

4.4. Liquid Chromatography Mass Spectrometry (LC-MS) Analysis

The isolated compounds were analysed by LC-MS, to determine their molecular mass, and their retention times compared with purchased standard compounds. The Agilent technologies 1260 Infinity coupled to 6530 Accurate mass Q-TOF LC-MS system was used (Agilent Technologies, Santa Clara, CA, USA). The gas temperature was 320 °C, with a dry gas flow rate of 11 L/min. Electrospray ionization (ESI) was operated at a voltage of 4000 V, an *m*/*z* range of 100–2000. The samples were injected at a volume of 5 µl and the run time was 15 min. The data was analysed by the Agilent mass-hunter qualitative analysis software.

4.5. NMR Spectroscopy

^{1}H NMR spectra of the isolated compounds in $CDCl_3$ or CD_3OD were obtained with a Bruker 1D (DPX-300) NMR spectrometer for 300 MHz (Bruker, Billerica, MA, USA), and Bruker 1D (DPX-500) for 500 MHz. ^{13}C NMR spectra were obtained at 125 MHz with a Bruker NMR spectrometer. ACD/Labs

10 Freeware (Advanced Chemistry Development Inc., Toronto, ON, Canada) was used in the Analysis of the NMR Spectra.

4.6. Purification and Isolation of Bioactive Compounds

The *n*-butanol fraction was submitted to silica gel column chromatography by eluting with ethyl acetate/methanol. Sub-fractions were subjected to high performance liquid chromatography (HPLC) on a semi-preparative column with a mobile phase composition of 0.1% trifluoroacetic acid (TFA) as solvent A and 80% acetonitrile with 0.1% TFA as solvent B. The gradient began with 100% of A for 5 min and increased to 80% B over 25 min. Then the gradient was increased to 100% B and maintained for 6 min. Palmatine (10 mg, **1**) was obtained as a yellow powder. LC positive ESI-MS, *m/z*: 352.1548 (Figure S1). ^{1}H NMR (500 MHz, CD_3OD), δ 9.76 (1H, s, H-8), 8.80 (1H, s, H-13), 8.12 (1H, d, J = 9.0 Hz, H-11), 8.01 (1H, d, J = 9.1 Hz, H-12), 7.67 (1H, s, H-1), 7.05 (1H, s, H-4), 4.92 (2H, t, J = 6.5 Hz, H-6), 4.21 (3H, s, OCH_3), 4.11 (3H, s, OCH_3), 3.99 (3H, s, OCH_3), 3.94 (3H, s, OCH_3), 3.28 (2H, t, J = 6.4 Hz, H-5). ^{13}C NMR (125 MHz), δ 152.5, 150.5, 149.5, 145.0, 144.4, 138.5, 133.9, 128.7, 126.7, 123.0, 121.9, 119.9, 119.1, 110.8, 108.6, 61.1 (OCH_3), 56.3 (OCH_3), 55.9 (OCH_3), 55.6 (CH_2), 55.3 (OCH_3), 26.4 (CH_2). These data are consistent with those reported [40,41].

The ethyl acetate fraction (RP-EA) of *R. parviflora* was also fractionated on silica gel column with a solvent gradient range of 100% *n*-hexane to 100% ethyl acetate to yield ten sub-fractions. Sub-fraction F2 was treated with hot methanol and recrystallized at a temperature of 4 °C to yield a white crystal (yield: 12.8%), urs-12-ene-24-oic acid, 3-oxo, methyl ester (**2**), referring as EA2 (Figure 1), by GC-MS and ^{1}H NMR. EI-MS, *m/z* (%): 468 $[M]^+$ (2), 453 (1), 218.2 (100), 203.2 (15), 189.2 (21), 178.1 (6), 161.1 (10), 135.1 (18), 21.1 (17), 95.1(22), 81.1 (18), 69.1 (20), 43.1 (48). ^{1}H NMR (300 MHz, $CDCl_3$) δ: 5.27 (brs), 3.49 (s, 3H, O-CH_3), 1.90-2.32 (m), 0.80-1.70 (m). Its ^{1}H NMR is consistent with those reported [42].

4.7. Cell Culture

The human ovarian cancer cell lines, OVCAR-4, OVCAR-8, A2780, cisplatin-resistant A2780 (A2780cis) were bought from American Type Culture Collection (Manassas, VA, USA), and human ovarian epithelial (HOE) cells immortalized using SV40 large T antigen (Catalogue number: T1074) were purchased from Applied Biological Materials Inc. (Richmond, BC, Canada). All cells were cultured in Roswell Park Memorial Institute (RPMI) 1640 medium supplemented with 10% fetal bovine serum (FBS), penicillin-streptomycin (50 U/ml) and glutamine (2 mM).

4.8. Cell Growth Assay

The SRB assay was used to evaluate the effects of the plant extracts and pure compounds on the growth of ovarian cancer and HOE cell lines after treatment of 72 h [16,20,43]. 100 mg/ml concentrations of the plant extracts were prepared using dimethyl sulfoxide (DMSO) (organic extracts) and media (aqueous extracts) as stock solutions. 20 mM of purified compounds (**1** and **2**) and carboplatin were prepared in DMSO and used in the assay. The 0.1% DMSO in growth media was added to the cells; referred to as vehicle-treated cells (control). Nine concentrations of the drugs were prepared by using a two-fold serial dilution. Each well of the 96-well plates were seeded with 80 µl of the ovarian cancer and HOE cells at a density of 2000 cells/well, except for OVCAR-4 which was plated at a density of 5000 cells/well. After 24 h, 20 µl of plant extracts (1 mg/ml) after further 100-fold dilution of the stock solution in the medium or pure compounds were added and the cell cultures were kept in the incubator at 37 °C under 5% CO_2 for 72 h in a humidified atmosphere. The medium was decanted and the cells fixed with 10% TCA on ice for 30 min and dried. The dried plates were stained with 0.4% SRB for 30 min, washed with 1% acetic acid and dried. 100 µL of Tris-base (10 mM) were added to the plates and shaken for 10 min to solubilise the protein-bound SRB dye. The absorbance at 570 nm was measured using a multi-mode microplate reader BioTEK Synergy 2 (Winooski, VT, USA). The recorded data was analysed by non-linear regression using the GraphPad PRISM software (GraphPad Software v6.0, San Diego, CA, USA) to fit a 4 parameter sigmoidal dose-response curve to determine IC_{50} values and the Hill coefficient. Based on the mean IC_{50}s obtained, the selectivity index (SI) was calculated for each bioactive compound and carboplatin using the following formula:

$$SI = \frac{IC_{50}\ (\text{HOE cell line})}{IC_{50}\ (\text{cancer cell line})} \quad (1)$$

The SI value obtained is an indication of the preferential selectivity in the cytotoxicity of the compound for cancer cells. A large value suggests that the compound would be more cytotoxic to cancer cells than HOE cells.

4.9. Caspase 3/7 Activity

Caspase 3/7 activity was measured in cells pre-treated with vehicle (0.1% DMSO in growth medium), carboplatin, palmatine (**1**) and urs-12-ene-24-oic acid, 3-oxo, methyl ester (**2**) by use of the caspase 3/7 Glo-reagent obtained from Promega (Southampton, UK). For the determination of caspase 3/7 activity, OVCAR-4 cells were seeded in 96-well plates at a density of 5000 cells/well, and exposed to the compounds at different concentrations for 48 h. Then 25 µl of caspase 3/7 Glo-reagent was carefully added to the cells in the dark. The foiled plates were placed on a rocker for 30 min before measurement of luminescence.

4.10. Western Blotting

The measurement of PARP cleavage was carried out as described previously [20]. Briefly, six-wells plates seeded with OVCAR-4 cells at a density of 300,000/well and treated with 0.1% DMSO vehicle, 10 µM or 20 µM of palmatine or 40 µM of carboplatin for 48 h. The cells were collected, trypsinized, washed with cold phosphate buffered saline (PBS) and lysed with radioimmunoprecipitation assay (RIPA) buffer consisting of 20 mM Hepes, 150 mM NaCl, 2 mM EDTA, 0.05 mM pepstatin, 0.12 mM leupeptin, 1 mM phenylmethylsulfonyl fluoride, 0.5% sodium deoxycholate and 1% NP40.

The protein concentrations were evaluated by the bicinchoninic acid (BCA) Assay. About 10 µg protein of each cell lysate sample was carefully added to NuPAGE sample buffer made up with 5% β-mercaptoethanol, before electrophoresis using sodium dodecyl sulphate polyacrylamide gel electrophoresis (SDS-PAGE), for 15 min at a temperature of 70 °C to denature the proteins. The denatured proteins were added to a 4–20% Tris-Glycine polyacrylamide gradient gel to separate the proteins from the cell lysates using 100 mM Hepes, 100 mM Tris and SDS (1%) as running buffer.

The separated proteins were transferred to Amersham Hybond P (GE Healthcare Life Sciences, Buckinghamshire, UK) 0.45 µm polyvinylidene difluoride membrane and incubated in transfer buffer with a composition of 200 mM glycine, 25 mM Tris, 10% methanol and 0.075% SDS for 1.5 h. The membrane was again incubated in Tris buffered saline with tween (TBST, 150 mM NaCl, 0.1% Tween 20, 50 mM Tris hydrochloride, and 5% skimmed milk powder, pH 7.4) for 1.5 h on a rocker at room temperature to achieve blocking of the membranes. The membrane was subsequently incubated overnight at 4 °C in the buffer with primary antibody against poly-ADP-ribose polymerase (Catalogue number: 9542, Cell Signaling Technology Inc., London, UK) (1:1000) and antibody against glyceraldehde-3-phosphate dehydrogenase (GAPDH) (Catalogue number: MAB374, Millipore, Watford, UK) (1:5000). The membrane was washed several times with TBST, before incubation in the buffer with IgG secondary antibody conjugated with horseradish peroxidase (HRP) (Catalogue number: 7074, Cell Signaling Technology Inc.) (1:2000) for 1 h at room temperature on a rocker. Following several washes in TBST, the membrane was analyzed on a FluorChem M Imager (ProteinSimple, San Jose, CA, USA), for the visualization of the protein bands using the UptiLight HRP chemiluminescent substrate according to the manufacturer's instructions (Uptima, San Jose, CA, USA).

4.11. Flow Cytometry

Flow cytometry analysis of annexin V/PI labelled treated cells were carried out as described as before [20,21]. Briefly, OVCAR-4 cells were seeded into 12-well plates and exposed to the treatment compounds at the indicated concentrations, followed by annexin V labelling of the cells using an annexin V-FITC kit procured from Miltenyi biotech (Bergisch Gladbach, Germany).

4.12. Statistical Analysis

The cytotoxicity, caspase activity and flow cytometry results were presented as mean values ± standard error of means (SEM). Statistical analysis was performed using one-way analysis of variance (ANOVA) and GraphPad Prism software v6.0 for the determination of statistical significance of difference between means. $p < 0.05$ were considered statistically significant.

5. Conclusions

In conclusion, this is the first report of the cytotoxic activities of *R. parviflora*, a medicinal plant, from folk medicine in Nigeria. Palmatine was isolated and identified from this unexplored plant, which provides a new source of palmatine. This study has also shown that palmatine possesses cytotoxicity, with apoptosis as the route of cell killing. Therefore, palmatine is a potential lead compound for the development of treatment of ovarian cancer, and merits further investigation.

Supplementary Materials: The following are available online at http://www.mdpi.com/2072-6651/11/4/237/s1, Figure S1: LC-MS chromatogram and MS of palmatine isolated from *Rutidea parviflora*. Figure S2: The effects of palmatine (20 μM) on cell morphology of OVCAR-4 cells monitored by light microscope for 48–72h.

Author Contributions: O.R.J.-A. collected the plant materials and performed the experimental work. O.R.J.-A. and W.-W.L. analyzed and interpreted the experimental data. All authors contributed to write the article.

Funding: This work was supported by Nigerian ETF (A PhD studentship to O.R.J.-A.).

Acknowledgments: We thank Falko Drijfhout, John Clews and Tim Claridge for LC-MS or NMR measurements.

Conflicts of Interest: All authors have no conflict of interest to disclose.

References

1. Torre, L.A.; Trabert, B.; DeSantis, C.E.; Miller, K.D.; Samimi, G.; Runowicz, C.D.; Gaudet, M.M.; Jemal, A.; Siegel, R.L. Ovarian cancer statistics, 2018. *CA Cancer J. Clin.* **2018**, *68*, 284–296. [CrossRef] [PubMed]
2. Ovarian Cancer Statistics. Available online: https://www.cancerresearchuk.org/health-professional/cancer-statistics/statistics-by-cancer-type/ovarian-cancer (accessed on 23 April 2019).
3. Domcke, S.; Sinha, R.; Levine, D.A.; Sander, C.; Schultz, N. Evaluating cell lines as tumour models by comparison of genomic profiles. *Nat. Commun.* **2013**, *4*, 2126. [CrossRef] [PubMed]
4. Jelovac, D.; Armstrong, D.K. Recent progress in the diagnosis and treatment of ovarian cancer. *CA Cancer J. Clin.* **2011**, *61*, 183–203.
5. Bowtell, D.D.; Bohm, S.; Ahmed, A.A.; Aspuria, P.J.; Bast, R.C., Jr.; Beral, V.; Berek, J.S.; Birrer, M.J.; Blagden, S.; Bookman, M.A.; et al. Rethinking ovarian cancer II: Reducing mortality from high-grade serous ovarian cancer. *Nat. Rev. Cancer* **2015**, *15*, 668–679. [CrossRef]
6. Petty, R.; Evans, A.; Duncan, I.; Kurbacher, C.; Cree, I. Drug resistance in ovarian cancer—The role of p53. *Pathol. Oncol. Res.* **1998**, *4*, 97–102. [CrossRef]
7. Vasey, P.A. Resistance to chemotherapy in advanced ovarian cancer: mechanisms and current strategies. *Br. J. Cancer* **2003**, *89*, S23–28. [CrossRef]
8. Binju, M.; Padilla, M.A.; Singomat, T.; Kaur, P.; Suryo Rahmanto, Y.; Cohen, P.A.; Yu, Y. Mechanisms underlying acquired platinum resistance in high grade serous ovarian cancer—A mini review. *Biochim. Biophys. Acta Gen. Subj.* **2019**, *1863*, 371–378. [CrossRef]
9. Franzese, E.; Centonze, S.; Diana, A.; Carlino, F.; Guerrera, L.P.; Di Napoli, M.; De Vita, F.; Pignata, S.; Ciardiello, F.; Orditura, M. PARP inhibitors in ovarian cancer. *Cancer Treat. Rev.* **2019**, *73*, 1–9. [CrossRef] [PubMed]

10. Sowemimo, A.; Van de Venter, M.; Baatjies, L.; Koekemoer, T. Cytotoxicity of some Nigerian plants used in traditional cancer treatment. *Planta Med.* **2010**, *76*, 1224–1225. [CrossRef]
11. Tariq, A.; Sadia, S.; Pan, K.; Ullah, I.; Mussarat, S.; Sun, F.; Abiodun, O.O.; Batbaatar, A.; Li, Z.; Song, D.; et al. A systematic review on ethnomedicines of anti-cancer plants. *Phytother. Res.* **2017**, *31*, 202–264. [CrossRef]
12. Salehi, B.; Zucca, P.; Sharifi-Rad, M.; Pezzani, R.; Rajabi, S.; Setzer, W.N.; Varoni, E.M.; Iriti, M.; Kobarfard, F.; Sharifi-Rad, J. Phytotherapeutics in cancer invasion and metastasis. *Phytother. Res.* **2018**, *32*, 1425–1449. [CrossRef] [PubMed]
13. Li, W.W.; Johnson-Ajinwo, O.R.; Uche, F.I. Potential of phytochemicals and their derivatives in the treatment of ovarian cancer. In *Handbook on Ovarian Cancer: Risk Factors, Therapies and Prognosis*; Collier, B.C., Ed.; Nova Science publishers: Hauppauge, NY, USA, 2015.
14. Newman, D.J.; Cragg, G.M. Natural Products as Sources of New Drugs from 1981 to 2014. *J. Nat. Prod.* **2016**, *79*, 629–661. [CrossRef]
15. Uche, F.I.; Li, W.W.; Richardson, A.; Greenhough, T.J. Anticancer activities of cyclotides from Viola yedeonsis Makino (Violaceae). *Planta Med.* **2014**, *80*, 818. [CrossRef]
16. Johnson-Ajinwo, O.R.; Richardson, A.; Li, W.W. Cytotoxic effects of stem bark extracts and pure compounds from Margaritaria discoidea on human ovarian cancer cell lines. *Phytomedicine* **2015**, *22*, 1–4. [CrossRef] [PubMed]
17. Johnson-Ajinwo, O.R.; Richardson, A.; Li, W.W. Identification and evaluation of anticancer compounds from three Nigerian plants used in traditional medicines. *Biochem. Pharmacol.* **2017**, *139*, 128. [CrossRef]
18. Uche, F.; Li, W.W.; Richardson, A.; Greenhough, T.J. Anti-ovarian cancer activities of alkaloids from Triclisia subcordata olive (Menispermecaea). *Planta Med.* **2014**, *80*, 813. [CrossRef]
19. Uche, F.I.; Drijfhout, F.; McCullagh, J.; Richardson, A.; Li, W.W. Cytotoxicity effects and apoptosis induction by cycleanine and tetrandrine. *Planta Med.* **2016**, 82. [CrossRef]
20. Uche, F.I.; Drijfhout, F.P.; McCullagh, J.; Richardson, A.; Li, W.W. Cytotoxicity Effects and Apoptosis Induction by Bisbenzylisoquinoline Alkaloids from Triclisia subcordata. *Phytother. Res.* **2016**, *30*, 1533–1539. [CrossRef]
21. Uche, F.I.; Abed, M.; Abdullah, M.I.; Drijfhout, F.P.; McCullagh, J.; Claridge, T.W.D.; Richardson, A.; Li, W.W. Isolation, identification and anti-cancer activity of minor alkaloids from Triclisia subcordata Oliv. *Biochem. Pharmacol.* **2017**, *139*, 112. [CrossRef]
22. Uche, F.I.; Abed, M.N.; Abdullah, M.I.; Drijfhout, F.P.; McCullagh, J.; Claridge, T.W.D.; Richardson, A.; Li, W.W. Isochondodendrine and 2 ′-norcocsuline: additional alkaloids from Triclisia subcordata induce cytotoxicity and apoptosis in ovarian cancer cell lines. *Rsc Adv.* **2017**, *7*, 44154–44161. [CrossRef]
23. Uche, F.I.; McCullagh, J.; Claridge, T.W.D.; Richardson, A.; Li, W.W. Synthesis of (aminoalkyl)cycleanine analogues: Cytotoxicity, cellular uptake, and apoptosis induction in ovarian cancer cells. *Bioorg. Med. Chem. Lett.* **2018**, *28*, 1652–1656. [CrossRef]
24. Johnson-Ajinwo, O.R.; Ullah, I.; Mbye, H.; Richardson, A.; Horrocks, P.; Li, W.W. The synthesis and evaluation of thymoquinone analogues as anti-ovarian cancer and antimalarial agents. *Bioorg. Med. Chem. Lett.* **2018**, *28*, 1219–1222. [CrossRef]
25. Burkill, H.M. *The useful plants of west tropical Africa*; Royal Botanic Gardens: London, UK, 1985.
26. Thanga Krishna Kumari, S.; Muthukumarasamy, S.; Mohan, V.R. GC-MS determination of bioactive components of Canscora perfoliata Lam. (Gentianaceae). *J. Appl. Pharmal. Sci.* **2012**, *2*, 210–214.
27. Whitacre, C.M.; Zborowska, E.; Willson, J.K.; Berger, N.A. Detection of poly(ADP-ribose) polymerase cleavage in response to treatment with topoisomerase I inhibitors: A potential surrogate end point to assess treatment effectiveness. *Clin. Cancer Res.* **1999**, *5*, 665–672.
28. Trucco, C.; Oliver, F.J.; de Murcia, G.; Menissier-de Murcia, J. DNA repair defect in poly(ADP-ribose) polymerase-deficient cell lines. *Nucleic Acids Res.* **1998**, *26*, 2644–2649. [CrossRef]
29. D'Amours, D.; Germain, M.; Orth, K.; Dixit, V.M.; Poirier, G.G. Proteolysis of poly(ADP-ribose) polymerase by caspase 3: kinetics of cleavage of mono(ADP-ribosyl)ated and DNA-bound substrates. *Radiat. Res.* **1998**, *150*, 3–10. [CrossRef]
30. Ding, P.L.; Chen, L.Q.; Lu, Y.; Li, Y.G. Determination of protoberberine alkaloids in Rhizoma Coptidis by ERETIC (1)H NMR method. *J. Pharm. Biomed. Anal.* **2012**, *60*, 44–50. [CrossRef]
31. Hambright, H.G.; Batth, I.S.; Xie, J.; Ghosh, R.; Kumar, A.P. Palmatine inhibits growth and invasion in prostate cancer cell: Potential role for rpS6/NFkappaB/FLIP. *Mol. Carcinog.* **2015**, *54*, 1227–1234. [CrossRef]

32. Wu, J.; Xiao, Q.; Zhang, N.; Xue, C.; Leung, A.W.; Zhang, H.; Tang, Q.J.; Xu, C. Palmatine hydrochloride mediated photodynamic inactivation of breast cancer MCF-7 cells: Effectiveness and mechanism of action. *Photodiagnosis Photodyn. Ther.* **2016**, *15*, 133–138. [CrossRef]
33. Chakravarthy, D.; Munoz, A.R.; Su, A.; Hwang, R.F.; Keppler, B.R.; Chan, D.E.; Halff, G.; Ghosh, R.; Kumar, A.P. Palmatine suppresses glutamine-mediated interaction between pancreatic cancer and stellate cells through simultaneous inhibition of survivin and COL1A1. *Cancer Lett.* **2018**, *419*, 103–115. [CrossRef]
34. Wu, J.; Xiao, Q.; Zhang, N.; Xue, C.; Leung, A.W.; Zhang, H.; Xu, C.; Tang, Q.J. Photodynamic action of palmatine hydrochloride on colon adenocarcinoma HT-29 cells. *Photodiagnosis Photodyn. Ther.* **2016**, *15*, 53–58. [CrossRef] [PubMed]
35. Padmapriya, K.; Barthwal, R. Structural and biophysical insight into dual site binding of the protoberberine alkaloid palmatine to parallel G-quadruplex DNA using NMR, fluorescence and Circular Dichroism spectroscopy. *Biochimie* **2018**, *147*, 153–169. [CrossRef] [PubMed]
36. Ma, W.K.; Li, H.; Dong, C.L.; He, X.; Guo, C.R.; Zhang, C.F.; Yu, C.H.; Wang, C.Z.; Yuan, C.S. Palmatine from Mahonia bealei attenuates gut tumorigenesis in ApcMin/+ mice via inhibition of inflammatory cytokines. *Mol. Med. Rep.* **2016**, *14*, 491–498. [CrossRef] [PubMed]
37. McCloud, T.G. High throughput extraction of plant, marine and fungal specimens for preservation of biologically active molecules. *Molecules* **2010**, *15*, 4526–4563. [CrossRef]
38. Li, W.W.; Barz, W. Structure and accumulation of phenolics in elicited Echinacea purpurea cell cultures. *Planta Med.* **2006**, *72*, 248–254. [CrossRef] [PubMed]
39. Li, W.W.; Barz, W. Biotechnological production of two new 8,4′-oxynorneolignans by elicitation of Echinacea purpurea cell cultures. *Tetrahedron Lett.* **2005**, *46*, 2973–2977. [CrossRef]
40. Zhou, S.Q.; Tong, R.B. A General, Concise Strategy that Enables Collective Total Syntheses of over 50 Protoberberine and Five Aporhoeadane Alkaloids within Four to Eight Steps. *Chem.-Eur. J.* **2016**, *22*, 7084–7089. [CrossRef] [PubMed]
41. Yuan, L.T.; Kao, C.L.; Huang, S.C.; Chen, C.T.; Li, H.T.; Chen, C.Y. Secondary Metabolites from the Stems of Mahonia oiwakensis. *Chem. Nat. Compd.* **2017**, *53*, 997–998. [CrossRef]
42. Shah, B.A.; Kumar, A.; Gupta, P.; Sharma, M.; Sethi, V.K.; Saxena, A.K.; Singh, J.; Qazi, G.N.; Taneja, S.C. Cytotoxic and apoptotic activities of novel amino analogues of boswellic acids. *Bioorg. Med. Chem. Lett.* **2007**, *17*, 6411–6416. [CrossRef]
43. Vichai, V.; Kirtikara, K. Sulforhodamine B colorimetric assay for cytotoxicity screening. *Nat. Protoc.* **2006**, *1*, 1112–1116. [CrossRef]

Article

Cytotoxic and Proapoptotic Activity of Sanguinarine, Berberine, and Extracts of *Chelidonium majus* L. and *Berberis thunbergii* DC. toward Hematopoietic Cancer Cell Lines

Anna Och [1], Daniel Zalewski [1], Łukasz Komsta [2], Przemysław Kołodziej [1], Janusz Kocki [3] and Anna Bogucka-Kocka [1,*]

1 Chair and Department of Biology and Genetics, Medical University of Lublin, 4a Chodźki St., 20-093 Lublin, Poland
2 Chair and Department of Medicinal Chemistry, Medical University of Lublin, 4 Jaczewskiego St., 20-090 Lublin, Poland
3 Department of Clinical Genetics, Chair of Medical Genetics, Medical University of Lublin, 11 Radziwiłłowska St., 20-080 Lublin, Poland
* Correspondence: anna.kocka@umlub.pl; Tel.: +48-81-448-7232

Received: 31 July 2019; Accepted: 21 August 2019; Published: 23 August 2019

Abstract: Isoquinoline alkaloids belong to the toxic secondary metabolites occurring in plants of many families. The high biological activity makes these compounds promising agents for use in medicine, particularly as anticancer drugs. The aim of our study was to evaluate the cytotoxicity and proapoptotic activity of sanguinarine, berberine, and extracts of *Chelidonium majus* L. and *Berberis thunbergii* DC. IC10, IC50, and IC90 doses were established toward hematopoietic cancer cell lines using trypan blue staining. Alterations in the expression of 18 apoptosis-related genes in cells exposed to IC10, IC50, and IC90 were evaluated using real-time PCR. Sanguinarine and *Chelidonium majus* L. extract exhibit significant cytotoxicity against all studied cell lines. Lower cytotoxic activity was demonstrated for berberine. *Berberis thunbergii* DC. extract had no influence on cell viability. Berberine, sanguinarine, and *Chelidonium majus* L. extract altered the expression of apoptosis-related genes in all tested cell lines, indicating the induction of apoptosis. The presented study confirmed the substantial cytotoxicity and proapoptotic activity of sanguinarine, berberine, and *Chelidonium majus* L. extract toward the studied hematopoietic cell lines, which indicates the utility of these substances in anticancer therapy.

Keywords: cytotoxicity; apoptosis; sanguinarine; berberine; *Chelidonium majus*; *Berberis thunbergii*; leukemia; anticancer

Key Contribution: In this study, we demonstrated cytotoxic and proapoptotic properties of sanguinarine, berberine and extract of *Chelidonium majus* L., which indicates the applicability of the examined substances in anticancer therapy.

---

1. Introduction

Sanguinarine and berberine are two of the most intensively investigated isoquinoline alkaloids in terms of their use in medicine. Sanguinarine is a benzophenantridine-type alkaloid occurring in *Papaveraceae*, *Ranunculaceae*, and *Berberidaceae* families. Berberine is a tertiary, protoberberine-derived alkaloid occurring in *Berberidaceae*, *Papaveraceae*, *Menispermaceae*, *Ranunculaceae*, and *Rutaceae* families. Due to their toxicity, sanguinarine and berberine play a defending role against viruses and fungi that are pathogenic towards plants [1,2].

Sanguinarine exhibits various pharmacological activities, including antibacterial [3], anti-inflammatory [4,5], anti-depressant [6], antinociceptive [7], antihypertensive [8], and antiplatelet [9] properties. It was previously demonstrated that sanguinarine inhibits acetylcholinesterase [10] and alfa-amylase [11], which broadens the potential clinical applications of this compound. Berberine is also an alkaloid with diverse biological activities [12], showing antimicrobial [13], anti-inflammatory, antioxidative [14], antidiabetic [15], cardioprotective [16], antidepressant [17], and neuroprotective effects [18].

The most intensively investigated of these is the anticancer activity of sanguinarine and berberine. Many pieces of evidence indicate that sanguinarine inhibits cell cycle and induce apoptosis in various types of cancer cells [19,20]. Sanguinarine induces apoptosis in receptor [21] and mitochondrial pathways [21–24]. In primary effusion lymphoma (PEL) cells exposed to sanguinarine, typical symptoms of the receptor-induced apoptosis were observed, including overexpression of DR5 receptors, activation of caspase-8, and truncation of BID protein. In turn, truncated BID protein mediated the mitochondrial pathway of apoptosis, which was evidenced by loss of mitochondrial membrane potential, release of cytochrome c to cytosol, and activation of caspase-3 and -9 demonstrated in PEL cells exposed to sanguinarine [21]. In high doses, sanguinarine also induces the death of cells via the process of necrosis [22,23]. The anticancer properties of sanguinarine also include inhibition of tumor invasiveness and angiogenesis through inhibition of matrix metalloproteinases activity and VEGF signaling [25–28].

Numerous studies have evidenced the anticancer activity of berberine [29,30]. Berberine was demonstrated to be non-toxic for normal cells and cytotoxic for cancer cells [31]. Berberine disrupts cell cycle, induces apoptosis, and inhibits angiogenesis [31–33]. Berberine enhances the radiosensitivity of cancer cells [34], but normal cells seem to be protected against radiation [35].

Clinical applications of sanguinarine and berberine are limited by their toxic effects [36]. In India, higher incidence of gall bladder cancer and epidemic dropsy was related to the consumption of mustard oil contaminated with sanguinarine [37,38]. The mechanism of cytotoxic and mutagenic activity of sanguinarine is an intercalation of DNA, binding to the tRNA molecules, causing induction of apoptosis and inhibition of oxidative phosphorylation and ATP synthesis [39–43]. Sanguinarine induces hepatotoxicity in animal models [44]. Berberine can intercalate DNA, but with much weaker effects than sanguinarine [45–47]. Berberine forms complexes with nuclear DNA and causes breaking in the double helix of DNA in a dose-dependent manner [48]. Berberine affects gene expression through binding to TATA boxes and the poly-adenine tails in mRNA [49]. This interaction is probably responsible for the neuroprotective effect of berberine in brain ischemia [50].

Despite the toxicity and mutagenicity of sanguinarine and berberine, the compounds are still extensively studied due to the possibility of synthesis of derivatives with reduced toxicity. In optimized doses, these alkaloids could exhibit potential therapeutic effects with limited side effects. A particularly important direction of research for this aspect is to determine the effect of the compound and mechanisms of action towards different types of cancer cells. Therefore, in this study we performed an assessment of cytotoxic and proapoptotic activities of sanguinarine and berberine on selected hematopoietic cell lines derived from various types of leukemia, including HL-60, HL-60/MX1, HL-60/MX2 (acute promyelocytic leukemia), J45.01 (acute T cell leukemia), U266B1 (myeloma), CCRF/CEM, and CEM/C1 (acute lymphoblastic leukemia). HL-60/MX1 and HL-60/MX2 cell lines are the multidrug-resistant derivatives of the HL-60 cells, therefore these cell lines were used to evaluate the effects of studied samples on cells resistant to anticancer treatment.

Previously a screening of alkaloid composition was performed for nine species: *Chelidonium majus* L., *Macleaya cordata* Willd., *Lamprocapnos spectabilis* (L.) Fukuhara, *Fumaria officinalis* L., *Glaucium flavum* Crantz, *Corydalis cava* (L.) Schweigg and Körte, *Berberis thunbergii* DC., *Meconopsis cambrica* (L.) Vig., and *Mahonia aquifolium* (Pursh) Nutt. [51]. The highest amount of sanguinarine was demonstrated in *Chelidonium majus* L. and the highest amount of berberine was found in *Berberis thunbergii* DC. [51].

Therefore, extracts of these two species were also included in this study for evaluation of their cytotoxic and proapoptotic activity toward selected hematopoietic cell lines.

Plant extracts are very complex mixtures of various compounds, which could exhibit antagonism or synergy of biological activity. The application of plant extracts in pharmacotherapy has a higher risk of side effects than pure compounds, the advantage of which is a more predictable therapeutic effect. In this study, we evaluated the cytotoxicity and proapoptotic activity of both pure alkaloids, sanguinarine, and berberine, as well as the extracts prepared from plants containing high amounts of these alkaloids, *Chelidonium majus* L. and *Berberis thunbergii* DC., respectively.

2. Results

2.1. Sanguinarine, Berberine, and Chelidonium majus L. Extract Exhibit Cytotoxic Activity against Hematopoietic Cell Lines

Cytotoxicity of sanguinarine, berberine, and extracts of *Chelidonium majus* L. and *Berberis thunbergii* DC. toward HL-60, HL-60/MX1, HL-60/MX2, CCRF/CEM, CEM/C1, J45.01, and U266B1 cell lines was evaluated by determination of IC10, IC50, and IC90 doses in trypan blue staining test (Table 1). Sanguinarine exhibited the highest cytotoxic activity against all study cell lines, with low variability in the IC10, IC50, and IC90 doses between individual cell lines. The strongest cytotoxic effect of sanguinarine was observed toward HL-60/MX2 cells. CCRF/CEM and U266B1 cell lines were the least sensitive to the exposure to this compound (Table 1).

Table 1. IC10, IC50, and IC90 inhibitory concentrations determined for sanguinarine, berberine, and *Chelidonium majus L.* extract toward seven tested hematopoietic cell lines. Note: SD = standard deviation, * = maximum concentration of exposure.

Cell Line	IC10	IC50	IC90
	Sanguinarine (μM ± SD)		
J45.01	0.10 ± 0.04	0.50 ± 0.04	1.00 ± 0.05
U266B1	0.80 ± 0.04	1.05 ± 0.05	1.80 ± 0.04
HL-60	0.20 ± 0.03	0.60 ± 0.06	1.80 ± 0.03
HL-60/MX1	0.15 ± 0.03	0.50 ± 0.04	1.80 ± 0.05
HL-60/MX2	0.06 ± 0.05	0.10 ± 0.05	1.20 ± 0.05
CCRF/CEM	0.50 ± 0.04	0.70 ± 0.03	1.20 ± 0.03
CEM/C1	0.30 ± 0.04	0.50 ± 0.04	1.00 ± 0.03
	Berberine (μM ± SD)		
J45.01	25.15 ± 3.15	80.15 ± 4.65	200.80 ± 4.65
U266B1	125.15 ± 2.68	240.45 ± 4.15	250.00 ± 1.10*
HL-60	50.32 ± 4.56	90.45 ± 5.83	250.00 ± 4.35*
HL-60/MX1	25.05 ± 2.13	110.05 ± 6.72	250.00 ± 3.15*
HL-60/MX2	75.25 ± 6.52	250.00 ± 2.15*	-
CCRF/CEM	50.40 ± 1.18	80.00 ± 2.13	130.25 ± 1.18
CEM/C1	50.25 ± 4.25	225.15 ± 5.25	250.00 ± 2.85*
	Chelidonium majus L. extract (μg/mL ± SD)		
J45.01	5.05 ± 2.15	12.25 ± 2.85	38.65 ± 5.23
U266B1	8.05 ± 3.45	21.50 ± 5.65	264.50 ± 4.12
HL-60	9.01 ± 2.35	13.82 ± 3.15	280.02 ± 6.15
HL-60/MX1	7.81 ± 6.18	20.15 ± 4.16	202.11 ± 4.32
HL-60/MX2	19.85 ± 5.68	64.50 ± 5.48	236.0 ± 4.82
CCRF/CEM	7.58 ± 2.89	10.75 ± 2.15	27.75 ± 1.63
CEM/C1	7.33 ± 5.48	13.25 ± 3.23	27.30 ± 1.89

Berberine was cytotoxic to all tested cell lines, but significantly less so than sanguinarine. The cell viability of the HL-60/MX2 cells exposed to berberine did not fall below 50% despite exposure to the maximum concentrations possible to be obtained. Therefore, IC50 dose could not be determined for this line due to the poor cytotoxicity of the compound. The maximum dose to which HL-60/MX2 cells were exposed (250 μM) was marked as IC50 with an asterisk (IC50*). The IC90 values of berberine were determined only for the CCRF/CEM cells at130 μM and J45.01 cells at200μM. For the remaining cell lines, the IC90 dose could not be determined due to insufficient cytotoxicity of berberine even at the maximum possible dose (250 μM) when the viability of the U266B1, CEM/C1, HL-60, and HL-60/MX1 cell lines did not fall below 90%. For this reason, the maximum dose to which the cells were exposed was marked as IC90*. The most potent cytotoxicity of berberine was found toward CCRF/CEM, J45.01, and HL-60/MX1 cells. The HL-60/MX2 line was the least sensitive to the compound (Table 1).

The *Chelidonium majus* L. extract showed differentiated cytotoxic effect depending on the study cell line. The strongest cytotoxic effect of *Chelidonium majus* L. extract was exerted on the J45.01, CCRF/CEM, and CEM/C1 cells, and the cells of the HL-60 and HL-60/MX2 lines were the least sensitive to this extract (Table 1).

The extract of *Berberis thunbergii* DC. showed no cytotoxicity toward any of the cell lines tested. The maximum concentration of this extract administered in the cytotoxicity tests was 145.5 μg/mL, because in this concentration the DMSO content in the assay is equal to 0.5%, which is the maximum concentration not having a cytotoxic effect on the cells. The viability of exposed cells remained in the range of 89–95% for all tested cell lines. In this respect, IC10, IC50, and IC90 were not determined for *Berberis thunbergii* DC. extract.

2.2. Sanguinarine, Berberine, and Chelidionium majus L. Extract Affect the Expression of Genes Related to Apoptosis

Gene expression analysis considered 18 genes associated with apoptosis and was performed in cells exposed to IC10, IC50, and IC90 doses of sanguinarine, berberine, and *Chelidonium majus* L. extract as well as to *Berb4eris thunbergii* DC. extract in 145.5 μg/mL concentration using a real-time PCR with relative quantification method. Analysis of these gene expression data revealed differences in levels of expression in exposed cells compared to controls (non-exposed cells) (Figures S1–S8). Standard deviation, logRQ, and mean values are presented in Tables S1–S10 in Supplementary Materials.

2.2.1. U266B1 Cell Line

Significant overexpression of *BAK1, BNIP3,* and *CASP9* was observed in U266B1 cells after exposure to sanguinarine, berberine, and *Chelidonium majus* L. extract. For IC10 doses, the highest gene expression modulatory activity was exhibited by sanguinarine, but in higher doses of tested samples, the most affecting expression of studied genes was berberine. A higher dose of berberine and a lower expression of *BCL2, BIK, BNIP2,* and *CASP3* were observed (Table 2, Figure S1).

2.2.2. CEM/C1 Cell Line

Exposure to IC10 doses of sanguinarine and *Chelidonium majus* L. extract caused an increase in expression of the majority of studied genes, with *BAK1* and *MCL1* reaching logRQ > 1. For IC50 and IC90 doses of tested samples, the highest expression was observed for *BAK1, BNIP3,* and *CASP9* in cells exposed to berberine. All genes (except *BIK*) were upregulated in cells exposed to sanguinarine in IC10 dose, but in cells exposed to IC90 dose of this alkaloid these genes (excluding *BAK1* and *MCL1*) were downregulated (Table 2, Figure S2).

2.2.3. CCRF/CEM Cell Line

Sanguinarine at IC10 dose caused a significant increase in expression of all studied genes, with *BAK1, BCL2, BCL2L2, BNIP3,* and *CASP9* reaching logRQ ≥ 2. In the case of IC50 and IC90 doses of this compound, upregulation of these genes was diminished. The distinctively high expression of the *TP53* gene was observed in cells exposed to IC50 and IC90 doses of sanguinarine, berberine, and *Chelidonium majus* L. extract (Table 2, Figure S3).

2.2.4. HL60 Cell Line

Sanguinarine, berberine, and *Chelidonium majus* L. extract did not cause substantial changes in gene expression, excluding *BCL2L2* and *TP53* genes. *BCL2L2* and *TP53* were highly expressed in cells exposed to IC10 of *Chelidonium majus* L. extract and sanguinarine, respectively. Berberine caused notable downregulation of *TP53* in each dose of exposure (Table 2, Figure S4).

Table 2. Differentially expressed genes associated with apoptosis in studied hematopoietic cell lines exposed to IC10, IC50, and IC90 doses of sanguinarine, berberine, and *Chelidonium majus* L. extract, as well as to *Berberis thunbergii* DC. extract at a concentration of 145.5 µg/mL. Table presents upregulated genes with logRQ > 0.5 and downregulated genes with logRQ < −0.5. The lack of genes meeting the assumed criteria was marked with "NA".

Cell Line	IC10	IC50	IC90
		Sanguinarine	
J45.01	↑(*BNIP3*), ↓(*BCL2L2, TP53*)	↑(*BCL2, BNIP3*), ↓(*BAD, TP53*)	↑(*BCL2, BNIP3*), ↓(*TP53*)
U266B1	↑(*BAK1, BNIP3, CASP8, MCL1, PMAIP1*)	↑(*BAK1, BCL2, BNIP3*)	↑(*BNIP3*)
HL-60	↑(*PMAIP1, TP53*)	↑(*BAD*)	NA
HL-60/MX1	↑(*BAK1, CASP8, PMAIP1*), ↓(*BCL2L2, BID*)	↑(*BCL2*), ↓(*TP53*)	↓(*BCL2L2, BNIP2, CASP3, TP53*)
HL-60/MX2	↑(*BAK1, BAX, BIK, CASP3, CASP8, MCL1, PMAIP1, TP53*)	↓(*BAD, BNIP2, PMAIP1, TP53*)	↓(*BAD, BNIP2, TP53*)
CCRF/CEM	(*B2M, BAD, BAK1, BAX, BCL2, BCL2L1, BCL2L2, BID, BIK, BNIP1, BNIP2, BNIP3, CASP3, CASP8, CASP9, MCL1, PMAIP1, TP53*)	↑(*BAK1, BNIP3, CASP8, PMAIP1, TP53*)	↑(*BAD, BAK1, BCL2, BNIP2, BNIP3, CASP8, CASP9, MCL1, PMAIP1, TP53*)
CEM/C1	↑(*BAD, BAK1, BAX, BCL2, BCL2L2, BNIP1, BNIP2, CASP3, CASP8, MCL1, PMAIP1, TP53*)	↑(*BAK1, MCL1*)	↓(*BNIP3*)
		Berberine	
J45.01	↑(*BAK1, BCL2, BCL2L2, BID, BNIP3, CASP9*)	↓(*BCL2L2*)	↑(*BNIP3, CASP9*)
U266B1	NA	↑(*BAK1, BNIP3, CASP9*), ↓(*BCL2L2, BNIP2, MCL1, TP53*)	↑(*BAK1, BAX, BNIP3, CASP9*), ↓(*BCL2, BIK, BNIP2*)
HL-60	↓(*BAK1, BNIP2, TP53*)	↓(*BAK1, BAX, BNIP2, BNIP3, TP53*)	↓(*TP53*)
HL-60/MX1	↑(*BAD, BAK1, BCL2, BNIP3, CASP9, TP53*), ↓(*BNIP2, MCL1*)	↑(*BCL2*), ↓(*BAK1, BNIP2, BNIP3, TP53*)	↑(*BAK1, BCL2, BNIP3, CASP9, PMAIP1*), ↓(*BNIP2, CASP8, MCL1*)
HL-60/MX2	↑(*BAD, BAK1, BAX, BCL2, BCL2L1, BCL2L2, BID, BIK, BNIP1, BNIP2, BNIP3, CASP3, CASP8, CASP9, MCL1, PMAIP1, TP53*)	↑(*BAX, BIK, CASP3, CASP9, MCL1, TP53*)	NA
CCRF/CEM	↑(*TP53*), ↓(*BAD, BAK1, BNIP2*)	↑(*TP53*), ↓(*BAD, BAK1*)	↑(*TP53*), ↓(*BAD, BAK1*)
CEM/C1	NA	NA	↑(*B2M, BAK1, BAX, BCL2, BNIP3, CASP9, PMAIP1*), ↓(*BNIP2*)
		Chelidonium majus L. extract	
J45.01	↑(*BAK1, BCL2, BNIP3, CASP8, CASP9, MCL1, PMAIP1*)	↑(*BNIP3*), ↓(*BAD, BCL2L2, TP53*)	↑(*BNIP3*), ↓(*TP53*)
U266B1	↑(*BAK1*)	↑(*BAK1, BCL2, BCL2L2, BNIP2*)	↑(*BAK1*)
HL-60	↑(*BAK1, BCL2L2, MCL1*)	↑(*BAK1*)	↓(*TP53*)
HL-60/MX1	↑(*BAK1, BCL2L2, MCL1, TP53*)	↑(*BAK1, TP53*)	↑(*TP53*)
HL-60/MX2	↑(*BAK1, BAX, MCL1*), ↓(*TP53*)	↑(*BAK1, BAX, CASP3, MCL1*), ↓(*TP53*)	↑(*BAK1, BAX, MCL1*)
CCRF/CEM	↑(*BCL2, BCL2L2, BNIP2, BNIP3, CASP8, MCL1, PMAIP1, TP53*)	↑(*BAK1, BCL2L2, BIK, BNIP2, TP53*)	↑(*TP53*), ↓(*BAD*)
CEM/C1	↑(*BAK1, BCL2L2, CASP8, MCL1, PMAIP3*)	↑(*BAK1*)	↑(*BAD, BAK1, BCL2L1, BCL2L2, BNIP2*)
		Berberis thunbergii DC. extract (145.5 µg/mL)	
J45.01		↑(*BAK1, BAX, BCL2, BID, BNIP1, BNIP3, CASP9*), ↓(*BNIP2, TP53*)	
U266B1		↓(*BNIP3*)	
HL-60		↑(*BAK1, BNIP3, CASP9, TP53*), ↓(*BCL2L1, BIK, BNIP2, CASP3, CASP8,* MCLI)	
HL-60/MX1		↑(*BCL2*), ↓(*BNIP2, TP53*)	
HL-60/MX2		↑(*BAX, BIK, CASP3, MCL1*), ↓(*TP53*)	
CCRF/CEM		NA	
CEM/C1		↑(*BCL2*), ↓(*BAD, BAK1, TP53*)	

2.2.5. HL60/MX1 Cell Line

This cell line was characterized by a high level of differentiation in gene expression levels after exposure to sanguinarine, berberine, and *Chelidonium majus* L. extract. The influence on gene expression levels was larger in cells exposed to IC10 doses of all tested samples and to all doses of berberine. *BAK1*, *BCL2*, and *TP53* were the most upregulated genes in cells exposed to IC10 dose of berberine. For IC50 dose, berberine caused a significant decrease in *TP53* expression level. Significant downregulation of this gene was also caused by sanguinarine in IC50 and IC90 doses (Table 2, Figure S5).

2.2.6. HL60/MX2 Cell Line

Berberine in IC10 dose caused a pronounced increase in expression of all studied genes, but for IC50 dose, the high expression remained only for *BAX*, *BIK*, *CASP3*, *MCL1*, and *TP53*. Sanguinarine at IC50 concentration induced upregulation of *BAD*, *BCL2L1*, *BCL2L2*, *BNIP2*, *CASP8*, *CASP9*, *PMAIP1*, and *TP53*, but for IC50 and IC90 doses of this compound these genes were significantly downregulated (Table 2, Figure S6).

2.2.7. J45.01 Cell Line

Upregulation of *BNIP3*, *BCL2*, *CASP9*, and *BIK* was observed in J45.01 cells after exposure to sanguinarine, berberine, and *Chelidonium majus* L. extract in IC10, IC50, and IC90 concentrations (Table 2, Figure S7).

After exposure to sanguinarine in IC10 dose, the highest expression of almost all studied genes (excluding *MCL1* and *TP53*) was observed in CCRF/CEM cells compared to other cell lines. In turn, the highest expression of *MCL1* and *TP53* was demonstrated in CEM/C1 and HL-60 cells, respectively. The highest induction of genes associated with apoptosis by IC90 dose of sanguinarine was also observed in the CCRF/CEM cell line. In HL60/MX1 and HL60/MX2 cells, sanguinarine caused the lowest alterations in the expression of the studied genes. In these cell lines, the highest increase in expression of studied genes was observed after exposure to IC10 of berberine. Expression of examined genes in cells exposed to a higher concentration of berberine and extracts of *Chelidonium majus* L. and *Berberis thunbergii* DC. was differentiated across studied cell lines. Exposure to *Berberis thunbergii* DC. extract (145.5 µg/mL) caused upregulation of *B2M* and downregulation of *BAD* and *BNIP2* in all studied cell lines (Table 2, Figure S8).

2.3. Clustering and PARAFAC Analysis

In order to assess the distribution of similarities in the doses of tested compounds in overall gene expression, a combined chemometric analysis of all values of gene expression changes obtained after exposure of the tested cell lines to sanguinarine, berberine, and *Chelidonium majus* L. extract at the IC10, IC50, and IC90 doses was performed (Figure 1).

Clustering with Euclidean distance revealed two subgroups, one of them containing sanguinarine with *Chelidonium majus* L. extract doses, and the other covering berberine doses (Figure 1A). This indicates that cellular responses in studied gene expression after exposure to berberine differ from the response to exposure to sanguinarine and *Chelidonium majus* L. extract.

PARAFAC analysis was performed on a tensor consisting of expression data for 18 genes, 7 cell lines, and 9 doses (IC10, IC50, and IC90 for sanguinarine, berberine, and *Chelidonium majus* L. extract). The PARAFAC univariate analysis explained only 28.3% of the variance, so bivariate analysis was performed, explaining 50.4% of the variance. The obtained results allowed the isolation of IC10 concentration of sanguinarine as an outlier (Figure 1B).

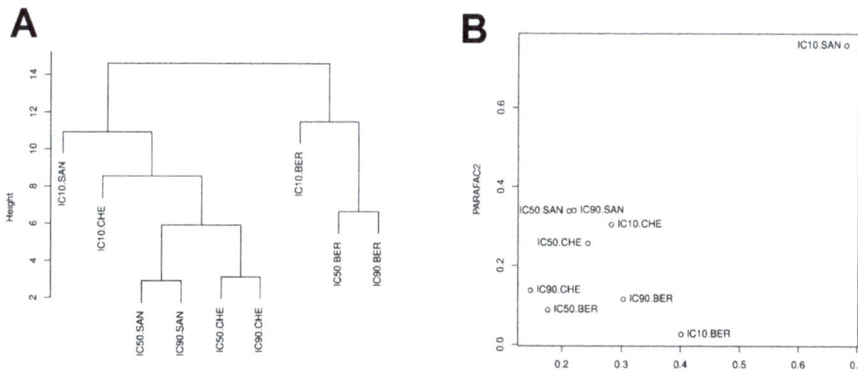

Figure 1. Distribution of similarities in apoptosis-associated gene expression in response to the IC10, IC50, and IC90 doses of sanguinarine, berberine, and *Chelidonium majus* L. extract, evaluated by the clustering analysis with Euclidean distance (**A**) and the PARAFAC analysis (**B**). Note: IC10.SAN = IC10 concentration of sanguinarine; IC50.SAN = IC50 concentration of sanguinarine; IC90.SAN = IC90 concentration of sanguinarine; IC10.BER = IC_{10} concentration of berberine; IC50.BER = IC50 concentration of berberine; IC90.BER = IC90 concentration of berberine; IC10.CHE = IC10 concentration of *Chelidonium majus* L. extract; IC50.CHE = IC50 concentration of *Chelidonium majus* L. extract; IC90.CHE = IC90 concentration of *Chelidonium majus* L. extract.

Clustering analysis based on Euclidean distances and PARAFAC analysis were performed to assess similarities between the expression of 18 examined genes in the studied cell lines after the exposure to sanguinarine, berberine, and *Chelidonium majus* L. extract, (Figure 2). After exposure to sanguinarine, expression of *TP53* significantly differ from the expression of other studied genes. After exposure to berberine, expression of *TP53*, *BAK1*, and *BNIP3* differed from the expression of other studied genes. After exposure to *Chelidonium majus* L. extract, expression of *BCL2L2*, *BAK1*, and *MCL1* was different from the expression of other studied genes (Figure 2).

In order to assess the biological response to exposure to sanguinarine, berberine, and *Chelidonium majus* L. extract, the similarity of cell lines in changes in expression of studied genes was also assessed by Euclidean clustering and PARAFAC analyses. CCRF/CEM, HL60/MX2, and HL60 cell lines differed from other cell lines after exposure to sanguinarine, berberine, and *Chelidonium majus* L. extract, respectively (Figure 3).

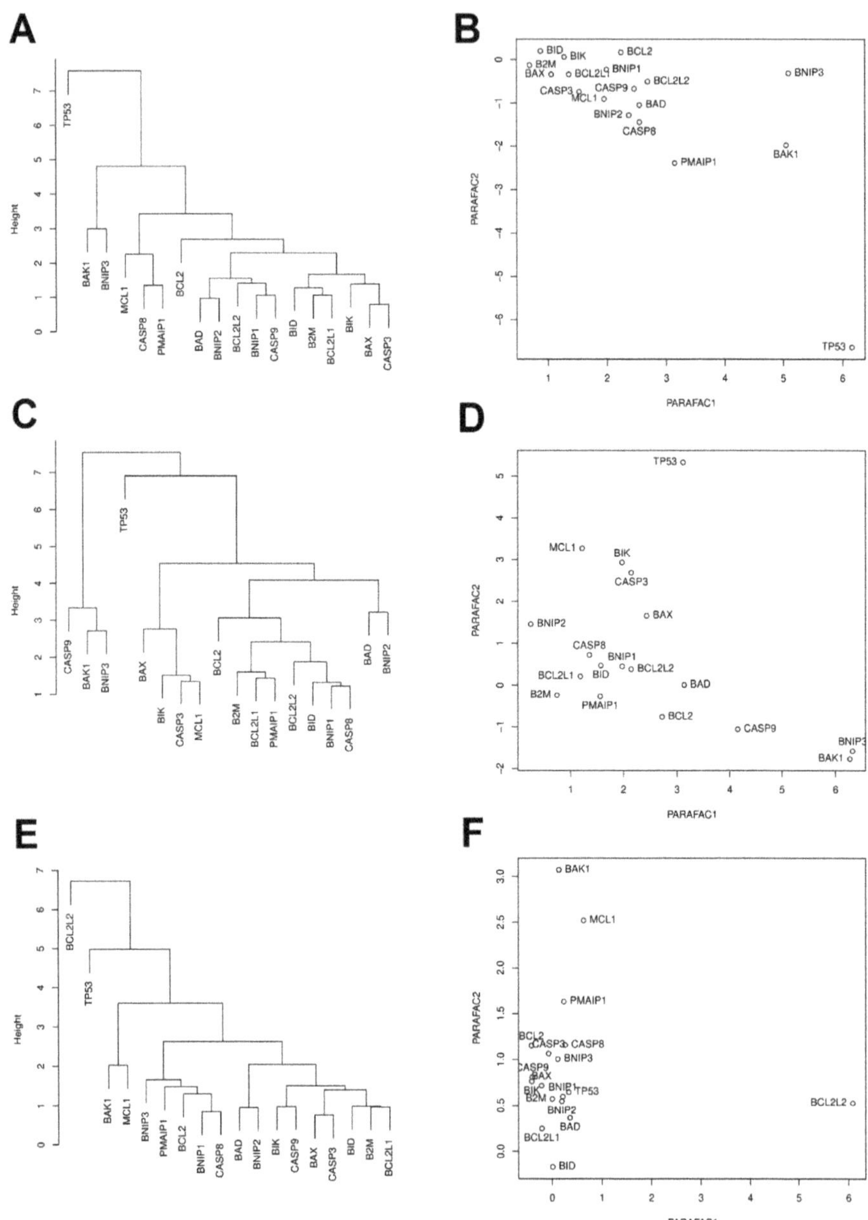

Figure 2. Distribution of similarities in expression of 18 apoptosis-associated genes after exposure to sanguinarine (**A**,**B**), berberine (**C**,**D**), and *Chelidonium majus* L. extract (**E**,**F**), evaluated by the clustering analysis with Euclidean distance (**A**,**C**,**E**) and the PARAFAC analysis (**B**,**D**,**F**). Note: IC10.SAN = IC10 concentration of sanguinarine; IC50.SAN = IC50 concentration of sanguinarine; IC90.SAN = IC90 concentration of sanguinarine; IC10.BER = IC10 concentration of berberine; IC50.BER = IC50 concentration of berberine; IC90.BER = IC90 concentration of berberine; IC10.CHE = IC10 concentration of *Chelidonium majus* L. extract; IC50.CHE = IC50 concentration of *Chelidonium majus* L. extract; IC90.CHE = IC90 concentration of *Chelidonium majus* L. extract.

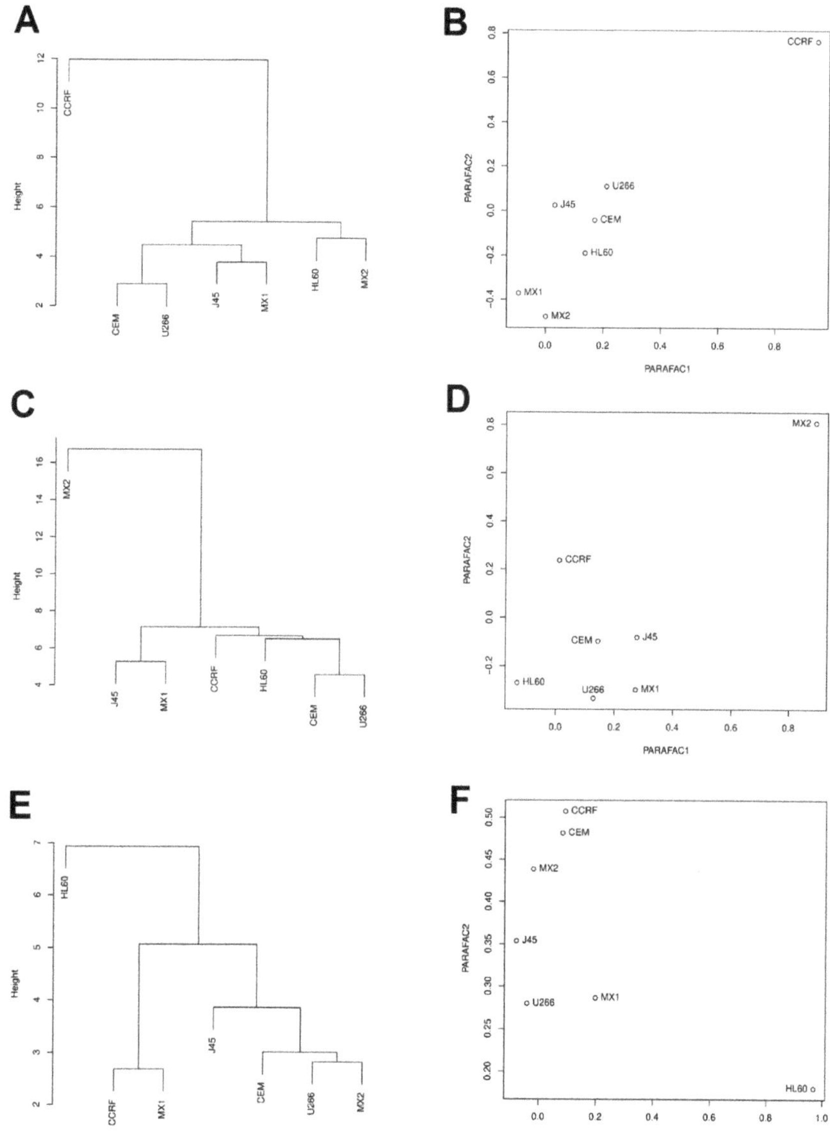

Figure 3. Distribution of similarities in cell lines response in expression of 18 apoptosis-related genes after exposure to sanguinarine (**A**,**B**), berberine (**C**,**D**), and *Chelidonium majus* L. extract (**E**,**F**), evaluated by the clustering analysis with Euclidean distance (**A**,**C**,**E**) and the PARAFAC analysis (**B**,**D**,**F**). Note: IC10.SAN = IC10 concentration of sanguinarine; IC50.SAN = IC50 concentration of sanguinarine; IC90.SAN = IC90 concentration of sanguinarine; IC10.BER = IC10 concentration of berberine; IC50.BER = IC50 concentration of berberine; IC90.BER = IC90 concentration of berberine; IC10.CHE = IC10 concentration of *Chelidonium majus* L. extract; IC50.CHE = IC50 concentration of *Chelidonium majus* L. extract; IC90.CHE = IC90 concentration of *Chelidonium majus* L. extract.

3. Discussion

In this study, the cytotoxicity and proapoptotic activity of sanguinarine, berberine, and extracts of *Chelidonium majus* L. and *Berberis thunbergii* DC. were investigated. IC10, IC50, and IC90 doses were determined in trypan blue staining tests. The influence of these doses on expression of 18 apoptosis-related genes was evaluated.

Sanguinarine exhibited high cytotoxic activity against all studied cell lines. IC50 concentrations of sanguinarine were established toward HL60 cells in previous studies, showing values of 0.37 μM [52] and 1.02 μM [53]. The value obtained in our study was similar and was 0.6 μM.

Inhibitory concentrations received for *Chelidonium majus* L. extract indicated the cytotoxicity of this extract against all tested cell lines. The molar concentration of sanguinarine in the IC50 doses of *Chelidonium majus* L. extract (0.114–0.686 μM) was similar to the IC50 dose of the sanguinarine solution (0.1–1.05 μM) [51]. These findings suggest that sanguinarine is the main compound responsible for the cytotoxic activity of *Chelidonium majus* L. extract, but sanguinarine does not occur in the highest amount in this plant [51,54].

Our research confirmed the cytotoxic activity of *Chelidonium majus* L. extract, as previously evidenced in HL60 cells [55]. The results broaden the wide spectrum of evidence supporting the potential clinical application of preparations containing *Chelidonium majus* L. [56].

Berberine also exhibited cytotoxic activity against all examined cell lines, however it was weaker than sanguinarine. IC10, IC50, and IC90 doses of berberine were significantly higher than determined for sanguinarine and was characterized by a high variation between individual cell lines. Due to low cytotoxicity, we were not able to determine the IC50 dose of berberine for HL-60/MX2 cells and IC90 doses for HL-60, HL-60/MX1, HL-60/MX2, CEM/C1, and U266B1 cell lines. Cytotoxic activity of berberine toward U266B2 cell line was previously evaluated by Hu and collaborators using Cell Counting Kit-8 [57]. The authors reported a statistically significant reduction in cell viability after 48 h exposure to 40–160 μM of berberine. In our study, reduction in cell viability was achieved after exposure to higher berberine concentration (IC50 = 240.45 μM), which probably was a result of the shorter exposure time in our study (24 h). The cytotoxicity of berberine toward CCRF/CEM cell line was investigated by Efferth and collaborators in a 96-h model using MTT test [58]. The IC50 dose of berberine determined in the cited study was equal to 26 μM. The IC50 dose of berberine determined in the current work was higher and amounted to 80.00 μM, which was probably caused by the application of the shorter 24-h model.

In our study, the cytotoxic activity of berberine was evaluated for the first time in relation to J45.01, HL60/MX1, and HL60/MX2 cell lines.

The demonstrated cytotoxicity of berberine toward tested cells suggests the potential activity of *Berberis thunbergii* DC. extract, which contained a relatively high amount of berberine [51]. However, this extract at the maximum dose of 145.5 μg/mL did not cause a decrease in the cell viability of any of the tested cell lines. The probable explanation of this phenomenon is that the content of berberine in the *Berberis thunbergii* DC. extract is insufficient to induce a cytotoxic effect, which is supported by relatively high values of IC10, IC50, and IC90 concentrations (25.15 μM–250.45 μM) determined for berberine.

In the next step of the study, the influence of sanguinarine, berberine, and extracts from *Chelidonium majus* L. and *Berberis thunbergii* DC. on the expression of 18 genes associated with apoptosis was tested in hematopoietic cell lines. The similarities between administered doses and the response of the cells after 24 h exposure to the tested samples were also investigated.

Exposure to sanguinarine, especially in IC10 dose, caused the strongest upregulation of studied genes associated with apoptosis. Upregulation of *BAK1* and *BNIP3* was the most frequently observed. In CCRF/CEM cell line, all studied genes were upregulated with logRQ > 0.5 after exposure to sanguinarine. Berberine also exhibited an ability to raise expression of studied genes, but to a lower extent than sanguinarine. All studied genes were upregulated in HL-60/MX2 cells after IC10 berberine treatment. The expression of *TP53* was differentiated across studied cell lines exposed to IC50 of berberine. In CEM/C1, CCRF/CEM, and HL-60/MX2 cells *TP53* was upregulated, but in other cell lines the downregulation of this gene was demonstrated. This indicates diverse *TP53*-mediated responses of studied cells after treatment with berberine. Upregulation of *BAK1* and *BCL2L2* was the most often observed effect in studied cell lines after treatment with *Chelidonium majus* L. extract.

PARAFAC analysis shows that IC10 dose of sanguinarine was an outlier among other doses of tested samples. Exposure to sanguinarine at this dose induced interesting changes in gene expression, which seem to be more unambiguous in interpretation and more beneficial than those induced after exposure to higher doses of the compound. Sanguinarine in IC10 doses induced an increase in *TP53* gene expression. This is a beneficial effect because the promotion of neoplastic transformation cancer is associated with a reduced expression of the *TP53*. These findings support evidence that sanguinarine has a strong antiproliferative effect on cells with *TP53* gene dysfunction [21]. In the case of cells exposed to the IC90 dose of sanguinarine, an adverse decrease in the expression of *TP53* was observed.

The presented results show that the effect of *Chelidonium majus* L. extract on the expression of genes related to the apoptosis process is very similar to the effect of sanguinarine in the IC10 dose. This suggests similarities in the mechanism of apoptosis induction between sanguinarine and *Chelidonium majus* L. extract. Upregulation of *CASP8* and *CASP3* observed in cells exposed to IC10 of sanguinarine and IC10, IC50, and IC90 of *Chelidonium majus* L. extract indicated induction of apoptosis in the extrinsic pathway. The intrinsic apoptosis is probably prevented due to the accumulation of sanguinarine close to the outer side of the inner mitochondrial membrane during its stimulation. Sanguinarine neutralizes the effects of stimulation of the mitochondrial membrane, inhibits the synthesis of ATP, and breaks the oxidative phosphorylation [40]. An increase in *CASP9* expression observed in cells exposed to *Chelidonium majus* L. extract suggests the intrinsic course of apoptosis, probably conjugated with the extrinsic apoptosis pathway.

Another effect that clearly indicates promotion of the process of apoptosis in cells exposed to both *Chelidonium majus* L. extract and low sanguinarine doses is the increase in the expressions of *BAK1*, *BAD*, and *BNIP3*. *BAK1* gene encodes the Bak protein, which participates in the formation of mitochondrial transmembrane channels and mediates the release of proapoptotic factors from the intra-mitochondrial space, including cytochrome c [22,23]. The active form of the BAD protein, encoded by the *BAD* gene, has the ability to form heterodimers with antiapoptotic Bcl-2 and Bcl-xL proteins, enhancing the proapoptotic activity of the BAX and Bak proteins [59]. The BNIP3 protein belongs to the group of BOP proteins (BH3-only proteins) belonging to the *BCL2* family. This protein is responsible for the neutralization of anti-apoptotic proteins after the onset of the mitochondrial permeabilizing membrane [60,61]. It can be concluded that the effect on the balance between the expression levels of pro- and antiapoptotic genes of the *BCL2* family is regulated by sanguinarine and *Chelidonium majus* L. extract via the BNIP3 protein. Sanguinarine seems to be the main factor inducing apoptosis of cells exposed to *Chelidonium majus* L., despite a lower amount of this compound compared to the amounts of other alkaloids in this plant [51,54].

It should be noted that in the cells exposed to the higher doses of sanguinarine, the expression of *BAK1*, *BAD*, and *BNIP3* was lower, suggesting the lower proapoptotic potential of sanguinarine in higher doses.

Both the PARAFAC analysis and the clustering with the Euclidean distances indicated that CCFR/CEM cell line clearly differed from other cell lines after exposure to all doses of sanguinarine. In this cell line, all doses of sanguinarine increased the expression of almost all analyzed genes. This is particularly important in the case of the *TP53* gene, whose expression was strongly elevated in the CCRF/CEM cells (for IC50 by 303% and for IC90 by 342%), but in other cell lines after exposure to the IC50 and IC90 doses of sanguinarine was reduced.

PARAFAC analysis and Euclidean clustering showed that the HL60 line is an outlier after exposure to all doses of *Chelidonium majus* L. extract. In these cells, as well as in the HL-60/MX1 cells, the anti-apoptotic *BCL2* expression was lowered. The remaining cell lines responded with an increase in *BCL2* gene expression after exposure to all doses of *Chelidonium majus* L. extract. This result indicates that the sensitivity of HL60 and HL60/MX1 cells to the proapoptotic action of *Chelidonium majus* L. extract is mediated by the downregulation of *BCL2*.

In this research, we evaluated the effect of berberine and *Berberis thunbergii* DC. extract for the transcription of 18 apoptosis-related genes in hematopoietic cancer cell lines. Berberine is a particularly interesting compound because it is characterized by a lack of toxicity in relation to normal cells, while the cytotoxicity towards cancer cells were reported [30]. Despite the proven proapoptotic activity of the compound [31,32], the role of individual genes in the regulation of programmed cell death exposed to berberine has been poorly investigated.

In the current study, a high increase in *CASP3* expression was observed in the CEM/C1, CCRF/CEM, and HL-60/MX2 cells after exposure to all doses of berberine. This result clearly indicates that the induction of cell death of these lines was in the apoptosis way. Moreover, an increase in *CASP9* expression, observed in the case of all cell lines exposed to berberine in the IC10, IC50, and IC90 doses, suggests the induction of programmed cell death with the intrinsic pathway.

The multidrug resistant HL-60/MX2 cell line was the most sensitivity to sanguinarine and was characterized by higher upregulation of proapoptotic genes compared to HL-60 and HL-60/MX1 cells, which exhibit weaker multidrug resistance potential (Table 1, Table 2). It may suggest that sanguinarine has an ability to break the multidrug resistance and induce apoptosis in HL-60/MX2 cells and the mechanisms determining the resistance may facilitate the cytotoxic effect of this alkaloid. This hypothesis is an interesting path for further investigations.

Euclidean clustering and the PARAFAC analysis showed that the cells of the HL-60/MX2 line exposed to berberine clearly differ from the other examined cell lines. HL-60/MX2 cell line differs in the DNA profile from the HL-60/MX1 clone by the presence of the 11 allele in the TPOX locus. This increases the multidrug resistance of the HL-60/MX2 cells, which may be responsible for the high resistance to the cytotoxic action of berberine and *Chelidonium majus* L. extract (Table 1). Despite low cytotoxicity, analysis of changes in gene expression in HL-60/MX2 cells exposed to berberine showed the greatest modulation among all tested samples. Berberine caused a significant increase in the expression of caspases *CASP3*, *CASP8*, and *CASP9*, as well as proapoptotic genes *BAX*, *BAK1*, and *BIK*, accompanied by downregulation of the anti-apoptotic *BCL2* and *BCL2L2*, as well as *BNIP1*, *BNIP3*, and *BNIP3*. It follows that gene regulation is moving towards the induction of apoptosis; however, this process is clearly slower than in the case of sanguinarine. For a more in-depth analysis of berberine-induced apoptosis, further experiments in the 72- and 96-h model should be carried out.

Exposure to the *Berberis thunbergii* DC. extract increased the expression of *BAX*, *BAK1*, *BIK*, and *CASP9* in the examined cell lines, however, was weaker when compared to berberine. This result indicates the possibility of inducing the internal pathway of apoptosis, however, similarly to berberine, signs of apoptosis could possibly be detected after exposure longer than 24 h. The berberine content in the tested dose of extract was lower than the corresponding doses of berberine (IC10, IC50, IC90), which may explain a lower increase in the expression of the same genes. It may also suggest the possibility of proapoptotic activity of other compounds present in the extract, whose identification requires further investigation. Our results confirmed that berberine and the *Berberis thunbergii* DC. extract have an activity promoting the apoptosis with minimal cytotoxicity in studied concentrations.

The limitation of this study is assessment of cytotoxicity and gene expression in exposed cells independently from cell population growth parameters, like doubling time or maximum growth rate. Gene expression was evaluated after 24 h of exposure, however, mRNA levels may change in shorter time (couple of hours, even minutes or seconds) and be a result of complex interaction with other RNAs. The influx and efflux time in relation to concentration of study samples could also affect the outcome. The influence of these variables on the sensitivity of the cells to cytotoxic activity of studied compounds should be elucidated in further investigations with optimized time points of the exposure.

Observations presented in current study may have a significant impact on the understanding of the regulation of apoptosis at the transcriptome level after exposure to sanguinarine, berberine and extracts from *Chelidonium majus* L. and *Berberis thunbergii* DC. Understanding the mechanisms that accompany the change in the expression of the studied genes is substantial when trying to determine the effects associated with the interaction of sanguinarine, berberine and the extracts studied for their anticancer activity. The obtained results significantly enrich the knowledge of the antineoplastic activity of substances and may accelerate their introduction into the clinical trial phase in the treatment of various types of leukemia.

4. Materials and Methods

4.1. Sanguinarine and Berberine Stock Solutions

The reference compounds sanguinarine and berberine were of analytical grade from Sigma-Aldrich Company (St. Louis, MO, USA). The purity of each compound was more than 98%, according to the manufacturer. 10 mM stock solution of sanguinarine in DMSO was prepared. For the cytotoxic study, a series of dilutions was prepared with concentrations 0.8, 0.6, 0.5, 0.4, 0.2, 0.1 and 0.01 mM. The final concentration of sanguinarine in cell suspensions were 1000-fold lower. Due to the limited solubility of berberine in DMSO (maximum 50 mM), berberine dilutions were prepared in 25 mM, 10 mM and 5 mM concentrations.

4.2. Plant Extracts Preparation

Plant material selection and procedure of extraction were carried out as previously described [51]. For cytotoxicity tests, 1 mL of methanolic extracts were evaporated in a nitrous atmosphere and resolved in 1 mL DMSO, obtaining 56 mg/mL concentration of *Chelidonium majus* L. extract and 29.1 mg/mL concentration of *Berberis thunbergii* DC. extract. Due to a harmful effect of DMSO to cells [62], to the cell suspension was added no more than 5 μL of sample per 1000 μL of cell suspension.

4.3. Cell Lines

HL-60 (CCL-240), HL-60/MX1 (CRL-2258), HL-60/MX2 (CRL-2257), CCRF/CEM (CCL-119) and CEM/C1 (CRL-2265) cell lines were from ATCC collection and were purchased from LGC Standards, UK. J45.01 (ECACC 93031145-1VL) and U266B1 (ECACC 85051003-1VL) cell lines were from ECACC collection and were purchased from Sigma-Aldrich Co. (St. Louis, MO, USA). Before experiments, 1 mL of cell suspension, containing 5–9 × 105 cells, was cultured in sterile 12-wells plates (3.8 cm^2 per well) in standard conditions (5% CO_2, 37 °C) for 24 h using Galaxy B incubator (RS Biotech, Irvine, UK).

4.4. Trypan Blue Staining

Sanguinarine and berberine stock solutions and examined extracts were added to cell cultures after 24 h incubation. After 24 h of exposure to studied samples, cell suspensions were centrifuged at 800 rpm for 10 min (Eppendorf 5810R centrifuge, Eppendorf AG., Hamburg, Germany). The supernatant was discarded, to the cells were added 1 mL PBS (Phosphate Buffered Saline, Biomed-Lublin WSiS, Lublin, Poland) and centrifugation was repeated. The supernatant was removed and the cells were resuspended in 50 μL of PBS. Subsequently, 10 μL of suspension was mixed with 10 μL of trypan blue solution (0.4% trypan blue in 0.81% sodium chloride and 0.06% potassium dihydrogen phosphate, Bio-Rad, Hercules, CA, USA) and incubated for 1 min in 37 °C. The total number of cells, as well as the percentages of normal and death cells, were counted on Counting Slides using Automated Cell Counter (Bio-Rad). The experiments were performed in triplicates on independent cell lines passages and the mean values were calculated. The curves presenting relationships between cell viability and the concentration of the studied samples were prepared in order to determine IC10, IC50 and IC90 doses for each of the studied cell lines.

4.5. RNA Isolation

After 24 h exposure to IC10, IC50 and IC90 doses of studied samples, cells were centrifuged at 800 rpm for 10 min and the supernatant was discarded. Total RNA was isolated from the cells using Chomczyński and Sacchi method [63] and TRI reagent (Sigma-Aldrich Co., St. Louis, MO, USA), according to the manufacturer procedure. Briefly, 500 μL of TRI reagent was added to the cells and the suspension was shaken for 10 min on IKA MS 3 Basic shaker (IKA WERKE Gmbh, Staufen, Germany). Subsequently, 50 μL of chloroform was added, the mixture was shaken for 10 min and centrifuged at 14,000 rpm for 15 min at 4 °C. The aqueous layer was collected to a new tube and 250 μL of isopropanol was added to the collected layer. The mixture was incubated for 15 min at room temperature and centrifuged (14,000 rpm, 4 °C, 10 min). The supernatant was discarded, RNA pellets were washed with 75% ethanol, dried and dissolved in 20 μL of RNAse-free water.

The quality and quantity of isolated RNA was assessed by NanoDrop2000 UV-VIS spectrophotometer with NanoDrop2000 Operating Software (Thermo Fisher Scientific Inc., Waltham, MA, USA). RNA samples with A260/A280 ratio ranged between 1.8 and 2.0 were intended to further investigations.

4.6. cDNA Synthesis

Synthesis of cDNA was performed using High Capacity cDNA Reverse Transcription Kit (Applied Biosystems, Foster City, CA, USA), according to the manufacturer procedure. Briefly, the following reaction mixture was assembled on ice: 2 μL of 10X RT Buffer, 0.8 μL of 25X dNTP Mix (100 mM), 2 μL of 10X RT Random Primers, 1 μL of MultiScribe Reverse Transcriptase (50 U/μL), 0.5 μL of RNase Inhibitor (40 U/μL), 1 μL of isolated RNA and 1 μL of DEPC-treated nuclease-free water. The reaction mixture was incubated in thermocycler Mastercycler Personal (Eppendorf AG., Hamburg, Germany) in the following conditions: 25 °C for 10 min, 37 °C for 120 min and 85 °C for 5 s. Obtained cDNA samples were stored in −20 °C.

4.7. Real-Time PCR

Gene expression analysis was carried out for 18 genes related to apoptosis (*B2M*, *BAD*, *BAK1*, *BAX*, *BCL2*, *BCL2L1*, *BCL2L2*, *BID*, *BIK*, *BNIP1*, *BNIP2*, *BNIP3*, *CASP3*, *CASP8*, *CASP9*, *MCL1*, *PMAIP1*, and *TP53*) using real-time PCR method. cDNA samples were amplified using a 7900HT Fast Real-Time PCR System (Applied Biosystems, Foster City, CA, USA). Reaction mixtures contained 1.25 μL of gene-specific TaqMan probe (Applied Biosystems, USA) described in Table A1, 12.5 μL of TaqMan Gene Expression Master Mix (Applied Biosystems, USA), and 11.25 μL of cDNA sample. PCR reactions were performed in μAmp Optical 96-Well Reaction Plates (Life Technologies Corporation, Carlsbad, CA, USA). The expression of GAPDH was used as an endogenous control. The reaction was conducted in the following conditions: 95 °C for 10 min, 40 cycles: 95 °C for 15 s, and 60 °C for 60 s. Gene expression levels (Ct) obtained in exposed cells were compared to the expression levels in no exposed cells (control) using the relative quantitation method (RQ = $2^{-\Delta\Delta Ct}$) [64,65]. The experiments were performed in quadruplicate and the mean values were calculated. Data was analyzed using SDS 2.4 Study software (Applied Biosystems, USA).

4.8. Statistical Analysis

Clustering analysis with Euclidean distance and parallel factor analysis (PARAFAC) were applied to reduce the large dimensionality of the data and to assess the similarities between the doses of the tested compounds, between the analyzed cell lines, and between changes in the expression of examined genes after exposure to the samples. The analysis was performed using R programming software. The analysis did not include data obtained after exposure to the *Berberis thunbergii* DC. extract due to the lack of cytotoxicity in relation to the tested cell lines.

Supplementary Materials: The following are available online at http://www.mdpi.com/2072-6651/11/9/485/s1, Figure S1: Expression profiles of 18 apoptosis-related genes in U266B1 cell line exposed to IC10 (upper panel), IC50 (middle panel) and IC90 (lower panel) of sanguinarine (blue), berberine (yellow) and *Chelidonium majus* L. extract (red), Figure S2: Expression profiles of 18 apoptosis-related genes in CEM/C1 cell line exposed to IC10 (upper panel), IC50 (middle panel) and IC90 (lower panel) of sanguinarine (blue), berberine (yellow) and *Chelidonium majus* L. extract (red), Figure S3: Expression profiles of 18 apoptosis-related genes in CCRF/CEM cell line exposed to IC10 (upper panel), IC50 (middle panel) and IC90 (lower panel) of sanguinarine (blue), berberine (yellow) and *Chelidonium majus* L. extract (red), Figure S4: Expression profiles of 18 apoptosis-related genes in HL60 cell line exposed to IC10 (upper panel), IC50 (middle panel) and IC90 (lower panel) of sanguinarine (blue), berberine (yellow) and *Chelidonium majus* L. extract (red), Figure S5: Expression profiles of 18 apoptosis-related genes in HL60/MX1 cell line exposed to IC10 (upper panel), IC50 (middle panel) and IC90 (lower panel) of sanguinarine (blue), berberine (yellow) and *Chelidonium majus* L. extract (red), Figure S6: Expression profiles of 18 apoptosis-related genes in HL60/MX2 cell line exposed to IC10 (upper panel), IC50 (middle panel) and IC90 (lower panel) of sanguinarine (blue), berberine (yellow) and *Chelidonium majus* L. extract (red), Figure S7: Expression profiles of 18 apoptosis-related genes in J45.01 cell line exposed to IC10 (upper panel), IC50 (middle panel) and IC90 (lower panel) of sanguinarine (blue), berberine (yellow) and *Chelidonium majus* L. extract (red), Figure S8: Expression profiles of 18 apoptosis-related genes in seven hematopoietic cell lines exposed to *Berberis thunbergii* DC. extract at a concentration of 145.5 μg/mL, Table S1: logRQ values for technical replicates, means and Standard Deviations calculated for changes in expression of 18 genes related to apoptosis (*B2M, BAD, BAK1, BAX, BCL2, BCL2L1, BCL2L2, BID, BIK, BNIP1, BNIP2, BNIP3, CASP3, CASP8, CASP9, MCL1, PMAIP1, TP53*) in seven hematopoietic cancer cell lines (CEM/C1, J45.01, CCRF/CEM, HL-60, HL-60/MX1, HL-60/MX2 and U266B1) after 24 h exposure to IC10 concentration of berberine, Table S2: logRQ values for technical replicates, means and Standard Deviations calculated for changes in expression of 18 genes related to apoptosis (*B2M, BAD, BAK1, BAX, BCL2, BCL2L1, BCL2L2, BID, BIK, BNIP1, BNIP2, BNIP3, CASP3, CASP8, CASP9, MCL1, PMAIP1, TP53*) in seven hematopoietic cancer cell lines (CEM/C1, J45.01, CCRF/CEM, HL-60, HL-60/MX1, HL-60/MX2 and U266B1) after 24 h exposure to IC50 concentration of berberine, Table S3: logRQ values for technical replicates, means and Standard Deviations calculated for changes in expression of 18 genes related to apoptosis (*B2M, BAD, BAK1, BAX, BCL2, BCL2L1, BCL2L2, BID, BIK, BNIP1, BNIP2, BNIP3, CASP3, CASP8, CASP9, MCL1, PMAIP1, TP53*) in seven hematopoietic cancer cell lines (CEM/C1, J45.01, CCRF/CEM, HL-60, HL-60/MX1 and U266B1) after 24 h exposure to IC90 concentration of berberine, Table S4: logRQ values for technical replicates, means and Standard Deviations calculated for changes in expression of 18 genes related to apoptosis (*B2M, BAD, BAK1, BAX, BCL2, BCL2L1, BCL2L2, BID, BIK, BNIP1, BNIP2, BNIP3, CASP3, CASP8, CASP9, MCL1, PMAIP1, TP53*) in seven hematopoietic cancer cell lines (CEM/C1, J45.01, CCRF/CEM, HL-60, HL-60/MX1, HL-60/MX2 and U266B1) after 24 h exposure to IC10 concentration of Chelidonium majus L. extract, Table S5: logRQ values for technical replicates, means and Standard Deviations calculated for changes in expression of 18 genes related to apoptosis (*B2M, BAD, BAK1, BAX, BCL2, BCL2L1, BCL2L2, BID, BIK, BNIP1, BNIP2, BNIP3, CASP3, CASP8, CASP9, MCL1, PMAIP1, TP53*) in seven hematopoietic cancer cell lines (CEM/C1, J45.01, CCRF/CEM, HL-60, HL-60/MX1, HL-60/MX2 and U266B1) after 24 h exposure to IC50 concentration of Chelidonium majus L. extract, Table S6: logRQ values for technical replicates, means and Standard Deviations calculated for changes in expression of 18 genes related to apoptosis (*B2M, BAD, BAK1, BAX, BCL2, BCL2L1, BCL2L2, BID, BIK, BNIP1, BNIP2, BNIP3, CASP3, CASP8, CASP9, MCL1, PMAIP1, TP53*) in seven hematopoietic cancer cell lines (CEM/C1, J45.01, CCRF/CEM, HL-60, HL-60/MX1, HL-60/MX2 and U266B1) after 24 h exposure to IC90 concentration of Chelidonium majus L. extract, Table S7: logRQ values for technical replicates, means and Standard Deviations calculated for changes in expression of 18 genes related to apoptosis (*B2M, BAD, BAK1, BAX, BCL2, BCL2L1, BCL2L2, BID, BIK, BNIP1, BNIP2, BNIP3, CASP3, CASP8, CASP9, MCL1, PMAIP1, TP53*) in seven hematopoietic cancer cell lines (CEM/C1, J45.01, CCRF/CEM, HL-60, HL-60/MX1, HL-60/MX2 and U266B1) after 24 h exposure to IC10 concentration of sanguinarine, Table S8: logRQ values for technical replicates, means and Standard Deviations calculated for changes in expression of 18 genes related to apoptosis (*B2M, BAD, BAK1, BAX, BCL2, BCL2L1, BCL2L2, BID, BIK, BNIP1, BNIP2, BNIP3, CASP3, CASP8, CASP9, MCL1, PMAIP1, TP53*) in seven hematopoietic cancer cell lines (CEM/C1, J45.01, CCRF/CEM, HL-60, HL-60/MX1, HL-60/MX2 and U266B1) after 24 h exposure to IC50 concentration of sanguinarine, Table S9: logRQ values for technical replicates, means and Standard Deviations calculated for changes in expression of 18 genes related to apoptosis (*B2M, BAD, BAK1, BAX, BCL2, BCL2L1, BCL2L2, BID, BIK, BNIP1, BNIP2, BNIP3, CASP3, CASP8, CASP9, MCL1, PMAIP1, TP53*) in seven hematopoietic cancer cell lines (CEM/C1, J45.01, CCRF/CEM, HL-60, HL-60/MX1, HL-60/MX2 and U266B1) after 24 h exposure to IC90 concentration of sanguinarine, Table S10: Mean values of logRQ calculated for changes in expression of 18 genes related to apoptosis (*B2M, BAD, BAK1, BAX, BCL2, BCL2L1, BCL2L2, BID, BIK, BNIP1, BNIP2, BNIP3, CASP3, CASP8, CASP9, MCL1, PMAIP1, TP53*) in seven hematopoietic cancer cell lines (CEM/C1, J45.01, CCRF/CEM, HL-60, HL-60/MX1, HL-60/MX2 and U266B1) after 24 h exposure to Berberis thunbergii DC. extract in concentration of 145.5 μg/mL.

Author Contributions: Conceptualization, J.K. and A.B.-K.; data curation, A.O., Ł.K., and A.B.-K.; formal analysis, A.O., D.Z., Ł.K., P.K., and A.B.-K.; funding acquisition, A.B.-K.; investigation, A.O., Ł.K., and A.B.-K.; methodology, A.O., Ł.K., and A.B.-K.; project administration, A.B.-K.; resources, J.K. and A.B.-K.; software, Ł.K.; visualization, A.O., D.Z., and Ł.K.; writing—original draft, A.O., D.Z., and P.K.; writing—review and editing, J.K. and A.B.-K.

Funding: This research was funded by statutory funds of the Medical University of Lublin (DS43) provided by the Polish Ministry of Science and Higher Education for Medical University of Lublin, Poland.

Acknowledgments: The research was performed using the equipment purchased within the project "The equipment of innovative laboratories doing research on new medicines used in the therapy of civilization and neoplastic diseases" within the Operational Program Development of Eastern Poland 2007–2013, Priority Axis I Modern Economy, Operations I.3 Innovation Promotion.

Conflicts of Interest: The authors declare no conflict of interest. The funders had no role in the design of the study; in the collection, analyses, or interpretation of data; in the writing of the manuscript, or in the decision to publish the results.

Appendix A

Table A1. The list of TaqMan probes (Applied Biosystems, Foster City, CA, USA) applied to the study.

Gene Symbol	Probe	Gene Name
BAK1	Hs00940249_m1	*BCL2* antagonist/killer 1
BAX	Hs00180363_m1	*BCL2* associated X, apoptosis regulator
BCL2	Hs00608023_m1	*BCL2* apoptosis regulator
MCL1	Hs01050896_m1	*MCL1* apoptosis regulator, *BCL2* family member
BCL2L1	Hs00236329_m1	*BCL2* like 1
BCL2L2	Hs00187848_m1	*BCL2* like 2
BID	Hs01026792_m1	*BH3* interacting domain death agonist
BIK	Hs00154189_m1	*BCL2* interacting killer
BNIP1	Hs00241824_m1	*BCL2* interacting protein 1
BNIP2	Hs00188939_m1	*BCL2* interacting protein 2
BNIP3	Hs00969291_m1	*BCL2* interacting protein 3
PMAIP1	Hs00560402_m1	Phorbol-12-myristate-13-acetate-induced protein 1
BAD	Hs00188930_m1	*BCL2* associated agonist of cell death
CASP3	Hs00234387_m1	Caspase 3
CASP8	Hs01018151_m1	Caspase 8
CASP9	Hs00154261_m1	Caspase 9
TP53	Hs01034249_m1	Tumor protein p53
GAPDH (Endogenous control)	Hs99999905_m1	Glyceraldehyde-3-phosphate dehydrogenase

References

1. Trujillo-Villanueva, K.; Rubio-Piña, J.; Monforte-González, M.; Ramírez-Benítez, E.; Vázquez-Flota, F. The sequential exposure to jasmonate, salicylic acid and yeast extract promotes sanguinarine accumulation in Argemone mexicana cell cultures. *Biotechnol. Lett.* **2012**, *34*, 379–385. [CrossRef] [PubMed]
2. Cho, H.Y.; Rhee, H.S.; Yoon, S.Y.; Park, J.M. Differential induction of protein expression and benzophenanthridine alkaloid accumulation in Eschscholtzia californica suspension cultures by methyl jasmonate and yeast extract. *J. Microbiol. Biotechnol.* **2008**, *18*, 255–262. [PubMed]
3. Obiang-Obounou, B.W.; Kang, O.H.; Choi, J.G.; Keum, J.H.; Kim, S.B.; Mun, S.H.; Shin, D.W.; Kim, K.W.; Park, C.B.; Kim, Y.G.; et al. The mechanism of action of sanguinarine against methicillin-resistant Staphylococcus aureus. *J. Toxicol. Sci.* **2011**, *36*, 277–283. [CrossRef] [PubMed]
4. Wang, Q.; Dai, P.; Bao, H.; Liang, P.; Wang, W.; Xing, A.; Sun, J. Anti-inflammatory and neuroprotective effects of sanguinarine following cerebral ischemia in rats. *Exp. Ther. Med.* **2017**, *13*, 263–268. [CrossRef] [PubMed]
5. Ma, Y.; Sun, X.; Huang, K.; Shen, S.; Lin, X.; Xie, Z.; Wang, J.; Fan, S.; Ma, J.; Zhao, X. Sanguinarine protects against osteoarthritis by suppressing the expression of catabolic proteases. *Oncotarget* **2017**, *8*, 62900–62913. [CrossRef] [PubMed]
6. Lee, S.S.; Kai, M.; Lee, M.K. Inhibitory effects of sanguinarine on monoamine oxidase activity in mouse brain. *Phytother. Res.* **2001**, *15*, 167–169. [CrossRef] [PubMed]
7. Jursky, F.; Baliova, M. Differential effect of the benzophenanthridine alkaloids sanguinarine and chelerythrine on glycine transporters. *Neurochem. Int.* **2011**, *58*, 641–647. [CrossRef] [PubMed]

8. Singh, R.; Mackraj, I.; Naidoo, R.; Gathiram, P. Sanguinarine downregulates AT1a gene expression in a hypertensive rat model. *J. Cardiovasc. Pharmacol.* **2006**, *48*, 14–21. [CrossRef] [PubMed]
9. Jeng, J.H.; Wu, H.L.; Lin, B.R.; Lan, W.H.; Chang, H.H.; Ho, Y.S.; Lee, P.H.; Wang, Y.J.; Wang, J.S.; Chen, Y.J.; et al. Antiplatelet effect of sanguinarine is correlated to calcium mobilization, thromboxane and cAMP production. *Atherosclerosis* **2007**, *191*, 250–258. [CrossRef]
10. Kuznetsova, L.P.; Nikol'skaya, E.B.; Sochilina, E.E.; Faddeeva, M.D. Inhibition of Human Blood Acetylcholinesterase and Butyrylcholinesterase by Some Alkaloids. *J. Evol. Biochem. Physiol.* **2002**, *38*, 35–39. [CrossRef]
11. Zajoncova, L.; Kosina, P.; Vicar, J.; Ulrichová, J.; Pec, P. Study of the inhibition of alpha-amylase by the benzo[c]phenanthridine alkaloids sanguinarine and chelerythrine. *J. Enzym. Inhib. Med. Chem.* **2005**, *20*, 261–267. [CrossRef] [PubMed]
12. Imenshahidi, M.; Hosseinzadeh, H. Berberine and barberry (*Berberis vulgaris*): A clinical review. *Phytother. Res.* **2019**, *33*, 504–523. [CrossRef] [PubMed]
13. Xu, Y.; Wang, Y.; Yan, L.; Liang, R.M.; Dai, B.D.; Tang, R.J.; Gao, P.H.; Jiang, Y.Y. Proteomic analysis reveals a synergistic mechanism of fluconazole and berberine against fluconazole-resistant Candida albicans: Endogenous ROS augmentation. *J. Proteome Res.* **2009**, *8*, 5296–5304. [CrossRef] [PubMed]
14. Li, Z.; Geng, Y.N.; Jiang, J.D.; Kong, W.J. Antioxidant and anti-inflammatory activities of berberine in the treatment of diabetes mellitus. *Evid. Based Complement. Altern. Med.* **2014**, *2014*, 289264. [CrossRef] [PubMed]
15. Chang, W.; Chen, L.; Hatch, G.M. Berberine as a therapy for type 2 diabetes and its complications: From mechanism of action to clinical studies. *Biochem. Cell Biol.* **2015**, *93*, 479–486. [CrossRef] [PubMed]
16. Wang, Y.; Liu, J.; Ma, A.; Chen, Y. Cardioprotective effect of berberine against myocardial ischemia/reperfusion injury via attenuating mitochondrial dysfunction and apoptosis. *Int. J. Clin. Exp. Med.* **2015**, *8*, 14513–14519. [PubMed]
17. Fan, J.; Zhang, K.; Jin, Y.; Li, B.; Gao, S.; Zhu, J.; Cui, R. Pharmacological effects of berberine on mood disorders. *J. Cell Mol. Med.* **2019**, *23*, 21–28. [CrossRef] [PubMed]
18. Kongkiatpaiboon, S.; Duangdee, N.; Prateeptongkum, S.; Chaijaroenkul, W. Acetylcholinesterase Inhibitory Activity of Alkaloids Isolated from Stephania venosa. *Nat. Prod. Commun.* **2016**, *11*, 1805–1806. [CrossRef] [PubMed]
19. Galadari, S.; Rahman, A.; Pallichankandy, S.; Thayyullathil, F. Molecular targets and anticancer potential of sanguinarine—A benzophenanthridine alkaloid. *Phytomedicine* **2017**, *34*, 143–153. [CrossRef] [PubMed]
20. Gaziano, R.; Moroni, G.; Buè, C.; Miele, M.T.; Sinibaldi-Vallebona, P.; Pica, F. Antitumor effects of the benzophenanthridine alkaloid sanguinarine: Evidence and perspectives. *World J. Gastrointest. Oncol.* **2016**, *8*, 30–39. [CrossRef] [PubMed]
21. Hussain, A.R.; Al-Jomah, N.A.; Siraj, A.K.; Manogaran, P.; Al-Hussein, K.; Abubaker, J.; Platanias, L.C.; Al-Kuraya, K.S.; Uddin, S. Sanguinarine-dependent induction of apoptosis in primary effusion lymphoma cells. *Cancer Res.* **2007**, *67*, 3888–3897. [CrossRef] [PubMed]
22. Weerasinghe, P.; Hallock, S.; Tang, S.C.; Liepins, A. Role of Bcl-2 family proteins and caspase-3 in sanguinarine-induced bimodal cell death. *Cell Biol. Toxicol.* **2001**, *17*, 371–381. [CrossRef] [PubMed]
23. Weerasinghe, P.; Hallock, S.; Tang, S.C.; Liepins, A. Sanguinarine induces bimodal cell death in K562 but not in high Bcl-2-expressing JM1 cells. *Pathol. Res. Pract.* **2001**, *197*, 717–726. [CrossRef] [PubMed]
24. Kim, S.; Lee, T.J.; Leem, J.; Choi, K.S.; Park, J.W.; Kwon, T.K. Sanguinarine-induced apoptosis: Generation of ROS, down-regulation of Bcl-2, c-FLIP, and synergy with TRAIL. *J. Cell Biochem.* **2008**, *104*, 895–907. [CrossRef] [PubMed]
25. Park, S.Y.; Jin, M.L.; Kim, Y.H.; Lee, S.J.; Park, G. Sanguinarine inhibits invasiveness and the MMP-9 and COX-2 expression in TPA-induced breast cancer cells by inducing HO-1 expression. *Oncol. Rep.* **2014**, *31*, 497–504. [CrossRef] [PubMed]
26. Basini, G.; Bussolati, S.; Santini, S.E.; Grasselli, F. Sanguinarine inhibits VEGF-induced angiogenesis in a fibrin gel matrix. *Biofactors* **2007**, *29*, 11–18. [CrossRef]
27. Basini, G.; Santini, S.E.; Bussolati, S.; Grasselli, F. Sanguinarine inhibits VEGF-induced Akt phosphorylation. *Ann. N. Y. Acad. Sci.* **2007**, *1095*, 371–376. [CrossRef]
28. Eun, J.P.; Koh, G.Y. Suppression of angiogenesis by the plant alkaloid, sanguinarine. *Biochem. Biophys. Res. Commun.* **2004**, *317*, 618–624. [CrossRef]

29. Zou, K.; Li, Z.; Zhang, Y.; Zhang, H.Y.; Li, B.; Zhu, W.L.; Shi, J.Y.; Jia, Q.; Li, Y.M. Advances in the study of berberine and its derivatives: A focus on anti-inflammatory and anti-tumor effects in the digestive system. *Acta Pharmacol. Sin.* **2017**, *38*, 157–167. [CrossRef]
30. Yan, K.; Zhang, C.; Feng, J.; Hou, L.; Yan, L.; Zhou, Z.; Liu, Z.; Liu, C.; Fan, Y.; Zheng, B.; et al. Induction of G1 cell cycle arrest and apoptosis by berberine in bladder cancer cells. *Eur. J. Pharmacol.* **2011**, *661*, 1–7. [CrossRef]
31. Goto, H.; Kariya, R.; Shimamoto, M.; Kudo, E.; Taura, M.; Katano, H.; Okada, S. Antitumor effect of berberine against primary effusion lymphoma via inhibition of NF-κB pathway. *Cancer Sci.* **2012**, *103*, 775–781. [CrossRef] [PubMed]
32. Liu, Z.; Liu, Q.; Xu, B.; Wu, J.; Guo, C.; Zhu, F.; Yang, Q.; Gao, G.; Gong, Y.; Shao, C. Berberine induces p53-dependent cell cycle arrest and apoptosis of human osteosarcoma cells by inflicting DNA damage. *Mutat. Res.* **2009**, *662*, 75–83. [CrossRef] [PubMed]
33. Jin, F.; Xie, T.; Huang, X.; Zhao, X. Berberine inhibits angiogenesis in glioblastoma xenografts by targeting the VEGFR2/ERK pathway. *Pharm. Biol.* **2018**, *56*, 665–671. [CrossRef] [PubMed]
34. Wang, J.; Kang, M.; Wen, Q.; Qin, Y.T.; Wei, Z.X.; Xiao, J.J.; Wang, R.S. Berberine sensitizes nasopharyngeal carcinoma cells to radiation through inhibition of Sp1 and EMT. *Oncol. Rep.* **2017**, *37*, 2425–2432. [CrossRef] [PubMed]
35. Liu, Y.; Yu, H.; Zhang, C.; Cheng, Y.; Hu, L.; Meng, X.; Zhao, Y. Protective effects of berberine on radiation-induced lung injury via intercellular adhesion molecular-1 and transforming growth factor-beta-1 in patients with lung cancer. *Eur. J. Cancer* **2008**, *44*, 2425–2432. [CrossRef] [PubMed]
36. Singh, N.; Sharma, B. Toxicological Effects of Berberine and Sanguinarine. *Front. Mol. Biosci.* **2018**, *5*, 21. [CrossRef] [PubMed]
37. Das, M.; Ansari, K.M.; Dhawan, A.; Shukla, Y.; Khanna, S.K. Correlation of DNA damage in epidemic dropsy patients to carcinogenic potential of argemone oil and isolated sanguinarine alkaloid in mice. *Int. J. Cancer* **2005**, *117*, 709–717. [CrossRef] [PubMed]
38. Dixit, R.; Srivastava, P.; Basu, S.; Srivastava, P.; Mishra, P.K.; Shukla, V.K. Association of mustard oil as cooking media with carcinoma of the gallbladder. *J. Gastrointest. Cancer* **2013**, *44*, 177–181. [CrossRef]
39. Hossain, M.; Kabir, A.; Suresh Kumar, G. Binding of the anticancer alkaloid sanguinarine with tRNA (phe): Spectroscopic and calorimetric studies. *J. Biomol. Struct. Dyn.* **2012**, *30*, 223–234. [CrossRef]
40. Adhami, V.M.; Aziz, M.H.; Mukhtar, H.; Ahmad, N. Activation of prodeath Bcl-2 family proteins and mitochondrial apoptosis pathway by sanguinarine in immortalized human HaCaT keratinocytes. *Clin. Cancer Res.* **2003**, *9*, 3176–3182.
41. Selvi, B.R.; Pradhan, S.K.; Shandilya, J.; Das, C.; Sailaja, B.S.; Shankar, G.N.; Gadad, S.S.; Reddy, A.; Dasgupta, D.; Kundu, T.K. Sanguinarine interacts with chromatin, modulates epigenetic modifications, and transcription in the context of chromatin. *Chem. Biol.* **2009**, *16*, 203–216. [CrossRef] [PubMed]
42. Maiti, M.; Kumar, G.S. Polymorphic nucleic Acid binding of bioactive isoquinoline alkaloids and their role in cancer. *J. Nucleic Acids* **2010**, *2010*, 593408. [CrossRef] [PubMed]
43. Ji, X.; Sun, H.; Zhou, H.; Xiang, J.; Tang, Y.; Zhao, C. The interaction of telomeric DNA and C-myc22 G-quadruplex with 11 natural alkaloids. *Nucleic Acid Ther.* **2012**, *22*, 127–136. [CrossRef] [PubMed]
44. Choy, C.S.; Cheah, K.P.; Chiou, H.Y.; Li, J.S.; Liu, Y.H.; Yong, S.F.; Chiu, W.T.; Liao, J.W.; Hu, C.M. Induction of hepatotoxicity by sanguinarine is associated with oxidation of protein thiols and disturbance of mitochondrial respiration. *J. Appl. Toxicol.* **2008**, *28*, 945–956. [CrossRef] [PubMed]
45. Ferraroni, M.; Bazzicalupi, C.; Bilia, A.R.; Gratteri, P. X-Ray diffraction analyses of the natural isoquinoline alkaloids Berberine and Sanguinarine complexed with double helix DNA d(CGTACG). *Chem. Commun. (Camb.)* **2011**, *47*, 4917–4919. [CrossRef] [PubMed]
46. Maiti, M.; Kumar, G.S. Molecular aspects on the interaction of protoberberine, benzophenanthridine, and aristolochia group of alkaloids with nucleic acid structures and biological perspectives. *Med. Res. Rev.* **2007**, *27*, 649–695. [CrossRef] [PubMed]
47. Li, X.L.; Hu, Y.J.; Wang, H.; Yu, B.Q.; Yue, H.L. Molecular spectroscopy evidence of berberine binding to DNA: Comparative binding and thermodynamic profile of intercalation. *Biomacromolecules* **2012**, *13*, 873–880. [CrossRef] [PubMed]
48. Hu, X.; Wu, X.; Huang, Y.; Tong, Q.; Takeda, S.; Qing, Y. Berberine induces double-strand DNA breaks in Rev3 deficient cells. *Mol. Med. Rep.* **2014**, *9*, 1883–1888. [CrossRef]

49. Yuan, Z.Y.; Lu, X.; Lei, F.; Chai, Y.S.; Wang, Y.G.; Jiang, J.F.; Feng, T.S.; Wang, X.P.; Yu, X.; Yan, X.J.; et al. TATA boxes in gene transcription and poly (A) tails in mRNA stability: New perspective on the effects of berberine. *Sci. Rep.* **2015**, *5*, 18326. [CrossRef]
50. Chai, Y.S.; Yuan, Z.Y.; Lei, F.; Wang, Y.G.; Hu, J.; Du, F.; Lu, X.; Jiang, J.F.; Xing, D.M.; Du, L.J. Inhibition of retinoblastoma mRNA degradation through Poly (A) involved in the neuroprotective effect of berberine against cerebral ischemia. *PLoS ONE* **2014**, *9*, e90850. [CrossRef]
51. Och, A.; Szewczyk, K.; Pecio, Ł.; Stochmal, A.; Załuski, D.; Bogucka-Kocka, A. UPLC-MS/MS Profile of Alkaloids with Cytotoxic Properties of Selected Medicinal Plants of the Berberidaceae and Papaveraceae Families. *Oxid. Med. Cell. Longev.* **2017**, *2017*, 9369872. [CrossRef] [PubMed]
52. Aburai, N.; Yoshida, M.; Ohnishi, M.; Kimura, K. Sanguinarine as a potent and specific inhibitor of protein phosphatase 2C in vitro and induces apoptosis via phosphorylation of p38 in HL60 cells. *Biosci. Biotechnol. Biochem.* **2010**, *74*, 548–552. [CrossRef] [PubMed]
53. Slaninová, I.; Slanina, J.; Táborská, E. Quaternary benzo[c]phenanthridine alkaloids—Novel cell permeant and red fluorescing DNA probes. *Cytometry A* **2007**, *71*, 700–708. [CrossRef] [PubMed]
54. Bogucka-Kocka, A.; Zalewski, D. Qualitative and quantitative determination of main alkaloids of *Chelidonium majus* L. using thin-layer chromatographic-densitometric method. *Acta Chromatogr.* **2017**, *29*, 385–397. [CrossRef]
55. Nadova, S.; Miadokova, E.; Alfoldiova, L.; Kopaskova, M.; Hasplova, K.; Hudecova, A.; Vaculcikova, D.; Gregan, F.; Cipak, L. Potential antioxidant activity, cytotoxic and apoptosis-inducing effects of *Chelidonium majus* L. extract on leukemia cells. *Neuro Endocrinol. Lett.* **2008**, *29*, 649–652. [PubMed]
56. Zielińska, S.; Jezierska-Domaradzka, A.; Wójciak-Kosior, M.; Sowa, I.; Junka, A.; Matkowski, A.M. Greater Celandine's Ups and Downs-21 Centuries of Medicinal Uses of *Chelidonium majus* From the Viewpoint of Today's Pharmacology. *Front. Pharmacol.* **2018**, *9*, 299. [CrossRef] [PubMed]
57. Hu, H.Y.; Li, K.P.; Wang, X.J.; Liu, Y.; Lu, Z.G.; Dong, R.H.; Guo, H.B.; Zhang, M.X. Set9, NF-κB, and μRNA-21 mediate berberine-induced apoptosis of human multiple myeloma cells. *Acta Pharmacol. Sin.* **2013**, *34*, 157–166. [CrossRef] [PubMed]
58. Efferth, T.; Davey, M.; Olbrich, A.; Rücker, G.; Gebhart, E.; Davey, R. Activity of drugs from traditional Chinese medicine toward sensitive and MDR1-or MRP1-overexpressing multidrug-resistant human CCRF-CEM leukemia cells. *Blood Cells Mol. Dis.* **2002**, *28*, 160–168. [CrossRef]
59. Bui, N.L.; Pandey, V.; Zhu, T.; Ma, L.; Basappa; Lobie, P.E. BAD phosphorylation as a target of inhibition in oncology. *Cancer Lett.* **2018**, *415*, 177–186. [CrossRef]
60. Macher-Goeppinger, S.; Keith, M.; Hatiboglu, G.; Hohenfellner, M.; Schirmacher, P.; Roth, W.; Tagscherer, K.E. Expression and Functional Characterization of the *BNIP3* Protein in Renal Cell Carcinomas. *Transl. Oncol.* **2017**, *10*, 869–875. [CrossRef]
61. Singh, A.; Azad, M.; Shymko, M.D.; Henson, E.S.; Katyal, S.; Eisenstat, D.D.; Gibson, S.B. The BH3 only Bcl-2 family member *BNIP3* regulates cellular proliferation. *PLoS ONE* **2018**, *13*, e0204792. [CrossRef] [PubMed]
62. Bishayee, A.; Rao, D.V.; Bouchet, L.G.; Bolch, W.E.; Howell, R.W. Protection by DMSO against cell death caused by intracellularly localized iodine-125, iodine-131 and polonium-210. *Radiat. Res.* **2000**, *153*, 416–427. [CrossRef]
63. Chomczynski, P.; Sacchi, N. The single-step method of RNA isolation by acid guanidinium thiocyanate-phenol-chloroform extraction: Twenty-something years on. *Nat. Protoc.* **2006**, *1*, 581–585. [CrossRef] [PubMed]
64. Livak, K.J.; Schmittgen, T.D. Analysis of relative gene expression data using real-time quantitative PCR and the 2(-Delta Delta C(T)) Method. *Methods* **2001**, *25*, 402–408. [CrossRef] [PubMed]
65. Schmittgen, T.D.; Livak, K.J. Analyzing real-time PCR data by the comparative C (T) method. *Nat. Protoc.* **2008**, *3*, 1101–1108. [CrossRef] [PubMed]

Article

Curine Inhibits Macrophage Activation and Neutrophil Recruitment in a Mouse Model of Lipopolysaccharide-Induced Inflammation

Jaime Ribeiro-Filho [1,*], Fagner Carvalho Leite [2], Andrea Surrage Calheiros [3], Alan de Brito Carneiro [3], Juliana Alves Azeredo [3], Edson Fernandes de Assis [3], Celidarque da Silva Dias [4], Márcia Regina Piuvezam [2] and Patrícia T. Bozza [3]

1. Laboratório de Investigação em Genética e Hematologia Translacional, Instituto Gonçalo Moniz, FIOCRUZ, Salvador 40296-710, Brazil
2. Laboratório de Imunofarmacologia, Departamento de Fisiologia e Patologia, UFPB, João Pessoa 58051-900, Brazil; fagnercarvalho.farm@gmail.com (F.C.L.); mrpiuvezam@ltf.ufpb.br (M.R.P.)
3. Laboratório de Imunofarmacologia, Instituto Oswaldo Cruz, FIOCRUZ, Rio de Janeiro 21040-360, Brazil; andrea.surrage@gmail.com (A.S.C.); alan.fiocruz@gmail.com (A.d.B.C.); jazeredo@ioc.fiocruz.br (J.A.A.); edassis@ioc.fiocruz.br (E.F.d.A.); pbozza@gmail.com (P.T.B.)
4. Laboratório de Fitoquímica, Departamento de Ciências Farmacêuticas, UFPB, João Pessoa 58051-900, Brazil; celidarquedias@ltf.ufpb.br

* Correspondence: jaime.ribeiro@fiocruz.br; Tel.: +55-71-3176-2226

Received: 18 September 2019; Accepted: 22 October 2019; Published: 3 December 2019

Abstract: Curine is a bisbenzylisoquinoline alkaloid (BBA) with anti-allergic, analgesic, and anti-inflammatory properties. Previous studies have demonstrated that this alkaloid is orally active at non-toxic doses. However, the mechanisms underlying its anti-inflammatory effects remain to be elucidated. This work aimed to investigate the effects of curine on macrophage activation and neutrophil recruitment. Using a murine model of lipopolysaccharide (LPS)-induced pleurisy, we demonstrated that curine significantly inhibited the recruitment of neutrophils in association with the inhibition of cytokines tumor necrosis factor (TNF-α), interleukin (IL)-1β, IL-6, monocyte chemotactic protein (CCL2/MCP-1) as well as leukotriene B_4 in the pleural lavage of mice. Curine treatment reduced cytokine levels and the expression of iNOS in in vitro cultures of macrophages stimulated with LPS. Treatment with a calcium channel blocker resulted in comparable inhibition of TNF-α and IL-1β production, as well as iNOS expression by macrophages, suggesting that the anti-inflammatory effects of curine may be related to the inhibition of calcium-dependent mechanisms involved in macrophage activation. In conclusion, curine presented anti-inflammatory effects that are associated with inhibition of macrophage activation and neutrophil recruitment by inhibiting the production of inflammatory cytokines, LTB_4 and nitric oxide (NO), and possibly by negatively modulating Ca^{2+} influx.

Keywords: Curine; alkaloid; macrophage; neutrophil; lipopolysaccharide

Key Contribution: This study attempts to contribute to the elucidation of mechanisms involved in the anti-inflammatory action of curine, an orally active alkaloid.

1. Introduction

Macrophages work as detectors of inflammatory signals, including those produced by the host and the derived from microorganisms, such as lipopolysaccharide (LPS) [1]. LPS signaling through TLR_4 induces macrophage activation by regulating intracellular pathways involved in cytokine, lipid mediator and oxygen reactive species (ROS) production, in a process regulated by calcium signaling [2,3]. The mediators released by activated macrophages play critical roles in neutrophil recruitment and activation, and therefore, influence the progress of immune responses as well as the development of many inflammatory diseases [4,5].

Curine (Figure 1A) is the principal bisbenzilisoquinoline alkaloid (BBA) obtained from *Chondrodendron platyphyllum* (Menispermaceae). Earlier studies reported that this alkaloid, as well as the structurally related compounds isocurine and 12-O-metilcurine, have promising pharmacological effects [6] corroborating ethnopharmacological data which points *C. phatyphyllum* as a plant with medicinal properties [7]. Studies have shown that BBA are bioactive natural compounds presenting anti-inflammatory, anti-allergic, and analgesic activities [8] and there is evidence that their mechanism of action involves a direct inhibition of calcium channels [6,9,10].

Studies carried out by our group have demonstrated the effects of curine treatment in an experimental model of allergic asthma. The oral administration of this compound to allergic mice significantly inhibited eosinophilic inflammation and airway hyper-responsiveness (AHR), which are critical hallmarks of the allergic response in this model. In addition, curine prevented lipid body formation and cytokine production in vivo, suggesting that it has an inhibitory role in eosinophil activation. A similarity between the anti-allergic effects of curine and verapamil (a calcium channel blocker) as well as inhibition of calcium-induced tracheal contraction by curine strongly suggested that its anti-allergic effects are associated with modulation of calcium-dependent responses [11]. These findings were affirmed by another study in which we demonstrated the anti-allergic effects of curine and verapamil in a mice model of mast cell activation. In addition to inhibiting the scratching behavior, the oral treatment with curine prevented the anaphylactic shock reaction in systemically-challenged mice. Additionally, these treatments inhibited the production of lipid mediators and cytokines associated with mast cell activation [12]. Importantly, an analysis of physical, behavioral, histological, hematologic and biochemical parameters revealed that the oral treatment with curine for seven consecutive days did not induce evident toxicity in mice [11]. Additionally, this alkaloid presented analgesic effects that were not associated with an activity in the central nervous system but involve anti-inflammatory mechanisms [13].

Accumulating evidence places curine as a potent anti-inflammatory and anti-allergic compound with low-toxicity. Through both in vivo and in vitro studies, we have described the general pharmacological properties of this alkaloid. However, the mechanisms underlying its anti-inflammatory effects are still poorly understood. Therefore, the objective of this study was to investigate the effects of curine on macrophage activation and neutrophil recruitment in a mouse model of LPS-induced inflammation. Here we analyze the impact of inflammatory mediator production modulation as well as the importance of calcium influx inhibition in the anti-inflammatory mechanisms of curine.

2. Results

2.1. Curine Inhibits Neutrophil Recruitment in LPS-Challenged Mice

An intrapleural administration of LPS was observed to induce a significant increase in the number of neutrophils in the pleural lavages of C57Bl/6 mice (Figure 1B). The oral treatment with curine (2.5 mg/kg) or dexamethasone (2 mg/kg) 1 h prior to the LPS challenge caused a significant reduction in neutrophil counts (Figure 1B) in comparison with the group of untreated and challenged mice, thus demonstrating the inhibitory role played by curine with regard to neutrophil recruitment during the pleural inflammation.

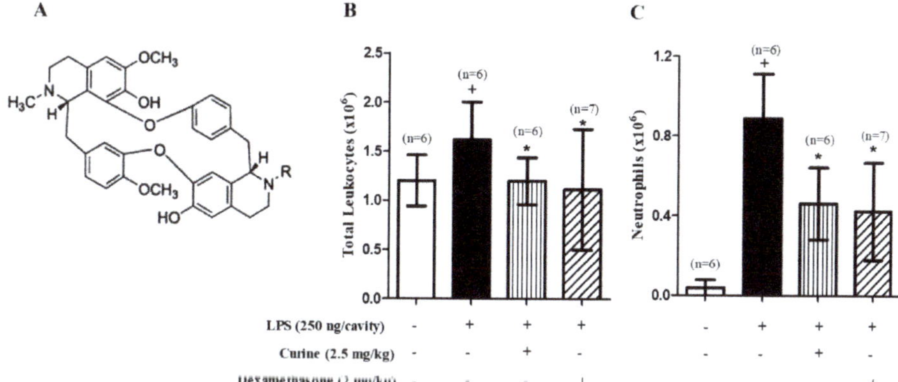

Figure 1. Effect of curine on neutrophil recruitment in lipopolysaccharide (LPS)-induced pleurisy. (**A**) The chemical structure of curine. Total leukocytes (**B**) and neutrophils (**C**) per pleural lavage of C57Bl/6 mice orally pre-treated with curine (2.5 mg/kg) or dexamethasone (2 mg/kg), counted under light microscopy 4h after LPS-challenge. Results are expressed as means ± SD from at least six animals. + significant difference ($p < 0.05$) from the unchallenged group; * significant difference ($p < 0.05$) from the untreated LPS-challenged group. Statistical significance was determined with one-way ANOVA and post hoc Tukey test.

2.2. Curine Inhibits Inflammatory Mediator Production in Vivo

Based on our finding of increased neutrophil recruitment in response to the LPS challenge, we analyzed the effect of curine treatment on the production of mediators involved in neutrophil recruitment and inflammation. Supernatants obtained from the pleural lavages of LPS-challenged mice presented increased levels of inflammatory mediators (Figure 2A–F), in comparison to unstimulated animals. Curine treatment was observed to significantly inhibit the production of interleukin (IL)-6, tumor necrosis factor (TNF)-α, monocyte chemotactic protein (MCP)-1/CCL2, keratinocyte-derived chemokine (KC/CXCL1) and leukotriene B_4 (LTB_4), thereby providing evidence of a link between the inhibitory effect of curine on neutrophil recruitment and associated inflammatory mediator production.

2.3. Curine Inhibits Macrophage Activation in Vitro

As activated macrophages are crucially involved in the production of mediators in early inflammatory events, we analyzed the direct effects of curine on macrophage activation by investigating its interference on cytokine production in vitro. Figure 3 shows that stimulation of peritoneal macrophage cultures with LPS increased the levels of IL-1β, IL-6 and TNF-α in comparison with control cells. Pre-treatment with curine at 1 or 10 μM significantly reduced the levels of these cytokines in the supernatants indicating that this BBA inhibits TLR-4 mediated macrophage activation in vitro (Figure 3).

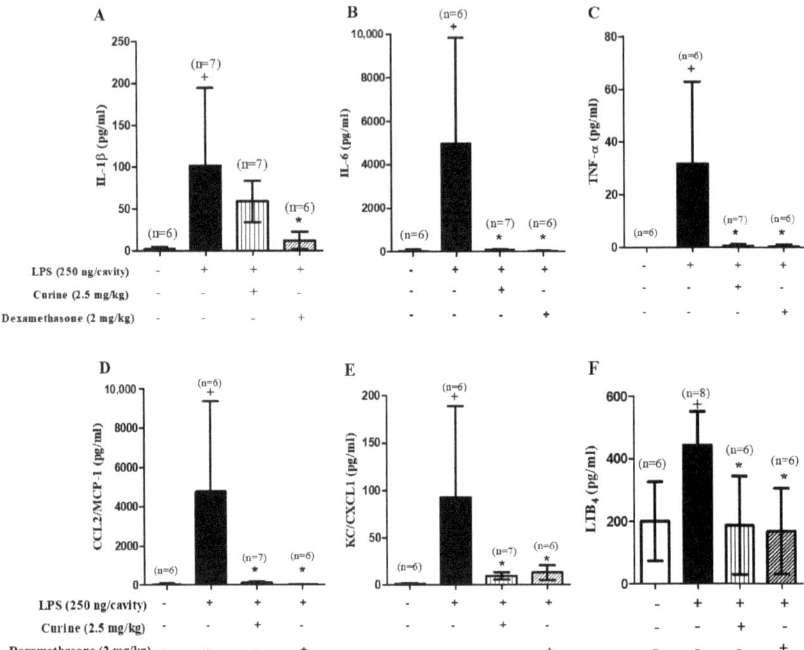

Figure 2. Effects of curine pre-treatment on in vivo cytokine production 4h after LPS challenge. Concentrations of interleukin (IL)-1β (**A**), IL-6 (**B**), tumor necrosis factor (TNF)-α (**C**), CCL2/monocyte chemotactic protein (MCP)-1 (**D**), keratinocyte-derived chemokine (KC/CXCL-1) (**E**) and leukotriene B_4 (LTB_4)(**F**) in the pleural lavages of C57Bl/6 mice orally pre-treated with curine (2.5 mg/kg) or dexamethasone (2 mg/kg). These results are expressed as the mean ± SD of at least 6 animals. + significant difference ($p < 0.05$) from the unchallenged group; * significant difference ($p < 0.05$) from the untreated LPS-challenged group. Statistical significance was determined with one-way ANOVA and post hoc Tukey test. Assay range of IL-1β: 10.36–60,631 pg/mL; IL-6: 0.74–12,053 pg/mL; TNF-α: 5.86–59,626 pg/mL; CCL2/MCP-1: 22.4–41,873 pg/mL; KC/CXCL1: 3.2–182 pg/mL and LTB_4: 3.9–500 pg/mL.

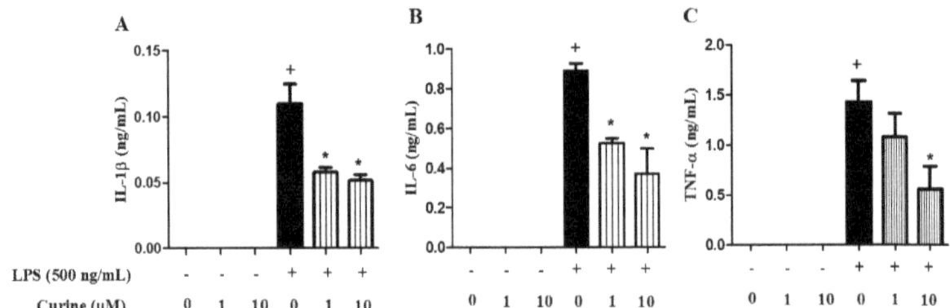

Figure 3. Effects of curine treatment on macrophage activation 4 h after LPS challenge. Concentrations of IL-1β (**A**), IL-6 (**B**) and TNF-α (**C**) in the supernatants of peritoneal macrophage cultures treated with curine (1 or 10 μM) 4h after the stimulus with LPS (500 ng/mL). Results are expressed as means ± Standard Error of Mean (SEM) of two experiments performed in triplicate. + significant difference ($p < 0.05$) from the unchallenged cells; * significant difference ($p < 0.05$) from the untreated LPS-challenged cells. Statistical significance was determined with one-way ANOVA and post hoc Tukey test. Assay range of IL-1β and IL-6: 15.6–1000 pg/mL and TNF-α: 31.3–2000 pg/mL.

2.4. Effects of Calcium Influx Inhibition on Macrophage Activation

Our group recently demonstrated that curine and verapamil presented anti-allergic effects that might be associated with calcium signaling modulation [11,12]. To evaluate the importance of calcium influx inhibition on macrophage activation, as well as its potential participation in curine anti-inflammatory mechanisms, we made a comparison between the effects of curine and verapamil on macrophage activation. As shown in Figure 4, treatment with curine or verapamil at the same concentration induced a similar inhibition in IL-1β (A) and TNF-α (B) production, which suggests that the effects of curine on macrophage activation might be dependent on calcium influx inhibition.

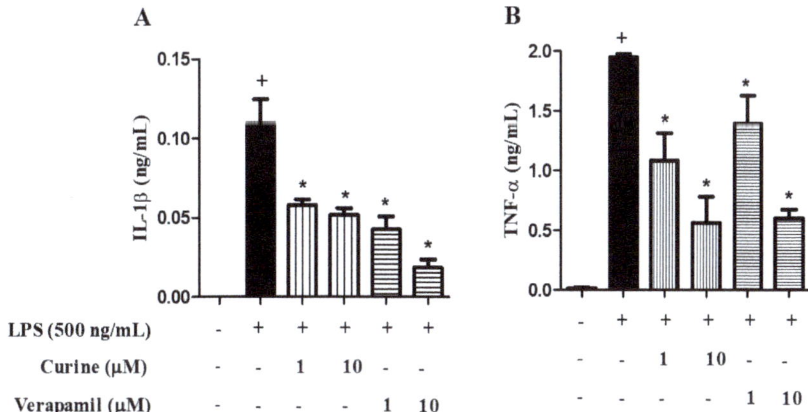

Figure 4. Effects of calcium influx inhibition on macrophage activation. Concentrations of IL-1β (**A**), TNF-α (**B**) in the supernatants of peritoneal macrophage cultures treated with curine or verapamil (1 or 10 µM) were evaluated 4 h after LPS challenge. Results are expressed as means ± SEM from two experiments performed in triplicate. + significant difference ($p < 0.05$) from the unchallenged cells; * significant difference ($p < 0.05$) from the untreated LPS-challenged cells. Statistical significance was determined with one-way ANOVA and post hoc Tukey test. Assay range of IL-1β: 15.6–1000 pg/mL and TNF-α: 31.3–2000 pg/mL.

2.5. Curine Inhibits Nitric Oxide (NO) Production by Regulating iNOS Expression in Macrophages

It has been demonstrated that LPS stimulates the synthesis of NO via IL-1β, TNF-α, and IFN-γ [14]. Figure 5A illustrates the effects of curine on NO production. The supernatants of LPS-stimulated peritoneal macrophages presented significantly increased concentrations of nitrite, which was significantly decreased by curine treatment. Moreover, while LPS stimulation was found to induce increased expression of iNOS by macrophages, treatment with curine or verapamil reduced the expression of this enzyme (Figure 5B). These findings suggest that the inhibition of NO production through the curine-mediated regulation of iNOS expression could be associated with calcium influx inhibition.

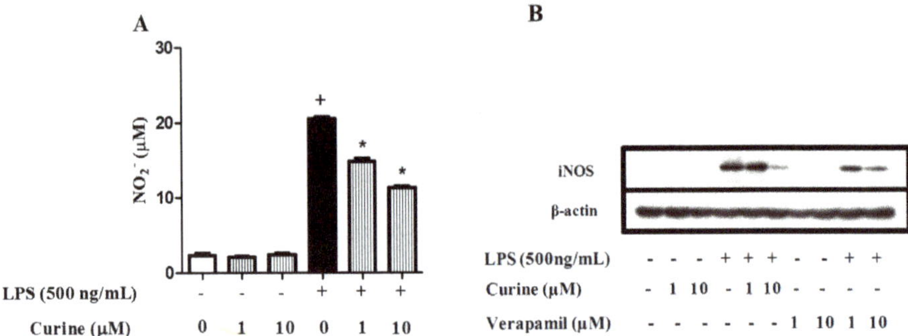

Figure 5. Effects of curine on nitric oxide (NO) production and iNOS expression. Concentrations of nitrite (**A**) in the supernatants of peritoneal macrophages treated with curine or verapamil (1 or 10 μM) 24 h after LPS challenge. iNOS expression (**B**) was analyzed by Western blotting 18 h after LPS stimulus. Results are expressed as means ± SEM from two experiments performed in triplicate. + significant difference ($p < 0.05$) from the unchallenged cells; * significant difference ($p < 0.05$) from the untreated LPS-challenged cells. Statistical significance was determined with one-way ANOVA and post hoc Tukey test.

3. Discussion

The bisbenzylisoquinoline alkaloids (BBA) constitute a group of secondary metabolites that exert numerous biological effects. The medicinal properties of BBA-rich plants and isolated compounds have been demonstrated in different experimental models, indicating that this class of substances presents promising anti-allergic and anti-inflammatory activities [15,16]. Our group found that curine, a BBA identified as the main constituent of *Chondrodendron platyphyllum* (Menispermaceae), is an orally active alkaloid with potent immunomodulatory effects and low toxicity, which therefore makes it a promising candidate in the development of new anti-inflammatory drugs [16].

In a worldwide context, questions have been raised concerning the efficacy and safety of currently available medications [17]. Although corticosteroids, non-steroidal anti-inflammatory drugs (NSAIDs) and other conventional drugs effectively relieve most inflammatory symptoms, in specific conditions these are not effective, or can cause significant side effects [18]. Accordingly, the development of novel, safe and effective drugs is imperative to improving anti-inflammatory therapy.

Although inflammatory diseases differ in various aspects, some evidence has consistently shown that macrophages and neutrophils perform essential functions in the initiation and development of many inflammatory conditions [19]. The present mouse model of LPS-induced inflammation, used to characterize the effects of curine on macrophage activation and neutrophil recruitment, demonstrated new anti-inflammatory properties of this alkaloid compound. Our findings indicate that orally administered curine inhibited the recruitment of neutrophils to the pleural cavity of LPS-challenged mice. Accordingly, curine treatment reduced levels of IL-6, TNF-α, CCL2/MCP-1, and LTB_4 in the pleural lavages of these animals, providing evidence of a link between the inhibitory effect of this alkaloid on neutrophil in association with the production of inflammatory mediators.

Neutrophils can rapidly migrate to sites of inflammation [20] in response to inflammatory signals such as chemokines and cytokines produced by resident cells [1]. The chemokine CXCL1 (also known as KC in mice) plays a critical role in neutrophil recruitment and activation by signaling via CXCR2 on these cells [21]. Previous studies have also demonstrated that LTB_4 acts as an essential chemotactic agent [22,23] by stimulating the recruitment of neutrophils via BLT_1 receptor activation [24]. It follows that the inhibitory effect of curine on KC/CCL1 and LTB_4 production might therefore directly impact neutrophil recruitment. Additionally, LPS-induced cytokines, including TNF-α and IL-β, can directly affect neutrophil recruitment by stimulating the expression of adhesion molecules, including selectins and integrins [25–27].

As activated macrophages are one of the most critical sources of mediator production in the early phase of inflammation [28], we hypothesized that the inhibition of neutrophil recruitment and cytokine production in the pleural lavage induced by curine might be associated with decreased macrophage activation. Our data show that the production of IL-6, IL-β, and TNF-α was inhibited in murine macrophages stimulated with curine in vitro, which indicates that this compound may regulate neutrophil recruitment by inhibiting the production of key inflammatory mediators in macrophages. On the other hand, the release of products involved in monocyte/macrophage influx and activation, such as MCP-1, by neutrophils can also affect macrophage function [29]. Here, MCP-1 production was found to be significantly inhibited by curine, which suggests that cross-talk between neutrophils and macrophages could be impaired by curine treatment.

Our previous work has demonstrated that curine can exert anti-allergic effects associated with the inhibition of calcium influx [11,12,16]. Using tracheal rings preparations kept in Ca^{2+}-free medium, depolarized with KCl and stimulated with cumulative addition of Ca^{2+}, Ribeiro-Filho and colleagues [11] demonstrated ex-vivo that curine pre-treatment significantly inhibited calcium-induced trachea contractile response, suggesting that curine inhibits the influx of calcium via blockade of voltage-dependent Ca^{2+} channels in the rat tracheal smooth muscle. In addition, Medeiros and collaborators [10] demonstrated that curine decreased intracellular Ca^{2+} transients in A7r5 cells, which indicated that this alkaloid can have a direct inhibitory effect on L-type Ca^{2+} channels in vascular smooth muscle cells. Here we hypothesized that calcium influx inhibition would impair macrophage activation, which could be partially responsible for the anti-inflammatory effects associated with curine treatment. To confirm this, we compared the impact of verapamil and curine on cytokine production by LPS-stimulated macrophages, as a parameter to evaluate macrophage activation. When administered under identical conditions (concentrations and time of pre-treatment), verapamil and curine demonstrated similar inhibitory effects, suggesting that the modulation of calcium influx is indeed a potential mechanism by which curine inhibits the inflammatory response.

Curine was also found to significantly decrease nitrite concentrations in the supernatants of macrophages stimulated with LPS, indicating that nitric oxide (NO) production was inhibited in vitro. This finding adds to the role played by curine in macrophage activation, since NO production is a hallmark of activated macrophages [14]. Importantly, recent reports have demonstrated that NO, in association with other ROS, is critically involved in neutrophil extracellular trap (NET) formation [30]. Curiously, it was also recently reported that neutrophils might participate in the resolution of inflammation mediated by reparative macrophages [5]. It has been well established that LPS and inflammatory cytokines stimulate NO production by inducing iNOS expression [31]. Here, we demonstrated that curine and verapamil inhibited the expression of iNOS in LPS-stimulated murine macrophages, which thereby provides evidence of the inhibitory effect of these drugs on NO production. This finding further corroborates the impact of curine on TNF-α and IL-1β production and lends support to the notion that curine negatively modulates macrophage activation through the inhibition of a calcium-dependent response. In fact, studies have shown that calcium-dependent signaling potentiates macrophage activation [32] and stimulates proinflammatory cytokine production [33] in these cells. Therefore, further studies addressing the role of curine in modulating calcium influx and associated signaling pathways in immune cells, including macrophages, will contribute to characterize the molecular mechanism of action of curine as an anti-inflammatory compound.

We previously reported that a single dose of curine administered orally as pre-treatment exhibited anti-inflammatory and analgesic effects in mice [13]. Our investigation of the analgesic effects of curine revealed that instead of acting by way of neurogenic mechanisms, curine acts through anti-inflammatory mechanisms associated with the inhibition of PGE_2 production. These findings are in line with our previous work which demonstrated that curine inhibited the synthesis of cysteinyl leukotrienes and PGD_2 by mast cells [12]. Taken together, these findings indicate that curine affects signaling pathways involved in lipid mediator synthesis, possibly due to interference in leukocyte activation. Finally, the present findings suggest that the anti-inflammatory and analgesic effects of curine are also related to the inhibition of critical mediators of inflammatory pain, including IL-1β, TNF-α, and NO [34–36].

In conclusion, curine was shown to exert anti-inflammatory effects associated with the inhibition of macrophage activation and neutrophil recruitment, as well as reduced production of inflammatory cytokines, LTB4, NO and the modulation of Ca^{2+} influx (Figure 6).

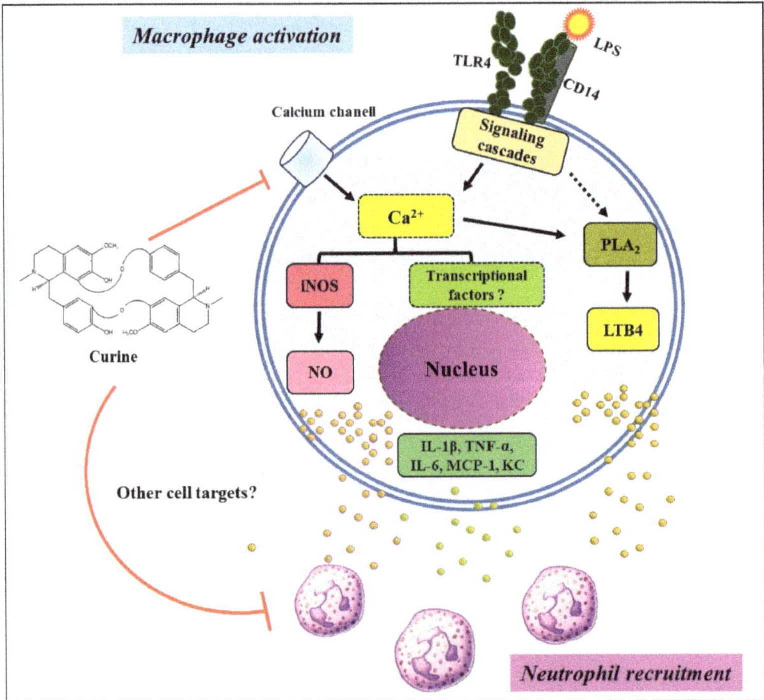

Figure 6. Schematic diagram illustrating a potential mechanism of action by curine on macrophage activation and neutrophil recruitment. The inhibition of calcium influx could be affecting signaling pathways associated with the synthesis of NO, LTB_4 and cytokines, which could impair neutrophil recruitment. Importantly, the possibility that curine may interfere with other cell targets cannot be ruled out.

4. Materials and Methods

4.1. Preparation of Curine Solution

Curine was purified from the total tertiary alkaloid fraction (TTA) obtained from the root bark of *Chondrodendron platyphyllum* Hil St. (Miers) as previously described [13]. The TTA was submitted to column chromatography followed by thin-layer chromatography (TLC) purification, from which curine was obtained in the form of a crystal. The chemical structure was analyzed by spectroscopy and comparison with the literature data [37]. The purity of curine was analyzed by NMR ^{13}C and NMR ^{1}H ($CDCl_3$, 400 MHz) data from the crystals when compared to the literature data [13] and the substance was considered spectroscopically pure. After purification, 1 mg of the crystal was dissolved in 50 μL of 1 N HCl and 500 μL of distilled water. The pH was adjusted (between 7–8) with 1 N NaOH and volume was adjusted to 1000 μL, with dilutions performed in phosphate-buffered saline (PBS). The use of *C. platyphyllum* in the present study was registered in the National System of Genetic Heritage Management and Associated Traditional Knowledge (SisGen, protocol A84A87E).

4.2. Animals

Male C57Bl/6 mice (age 6–8 weeks) weighing 20–30 g obtained from the Oswaldo Cruz (Fiocruz, Rio de Janeiro, Brazil) were maintained with food and water ad libitum in cages at room temperature ranging from 22 to 24 °C under a 12 h light/dark cycle. This study was carried out in accordance with the recommendations established by the Brazilian National Council for the Control of Animal Experimentation (CONCEA). All experimental protocols were approved by the Animal Welfare Committee of the Oswaldo Cruz Foundation (CEUA/FIOCRUZ-RJ, protocol #L-002/08).

4.3. Treatments

For in vivo experimentation, animals (6–8 per group) were randomly assigned by body weight for single oral pre-treatment with curine (2.5 mg/kg), dexamethasone (2 mg/kg) or PBS (negative control). All treatments were performed by simple awake gavage [38,39] 1 h prior to LPS challenge. Briefly, treatments (0.1 mL/10 g body weight) were administered using a 31 mm long, 1 mm diameter reusable curved stainless-steel feed needle containing a 1.7 mm ball at the tip (Bronther, Ribeirão Preto, SP, Brazil). The gavage needle was gently inserted into the oral cavity, ensuring the correct passage through the esophagus. For in vitro experiments, cells were treated with curine or verapamil 1 or 10 μM) or PBS 1 h before stimulation. Curine dosage (2.5 mg/kg) was based on results obtained by Ribeiro-Filho and colleagues [11].

4.4. LPS-Induced Pleurisy

Male C57Bl/6 mice (n = 6–8) were orally pre-treated with curine (2.5 mg/kg) or dexamethasone (2 mg/Kg) 1 h prior to the pleurisy protocol [40]. Animals were anesthetized with isoflurane (Forane™, Abbott, São Paulo, SP, Brazil) and challenged through an intrathoracic (i.t) injection of LPS (250 ng/cavity) dissolved in 100 μL of PBS. A group of mice receiving the same volume of PBS was used as the control. Four hours following the LPS injection, the animals were euthanized by CO_2 and the pleura was surgically exposed. Pleural lavage was collected by washing the pleural cavity with 1 mL of heparinized PBS (20 U/mL).

4.5. Leukocyte Counting

Leukocytes were counted under light microscopy after diluting the pleural lavage samples in Turk fluid (2% acetic acid). Differential counts were performed under an objective lens at 100× magnification after staining by the May–Grunwald–Giemsa method.

4.6. Peritoneal Macrophage Cultures

Peritoneal macrophages from C57Bl/6 mice were obtained four days after the injection of 4% thioglycollate. The peritoneal cavity was washed with RPMI 1640 medium supplemented with 100 U/mL penicillin and 100 μg/mL streptomycin (Thermo Fisher Scientific, Waltham, MA, USA). Cells were adjusted to 2×10^6/mL and plated on 24-well culture plates (500 μL) at 37 °C under 4% CO_2 overnight. Following incubation, cells were pre-treated with curine, or alternatively with verapamil, (both at 1 or 10 μM) and stimulated with LPS (500 ng/mL) 1 h later. Of note, curine pre-treatment (at either 1 or 10 μM) did not affect cell viability, which was higher than 90% in all experiments.

4.7. Cytokine and LTB_4 Analysis

Samples of the pleural lavage were centrifuged at 500 g for 8 min at 4 °C to obtain the supernatants. The concentrations of IL-1β, IL-6, TNF-α, CCL2/MCP-1 and KC/CXCL-1 in these supernatants were determined using a multiplex fluorescent microbead immunoassay (Bio-Rad Laboratories, Hercules, CA, USA). Cytokine levels were quantified using a Luminex technology (Bio-Plex Workstation; Bio-Rad Laboratories, Hercules, CA, USA). Data analysis was performed using Bio-Plex software (Bio-Rad Laboratories, Hercules, CA, USA). The concentrations of LTB_4 in the pleural lavages, as well as cytokines in the supernatants of macrophage cultures, were analyzed using ELISA kits in accordance with the manufacturer's instructions (Cayman Chemical, Ann Arbor, MI, USA: LTB_4 and R&D Systems, Minneapolis, MN, USA: Cytokines). All analyses were performed 4 h after stimulation with LPS.

4.8. Nitrite Quantification

For NO_2^- determination, 100 µL of macrophage culture supernatant was removed 24 h after LPS stimulus and incubated with Griess reagent (1% sulfanilamide and 0.1% naphthylenediamine hydrochloride in 2.5% H_3PO_4) for 10 min at room temperature. The readings were performed at 540 nm using a spectrophotometer (Titertek Multiscan, Flow Laboratories, Eflab Oy, Helsinki, Finland). Nitrite concentrations were calculated using a standard reference curve obtained from $NaNO_2$ (1–200 µM in culture medium).

4.9. SDS-PAGE and Western Blotting

Eighteen hours after LPS stimulus, Western blot was used to analyze iNOS expression. Briefly, cells were washed in PBS buffer and homogenized with 10 mM Tris-HCl buffer (pH 7.4), 150 mM NaCl, 0.5% triton X-100, 10% glycerol (*v*/*v*), 0.1 mM EDTA, 1 mM Dithiothreitol (DTT) and a cocktail of protease inhibitors (Roche Diagnostics GmbH, Mannheim, Germany). Proteins from the cell homogenate were separated by polyacrylamide gels in the presence of 10% SDS at a constant current of 16 mA. Full-range rainbow molecular weight markers (RPN800E, GE Healthcare Life Sciences, Piscataway, NJ, USA) were used as a relative molecular mass standard. After running the gels, samples were transferred at 200 mA (2.7 mA/cm^2) to a nitrocellulose membrane using 25 mM Tris–HCl and 192 mM glycine, pH 8.3, at 4 °C for 120 min. The membranes were then blocked with Tris-buffered saline (TBS)-0.5 tween 20 and 5% milk for 1 h at room temperature, incubated with a polyclonal antibody (1:1000) against iNOS (BD-610333) for 18 h at 4 °C, followed by incubation with a secondary antibody (anti-rabbit IgG-HRP, PI.1000, Vector Laboratories) for 1 h at room temperature. Reactions were developed using a Super Signal West Pico Chemiluminescent Substrate (Thermo Fisher Scientific, Rockford, IL, USA).

4.10. Statistical Analyses

Data were analyzed by one-way ANOVA followed by Tukey's post-test using GraphPad Prism software version 5.02 (GraphPad, San Diego, CA, USA, 2016). Values of in vivo experiments are expressed as means ± SD and values of in vitro assays as means ± Standard Error of Mean (SEM). Statistical significance was considered when $p < 0.05$.

Author Contributions: Conceptualization: J.R.-F., P.T.B.; experiments were conducted by: J.R.-F., F.C.L., A.S.C., A.d.B.C., J.A.A. and E.F.d.A.; data analysis: J.R.-F., P.T.B. and M.R.P.; contribution with reagents/materials/analytical tools: C.d.S.D.; writing of manuscript: J.R.-F.; critical review: P.T.B. and M.R.P.

Funding: This work was supported by PRONEX/MCT, CNPq, FAPERJ and INCT-Cancer.

Acknowledgments: The authors would like to thank Cristiane Zanon de Sousa for technical assistance and are grateful to Andris K. Walter for English language revision and copyediting assistance.

Conflicts of Interest: The authors declare no conflicts of interest.

References

1. Borregaard, N. Neutrophils, from marrow to microbes. *Immunity* **2010**, *33*, 657–670. [CrossRef] [PubMed]
2. Dean, J.L.E.; Brook, M.; Clark, A.R.; Saklatvala, J. p38 Mitogen-activated Protein Kinase Regulates Cyclooxygenase-2 mRNA Stability and Transcription in Lipopolysaccharide-treated Human Monocytes. *J. Biol. Chem.* **1999**, *274*, 264–269. [CrossRef] [PubMed]
3. Kim, E.; Ha, I.; Kim, K.; Park, D.; Lee, M.Y. Production of TLRs triggered pro-inflammatory cytokines through calcium dependent and independent pathways in HaCaT cells. *FASEB J.* **2017**, *31*, 184.6.
4. Kumar, K.P.; Nicholls, A.J.; Wong, C.H.Y. Partners in crime: neutrophils and monocytes/macrophages in inflammation and disease. *Cell Tissue Res.* **2018**, *371*, 551–565. [CrossRef] [PubMed]

5. Yang, W.; Tao, Y.; Wu, Y.; Zhao, X.; Ye, W.; Zhao, D.; Fu, L.; Tian, C.; Yang, J.; He, F.; et al. Neutrophils promote the development of reparative macrophages mediated by ROS to orchestrate liver repair. *Nat. Commun.* **2019**, *10*, 1076. [CrossRef] [PubMed]
6. Dias, C.S.; Barbosa-Filho, J.M.; Lemos, V.S.; Cortes, S.F. Mechanisms involved in the vasodilator effect of curine in rat resistance arteries. *Planta Med.* **2002**, *68*, 1049–1051. [CrossRef]
7. Correa, P.M. *Dicionário Das Plantas Úteis do Brasil e das Exóticas Cultivadas*; Instituto Brasileiro de Desenvolvimento Florestal, Ministério da Agricultura: Brasília, Brasil, 1984.
8. Souto, A.L.; Tavares, J.F.; Silva, M.S.; Diniz, M.F.; Athayde-Filho, P.F.; Barbosa-Filho, J.M. Anti-inflammatory activity of alkaloids: An update from 2000 to 2010. *Molecules* **2011**, *16*, 8515–8534. [CrossRef]
9. Guedes, D.N.; Barbosa-Filho, J.M.; Lemos, V.S.; Côrtes, S.F. Mechanism of the vasodilator effect of 12-O-methylcurine in rat aortic rings. *J. Pharm. Pharmacol.* **2002**, *54*, 853–858. [CrossRef]
10. Medeiros, M.A.; Pinho, J.F.; de-Lira, D.P.; Barbosa-Filho, J.M.; Araújo, D.A.; Cortes, S.F.; Lemos, V.S.; Cruz, J.S. Curine, abisbenzylisoquinoline alkaloid, blocks L-type Ca^{2+} channels and decreases intracellular Ca^{2+} transients in A7r5 cells. *Eur. J. Pharmacol.* **2011**, *669*, 100–107. [CrossRef]
11. Ribeiro-Filho, J.; Calheiros, A.S.; Vieira-de-Abreu, A.; Carvalho, K.I.M.; Mendes, D.S.; Bandeira-Melo, C.; Martins, M.A.; Dias, C.S.; Piuvezam, M.R.; Bozza, P.T. Curine inibits eosinophil activation and airway hyper-responsiveness in a mouse model of allergic asthma. *Toxicol. Appl. Pharmacol.* **2013**, *273*, 19–26. [CrossRef]
12. Ribeiro-Filho, J.; Leite, F.C.; Costa, H.F.; Calheiros, A.S.; Torres, R.C.; de Azevedo, C.T.; Martins, M.A.; Dias, C.S.; Bozza, P.T.; Piuvezam, M.R. Curine inhibits mast cell-dependent responses in mice. *J. Ethnopharmacol.* **2014**, *155*, 1118–1124. [CrossRef] [PubMed]
13. Leite, F.C.; Ribeiro-Filho, J.; Costa, H.F.; Salgado, P.R.; Calheiros, A.S.; Carneiro, A.B.; de Almeida, R.N.; Dias Cda, S.; Bozza, P.T.; Piuvezam, M.R. Curine, an Alkaloid Isolated from Chondrodendron platyphyllum Inhibits Prostaglandin E_2 in Experimental Models of Inflammation and Pain. *Planta Med.* **2014**, *80*, 1072–1078. [CrossRef]
14. Chang, C.I.; Liao, J.C.; Kuo, L. Arginase modulates nitric oxide production in activated macrophages. *Am. J. Physiol.* **1998**, *274*, 342–348. [CrossRef]
15. Barbosa-Filho, J.M.; Piuvezam, M.R.; Moura, M.D.; Silva, M.S.; Batista-Lima, K.V.; Leitão-da-Cunha, E.V.; Fechine, I.M.; Takemura, O.S. Anti-inflammatory activity of alkaloids: A twenty-century review. *Braz. J. Pharmacogn.* **2006**, *16*, 109–139. [CrossRef]
16. Ribeiro-Filho, J.; Piuvezam, M.R.; Bozza, P.T. Anti-allergic properties of curine, a bisbenzylisoquinoline alkaloid. *Molecules* **2015**, *20*, 4695–4707. [CrossRef]
17. Cosendey, M.A.E.; Bermudez, J.A.Z.; Reis, A.L.A.; Silva, H.F.; Oliveira, M.A.; Luiza, V.L. Assistência farmacêutica na atenção básica de saúde: A experiência de três estados brasileiros. *Cad. Saúde Pública* **2000**, *16*, 171–182. [CrossRef]
18. Rainsford, K.D. Anti-inflammatory drugs in the 21st century. *Sub-Cel. Biochem.* **2007**, *42*, 3–27.
19. Takeda, K.; Akira, S. Toll-like receptors in innate immunity. *Int. Immunol.* **2005**, *17*, 1–14. [CrossRef]
20. Amulic, B.; Cazalet, C.; Hayes, G.L.; Metzler, K.D.; Zychlinsky, A. Neutrophil Function: From Mechanisms to Disease. *Ann. Rev. Immunol.* **2012**, *30*, 459–489. [CrossRef]
21. Sawant, K.V.; Poluri, K.M.; Dutta, A.K.; Sepuru, K.M.; Troshkina, A.; Garofalo, R.P.; Rajarathnam, K. Chemokine CXCL1 mediated neutrophil recruitment: Role of glycosaminoglycan interactions. *Sci. Rep.* **2010**, *6*, 33123. [CrossRef]
22. Palmblad, J.; Malmsten, C.L.; Udén, A.M.; Rådmark, O.; Engstedt, L.; Samuelsson, B. Leukotriene B_4 is a potent and stereospecific stimulator of neutrophil chemotaxis and adherence. *Blood* **1981**, *58*, 658–661. [CrossRef] [PubMed]
23. Afonso, P.V.; Janka-Junttila, M.; Lee, Y.J.; McCann, C.P.; Oliver, C.M.; Aamer, K.A.; Losert, W.; Cicerone, M.T.; Parent, C.A. LTB_4 is a signal-relay molecule during neutrophil chemotaxis. *Dev. Cell* **2012**, *22*, 1079–1091. [CrossRef] [PubMed]
24. Monteiro, A.P.; Pinheiro, C.S.; Luna-Gomes, T.; Alves, L.R.; Maya-Monteiro, C.M.; Porto, B.N.; Barja-Fidalgo, C.F.; Benjamim, C.F.; Peters-Golden, M.; Bandeira-Melo, C.; et al. Leukotriene B_4 mediates neutrophil migration induced by heme. *J. Immunol.* **2011**, *186*, 6562–6567. [CrossRef] [PubMed]
25. Doukas, J.; Pober, J.S. IFN-gamma enhances endothelial activation induced by tumor necrosis factor but not IL-1. *J. Immunol.* **1990**, *145*, 1727–1733.

26. Marfaing-Koka, A.; Devergne, O.; Gorgone, G.; Portier, A.; Schall, T.J.; Galanaud, P.; Emilie, D. Regulation of the production of the RANTES chemokine by endothelial cells. Synergistic induction by IFN-gamma plus TNF-alpha and inhibition by IL-4 and IL-13. *J. Immunol.* **1995**, *154*, 1870–1878. [PubMed]
27. Libby, P. Inflammatory Mechanisms: The Molecular Basis of Inflammation and Disease. *Nutr. Rev.* **2007**, *65*, 140–146. [CrossRef]
28. Fujiwara, N.; Kobayashi, K. Macrophages in inflammation and Allergy. *Curr. Drug Targets* **2005**, *4*, 281–286. [CrossRef]
29. Soehnlein, O.; Weber, C.; Lindbom, L. Neutrophil granule proteins tune monocytic cell function. *Trends Immunol.* **2009**, *30*, 538–546. [CrossRef]
30. Patel, S.; Kumar, S.; Jyoti, A.; Srinag, B.S.S.; Keshari, R.S.; Saluja, R.; Verma, A.; Mitra, K.; Barthwal, M.K.; Krishnamurthy, H.; et al. Nitric oxide donors release extracellular traps from human neutrophils by augmenting free radical generation. *Nitric Oxide* **2010**, *22*, 226–234. [CrossRef]
31. Moncada, S.; Palmer, R.M.J.; Higgs, E.A. Nitric oxide: Physiology, pathophysiology, and pharmacology. *Pharmacol. Rev.* **1991**, *43*, 109–142.
32. Zumerle, S.; Calı, B.; Munari, F.; Angioni, R.; Virgilio, F.D.; Molon, B.; Viola, A. Intercellular Calcium Signaling Induced by ATP Potentiates Macrophage Phagocytosis. *Cell Rep.* **2019**, *27*, 1–10. [CrossRef] [PubMed]
33. Desai, B.M.; Leitinger, N. Purinergic and calcium signaling in macrophage function and plasticity. *Front. Immunol.* **2010**, *5*, 1–8. [CrossRef] [PubMed]
34. Cunha, F.Q.; Poole, S.; Lorenzetti, B.B.; Ferreira, S.H. The pivotal role of tumour necrosis factor alpha in the development of inflammatory hyperalgesia. *Br. J. Pharmacol.* **1992**, *107*, 660–664. [CrossRef] [PubMed]
35. Ferreira, S.H.; Lorenzetti, B.B.; Bristow, A.F.; Poole, S. Interleukin-1β as a potent hyperalgesic agent antagonized by a tripeptide analogue. *Nature* **1988**, *334*, 698–700. [CrossRef]
36. Cury, Y.; Picolo, G.; Gutierrez, V.P.; Ferreira, S.H. Pain and analgesia: The dual effect of nitric oxide in the nociceptive system. *Nitric Oxide* **2011**, *25*, 243–254. [CrossRef]
37. Mambu, L.; Martin, M.T.; Razafimahefa, D.; Ramanitrahasimbola, D.; Rasoanaivo, P.; Frappier, F. Spectral characterisation and antiplasmodial activity of bisbenzylisoquinolines from *Isolona ghesquiereina*. *Planta Med.* **2000**, *66*, 537–540. [CrossRef]
38. Brasil. Conselho Nacional de Controle de Experimentação animal (CONCEA). Procedimentos–Roedores E Lagomorfos Mantidos Em Instalações de Instituições de Ensino Ou Pesquisa Científica. Resolução Normativa N° 33, de 18 de Novembro de 2016. Available online: http://www.in.gov.br/materia/-/asset_publisher/Kujrw0TZC2Mb/content/id/22073702/do1-2016-11-21-resolucao-normativa-n-33-de-18-de-novembro-de-2016-22073453 (accessed on 15 October 2019).
39. McIntyre, K. Rodent gavage technique concerns: Avoiding excess mortality. *Contemp. Top. Lab. Anim. Sci.* **2001**, *40*, 7.
40. Laranjeira, A.P.; Silva, A.R.; Gomes, R.N.; Penido, C.; Henriques, M.G.M.O.; Castro-Faria-Neto, H.C.; Bozza, P.T. Mechanisms of allergen- and LPS-induced bone marrow eosinophil mobilization and eosinophil accumulation into the pleural cavity: A role for CD11b/CD18 complex. *Inflamm. Res.* **2001**, *50*, 309–316. [CrossRef]

Article

Alkaloid Lindoldhamine Inhibits Acid-Sensing Ion Channel 1a and Reveals Anti-Inflammatory Properties

Dmitry I. Osmakov [1,2], Sergey G. Koshelev [1], Victor A. Palikov [3], Yulia A. Palikova [3], Elvira R. Shaykhutdinova [3], Igor A. Dyachenko [3], Yaroslav A. Andreev [1,2] and Sergey A. Kozlov [1,*]

1 Shemyakin-Ovchinnikov Institute of Bioorganic Chemistry, Russian Academy of Sciences, 117997 Moscow, Russia; osmadim@gmail.com (D.I.O.); sknew@yandex.ru (S.G.K.); yaroslav.andreev@yahoo.com (Y.A.A.)
2 Institute of Molecular Medicine, Sechenov First Moscow State Medical University, 119991 Moscow, Russia
3 Branch of the Shemyakin-Ovchinnikov Institute of Bioorganic Chemistry, Russian Academy of Sciences, 6 Nauki Avenue, 142290 Pushchino, Russia; viktorpalikov@mail.ru (V.A.P.); yuliyapalikova@bibch.ru (Y.A.P.); shaykhutdinova@bibch.ru (E.R.S); dyachenko@bibch.ru (I.A.D.)
* Correspondence: serg@ibch.ru

Received: 29 July 2019; Accepted: 14 September 2019; Published: 18 September 2019

Abstract: Acid-sensing ion channels (ASICs), which are present in almost all types of neurons, play an important role in physiological and pathological processes. The ASIC1a subtype is the most sensitive channel to the medium's acidification, and it plays an important role in the excitation of neurons in the central nervous system. Ligands of the ASIC1a channel are of great interest, both fundamentally and pharmaceutically. Using a two-electrode voltage-clamp electrophysiological approach, we characterized lindoldhamine (a bisbenzylisoquinoline alkaloid extracted from the leaves of *Laurus nobilis* L.) as a novel inhibitor of the ASIC1a channel. Lindoldhamine significantly inhibited the ASIC1a channel's response to physiologically-relevant stimuli of pH 6.5–6.85 with IC_{50} range 150–9 μM, but produced only partial inhibition of that response to more acidic stimuli. In mice, the intravenous administration of lindoldhamine at a dose of 1 mg/kg significantly reversed complete Freund's adjuvant-induced thermal hyperalgesia and inflammation; however, this administration did not affect the pain response to an intraperitoneal injection of acetic acid (which correlated well with the function of ASIC1a in the peripheral nervous system). Thus, we describe lindoldhamine as a novel antagonist of the ASIC1a channel that could provide new approaches to drug design and structural studies regarding the determinants of ASIC1a activation.

Keywords: acid-sensing ion channel subtype 1a; bisbenzylisoquinoline alkaloid; lindoldhamine; nociception; inflammation

Key Contribution: Bisbenzylisoquinoline alkaloid lindoldhamine (LIN) possesses strong inhibitory action on ASIC1a at physiologically-relevant pH stimuli of 6.5–6.85. LIN possesses a strong anti-inflammatory, but no analgesic effect.

1. Introduction

Acid-sensing ion channels (ASICs) belong to the family of amiloride-sensitive degenerin/epithelial Na+ channels [1]. ASICs are widely expressed in the peripheral sensory and central neurons, where they serve an important function in the transmission of signals associated with local pH changes, both during normal neuronal activity and in several pathological conditions that cause significant extracellular acidosis [1–3].

In mammals, ASICs have six isoforms: 1a, 1b, 2a, 2b, 3, and 4. In the neurons of the central nervous system, ASIC1a, ASIC2, and ASIC4 are the dominant isoforms; these all have a wide distribution in the brain and spinal cord. ASIC3 and ASIC1b are widely distributed in the neurons of the peripheral nervous system, but also have been found in the brain [4]. In total, ASIC1a is one of the widely-expressed primary acid sensors in mammalians in the brain and primary sensory neurons of the dorsal root ganglion [5]. Together with the ASIC3 channel, ASIC1a is the most sensitive to protons with a pH for half-maximal activation values of 6.4–6.7. Activation occurs within a few milliseconds and is accompanied by desensitization lasting several seconds [6,7].

ASIC1a plays an important role in diverse physiological and pathological processes. This channel is directly involved in synaptic plasticity, learning, nerve-stimulation transmission, fear and anxiety, epilepsy, ischemic processes, neuronal death, and pain sensations [8]. When the central nervous system is under ischemic conditions, ASIC1a is one of the main factors in the dysregulation of ion homeostasis, which leads to neuronal death [2,9]. Sustained activation of ASIC1a leads to an excessive influx of cations, which causes ischemic brain damage. During ischemic diseases, glycolysis is increased, resulting in lactic-acid accumulation and subsequent tissue acidosis. Either the blockade or the transgenic knockdown of ASIC1a induces neuroprotective effects in models of ischemia; thus, ASIC1a is an important target of therapies for ischemic brain damage [10].

A limited number of molecules are known to affect ASIC1a. Endogenous neuropeptides dynorphins, RF-amide peptides, and endogenous cationic polyamine spermine potentiate ASIC1a's response to acidification by inhibiting the desensitization of the channel, which leads to the neuronal damage in ischemic stroke models [11,12]. Endogenous opioid peptide nocistatin activates ASIC1a at physiological pH and decreases the sensitivity of the channel to the protons [13]. The snake toxin MitTx at nanomolar concentrations activates the ASIC1a channel and induces pain behavior [14]. Other snake toxins known as mambalgins inhibit that channel with a half-maximal effective concentration (IC_{50}) of 55 нM. The spider toxin PcTx1 also inhibits ASIC1a (IC_{50} 1 nM) by promoting the desensitization of the channel. The use of mambalgins and PcTx1 in pain models in vivo demonstrates their significant antinociceptive effect [15,16]. Low molecular weight compounds used in medical practice, such as amiloride (diuretic), diarylamidines (anti-infective drugs), ibuprofen (a non-steroidal anti-inflammatory drug), and chloroquine (the anti-malarial drug) also inhibit ASIC1a when applied at high micromolar concentrations [3,17].

Medicinal herbs have been of therapeutic interest for centuries and have resulted in considerable contributions to the development of new drugs. Medicinal plants are useful sources for the isolation of novel, biologically-active molecules and the subsequent synthesis, modification, and optimization of those molecules' biological properties [18]. Alkaloids are a diverse natural group of biologically-active, secondary metabolites; alkaloids have attracted much attention due to their toxic and medicinal properties. One of these interesting and promising alkaloids is the dendrogenins, which elicit the proliferation of neuronal cells [19]. The other perspective compound, a plant alkaloid, sinomenine, extracted from the roots of *Sinomenium acutum*, effectively inhibits ASIC1a (IC_{50} ~1 μM) and possesses anti-inflammatory, antinociceptive, and neuroprotective effects [20]. However, there are no data on the effect of this compound on other ASIC isoforms. Here, we describe a plant bisbenzylisoquinoline alkaloid named lindoldhamine (LIN) as a potent inhibitor of ASIC1a activation by physiologically-relevant acidification. LIN was extracted from the leaves of *Laurus nobilis* belonging to the family Lauraceae. Representatives of this family synthesize an abundant variety of alkaloids with 22 structurally-distinct moieties (more comprehensive information can be found in the review [21]). Scholars have described LIN as an acetylcholinesterase inhibitor with an IC_{50} value of 3.5 μM [22]. LIN has demonstrated trypanocidal activity due to inhibition of trypanothione reductase (IC_{50} 27 μM) [23]. At a concentration of 100 μM, LIN completely inhibits platelet aggregation induced by collagen, arachidonic acid, and platelet-activating factor [24]. It has recently been shown that LIN acts as a proton-independent activator of ASIC3 (with EC_{50} values of 1.5 mM and 3.2 mM for human and rat ASIC3 channels, respectively). Besides, LIN potentiated the transient component of the human, but not rat ASIC3 current and restored it from desensitization, with EC_{50} values of 3.7 and 16.2 μM, respectively [25].

To assess the pharmacological potential of LIN, we tested it on the ASIC1a channel. We found that LIN stimulus-dependently inhibited the acid-induced currents of ASIC1a and produced anti-inflammatory effects in mice. These results demonstrated a good correlation between in vitro and in vivo data, as well as the participation of ASIC1a in some pathological processes occurring in the organism.

2. Results

2.1. LIN Proton-Dependently Inhibits the ASIC1a Current

Bisbenzylisoquinoline alkaloid LIN was extracted from the leaves of *Laurus nobilis* as described previously [25] (Figure 1A,B). Using two-electrode voltage-clamp recordings from *Xenopus laevis* oocytes that express rat ASIC1a, we found that, at the holding membrane potential of −50 mV, an LIN concentration of 300 μM effectively inhibited ASIC1a during channel activations with mild acidic stimuli (pH above 6.0). LIN inhibited ASIC1a by 98.1 ± 0.3%, 93.7 ± 0.9%, and 69.3 ± 2.3% for pHs of 6.85, 6.7, and 6.5, respectively (Figure 1C). LIN's inhibitory power tended to decrease as the pH dropped from 7.4 to 6.0 (34 ± 4% inhibition), and the inhibition was non-significant for stimuli with a pH less than 5.5.

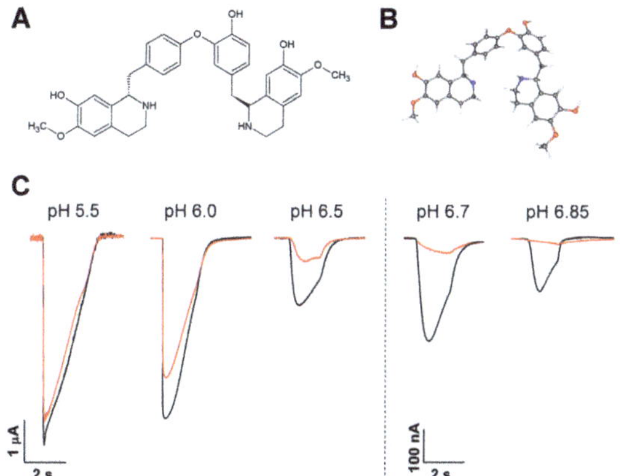

Figure 1. Lindoldhamine (LIN)'s effects on rat ASIC1a. (**A**) LIN's chemical structure. (**B**) LIN's predicted (cambered and flex) 3D structure (from PubChem; CID 10370752). Carbon atoms are in grey, hydrogen atoms in white, oxygen atoms\in red, and nitrogen atoms in blue. (**C**) The inhibitory effect that LIN (300 μM) had on the ASIC1a channel's activation by fast external solution acidification from pH 7.4 to the corresponding pH stimulus. The black traces are the control current; the red traces are the current obtained after 15 s of LIN pre-application. All presented traces are for a single cell.

To evaluate the effectiveness of LIN's ASIC1a inhibition, we measured the dose-dependence of LIN's inhibitory effect with various activating stimuli (Figure 2).

As expected, LIN very effectively inhibited ASIC1a during mild acidic stimulations, and its effectiveness (IC_{50} value) significantly decreased as the acidity of the stimulus increased. An important observation is that the stable dose-response curve shifted from a sub-micromolar IC_{50} value to one measuring hundreds of micromoles for pH values below 6.0. When applied before the pH dropped to above 6.0, LIN was able to inhibit the current completely; however, an application before a pH 6.0 stimulus dramatically changed LIN's mode of action, causing the inhibitory effect to reach saturation at 52 ± 11%. For a stimulus with the more acidic pH of 5.5, the dose-dependent curve no longer fit well. The values of the dose-dependent parameters are summarized in Table 1.

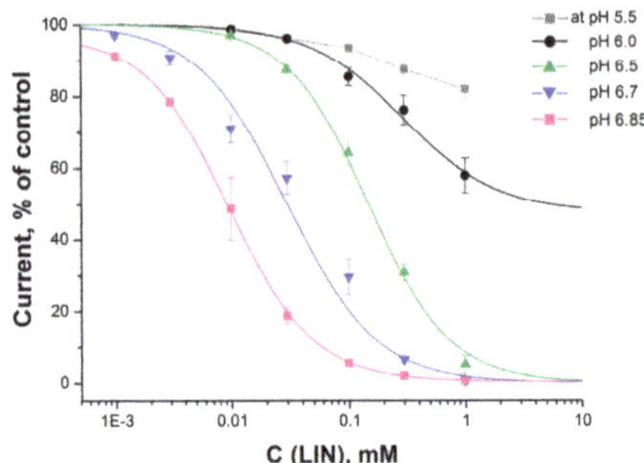

Figure 2. Dose-response curves for LIN's inhibitory activity on ASIC1a currents. The colored curves denote the corresponding pH stimuli that activate the channel, relative to the conditioning pH of 7.4. Each point is the mean ± SEM of five measurements. The data were fitted using the F_1 logistic equation (see the Materials and Methods Section for details).

Table 1. Calculated IC_{50} and Hill coefficient (nH) values of LIN's inhibitory action for activating stimuli, by pH.

	pH 6.85	pH 6.7	pH 6.5	pH 6.0
IC_{50}, µM	9 ± 0.3	25 ± 6	150 ± 10	296 ± 146
n_H	1.22 ± 0.03	1.29 ± 0.12	1.25 ± 0.05	1.07 ± 0.15

2.2. *LIN's Effect on the pH Dependence of the ASIC1a Activation*

In the presence of 0.3 mM of LIN, the proton dependence of the ASIC1a activation demonstrated an acidic shift (Figure 3). The amplitude's fit with the F_2 equation (see the formula in the Section 4 Materials and Methods) indicated that LIN shifted both the $pH_{50}1$ and $pH_{50}2$ values towards a more acidic area, with $pH_{50}1$ shifting from 6.72 ± 0.02 for the control to 6.47 ± 0.03 with LIN ($p < 0.001$) and with $pH_{50}2$ shifting from 6.49 ± 0.04 for the control to 6.23 ± 0.09 with LIN ($p < 0.05$).

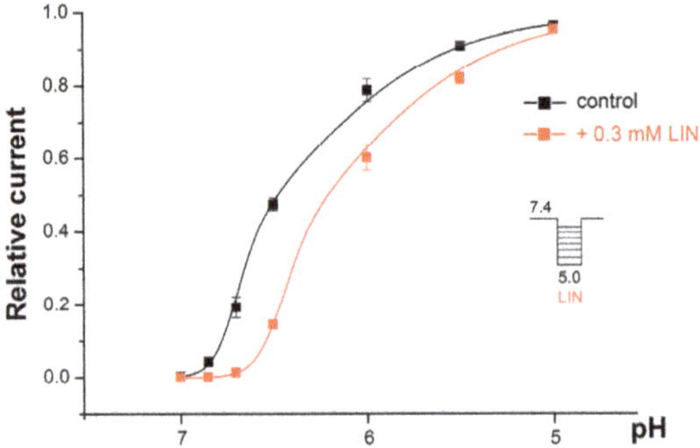

Figure 3. The pH dependence of the ASIC1a activation by protons alone (black line) and with the co-application of protons and 0.3 mM of LIN (red line). The ASIC1a channel was held at pH 7.4 and then activated by various acidic stimuli. The relative current is the amplitude of the peak current that the acidic stimuli evoked, normalized to the maximum amplitude (which we calculated for each cell via individual fitting). The data are presented as the mean ± SEM ($n = 5$).

The Hill coefficient n_H showed a strong tendency to decrease in the presence of LIN, although this change had no statistical significance (6.6 ± 0.5 for the control and 5.7 ± 0.7 for 0.3 mM LIN ($p = 0.14$)).

2.3. LIN Produced an Anti-Inflammatory Effect, but Failed to Decrease the Response to Acid

We studied the anti-inflammatory effect using the complete Freund's adjuvant (CFA)-induced inflammation model. The intraplantar injection of CFA in the hind paw caused hyperalgesia (i.e., increased sensitivity to noxious mechanical and thermal stimuli) and swelling due to the inflammatory process. We chose a 1 mg/kg dose so that the LIN would reach a blood plasma concentration of about 40 μM, as estimated with regard to blood volume (~8% of weight) and blood plasma content (55–65% of that volume [26]).

The group of mice treated with saline 24 h after the CFA injection showed significantly higher paw diameter (3.26 ± 0.03 mm) relative to that of the control group (2.09 ± 0.03 mm), which we treated with saline twice. Intravenous administration of LIN at a dose of 1 mg/kg decreased paw edema significantly, relative to that of the control group by 25%, 21%, and 19% 2, 4, and 24 h after administration, respectively (Figure 4A).

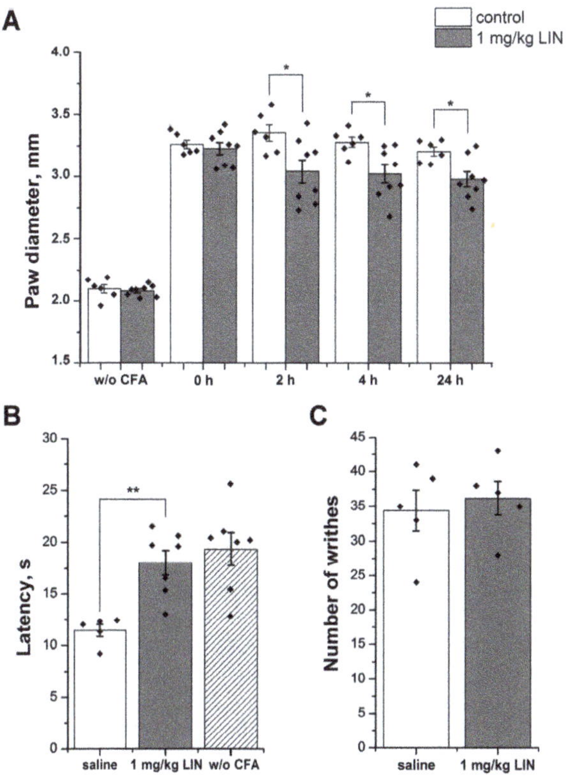

Figure 4. LIN activity in animal models. (**A**) LIN's anti-inflammatory effect. We induced paw edema via complete Freund's adjuvant (CFA) injection and estimated values before the administration of the CFA and the testing compound. LIN (1 mg/kg, i.v.) significantly reduced edema, even 24 h after injection. (**B**) LIN's reversal of thermal hyperalgesia. LIN (1 mg/kg, i.v., 30 min before testing) significantly reversed CFA-induced thermal hyperalgesia and prolonged the withdrawal latency for an inflamed hind paw placed on a hot plate. (**C**) LIN's effect on the writhing test. Pretreatment of mice with LIN (1 mg/kg, i.v., 30 min before testing) did not have any significant effect on the results of the writhing test, which involved the intraperitoneal administration of acetic acid. The results are presented as the mean ± SEM ($n = 5$–8); the p-values of the LIN group vs. the saline group are based on an analysis of variance and on Tukey's test. * $p < 0.05$, ** $p < 0.01$.

CFA injection provoked a significant decrease of inflamed hind paw withdrawal latency to thermal stimuli (11.44 ± 0.59 s) in comparison with the group treated with saline twice (19.34 ± 1.56 s). Intravenous administration of LIN (1 mg/kg) reversed 83.4% of thermal hyperalgesia. The latency of the response to the thermal stimulus for the LIN-treated group was 18.03 ± 1.18 s (Figure 4B).

The intraperitoneal injection of acetic acid provoked a pain behavior in the mice, as characterized by abdominal contractions (acetic acid-induced writhes). LIN pretreatment had no significant analgesic effect on the mice (Figure 4C), as the number of writhes was 34.4 ± 2.9 for the saline control group and 36.2 ± 2.4 for the LIN pretreated group.

3. Discussion

Bisbenzylisoquinoline alkaloids represent a fairly large family of isoquinoline alkaloids, in the abundance occurring in plants of the Lauraceae family. The pharmacology of bisbenzylisoquinoline alkaloids is also quite extensive and has remarkable medicinal relevance. Most of them are potent anticancer agents and possess anti-inflammatory, antiplasmodial, and antiviral activity. The most important and widely-studied compounds are primarily tetrandrine, curine, and curine-related compounds, cepharanthine, cycleanine, fangchinoline, and neferine [27]. For example, tetrandrine selectively inhibited the alpha7 subtype of neuronal nicotinic acetylcholine receptors and possessed an anti-inflammatory, antifibrotic, anticancer, and neuroprotective properties [28,29], Bisbenzylisoquinoline alkaloid lindoldhamine (LIN) has not been widely studied yet. To date, LIN has been shown to inhibit acetylcholinesterase, trypanothione reductase, platelet aggregation, as well as to activate and potentiate the ASIC3 channel [22–25].

In this study, we show the action LIN on the ASIC1a channel in vitro, as well as its effect in two pain models in vivo. We found that LIN effectively inhibited ASIC1a currents activated by mild acidic stimuli. LIN's inhibition efficiency correlated with the strength of the applied acid stimulus. For the minimal activation stimulus of pH 6.85, LIN completely eliminated the current, with an IC_{50} value of 9 μM. LIN also completely inhibited the current for two other low-scale stimuli (pHs 6.7 and 6.5), but its inhibitory effectiveness was more than 10-times lower. Medium acidification (pH 6.0 or lower) produced a greater ion current that the LIN could not completely inhibit; therefore, as the activation stimuli increased, the maximum inhibition decreased, and the IC_{50} value increased. We measured the inhibitory effect under pre-incubation conditions when the ligand had a preference for the channel binding, so we can assume that the LINs' binding site at least partially overlaps with some of the channel's numerous protonation sites.

A fast pH drop is essential to open the ASIC1a channel, and the rate of acidification was very fast in our experiments. Therefore, we were able to realize LIN's inhibitory effect through the competition for overlapping protonation sites. LIN was most efficacious for weak, physiologically-relevant acidification (Figures 2 and 3), but when the acidic stimuli had a pH below 6.0, the protons could simultaneously affect approximately 16 binding sites [30,31], resulting in a conversion of the ion channel into its so-called pre-activation form [32]. The LIN likely could not bind with the pre-activated form of ASIC1a, thus causing it to lose its affinity and its inhibitory activity such that it only affected a small population of the channels still not be pre-activated. As a result, LIN's overall effect was to shift the proton dependence curve of ASIC1a activation towards a more acidic region (Figure 3). Other well-known ASIC1a inhibitors, including 2-guanidine-4-methylquinazoline (GMQ) and mambalgins, have the same effect [33,34].

For rat ASIC1a, some researchers have described proton activation as having non-sigmoidal dependence [35,36] that fit well by equation F_2 (Materials and Methods). This equation describes the simultaneous protonation of a pool of "highly-cooperative" sites ($pH_{50}1$) and additional "non-cooperative" site ($pH_{50}2$) within the channel. LIN significantly decreased both the $pH_{50}1$ and $pH_{50}2$ parameters by 0.25 units, which indicated that it equally impacted both the highly-cooperative sites and the non-cooperative site. We thus assumed that LIN influenced more than one protonation site in ASIC1a, either through the overlap of binding sites or LIN could affect these sites allosterically.

However, to date, there are not enough data to verify with which protonation sites within ASIC1a LIN interacts.

We evaluated LIN's effects in vivo to confirm that it is an ASIC1a inhibitor. We used two tests: the acetic acid-induced writhing test (i.e., strong external acidification) and the CFA-induced inflammation test (i.e., inflammation-associated acidification).

In the CFA-induced inflammation test, LIN had a significant anti-inflammatory effect, as it reduced both thermal hyperalgesia and paw edema. Although the ASIC1a channel is widely represented in the neurons of the central nervous system and plays a significant role in central nociception in the neurons of the dorsal root ganglion, ASIC1a expression is elevated during inflammation and upregulated by inflammatory mediators [37,38]. In addition, in the neurons of the spinal dorsal horn, the ASIC1a channel contributed to inflammatory hypersensitivity to pain [39]. During the inflammation process, the pH of the inflamed area was 0.5–1 units lower than the physiological pH [40,41]; this amount is significant and is sufficient to activate the ASIC1a that are most sensitive to protons [6]. LIN's effectiveness confirms that ASIC1a plays an important role in CFA-induced inflammation; the data for this model correlated well with the in vitro data, which showed that LIN completely inhibited the channel at stimuli with a pH from 6.85–6.5.

In the acetic acid-induced writhing test, LIN did not have an analgesic effect. This may be because LIN cannot inhibit the ASIC1a channel for stimuli with pH below 6.0. An alternative assumption is that the ASIC1a channel does not participate in the development of acid-induced pain processes, confirming what was reported earlier [42].

LIN and other isoquinoline compounds [25,43] potentiated human ASIC3, but did not affect rat ASIC3; therefore, the animal acetic acid-induced writhing model could not produce the combined effect of ASIC3 potentiation and ASIC1a inhibition. In humans, the combined effect on ASIC1a and ASIC3 could have one of two results. First, the opposite effects could counteract this, resulting in zero final effect; this has been shown for high doses of APETx2, a well-known ASIC3 antagonist [44,45]. The other option is the significant enhancement of the analgesic effect because of desensitizing action on neurons of potentiators or weak activators, as was previously shown on the modulators of transient receptor potential vanilloid 1 (TRPV1) and transient receptor potential ankyrin 1 (TRPA1) channels [46–49].

ASIC1a's hyperactivity is associated with pathological processes in the central nervous system. Either a knockout or a pharmacological blockade of ASIC1a prevents the development of certain neurodegenerative diseases [8]. When combined with the fact that LIN effectively inhibits acetylcholinesterase activity and thus is a therapeutic option for preventing the progression of neurodegenerative diseases [22], this leads to a new perspective in which LIN can be used as a multipotent drug in the treatment of neurodegenerative diseases.

In summary, we demonstrated that LIN effectively inhibited ASIC1a activity under mild acidosis, but that this effectiveness decreased significantly as the acidity of the stimulus increased. These results are in good agreement with those for in vivo pain models. LIN effectively demonstrated anti-inflammatory activity, but did not have an analgesic effect against acetic acid-induced pain. These results confirm the importance of the ASIC1a channel in the development of pain processes during inflammation and together with LIN's previously-reported activity open new perspectives for the development of drugs to treat neurodegenerative diseases.

4. Materials and Methods

4.1. Chemical Reagents and Compounds

Lindoldhamine isolated from dried leaves of *Laurus nobilis* was used in all experiments. We obtained LIN using the isolation procedure [25] described previously. All reagents were obtained from Sigma-Aldrich (Steinheim, Germany). A working solution of the ligand was prepared in the ND96 buffer (96 mM NaCl, 2 mM KCl, 1 mM $MgCl_2$, and 5 mM HEPES

(4-(2-hydroxyethyl)-1-piperazineethanesulfonic acid) titrated to pH 7.4 with NaOH) immediately before the experiments.

4.2. Ethics Statement

This study strictly complied with the World Health Organization's International Guiding Principles for Biomedical Research Involving Animals. All experiments were approved by the Institutional Policy on the Use of Laboratory Animals of the Shemyakin-Ovchinnikov Institute of Bioorganic Chemistry Russian Academy of Sciences (Protocol Number 267/2018; date of approval: 2 October 2018) and by the Institutional Commission for the Control and Use of Laboratory Animals of the Branch of the Shemyakin-Ovchinnikov Institute of Bioorganic Chemistry of the Russian Academy of Sciences (identification code: 688/19; date of approval: 10 January 2019).

4.3. Electrophysiological Experiments

Electrophysiological experiments were performed on *X. laevis* oocytes expressing rat acid-sensing ion channel isoform 1a (ASIC1a channel). Unfertilized oocytes were harvested from female *X. laevis*. Tricaine methanesulfonate (MS222) (0.17% solution) was used to anaesthetize frogs, which, after surgery, were kept in a separate tank until they had completely recovered from the anesthesia. The follicle cell layers were removed from the oocytes by treatment at room temperature for 1.5–2 h with 1 mg/mL of collagenase in ND96 medium (96 mM NaCl, 2 mM KCl, 1 mM $MgCl_2$, and 5 mM HEPES (4-(2-hydroxyethyl)-1-piperazineethanesulfonic acid) titrated to pH 7.4 with NaOH) lacking calcium. Selected healthy Stage IV and V oocytes were injected with 2.5–5 ng of cRNA, synthesized from PCi plasmids containing the rat ASIC1a isoform, using the Nanoliter 2000 microinjection system (World Precision Instruments, USA). The injected oocytes were kept for 48–72 h at 17 °C and then for up to 6 days at 15 °C in ND96 medium supplemented with an antibiotic (gentamycin, 50 µg/mL). Two-electrode voltage-clamp recordings were carried out at a holding potential of −50 mV using the GeneClamp500 amplifier (Axon Instruments, Inverurie, U.K.). Microelectrodes were filled with 3 M KCl. The data were filtered and digitized at 20 Hz and 100 Hz, respectively, using the L780 AD converter (L-Card, Moscow, Russia). The external bath solution was ND96; the pH was adjusted to 7.4. For the activating test ND96 solutions with $6.5 \leq pH \leq 7.0$ and pH <6.5, HEPES was replaced by 10 mM MOPS (3-(N-morpholino)propanesulfonic acid) and 10 mM MES (2-(N-morpholino)ethanesulfonic acid), respectively. A computer-controlled valve system was used to achieve a streamflow of about 1 mL/min and a solution exchange rate about 60 mL/min in the recording chamber.

4.4. In Vivo Assay

4.4.1. Animals

Male adult CD-1 mice (20–25 g, 31 mice) were maintained at a normal 12-h light-dark cycle with water and food available ad libitum.

4.4.2. Complete Freund's Adjuvant-Induced Inflammation and Thermal Hyperalgesia

The oil/saline (1:1) CFA emulsion was injected into the dorsal surface of the left hind paw of mice (20 µL/paw) 24 h before the measurement, which caused the development of inflammation and thermal hyperalgesia of the paw. Saline (20 µL) was injected into control mice. The inflamed paw withdrawal latencies to thermal stimulation (53 °C) were measured 30 min after the LIN or saline injection. The paw diameter was evaluated using a digital caliper before the CFA injection, before LIN and saline administration, and 2, 4, and 24 h after the administration.

4.4.3. Acetic Acid-Induced Writhing (Abdominal Constriction Test of Visceral Pain)

Mice were divided into separate groups, and acetic acid in saline (0.6%, 10 mL/kg) was injected intraperitoneally 30 min after the intravenous administration of the LIN or saline for the control group.

Mice were directly located inside transparent glass cylinders, and the number of writhes was registered for 30 min.

4.5. Data and Statistical Analysis

Analysis of electrophysiological data was performed using OriginPro 8.6 software (OriginPro 8.6.0 (32 bit) version, OriginLab Corporation, Northampton, MA, USA, 2011). Curves were fitted using the following logistic equations:

$$\text{(a) } F_1(x) = ((a1 - a2)/(1 + (x/x0)\hat{}n_H)) + a2 \quad (1)$$

where $F_1(x)$ is the response value for a given LIN concentration; x is the concentration of LIN; a1 is the control response value (fixed at 100%); a2 is the response value at maximal inhibition (% of the control); x0 is the IC_{50} value; and n_H is the Hill coefficient;

$$\text{(b) } F_2(x) = A/((1 + (x/[pH_{50}1])\hat{}n_H) \times (1 + (x/[pH_{50}2])) \quad (2)$$

where $[pH_{50}1]$ is the half-maximal concentration of protons binding to the highly-cooperative sites (or half probability of "high-cooperative" sites' occupancy); $[pH_{50}2]$ is the half-maximal concentration of protons binding to the non-cooperative site (or half probability of "non-cooperative" site occupancy); n_H is the Hill coefficient; and A is the current's maximum amplitude.

The data significance in animal tests was determined by the analysis of variance (ANOVA) followed by Tukey's post-hoc test. Data are presented as the mean ± SEM.

Author Contributions: Conceptualization, Y.A.A. and S.A.K.; Formal analysis, D.I.O., S.G.K., I.A.D., Y.A.A. and S.A.K.; Investigation, D.I.O., S.G.K., V.A.P., Y.A.P. and E.R.S.; Methodology, D.I.O., S.G.K. and I.A.D.; Supervision, S.A.K.; Writing—original draft, D.I.O., Y.A.A. and S.A.K.; Writing—review & editing, Y.A.A. and S.A.K.

Funding: This research was funded by Russian Science Foundation grant 18-14-00138; animal experiments were partially carried out by V.A.P., Y.A.P. and E.R.S. using the equipment provided by the IBCh core facility (CKP IBCh, supported by Russian Ministry of Education and Science, grant RFMEFI62117X0018).

Acknowledgments: We are grateful to Sylvie Diochot (Institut de Pharmacologie Moléculaire et Cellulaire, Valbonne, France) for PCi plasmid containing cDNA of rat ASIC1a.

Conflicts of Interest: The authors declare no conflict of interest.

References

1. Benson, C.J.; Xie, J.; Wemmie, J.A.; Price, M.P.; Henss, J.M.; Welsh, M.J.; Snyder, P.M. Heteromultimers of DEG/ENaC subunits form H+-gated channels in mouse sensory neurons. *Proc. Natl. Acad. Sci. USA* **2002**, *99*, 2338–2343. [CrossRef] [PubMed]
2. Friese, M.A.; Craner, M.J.; Etzensperger, R.; Vergo, S.; Wemmie, J.A.; Welsh, M.J.; Vincent, A.; Fugger, L. Acid-sensing ion channel-1 contributes to axonal degeneration in autoimmune inflammation of the central nervous system. *Nat. Med.* **2007**, *13*, 1483–1489. [CrossRef] [PubMed]
3. Osmakov, D.I.; Andreev, Y.A.; Kozlov, S.A. Acid-Sensing Ion Channels and Their Modulators. *Biochemistry* **2014**, *79*, 1528–1545. [CrossRef] [PubMed]
4. Schuhmacher, L.-N.; Smith, E.S.J. Expression of acid-sensing ion channels and selection of reference genes in mouse and naked mole rat. *Mol. Brain* **2016**, *9*, 97. [CrossRef] [PubMed]
5. Rash, L.D. Acid-Sensing Ion Channel Pharmacology, Past, Present, and Future. In *Advances in Pharmacology*; Geraghty, D.P., Rash, L.D., Eds.; Elsevier: Amsterdam, The Netherlands, 2017; Volume 79, pp. 35–66.
6. Gründer, S.; Pusch, M. Biophysical properties of acid-sensing ion channels (ASICs). *Neuropharmacology* **2015**, *94*, 9–18. [CrossRef] [PubMed]
7. Osmakov, D.I.; Koshelev, S.G.; Lyukmanova, E.N.; Shulepko, M.A.; Andreev, Y.A.; Illes, P.; Kozlov, S.A. Multiple Modulation of Acid-Sensing Ion Channel 1a by the Alkaloid Daurisoline. *Biomolecules* **2019**, *9*, 336. [CrossRef] [PubMed]

8. Wemmie, J.A.; Taugher, R.J.; Kreple, C.J. Acid-sensing ion channels in pain and disease. *Nat. Rev. Neurosci.* **2013**, *14*, 461–471. [CrossRef]
9. Coryell, M.W.; Wunsch, A.M.; Haenfler, J.M.; Allen, J.E.; McBride, J.L.; Davidson, B.L.; Wemmie, J.A. Restoring Acid-Sensing Ion Channel-1a in the Amygdala of Knock-Out Mice Rescues Fear Memory but Not Unconditioned Fear Responses. *J. Neurosci.* **2008**, *28*, 13738–13741. [CrossRef]
10. Leng, T.; Shi, Y.; Xiong, Z.-G.; Sun, D. Proton-sensitive cation channels and ion exchangers in ischemic brain injury: New therapeutic targets for stroke? *Prog. Neurobiol.* **2014**, *115*, 189–209. [CrossRef]
11. Duan, B.; Wang, Y.-Z.; Yang, T.; Chu, X.-P.; Yu, Y.; Huang, Y.; Cao, H.; Hansen, J.; Simon, R.P.; Zhu, M.X.; et al. Extracellular Spermine Exacerbates Ischemic Neuronal Injury through Sensitization of ASIC1a Channels to Extracellular Acidosis. *J. Neurosci.* **2011**, *31*, 2101–2112. [CrossRef]
12. Vick, J.S.; Askwith, C.C. ASICs and neuropeptides. *Neuropharmacology* **2015**, *94*, 36–41. [CrossRef] [PubMed]
13. Osmakov, D.I.; Koshelev, S.G.; Ivanov, I.A.; Andreev, Y.A.; Kozlov, S.A. Endogenous Neuropeptide Nocistatin Is a Direct Agonist of Acid-Sensing Ion Channels (ASIC1, ASIC2 and ASIC3). *Biomolecules* **2019**, *9*, 401. [CrossRef]
14. Bohlen, C.J.; Chesler, A.T.; Sharif-Naeini, R.; Medzihradszky, K.F.; Zhou, S.; King, D.; Sánchez, E.E.; Burlingame, A.L.; Basbaum, A.I.; Julius, D. A heteromeric Texas coral snake toxin targets acid-sensing ion channels to produce pain. *Nature* **2011**, *479*, 410–414. [CrossRef] [PubMed]
15. Mazzuca, M.; Heurteaux, C.; Alloui, A.; Diochot, S.; Baron, A.; Voilley, N.; Blondeau, N.; Escoubas, P.; Gélot, A.; Cupo, A.; et al. A tarantula peptide against pain via ASIC1a channels and opioid mechanisms. *Nat. Neurosci.* **2007**, *10*, 943–945. [CrossRef] [PubMed]
16. Diochot, S.; Baron, A.; Salinas, M.; Douguet, D.; Scarzello, S.; Dabert-Gay, A.-S.; Debayle, D.; Friend, V.; Alloui, A.; Lazdunski, M.; et al. Black mamba venom peptides target acid-sensing ion channels to abolish pain. *Nature* **2012**, *490*, 552–555. [CrossRef] [PubMed]
17. Baron, A.; Lingueglia, E. Pharmacology of acid-sensing ion channels—Physiological and therapeutical perspectives. *Neuropharmacology* **2015**, *94*, 19–35. [CrossRef] [PubMed]
18. Atanasov, A.G.; Waltenberger, B.; Pferschy-Wenzig, E.-M.; Linder, T.; Wawrosch, C.; Uhrin, P.; Temml, V.; Wang, L.; Schwaiger, S.; Heiss, E.H.; et al. Discovery and resupply of pharmacologically active plant-derived natural products: A review. *Biotechnol. Adv.* **2015**, *33*, 1582–1614. [CrossRef]
19. Khalifa, S.A.M.; De Medina, P.; Erlandsson, A.; El-Seedi, H.R.; Silvente-Poirot, S.; Poirot, M. The novel steroidal alkaloids dendrogenin A and B promote proliferation of adult neural stem cells. *Biochem. Biophys. Res. Commun.* **2014**, *446*, 681–686. [CrossRef]
20. Wu, W.-N.; Wu, P.-F.; Chen, X.-L.; Zhang, Z.; Gu, J.; Yang, Y.-J.; Xiong, Q.-J.; Ni, L.; Wang, F.; Chen, J.-G. Sinomenine protects against ischaemic brain injury: Involvement of co-inhibition of acid-sensing ion channel 1a and L-type calcium channels. *Br. J. Pharmacol.* **2011**, *164*, 1445–1459. [CrossRef]
21. Silva Teles, M.M.R.; Vieira Pinheiro, A.A.; Da Silva Dias, C.; Fechine Tavares, J.; Barbosa Filho, J.M.; Leitão Da Cunha, E.V. Alkaloids of the Lauraceae. In *The Alkaloids: Chemistry and Biology*; Knölker, H.-J., Ed.; Elsevier: Amsterdam, The Netherlands, 2019; Volume 82, pp. 147–304.
22. Murebwayire, S.; Ingkaninan, K.; Changwijit, K.; Frédérich, M.; Duez, P. Triclisia sacleuxii (Pierre) Diels (Menispermaceae), a potential source of acetylcholinesterase inhibitors. *J. Pharm. Pharmacol.* **2009**, *61*, 103–107. [CrossRef]
23. Fournet, A.; Inchausti, A.; Yaluff, G.; De Arias, A.R.; Guinaudeau, H.; Bruneton, J.; Breidenbach, M.A.; Karplus, P.A.; Faerman, C.H. Trypanocidal Bisbenzylisoquinoline Alkaloids are Inhibitors of Trypanothione Reductase. *J. Enzyme Inhib.* **1998**, *13*, 1–9. [CrossRef] [PubMed]
24. Chia, Y.-C.; Chang, F.-R.; Wu, C.-C.; Teng, C.-M.; Chen, K.-S.; Wu, Y.-C. Effect of Isoquinoline Alkaloids of Different Structural Types on Antiplatelet Aggregation in Vitro. *Planta Med.* **2006**, *72*, 1238–1241. [CrossRef] [PubMed]
25. Osmakov, D.I.; Koshelev, S.G.; Andreev, Y.A.; Dubinnyi, M.A.; Kublitski, V.S.; Efremov, R.G.; Sobolevsky, A.I.; Kozlov, S.A. Proton-independent activation of acid-sensing ion channel 3 by an alkaloid, lindoldhamine, from Laurus nobilis. *Br. J. Pharmacol.* **2018**, *175*, 924–937. [CrossRef] [PubMed]
26. Morton, D.B.; Abbot, D.; Barclay, R.; Close, B.S.; Ewbank, R.; Gask, D.; Heath, M.; Mattic, S.; Poole, T.; Seamer, J.; et al. Removal of blood from laboratory mammals and birds: First Report of the BVA/FRAME/RSPCA/UFAW Joint Working Group on Refinement. *Lab. Anim.* **1993**, *27*, 1–22. [CrossRef]

27. Weber, C.; Opatz, T. Bisbenzylisoquinoline Alkaloids. In *The Alkaloids: Chemistry and Biology*; Knölker, H.-J., Ed.; Elsevier: Amsterdam, The Netherlands, 2019; Volume 81, pp. 1–114.
28. Bhagya, N.; Chandrashekar, K.R. Tetrandrine—A molecule of wide bioactivity. *Phytochemistry* **2016**, *125*, 5–13. [CrossRef] [PubMed]
29. Liu, T.; Liu, X.; Li, W. Tetrandrine, a Chinese plant-derived alkaloid, is a potential candidate for cancer chemotherapy. *Oncotarget* **2016**, *7*, 40800–40815. [CrossRef] [PubMed]
30. Ishikita, H. Proton-Binding Sites of Acid-Sensing Ion Channel 1. *PLoS ONE* **2011**, *6*, e16920. [CrossRef]
31. Vullo, S.; Bonifacio, G.; Roy, S.; Johner, N.; Bernèche, S.; Kellenberger, S. Conformational dynamics and role of the acidic pocket in ASIC pH-dependent gating. *Proc. Natl. Acad. Sci. USA* **2017**, *114*, 3768–3773. [CrossRef] [PubMed]
32. Laurent, B.; Murail, S.; Shahsavar, A.; Sauguet, L.; Delarue, M.; Baaden, M. Sites of Anesthetic Inhibitory Action on a Cationic Ligand-Gated Ion Channel. *Structure* **2016**, *24*, 595–605. [CrossRef] [PubMed]
33. Alijevic, O.; Kellenberger, S. Subtype-specific Modulation of Acid-sensing Ion Channel (ASIC) Function by 2-Guanidine-4-methylquinazoline. *J. Biol. Chem.* **2012**, *287*, 36059–36070. [CrossRef]
34. Besson, T.; Lingueglia, E.; Salinas, M. Pharmacological modulation of Acid-Sensing Ion Channels 1a and 3 by amiloride and 2-guanidine-4-methylquinazoline (GMQ). *Neuropharmacology* **2017**, *125*, 429–440. [CrossRef] [PubMed]
35. Sherwood, T.W.; Askwith, C.C. Endogenous arginine-phenylalanine-amide-related peptides alter steady-state desensitization of ASIC1a. *J. Biol. Chem.* **2008**, *283*, 1818–1830. [CrossRef] [PubMed]
36. Stephan, G.; Huang, L.; Tang, Y.; Vilotti, S.; Fabbretti, E.; Yu, Y.; Nörenberg, W.; Franke, H.; Gölöncsér, F.; Sperlágh, B.; et al. The ASIC3/P2X3 cognate receptor is a pain-relevant and ligand-gated cationic channel. *Nat. Commun.* **2018**, *9*, 1354. [CrossRef] [PubMed]
37. Voilley, N.; de Weille, J.; Mamet, J.; Lazdunski, M. Nonsteroid Anti-Inflammatory Drugs Inhibit Both the Activity and the Inflammation-Induced Expression of Acid-Sensing Ion Channels in Nociceptors. *J. Neurosci.* **2001**, *21*, 8026–8033. [CrossRef] [PubMed]
38. Mamet, J.; Baron, A.; Lazdunski, M.; Voilley, N. ProInflammatory Mediators, Stimulators of Sensory Neuron Excitability via the Expression of Acid-Sensing Ion Channels. *J. Neurosci.* **2002**, *22*, 10662–10670. [CrossRef] [PubMed]
39. Duan, B.; Wu, L.-J.; Yu, Y.-Q.; Ding, Y.; Jing, L.; Xu, L.; Chen, J.; Xu, T.-L. Upregulation of Acid-Sensing Ion Channel ASIC1a in Spinal Dorsal Horn Neurons Contributes to Inflammatory Pain Hypersensitivity. *J. Neurosci.* **2007**, *27*, 11139–11148. [CrossRef]
40. Steen, K.; Steen, A.; Reeh, P. A dominant role of acid pH in inflammatory excitation and sensitization of nociceptors in rat skin, in vitro. *J. Neurosci.* **1995**, *15*, 3982–3989. [CrossRef]
41. Lardner, A. The effects of extracellular pH on immune function. *J. Leukoc. Biol.* **2001**, *69*, 522–530. [CrossRef]
42. Sluka, K.A.; Price, M.P.; Breese, N.M.; Stucky, C.L.; Wemmie, J.A.; Welsh, M.J. Chronic hyperalgesia induced by repeated acid injections in muscle is abolished by the loss of ASIC3, but not ASIC1. *Pain* **2003**, *106*, 229–239. [CrossRef]
43. Osmakov, D.I.; Koshelev, S.G.; Andreev, Y.A.; Kozlov, S.A. Endogenous isoquinoline alkaloids agonists of acid-sensing ion channel type 3. *Front. Mol. Neurosci.* **2017**, *10*, 282. [CrossRef]
44. Karczewski, J.; Spencer, R.H.; Garsky, V.M.; Liang, A.; Leitl, M.D.; Cato, M.J.; Cook, S.P.; Kane, S.; Urban, M.O. Reversal of acid-induced and inflammatory pain by the selective ASIC3 inhibitor, APETx2. *Br. J. Pharmacol.* **2010**, *161*, 950–960. [CrossRef] [PubMed]
45. Andreev, Y.; Osmakov, D.; Koshelev, S.; Maleeva, E.; Logashina, Y.; Palikov, V.; Palikova, Y.; Dyachenko, I.; Kozlov, S. Analgesic Activity of Acid-Sensing Ion Channel 3 (ASIC3) Inhibitors: Sea Anemones Peptides Ugr9-1 and APETx2 versus Low Molecular Weight Compounds. *Mar. Drugs* **2018**, *16*, 500. [CrossRef] [PubMed]
46. Frias, B.; Merighi, A. Capsaicin, Nociception and Pain. *Molecules* **2016**, *21*, 797. [CrossRef] [PubMed]
47. Materazzi, S.; Benemei, S.; Fusi, C.; Gualdani, R.; De Siena, G.; Vastani, N.; Andersson, D.A.; Trevisan, G.; Moncelli, M.R.; Wei, X.; et al. Parthenolide inhibits nociception and neurogenic vasodilatation in the trigeminovascular system by targeting the TRPA1 channel. *Pain* **2013**, *154*, 2750–2758. [CrossRef] [PubMed]

48. Logashina, Y.A.; Mosharova, I.V.; Korolkova, Y.V.; Shelukhina, I.V.; Dyachenko, I.A.; Palikov, V.A.; Palikova, Y.A.; Murashev, A.N.; Kozlov, S.A.; Stensvåg, K.; et al. Peptide from Sea Anemone Metridium senile Affects Transient Receptor Potential Ankyrin-repeat 1 (TRPA1) Function and Produces Analgesic Effect. *J. Biol. Chem.* **2017**, *292*, 2992–3004. [CrossRef] [PubMed]
49. Logashina, Y.A.; Solstad, R.G.; Mineev, K.S.; Korolkova, Y.V.; Mosharova, I.V.; Dyachenko, I.A.; Palikov, V.A.; Palikova, Y.A.; Murashev, A.N.; Arseniev, A.S.; et al. New Disulfide-Stabilized Fold Provides Sea Anemone Peptide to Exhibit Both Antimicrobial and TRPA1 Potentiating Properties. *Toxins (Basel)* **2017**, *9*, 154. [CrossRef] [PubMed]

 toxins

Article

Dehydrocrenatidine Inhibits Voltage-Gated Sodium Channels and Ameliorates Mechanic Allodia in a Rat Model of Neuropathic Pain

Fang Zhao, Qinglian Tang, Jian Xu, Shuangyan Wang, Shaoheng Li, Xiaohan Zou * and Zhengyu Cao *

Jiangsu Key Laboratory of TCM Evaluation and Translational Research, School of Traditional Chinese Pharmacy, China Pharmaceutical University, Nanjing 211198, China; 1620184499@cpu.edu.cn (F.Z.); cputangqinglian@163.com (Q.T.); xujian0118@163.com (J.X.); shuangyanwcpu@163.com (S.W.); lsh199242@163.com (S.L.)

* Correspondence: 1620174411@cpu.edu.cn (X.Z.); zycao@cpu.edu.cn (Z.C.); Tel.: +86-25-8618-5955 (Z.C. & X.Z.); Fax: +86-25-8618-5158 (Z.C. & X.Z.)

Received: 27 March 2019; Accepted: 15 April 2019; Published: 18 April 2019

Abstract: *Picrasma quassioides* (D. Don) Benn, a medical plant, is used in clinic to treat inflammation, pain, sore throat, and eczema. The alkaloids are the main active components in *P. quassioides*. In this study, we examined the analgesic effect of dehydrocrenatidine (DHCT), a β-carboline alkaloid abundantly found in *P. quassioides* in a neuropathic pain rat model of a sciatic nerve chronic constriction injury. DHCT dose-dependently attenuated the mechanic allodynia. In acutely isolated dorsal root ganglion, DHCT completely suppressed the action potential firing. Further electrophysiological characterization demonstrated that DHCT suppressed both tetrodotoxin-resistant (TTX-R) and sensitive (TTX-S) voltage-gated sodium channel (VGSC) currents with IC_{50} values of 12.36 μM and 4.87 μM, respectively. DHCT shifted half-maximal voltage ($V_{1/2}$) of inactivation to hyperpolarizing direction by ~16.7 mV in TTX-S VGSCs. In TTX-R VGSCs, DHCT shifted $V_{1/2}$ of inactivation voltage to hyperpolarizing direction and $V_{1/2}$ of activation voltage to more depolarizing potential by ~23.9 mV and ~12.2 mV, respectively. DHCT preferred to interact with an inactivated state of VGSCs and prolonged the repriming time in both TTX-S and TTX-R VGSCs, transiting the channels into a slow inactivated state from a fast inactivated state. Considered together, these data demonstrated that the analgesic effect of DHCT was likely though the inhibition of neuronal excitability.

Keywords: dehydrocrenatidine; neuropathic pain; voltage-gated sodium channels

Key Contribution: DHCT attenuates mechanic allodynia in CCI neuropathic pain model; DHCT suppresses neuroexcitability through inhibition of VGSCs in DRG neurons.

---

1. Introduction

Neuropathic pain affects 6–8% of the population, among which 27% of patients are in a condition of chronic pain [1]. Despite the existence of seven categories of pain drugs, only 30% of patients get adequate relief. In addition, currently used analgesics produce significant side effects, such as addiction and sedation, which negatively affect life quality of the patients [2].

Neuropathic pain results from disorders of peripheral and/or central nervous systems. The pathological mechanisms are complex. Sensitization of neurons leads to enhanced neuronal excitability therefore representing the major cause of abnormal nociception [3]. Alterations in the activities and/or expression levels of ion channels, such as calcium, sodium, and potassium channels, have been demonstrated in neurons from neuropathic pain rodent models and contribute to the hyper-excitability of injured neurons [4–7]. Voltage-gated sodium channels (VGSCs) are responsible for the rising phase of action potential (AP) generation, therefore controlling neuronal excitability [8–10]. The altered expression levels and sensitivities of VGSCs in dorsal root ganglion (DRG) neurons in painful neuropathy result in hyper-excitability of neurons [11–13].

Picrasma quassioides (D. Don) Benn, a medical plant, has been used in clinics to treat diseases, such as inflammation, pain, sore throat, gastroenteritis, and eczema [14,15]. Phytochemical investigations revealed the presence of many types of chemical constituents such as alkaloids (mainly β-carboline and cathinone alkaloids), bitter principles (nigakihemiacetal A–F, nigakilactone A–N, picrasin C–G, picraqualides A–D and picrasinoside A–G), and triterpenoids in this medical plant [14,16–19]. Pharmacological studies demonstrate that the ethanolic extract of *P. quassioides* stems displays anti-cancer, anti-diabetes, and anti-hypertensive activities [14,17,20]. The total alkaloids extracted from *P. quassioides* are the main active components for antipyretic, anti-inflammatory, anti-bacterial, and anti-tumor activities [20,21]. In addition, these alkaloids are also effective in the peripheral nervous system and central nervous system. The total alkaloid fraction suppresses the sympathetic nerve firing therefore displaying an anti-hypertension effect [22].

Dehydrocrenatidine (DHCT) is one of the most abundant β-carboline alkaloids found in the *P. quassioides*. The levels of DHCT in the stems of *P. quassiodes* can reach as high as 2.72% [23]. DHCT has been reported to suppress lipopolysaccharide (LPS)-stimulated NO production and secretion of pro-inflammatory cytokines including tumor necrosis factor α (TNF-α) and interleukin-6 (IL-6) through inhibition of inducible nitric oxide synthase (iNOS) and cyclooxygenase-2 (COX-2) activities, as well as the expression levels in macrophages [15]. Recently, a study has also demonstrated that DHCT is a specific janus kinase (JAK)-specific inhibitor that induces DU145 and MDA-MB-468 cell apoptosis [24]. In addition, several synthesized β-carboline alkaloids possess anxiolytic and anticonvulsant, sedative effects through interaction with 5-hydroxytryptamine, dopamine, and benzodiazepine receptors suggesting their effects on neuronal excitability [25–29]. However, whether and how DHCT is able to affect neuronal excitability has never been explored, despite its high concentration in *P. quassioides*, which is used for treatment of pain in the clinic.

In the current study, we evaluated the analgesic activity of DHCT, a representative of β-carboline alkaloid from *P. quassioides*, in a rat neuropathic pain model. We demonstrate that DHCT ameliorates neuropathic pain. Mechanistically, DHCT suppresses VGSC Na^+ currents and AP firing in DRG neurons.

2. Results

2.1. Dehydrocrenatidine Ameliorated Mechanic Allodynia in a Neuropathic Pain Model of Sciatic Nerve Partial Ligation

As a representative of β-carboline alkaloid in *P. quassioides*, the analgesic effect of DHCT (Figure 1A) was investigated in a chronic constrictive injury (CCI) neuropathic pain model. Consistent with previous reports, after ligation of the sciatic nerve for 14 days, the paw withdrawal threshold (PWT) in CCI rats reached 1.90 ± 0.31 g, which was significantly lower than that observed in the sham group (15.10 ± 1.50 g, $n = 9$, $p < 0.01$) (Figure 1B). Intrathecal (*i.t.*) administration of DHCT (150 or 250 μg/kg) significantly attenuated CCI-induced mechanical allodynia (Figure 1B,C). The analgesic effect was peaked at 1.5 h post administration and lasted at least for 3 h (Figure 1B). At the highest dose (250 μg/kg) investigated, DHCT produced comparable efficacy with the positive control, morphine (50 μg/kg) (Figure 1B).

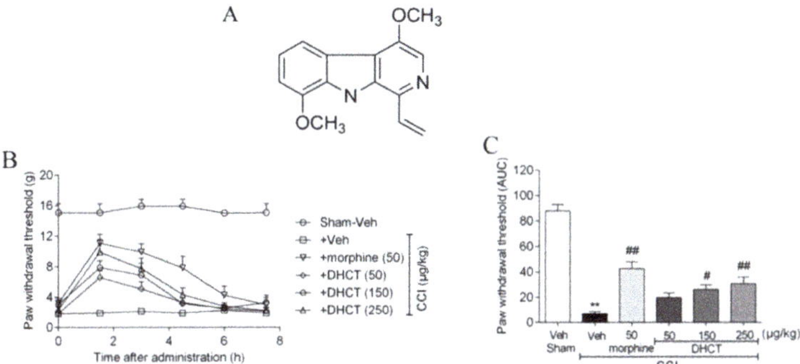

Figure 1. Dehydrocrenatidine (DHCT) attenuated sciatic nerve partial ligation-induced mechanical allodynia. (**A**) Chemical structure of DHCT. (**B**) Mechanical stimuli thresholds recorded after administration of vehicle (5% DMSO, 5% Tween-80, and 90% normal saline), DHCT (50, 150, and 250 μg/kg) or morphine (50 μg/kg) every 1.5 h, respectively. (**C**) Quantification of DHCT analgesic effects by using area under the curve (the area of paw withdrawal thresholds during 0–4.5 h after intrathecal administration). DHCT displayed significant analgesic effects in the rat chronic constriction neuropathic pain model. Each data point represents the mean ± SEM ($n = 9$). ** $p < 0.01$, model vs. sham group; # $p < 0.05$; ## $p < 0.01$, drugs vs. model group.

2.2. DHCT Suppressed Action Potential Generation in Acutely Dissociated Rat DRG Neurons

Given the analgesic effect of DHCT on CCI neuropathic pain model, we next examined whether DHCT was capable of suppressing neuronal excitability. Injection of a 30-pA (1 s) or 200-pA (1 s) current evoked repetitive APs (Figure 2A–C lower trace) in small or medium diameter DRG neurons. Bath application of DHCT (10 μM) for 1 min before current injections suppressed the firing frequency of APs elicited by injections of 30-pA and 200-pA currents in DRG neurons (Figure 2).

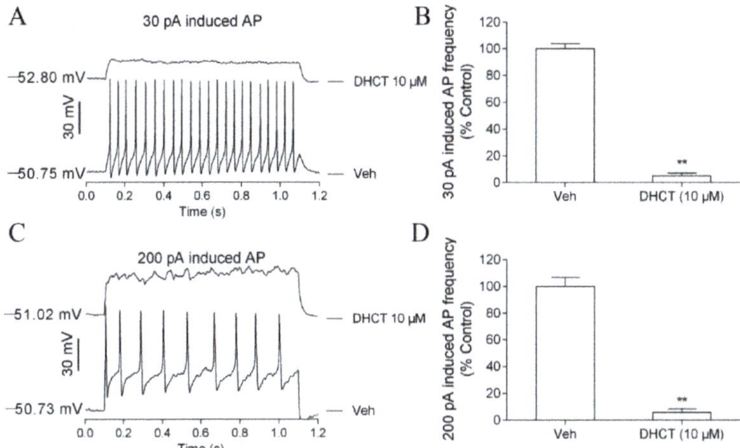

Figure 2. DHCT suppressed current-evoked action potential in dorsal root ganglion (DRG) neurons. (**A**) Representative traces of action potential (APs) evoked by an injection of a 30-pA (1 s) current in the absence and presence of 10 μM of DHCT. (**B**) Quantification of DHCT (10 μM) suppressed 30-pA induced APs in acutely dissociated rat DRG neurons. (**C**) Representative traces of APs evoked by an injection of a 200-pA (1 s) current in the absence and presence of 10 μM of DHCT. (**D**) Quantification of DHCT (10 μM) suppressed 200-pA induced APs in acutely dissociated rat DRG neurons. Each data point represents mean ± SEM. T-test was used to compared the statistical significance between Veh (0.1% DMSO) and DHCT (10 μM) groups. ** $p < 0.01$, $n = 13$.

2.3. DHCT Suppressed both Tetrodotoxin-Sensitive (TTX-S) and TTX-Resistant (TTX-R) VGSC Na^+ Currents in DRG Neurons

Given the inhibitory effect on AP generation in acutely dissociated rat DRG neurons and the pivotal role of VGSCs on neuroexcitability, we next examined the activity of DHCT on VGSCs in DRG neurons. Sodium currents were evoked by a 50-ms depolarization pulse to 0 mV from a holding potential of −100 mV. In medium- and large-diameter DRG neurons, membrane depolarization induced fast activated and fast inactivated Na^+ currents consistent with TTX-S VGSC currents [30]. Tetrodotoxin (TTX) (0.1–10 nM) concentration-dependently suppressed VGSC currents (Figure S1A,B). At a concentration of 10 nM, TTX produced a nearly complete suppression on VGSC Na^+ currents suggesting that medium- and large-diameter DRG neurons mainly express TTX-S VGSCs (Figure S1). DHCT concentration-dependently suppressed the Na^+ current in medium- and large-diameter DRG neurons with an IC_{50} value of 23.47 μM (14.59–37.74 μM, 95% confidential interval (CI)) (Figure 3A,B). To examine the voltage-dependence of DHCT action in TTX-S VGSCs, Na^+ currents were elicited by depolarization steps from −90 to +50 mV in a 5-mV increment in the presence or absence of DHCT (30 μM). DHCT suppressed the Na^+ peak currents evoked by different depolarization potentials (Figure 3C,D). The current–voltage (I-V) relationships in the presence and absence of 30 μM DHCT were illustrated in Figure 3E. Bath application of DHCT (30 μM) dramatically shifted the half-maximal voltage ($V_{1/2}$) of steady-state inactivation from −40.07 mV to −56.73 mV, but not activation $V_{1/2}$ of TTX-S VGSCs (Figure 3F).

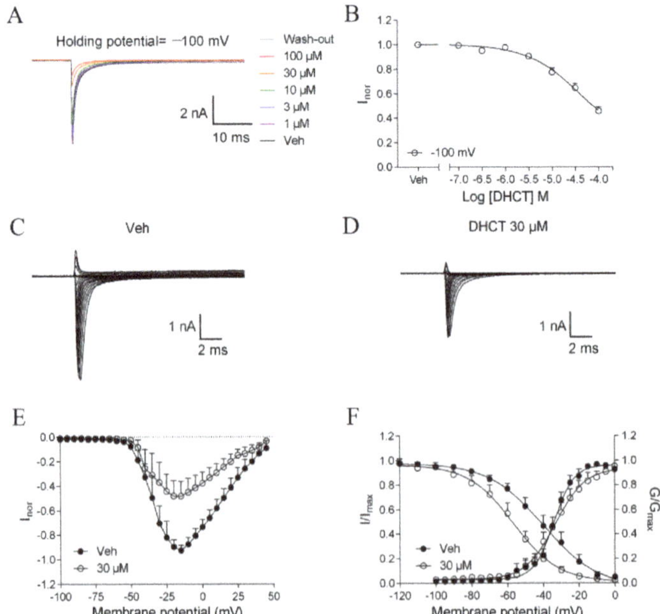

Figure 3. DHCT suppressed tetrodotoxin-sensitive (TTX-S) Na^+ currents in medium- and large-diameter -diameter DRG neurons. (**A**) Representative traces of DHCT suppressing TTX-S Na^+ currents in medium- and large-diameter DRG neurons. (**B**) Concentration–response curve of DHCT suppressed TTX-S Na^+ currents. Sodium currents were evoked by a 50-ms depolarization pulse to 0 mV from a holding potential of −100 mV. (**C**,**D**) Representative TTX-S Na^+ currents evoked by different depolarizing voltages from −90 to +50 mV in a 5-mV step in the absence (C) and presence (D) of DHCT (30 μM). (**E**) Current–voltage (I-V) relationships of TTX-S VGSCs in the presence or absence of DHCT. (**F**) Steady-state activation and inactivation curves of TTX-S Na^+ currents in the absence and presence of DHCT (30 μM). Each data point represents mean ± SEM (n = 5).

In small-diameter DRG neurons, in the presence of 300 nM TTX, depolarization induced slow activated and slow inactivated Na^+ currents (Figure 4A). Bath application of DHCT (30 μM) suppressed TTX-R Na^+ currents with an IC_{50} value of 7.42 μM (4.91–11.22 μM, 95% CI) (Figure 4A,B). DHCT (30 μM) suppressed the TTX-R Na^+ peak currents evoked by different depolarization potentials (Figure 4C–E). DHCT shifted the $V_{1/2}$ for activation of TTX-R VGSCs from −57.32 mV to −45.08 mV and $V_{1/2}$ for inactivation from −21.22 mV to −34.08 mV (Figure 4F).

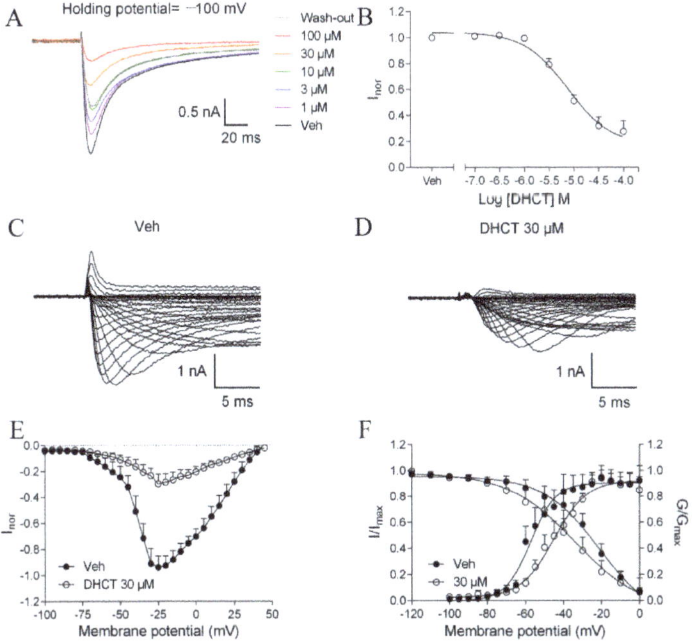

Figure 4. DHCT suppressed tetrodotoxin-resistant (TTX-R) Na^+ currents in small-diameter DRG neurons in the presence of 300 nM TTX. (**A**) Representative traces of DHCT suppressing TTX-R Na^+ currents. (**B**) Concentration–response curve of DHCT suppressed TTX-R Na^+ currents. Sodium currents were evoked by a 50-ms depolarization pulse to 0 mV from a holding potential of −100 mV. (**C**,**D**) Representative TTX-R Na^+ currents evoked by different depolarizing voltages from −90 to +50 mV in a 5-mV step in the absence (**C**) and presence (**D**) of DHCT (30 μM). (**E**) Current–voltage (I-V) relationships of TTX-R VGSCs in the presence or absence of 30 μM DHCT. (**F**) Steady-state activation and inactivation curves of TTX-R Na^+ currents in the absence and presence of DHCT (30 μM). Each data point depicts mean ± SEM (n = 5).

2.4. DHCT Preferred to Interact with Inactivated State of VGSCs

Many VGSC gating modifiers affect VGSC gating by preferentially interacting with an inactivated state of VGSCs [31–33]. Given the robust shift on $V_{1/2}$ for inactivation of DHCT on both TTX-S and TTX-R VGSCs, we next investigated whether DHCT preferentially interacted with inactivated state of VGSCs. When Na^+ currents were elicited by a depolarizing voltage step to 0 mV from −120 mV (950 ms), at which channels were predominantly in a resting state, DHCT concentration-dependently suppressed the TTX-S Na^+ currents in medium- and large-diameter DRG neurons with an IC_{50} value of 57.25 μM (27.31–120.00 μM, 95% CI) (Figure 5A–C). When a test pulse depolarized to 0 mV from a 950-ms conditioning voltage of −60 mV, at which, approximately 60% of TTX-S sodium channels were in an inactivated state, DHCT suppressed the TTX-S Na^+ currents with an IC_{50} value of 12.36 μM (8.91–17.15 μM, 95% CI) (Figure 5B,C). Similarly, DHCT also suppressed the TTX-R Na^+ currents in small-diameter DRG neurons in the presence of 300 nM TTX. The IC_{50} values were 19.97 μM (8.69–45.88 μM, 95% CI) and 4.87 μM (3.42–6.94 μM, 95% CI) when the holding potentials were −120 mV and −60 mV (~20% channels are in activated state), respectively (Figure 5D–F). The potency differences between distinct holding potentials suggested that DHCT preferentially bound to an inactivated state of VGSCs. The inactivated state dependence of DHCT was consistent with the hyperpolarized shifts of steady-state inactivation in both TTX-S and TTX-R VGSCs.

DHCT appeared to prolong the repriming time from an inactivated state when DRG neurons were held at a holding potential of 0 mV for 100 ms or 10 s, followed by a variable recovery interval at −120 mV before eliciting a current with a 20-ms pulse to 0 mV. The normalized current was plotted versus the recovery duration to determine the time constant (τ) of recovery from an inactivated state using first-, or second-order exponential growth equations [34]. When the precondition holding potential was set to be 0 mV for 100 ms, the recovery of TTX-S Na^+ currents displayed a monophasic profile and was best fit by a first-order exponential growth equation with a τ value of 1.43 ms (1.19–1.71 ms, 95% CI), which was consistent with a fast inactivation recovery. In the presence of DHCT (30 μM), recovery of Na^+ currents displayed a biphasic response and was best fit by a second-order exponential growth equation with τ values of 2.11 ms (1.15–5.11 ms, 95% CI) and 410.20 ms (23.88–7046.93 ms, 95% CI), respectively (Figure 6A–C). Similarly, in small-diameter DRG neurons in the presence of 300 nM TTX, bath application of DHCT also shifted a monophasic profile with a τ value of 3.80 ms (3.42–4.23 ms, 95% CI) to a biphasic response with τ values of 5.03 ms (3.16–8.02 ms, 95% CI) and 237.68 ms (89.54–632.41 ms, 95% CI), respectively (Figure 6D–F) suggesting that DHCT exposure transited both TTX-S and TTX-R Na^+ currents from a fast-inactivated state to a slow-inactivated state. When the precondition holding potential was set to be 0 mV for 10 s, recovery of TTX-S Na^+ currents displayed monophasic profiles in the absence and presence of DHCT with τ values of 205.9 ms (137.8–307.7 ms, 95% CI, n = 6) and 358.8 ms (254.1–506.7 ms, 95% CI, n = 6), respectively (Figure 6G–I). Similarly, recovery of TTX-R Na^+ currents in the absence and presence of DHCT both displayed monophasic profiles with τ values of 1524.05 ms (1106.62–2094.11 ms, 95% CI, n = 6) and 2218.20 ms (1774.19–2779.71 ms, 95% CI, n = 6), respectively (Figure 6J–L). The prolonged recovery time constants in the presence of DHCT suggested that DHCT was able to stabilize the inactivation state of VGSCs.

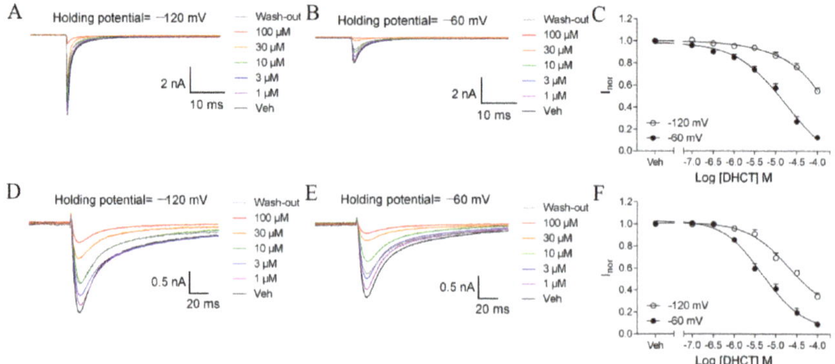

Figure 5. DHCT preferentially interacted with an inactivated state of TTX-S and TTX-R VGSCs in DRG neurons. (**A**) Representative TTX-S Na^+ current traces in the absence and presence of different concentrations of DHCT evoked by a depolarization potential to 0 mV from a holding potential of −120 mV. (**B**) Representative TTX-S Na^+ current traces in the absence and presence of different concentrations of DHCT evoked by a depolarization potential to 0 mV from a holding potential of −60 mV. (**C**) Concentration–response curves of DHCT suppressing TTX-S Na^+ currents at the holding potentials of −120 mV and −60 mV, respectively. (**D**) Representative TTX-R Na^+ current traces in the absence and presence of different concentrations of DHCT evoked by a depolarization potential to 0 mV from a holding potential of −120 mV. (**E**) Representative TTX-R Na^+ current traces in the absence

and presence of different concentrations of DHCT evoked by a depolarization potential to 0 mV from a holding potential of −60 mV. (**F**) Concentration–response curves of DHCT suppressing TTX-R Na^+ currents in DRG neurons at the holding potentials of −120 mV and −60 mV, respectively. Each data point represents mean ± SEM (n = 5–9).

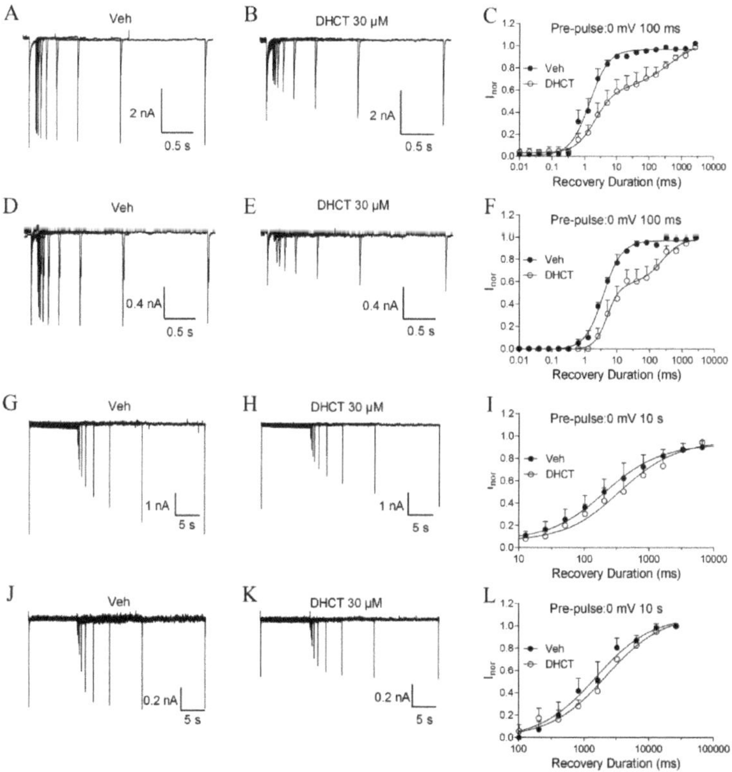

Figure 6. DHCT exposure transited both TTX-S and TTX-R Na^+ currents from a fast-inactivated state to a slow-inactivated state. (**A**,**B**) Representative traces of TTX-S Na^+ currents when the holding potential was set at 0 mV for 100 ms, followed by a recovery interval of variable duration to −120 mV before testing for an active current with a 20-ms pulse to 0 mV in the absence (**A**) and presence (**B**) of DHCT (30 μM). (**C**) Recovery curves of TTX-S Na^+ currents from fast-inactivation in the absence and presence of DHCT. (**D**,**E**) Representative traces of TTX-R Na^+ currents when the holding potential was set at 0 mV for 100 ms, followed by a recovery interval of variable duration to −120 mV before testing for an active current with a 20-ms pulse to 0 mV in the absence (**D**) and presence (**E**) of DHCT (30 μM). (**F**) Recovery curves of TTX-R Na^+ currents from fast-inactivation in the absence and presence of DHCT. (**G**,**H**) Representative traces of TTX-S Na^+ currents when the cells were pre-conditioned at 0 mV for 10 s, followed by a recovery interval of variable duration to −120 mV before testing for an active current with a 20-ms pulse to 0 mV in the absence (**G**) and presence (**H**) of DHCT (30 μM). (**I**) Recovery curves of TTX-S Na^+ currents from slow-inactivation in the absence and presence of DHCT. (**J**,**K**) Representative traces of TTX-R Na^+ currents when the cells were pre-conditioned at 0 mV for 10 s, followed by a recovery interval of variable duration to −120 mV before testing for an active current with a 20-ms pulse to 0 mV in the absence (**J**) and presence (**K**) of DHCT (30 μM). (**L**) Recovery curves of TTX-R Na^+ currents from slow-inactivation in the absence and presence of DHCT. Each data point represents mean ± SEM (n = 6–8).

3. Discussion

Although the mechanisms of neuropathic pain are complex, sensitization of nociceptors has been proposed to be the major cause for abnormal nociception [3]. VGSCs are responsible for the rising phase of the APs [8,35]. Sensitization of VGSC through various mechanisms has been demonstrated in the nociceptors from neuropathic rodent models and proposed to contribute to the hyper-excitability of injured neurons [4–7]. Discovery of VGSC gating modifiers, especially those displaying selectivity in VGSC subtypes ($Na_V1.7$ and $Na_V1.8$) is of high interest for pain therapy [36,37].

In this study, we investigated the analgesic effect of DHCT, a representative of β-carboline alkaloids found in the *P. quassiodes*, which was used clinically to treat various diseases including inflammation and pain as a traditional Chinese medicine. We demonstrated that DHCT displayed comparable efficacy with morphine in attenuation of mechanical allodynia in a sciatic nerve partial ligation neuropathic pain model. We further demonstrated that DHCT suppressed the neuronal excitability and inhibited the VGSC (both TTX-R and TTX-S) currents in DRG neurons. Considered together, these data demonstrated that the analgesic effect of DHCT was possibly through inhibition of VGSCs, one of the major mechanisms underlining neuropathic pain onset and progression [38]. It should be noted that CCI-injured DRG neurons displayed altered expression levels of sodium channel subtypes, such as $Na_V1.3$, $Na_V1.8$, and $Na_V1.9$ [39,40]. Whether DHCT is also capable of suppressing AP firing in CCI-injured neurons needs further examination. Recently, DHCT has been demonstrated to be a Jasus kinase (JAK) inhibitor [24]. Many studies have revealed that blockade of JAK/Signal transducers and activators of transcription (STAT) signaling pathways decreased microglia activation [41,42], reduced the release of cytokines [43–45], and attenuated both mechanical allodynia and thermal hyperalgesia in a spinal nerve ligation neuropathic pain model [46]. Therefore, the analgesic effect of DHCT was possibly through dual inhibitions of VGSCs and JAK-mediated microglia activation, two major mechanisms underlining in neuropathic pain onset and progression.

In DRG neurons, DHCT suppressed AP generation, suggesting that DHCT was able to inhibit the neuronal excitability. Various β-carboline alkaloids produced either a sedative, tremorgenic, anxiogenic, or convulsant effect by binding to benzodiazepine (BZ) receptors acting as full, partial agonists, antagonists, or inverse agonists [19,22,26–28]. Whether DHCT interacted with BZ receptors and contributed to the neuronal excitability is currently unknown. In the current study, we observed an inhibitory effect of DHCT on Na^+ currents in cultured DRG neurons. DHCT suppressed TTX-S and TTX-R Na^+ currents at −60 mV with an IC_{50} value of 12.36 μM and 4.87 μM respectively, in DRG neurons, suggesting that TTX-R VGSCs were more sensitive to DHCT exposure. DHCT at 10 μM completely suppressed the APs evoked by a 30-pA or 200-pA current injection. It should be mentioned that the resting membrane potential of DRG neuron was approximately −50 mV [47]. Considering that DHCT preferentially bound to inactivated state of VGSCs, the potency of DHCT to suppress VGSCs Na^+ currents in DRG neurons was therefore comparable to that on inhibition of AP firing. Therefore, it was likely that suppression of VGSCs was responsible for DHCT action in neuronal excitability.

Inhibition of both TTX-S and TTX-R Na^+ currents by DHCT in DRG neurons was state-dependent. The IC_{50} value for DHCT suppression of VGSC activity was around 5-fold (TTX-S) and 4-fold (TTX-R) more potent in an inactivated state (−60 mV holding potential) than that of a resting state (−120 mV holding potential). This state-dependence of DHCT action was further confirmed by prolonged repriming time of Na^+ current recovery. DHCT produced a depolarization shift of $V_{1/2}$ on inactivation in TTX-S VGSCs. In addition to the depolarized shifts of inactivation voltages, DHCT also robustly shifted voltage-dependent activation to hyperpolarized potentials in TTX-R VGSCs. The electrophysiological property of DHCT appeared to be distinct to the VGSC pore blocker, TTX [48]. Most of VGSC inhibitors such as lidocaine bound to the local anesthetic sites, which were comprised of amino acids of Ile/Phe/Met (IFM) (domain III–IV (DIII–DIV linker)), and the residues located at IS6, IIS6, and IVS6 segments [49,50]. Although both lidocaine and DHCT produced a hyperpolarized shift on voltage-dependent inactivation, lidocaine had no effect on voltage-dependent activation [51,52]. In addition, the kinetics of DHCT prolonged repriming time from fast-inactivated state of VGSCs was also distinct to that of lidocaine [34]. Considered together, these data suggested that DHCT binding sites on VGSCs were distinct to TTX, and were not identical to that of local anesthetics. In the current study, we are not able to clarify why DHCT distinctly affected the gating kinetics between TTX-S and TTX-R VGSCs. Gating modifiers produce conformational changes and the coupling between the different states of VGSCs. In general, activators binding to the voltage sensor (VS) of DII affect activation, while activators binding to the VS of DIV generally affect inactivation [53]. Whether DHCT also bound to VSs of both DII and DIV of TTX-R VGSCs requires further exploration.

In summary, we demonstrated that DHCT suppressed the mechanic allodia in a rat model of neuropathic pain. We also demonstrated that DHCT preferentially binds to the inactivated state of VGSCs and therefore suppresses the VGSC currents and neuronal excitability. Our data provided novel information on the mode of actions for DHCT, and to an extent, for β-carboline alkaloids on neuronal excitability.

4. Materials and Methods

4.1. Materials

DHCT was generously provided by Professor Dequan Yu (Chinese Academy of Medical Sciences and Peking Union Medical College, Beijing, China) and purified as described previously [54]. The purity was determined to be over 95% using HPLC. Penicillin, streptomycin, trypsin, soybean trypsin inhibitor, L-glutamine, fetal bovine serum (FBS), Dulbecco minimum essential medium (DMEM), and Neurobasal medium were obtained from Life Technology (Grand Island, NY, USA). Poly-D-lysine (PDL), deoxyribonuclease I, TTX, nerve growth factor (NGF), and all inorganic chemicals were purchased from Sigma-Aldrich (St. Louis, MO, USA).

4.2. Animal Care

Animal protocols were approved by Institutional Animal Care and Use Committee of China Pharmaceutical University (12100000466006834N, SYXK 2016-0011, 27 January 2016). Efforts were made to minimize animal suffering and reduce the number of experimental animals. Male Sprague-Dawley (SD) rats (200–220 g) were purchased from Qing-Long-Shan Laboratory Animal Center (Nanjing, Jiangsu, China). All the animals were maintained in a temperature-controlled (23 ± 2 °C) vivarium at a 12-h light/dark cycle provided with food and water ad libitum.

4.3. Acutely Dissected Rat Dorsal Root Ganglion Neurons

Acutely dissociated rat dorsal root ganglion neurons (DRG) were obtained from native male SD rats (200–220 g) as described previously [47,55]. In brief, the dissected DRG tissue was digested with 0.05% trypsin and 0.03% collagenase A for 25 min at 37 °C under gentle shaking. Cells were resuspended in neurobasal medium supplemented with 2% NS21 [56], 1 mM L-glutamine, 50 ng/mL NGF, 1% HEPES, and 5% FBS, and plated onto poly-D-lysine (PDL) (50 μg/mL) pre-coated 35-mm Petri dishes (Corning, NY, USA) at a density of 1000 cells/dish for patch clamp experiments after 6 h culture in vitro.

4.4. Patch Clamp Recording in DRG Neurons

Sodium currents in rat DRG neurons were recorded using a whole-cell patch clamp with an EPC-10 amplifier (HEKA Electronics, Germany) as described previously [57]. It has been reported that medium- (30–50 μm) to large-diameter (>50 μm) DRG neurons mainly express TTX-S VGSCs, while small-diameter (<30 μM) DRG neurons express both TTX-S and TTX-R VGSCs [30,58]. We therefore recorded TTX-S Na^+ currents in medium- and large-diameter DRG neurons. Small-diameter DRG neurons were chosen to record TTX-R Na^+ currents with bath application of 300 nM TTX. Current patch was recorded in small- and medium-diameter DRG neurons.

For voltage-clamp recordings, the extracellular solution was (in mM): NaCl 140, KCl 3, $CaCl_2$ 1, $MgCl_2$ 1, and HEPES 10 (pH adjusted to 7.4 with NaOH). The intracellular solution was (in mM): CsF 140, EGTA 1.1, NaCl 10, and HEPES 10 (pH adjusted to 7.3 with CsOH). Pipettes (1–3 MΩ) were pulled from 1.5-mm capillary tubing and filled with intracellular solution. For current clamp recordings, the intracellular solution contained (in mM): KCl 140, $MgCl_2$ 5, $CaCl_2$ 2.5, EGTA 5, ATP 4, GTP 0.3, and HEPES 10 (pH adjusted to 7.3 with KOH). The external solution contained (in mM): NaCl 140, $MgCl_2$ 1, KCl 5, $CaCl_2$ 2, glucose 10, and HEPES 10 (pH adjusted to 7.3 with NaOH). Voltage errors were minimized using 80% series resistance compensation, and the capacitance artifact was cancelled using computer-controlled circuitry of a patch clamp amplifier. Patchmaster (HEKA Electronics, Lambrecht/Pfalz, Germany, 2016) and OriginPro (Version 8, OriginLab, Northampton, MA, USA, 2018) were used to collect and analyze the experimental data, respectively. All data were presented as mean ± SEM, and each independent experimental cell was counted as an n number.

Concentration–response curves were fitted using a nonlinear logistic equation using prism software (Version 5.0, GraphPad Software Inc., San Diego, CA, USA, 2010). To study the effect of DHCT (30 μM) on current–voltage (I-V) relationships, the Na^+ currents were triggered by depolarized pulses from −100 mV to +50 mV in a 5-mV step in the absence and presence of DHCT. The peak current recorded after each voltage step was normalized into conductance (G) according to the equation: $I = G (V - V_{rev})$, where V_{rev} is the reversal potential of the sodium current. A steady-state activation curve was fitted using the Boltzmann equation: $G/G_{max}=1/(1 + exp(V_{1/2} - V)/k)$, where V is the membrane potential of the conditioning step, $V_{1/2}$ is the membrane potential at half-maximal activation, and *k* is the slope factor. To study voltage dependent steady-state inactivation, Na^+ currents were evoked by a 35-ms depolarizing pulse of 0 mV from −100 mV holding potential and pre-pulse ranging from −120 mV to 0 mV in a 10-mV step (100 ms). A steady-state inactivation curve was fitted using the Boltzmann equation: $I/I_{max}=1/(1 + exp(V_{1/2} - V)/k)$, where V is the membrane potential of the conditioning step, $V_{1/2}$ is the membrane potential at half-maximal inactivation, and *k* is the slope factor. To quantify the affinity toward resting state or inactivating state, cells were held at −120 mV for 950 ms or −60 mV for 950 ms, then depolarized to 0 mV for 50 ms to induce Na^+ current. To study the recovery time constant from fast and slow inactivation of VGSCs, DRG neurons were held at a holding potential of 0 mV for 100 ms or 10 s, followed by a variable recovery interval to −120 mV before testing for active current with a 20-ms pulse to 0 mV. The normalized amount of current measured was plotted versus the recovery duration to determine the τ value of recovery from inactivated state using first-, or second-order exponential growth equations as follows: $y = y_0 + A_1e^{x/t1}$ and $y = y_0 + A_1e^{x/t1} + A_2e^{x/t2}$, respectively. For both equations y_0 = the initial value at time 0, A_1 = weight factor (of total recovered) for fraction recovered with time constant t_1, A_2 = weight factor (of total recovered) for fraction recovered at t_2, x = the growth constant, t_1 = time constant for recovery of fast inactivated channels, and t_2 = time constant for recovery of potentially slow inactivated channels [34].

4.5. Chronic Constrictive Injury (CCI) Neuropathic Pain Model

Male rats (SD, 200–220 g) were randomly divided into 5 groups with 9 rats in each group. After anesthetized with 1% sodium pentobarbital (40 mg/kg), surgery was applied and sciatic nerves were exposed. Approximately one-fifth to one-fourth of the nerve was tied off. The sham group experienced surgical nerve exposing, but not ligating. After 14 days, animals were tested for development of mechanical allodynia by being placed on a wire mesh (20 × 15 × 20 cm, L × W × H) platform. After acclimatization for 15 min, a mechanical baseline was determined by applying a series of calibrated Von Frey fibers perpendicularly to the plantar surface of the hind paw of each rat. Positive responses included an abrupt withdrawal of the hind paw from the stimulus [59]. After a positive response was noted, a weaker fiber was applied. This was repeated until a 50% threshold for withdrawal could be observed.

4.6. Data Analysis

All data points were presented as mean ± SEM. The concentration–response curves were generated using GraphPad Prism 5.0 software using a non-linear logistic equation. Statistical significance between groups was calculated using T-test (two groups) or one-way ANOVA analysis when multiple groups were analyzed. p values below 0.05 were considered to be statistically significant.

Supplementary Materials: The following are available online at http://www.mdpi.com/2072-6651/11/4/229/s1, Figure S1: TTX suppressed TTX-S Na^+ currents in medium- and large-diameter DRG neurons.

Author Contributions: Conceptualization, X.Z. and Z.C.; Data curation, F.Z., S.W. and S.L.; Investigation, F.Z., Q.T. and J.X.; Project administration, X.Z. and Z.C.; Writing–original draft, F.Z.; Writing–review & editing, X.Z. and Z.C.

Funding: This work was supported by the National Natural Science Foundation of China (21777192, 81473539, 81603389), China Postdoctoral Science Foundation (2018M630645, 2018M642371), Jiangsu Provincial Natural Science Foundation (BK20160754; BK20160764), National Key R&D program of China (2018YFF0215200), National Science and Technology Major Projects for "Major New Drugs Innovation and Development" (2018ZX09101003-004-002), the National Key Laboratory of Natural Medicines, China Pharmaceutical University (SKLNMZZCX201825) and the "Double First-Class" project by China Pharmaceutical University (CPU2018GY18).

Acknowledgments: DHCT was generously provided by Ruoyun Chen and Dequan Yu (Chinese Academy of Medical Sciences and Peking Union Medical College).

Conflicts of Interest: Authors declare no conflicts of interest.

References

1. Mulvey, M.R.; Bennett, M.I.; Liwowsky, I.; Freynhagen, R. The role of screening tools in diagnosing neuropathic pain. *Pain Manag.* **2014**, *4*, 233–243. [CrossRef] [PubMed]
2. Melnikova, I. Pain market. *Dressnature Rev. Drug Discov.* **2010**, *9*, 589–590. [CrossRef] [PubMed]
3. Willis, W.D. Role of neurotransmitters in sensitization of pain responses. *Ann. N. Y. Acad. Sci.* **2001**, *933*, 142–156. [CrossRef] [PubMed]
4. Minett, M.S.; Nassar, M.A.; Clark, A.K.; Passmore, G.; Dickenson, A.H.; Wang, F.; Malcangio, M.; Wood, J.N. Distinct Nav1.7-dependent pain sensations require different sets of sensory and sympathetic neurons. *Nat. Commun.* **2012**, *3*, 85–100. [CrossRef] [PubMed]
5. Jay, G.W.; Barkin, R.L. Neuropathic pain: Etiology, pathophysiology, mechanisms, and evaluations. *Dis. Month* **2014**, *60*, 6–47. [CrossRef] [PubMed]
6. Amir, R.; Argoff, C.E.; Bennett, G.J.; Cummins, T.R.; Durieux, M.E.; Gerner, P.; Gold, M.S.; Porreca, F.; Strichartz, G.R. The role of sodium channels in chronic inflammatory and neuropathic pain. *J. Pain* **2006**, *7*, S1–S29. [CrossRef] [PubMed]
7. Brederson, J.D.; Kym, P.R.; Szallasi, A. Targeting TRP channels for pain relief. *Eur. J. Pharmacol.* **2013**, *716*, 61–76. [CrossRef]
8. Bean, B.P. The action potential in mammalian central neurons. *Nat. Rev. Neurosci.* **2007**, *8*, 451–465. [CrossRef]
9. Heyer, E.J.; Macdonald, R.L. Calcium- and sodium-dependent action potentials of mouse spinal cord and dorsal root ganglion neurons in cell culture. *J. Neurophysiol.* **1982**, *47*, 641–655. [CrossRef]
10. Fletcher, A. Action potential: Generation and propagation. *Anaesth. Intensive Care Med.* **2008**, *9*, 251–255. [CrossRef]
11. Thakor, K.; Lin, A.; Matsuka, Y.; Meyer, M.; Ruangsri, S.; Nishimura, I.; Spigelman, I. Nav1.8, neuropathic pain. *Mol. Pain* **2009**, *5*, 14. [PubMed]
12. Hong, C.J.; Hsueh, Y.P. Cask associates with glutamate receptor interacting protein and signaling molecules. *Biochem. Biophys. Res. Commun.* **2006**, *351*, 771–776. [CrossRef] [PubMed]
13. Hains, B.C.; Klein, J.P.; Saab, C.Y.; Craner, M.J.; Black, J.A.; Waxman, S.G. Upregulation of sodium channel Nav1.3 and functional involvement in neuronal hyperexcitability associated with central neuropathic pain after spinal cord injury. *J. Neurosci. Off. J. Soc. Neurosci.* **2003**, *23*, 8881–8892. [CrossRef]
14. Zhao, F.; Gao, Z.; Jiao, W.; Chen, L.; Chen, L.; Yao, X. In vitro anti-inflammatory effects of beta-carboline alkaloids, isolated from *Picrasma quassioides*, through inhibition of the inos pathway. *Planta Med.* **2012**, *78*, 1906–1911. [CrossRef] [PubMed]

15. Zhao, F.; Chen, L.; Bi, C.; Zhang, M.; Jiao, W.; Yao, X. In vitro anti-inflammatory effect of picrasmalignan a by the inhibition of iNOS and COX-2 expression in LPS-activated macrophage RAW 264.7 cells. *Mol. Med. Rep.* **2013**, *8*, 1575–1579. [CrossRef] [PubMed]
16. Jiao, W.H.; Gao, H.; Li, C.Y.; Zhou, G.X.; Kitanaka, S.; Ohmura, A.; Yao, X.S. β-carboline alkaloids from the stems of *Picrasma quassioides*. *Magn. Reson. Chem.* **2010**, *48*, 490–495. [PubMed]
17. Xu, J.; Xiao, D.; Lin, Q.H.; He, J.F.; Liu, W.Y.; Xie, N.; Feng, F.; Qu, W. Cytotoxic tirucallane and apotirucallane triterpenoids from the stems of *Picrasma quassioides*. *J. Natl. Prod.* **2016**, *79*, 1899–1910. [CrossRef] [PubMed]
18. Hikino, H.; Ohta, T.; Takemoto, T. Stereostructure of picrasin d and e, simaroubolides of *Picrasma quassioides*. *Chem. Pharm. Bull.* **2008**, *19*, 212–213. [CrossRef]
19. Jiao, W.H.; Gao, H.; Li, C.Y.; Zhao, F.; Jiang, R.W.; Wang, Y.; Zhou, G.X.; Yao, X.S. Quassidines A–D, Bis-β-carboline Alkaloids from the Stems of *Picrasma quassioides*. *J. Natl. Prod.* **2010**, *73*, 167–171. [CrossRef]
20. Yin, Y.; Heo, S.I.; Roh, K.S.; Wang, M.H. Biological activities of fractions from methanolic extract of *Picrasma quassioides*. *J. Plant Biol.* **2009**, *52*, 325–331. [CrossRef]
21. Jiao, W.H.; Gao, H.; Zhao, F.; Lin, H.W.; Pan, Y.M.; Zhou, G.X.; Yao, X.S. Anti-inflammatory alkaloids from the stems of *Picrasma quassioides* bennet. *Chem. Pharm. Bull.* **2011**, *59*, 359–364. [CrossRef]
22. Haghparast, A.; Farzin, D.; Ordikhani-Seyedlar, M.; Motaman, S.; Kermani, M.; Azizi, P. Effects of apomorphine and β-carbolines on firing rate of neurons in the ventral pallidum in the rats. *Behav. Brain Res.* **2012**, *227*, 109–115. [CrossRef]
23. Luo, S.R.; Guo, R.; Yang, J.S. Studies on the determination of alkaloids in *Prcrasma quassiodies* (d.Don) benn. *Acta Pharm. Sin.* **1988**, *23*, 4.
24. Zhang, J.; Zhu, N.; Du, Y.; Bai, Q.; Chen, X.; Nan, J.; Qin, X.; Zhang, X.; Hou, J.; Wang, Q. Dehydrocrenatidine is a novel janus kinase inhibitor. *Mol. Pharmacol.* **2015**, *87*, 572. [CrossRef]
25. Hevers, W.; Lüddens, H. The diversity of GABAA receptors. *Mol. Pharmacol.* **1998**, *18*, 35–86. [CrossRef]
26. Hollinshead, S.P.; Trudell, M.L.; Skolnick, P.; Cook, J.M. Structural requirements for agonist actions at the benzodiazepine receptor: Studies with analogs of 6-(benzyloxy)-4-(methoxymethyl)-.Beta.-carboline-3-carboxylic acid ethyl ester. *J. Med. Chem.* **1990**, *33*, 1062–1069. [CrossRef]
27. Grella, B.; Teitler, M.; Smith, C.; Herrickdavis, K.; Glennon, R.A. Binding of β-carbolines at 5-HT(2) serotonin receptors. *Bioorganic Med. Chem. Lett.* **2003**, *13*, 4421–4425. [CrossRef]
28. Glennon, R.A.; Grella, B.; Tyacke, R.J.; Lau, A.; Westaway, J.; Hudson, A.L. Binding of β-carbolines at imidazoline I2 receptors: A structure–affinity investigation. *Bioorganic Med. Chem. Lett.* **2004**, *14*, 999–1002. [CrossRef]
29. Abdel-Fattah, A.F.; Matsumoto, K.; Gammaz, H.A.; Watanabe, H. Hypothermic effect of harmala alkaloid in rats: Involvement of serotonergic mechanism. *Pharmacol. Biochem. Behav.* **1995**, *52*, 421–426. [CrossRef]
30. Ho, C.; O'Leary, M.E. Single-cell analysis of sodium channel expression in dorsal root ganglion neurons. *Mol. Cell. Neurosci.* **2011**, *46*, 159. [CrossRef]
31. O'Reilly, A.O.; Eberhardt, E.; Weidner, C.; Alzheimer, C.; Wallace, B.A.; Lampert, A. Bisphenol a binds to the local anesthetic receptor site to block the human cardiac sodium channel. *PLoS ONE* **2012**, *7*, e41667. [CrossRef]
32. Gaudioso, C.; Hao, J.; Martin-Eauclaire, M.F.; Gabriac, M.; Delmas, P. Menthol pain relief through cumulative inactivation of voltage-gated sodium channels. *Pain* **2012**, *153*, 473–484. [CrossRef]
33. Errington, A.C.; Stöhr, T.; Heers, C.; Lees, G. The investigational anticonvulsant lacosamide selectively enhances slow inactivation of voltage-gated sodium channels. *Mol. Pharmacol.* **2008**, *73*, 157–169. [CrossRef]
34. Sheets, P.L.; Jarecki, B.W.; Cummins, T.R. Lidocaine reduces the transition to slow inactivation in Nav1.7 voltage-gated sodium channels. *Br. J. Pharmacol.* **2011**, *164*, 719–730. [CrossRef]
35. Hildebrand, M.E.; Mezeyova, J.; Smith, P.L.; Salter, M.W.; Tringham, E.; Snutch, T.P. Identification of sodium channel isoforms that mediate action potential firing in lamina I/II spinal cord neurons. *Mol. Pain* **2011**, *7*, 1–13. [CrossRef]
36. Zhao, F.; Li, X.; Jin, L.; Zhang, F.; Inoue, M.; Yu, B.; Cao, Z. Development of a rapid throughput assay for identification of hNav1.7 antagonist using unique efficacious sodium channel agonist, antillatoxin. *Mar. Drugs* **2016**, *14*, 36. [CrossRef]
37. Wu, Y.; Ma, H.; Zhang, F.; Zhang, C.L.; Zou, X.; Cao, Z. Selective voltage-gated sodium channel peptide toxins from animal venom: Pharmacological probes and analgesic drug development. *ACS Chem. Neurosci.* **2018**, *9*, 187–197. [CrossRef]

38. Dibhajj, S.D.; Yang, Y.; Black, J.A.; Waxman, S.G. The Nav1.7 sodium channel: From molecule to man. *Nat. Rev. Neurosci.* **2012**, *14*, 49–62. [CrossRef]
39. Decosterd, I.; Ji, R.R.; Abdi, S.; Tate, S.; Woolf, C.J. The pattern of expression of the voltage-gated sodium channels Na(v)1.8 and Na(v)1.9 does not change in uninjured primary sensory neurons in experimental neuropathic pain models. *Pain* **2015**, *96*, 269–277. [CrossRef]
40. Wang, J.; Ou, S.W.; Wang, Y.J. Distribution and function of voltage-gated sodium channels in the nervous system. *Channels* **2017**, *11*, 534. [CrossRef]
41. Kim, H.Y.; Park, E.J.; Joe, E.H.; Jou, I. Curcumin suppresses janus kinase-STAT inflammatory signaling through activation of Src homology 2 domain-containing tyrosine phosphatase 2 in brain microglia. *J. Immunol.* **2003**, *171*, 6072. [CrossRef]
42. Natarajan, C.; Sriram, S.; Muthian, G.; Bright, J.J. Signaling through JAK2-STAT5 pathway is essential for IL-3. *Glia* **2004**, *45*, 188–196.
43. Yin, L.; Dai, Q.; Jiang, P.; Lin, Z.; Dai, H.; Yao, Z.; Hua, L.; Ma, X.; Qu, L.; Jiang, J. Manganese exposure facilitates microglial JAK2-STAT3 signaling and consequent secretion of TNF-a and IL-1β to promote neuronal death. *Neurotoxicology* **2018**, *64*, 195–203. [CrossRef]
44. Chi, Z.; Jie, Z.; Jing, Z.; Li, H.; Zhao, Z.; Liao, Y.; Wang, X.; Su, J.; Sang, S.; Yuan, X. Neuroprotective and anti-apoptotic effects of valproic acid on adult rat cerebral cortex through erk and akt signaling pathway at acute phase of traumatic brain injury. *Brain Res.* **2014**, *1555*, 1.
45. Aedín, M.; James, B.; Marina, L. Lps-induced release of il-6 from glia modulates production of IL-1β in a JAK2-dependent manner. *J. Neuroinflammation* **2012**, *9*, 126.
46. Dominguez, E.; Rivat, C.; Pommier, B.; Mauborgne, A.; Pohl, M. JAK/STAT3 pathway is activated in spinal cord microglia after peripheral nerve injury and contributes to neuropathic pain development in rat. *J. Neurochem.* **2008**, *107*, 50–60. [CrossRef]
47. Cummins, T.R.; Rush, A.M.; Estacion, M.; Dibhajj, S.D.; Waxman, S.G. Voltage-clamp and current-clamp recordings from mammalian drg neurons. *Nat. Protocols* **2009**, *4*, 1103–1112. [CrossRef]
48. Chen, R.; Chung, S.H. Mechanism of tetrodotoxin block and resistance in sodium channels. *Biochem. Biophys. Res. Commun.* **2014**, *446*, 370–374. [CrossRef]
49. Ulbricht, W. Sodium channel inactivation: Molecular determinants and modulation. *Physiol. Rev.* **2005**, *85*, 1271. [CrossRef]
50. Catterall, W.A.; Swanson, T.M. Structural basis for pharmacology of voltage-gated sodium and calcium channels. *Mol. Pharmacol.* **2015**, *88*, 141. [CrossRef]
51. Wang, Y.; Mi, J.; Lu, K.; Lu, Y.; Wang, K. Comparison of gating properties and use-dependent block of Nav1.5 and Nav1.7 channels by anti-arrhythmics mexiletine and lidocaine. *PLoS ONE* **2015**, *10*, e0128653. [CrossRef]
52. Chevrier, P.; Vijayaragavan, K.; Chahine, M. Differential modulation of Nav1.7 and Nav1.8 peripheral nerve sodium channels by the local anesthetic lidocaine. *Br. J. Pharmacol.* **2004**, *142*, 576–584. [CrossRef]
53. Capes, D.L.; Arcisio-Miranda, M.; Jarecki, B.W.; French, R.J.; Chanda, B. Gating transitions in the selectivity filter region of a sodium channel are coupled to the domain IV voltage sensor. *Proc. Natl. Acad. Sci. USA* **2012**, *109*, 2648–2653. [CrossRef]
54. Yang, Y.; Luo, S.; Shen, X.; Li, Y. [chemical investigation of the alkaloids of Ku-Mu [*picrasma quassioides* (d. Don) benn.] (author's transl)]. *Yao Xue Xue Bao* **1979**, *14*, 167–177.
55. Dibhajj, S.D.; Choi, J.S.; Macala, L.J.; Tyrrell, L.; Black, J.A.; Cummins, T.R.; Waxman, S.G. Transfection of rat or mouse neurons by biolistics or electroporation. *Nat. Protocols* **2009**, *4*, 1118. [CrossRef]
56. Chen, Y.; Stevens, B.; Chang, J.; Milbrandt, J.; Barres, B.A.; Hell, J.W. NS21: Re-defined and modified supplement B27 for neuronal cultures. *J. Neurosci. Methods* **2008**, *171*, 239–247. [CrossRef]
57. He, Y.; Zou, X.; Li, X.; Chen, J.; Liang, J.; Fan, Z.; Yu, B.; Cao, Z. Activation of sodium channels by α-scorpion toxin, BmK NT1, produced neurotoxicity in cerebellar granule cells: An association with intracellular Ca^{2+} overloading. *Arch. Toxikol.* **2016**, *91*, 935–948. [CrossRef]

58. Roy, M.L.; Narahashi, T. Differential properties of tetrodotoxin-sensitive and tetrodotoxin-resistant sodium channels in rat dorsal root ganglion neurons. *J. Neurosci.* **1992**, *12*, 2104–2111. [CrossRef]
59. Austin, P.J.; Wu, A.; Moalem-Taylor, G. Chronic constriction of the sciatic nerve and pain hypersensitivity testing in rats. *J. Vis. Exp. Jove* **2012**, *61*, e3393. [CrossRef]

Article

Arecoline Promotes Migration of A549 Lung Cancer Cells through Activating the EGFR/Src/FAK Pathway

Chih-Hsiang Chang [1,†], Mei-Chih Chen [2,3,†], Te-Huan Chiu [1], Yu-Hsuan Li [1], Wan-Chen Yu [1], Wan-Ling Liao [1], Muhammet Oner [1], Chang-Tze Ricky Yu [4], Chun-Chi Wu [5], Tsung-Ying Yang [6], Chieh-Lin Jerry Teng [7], Kun-Yuan Chiu [8], Kun-Chien Chen [6], Hsin-Yi Wang [9], Chia-Herng Yue [10], Chih-Ho Lai [11], Jer-Tsong Hsieh [12] and Ho Lin [1,13,14,*]

1 Department of Life Sciences, National Chung Hsing University, Taichung 40227, Taiwan; milkpa@gmail.com (C.-H.C.); atheism1989@gmail.com (T.-H.C.); lihsuancc@gmail.com (Y.-H.L.); sl91123j@yahoo.com.tw (W.-C.Y.); dowdow4820@gmail.com (W.-L.L.); muhammet.oner053@gmail.com (M.O.)

2 Medical Center for Exosomes and Mitochondria Related Diseases, China Medical University Hospital, Taichung 40447, Taiwan; midyjack@gmail.com

3 Department of Nursing, Asia University, Taichung 41345, Taiwan

4 Department of Applied Chemistry, National Chi Nan University, Nantou 54561, Taiwan; ctyu@ncnu.edu.tw

5 Institute of Medicine, Chung Shan Medical University, Taichung 40201, Taiwan; daniel@csmu.edu.tw

6 Division of Chest Medicine, Taichung Veterans General Hospital, Taichung 40705, Taiwan; jonyin@gmail.com (T.-Y.Y.); ckjohn@mail.vghtc.gov.tw (K.-C.C.)

7 Division of Hematology/Medical Oncology, Taichung Veterans General Hospital, Taichung 40705, Taiwan; drteng@vghtc.gov.tw

8 Division of Urology, Taichung Veterans General Hospital, Taichung 40705, Taiwan; chiu37782002@yahoo.com

9 Department of Nuclear Medicine, Taichung Veterans General Hospital, Taichung 40705, Taiwan; hywang@vghtc.gov.tw

10 Department of Surgery, Tung's Taichung Metro Harbor Hospital, Taichung 435, Taiwan; ericchyue@gmail.com

11 Department of Microbiology and Immunology, Chang Gung Medical University, Taoyuan 33302, Taiwan; chlai@mail.cgu.edu.tw

12 Department of Urology, University of Texas Southwestern Medical Center, TX 75390, USA; JT.Hsieh@UTSouthwestern.edu

13 Rong Hsing Research Center for Translational Medicine, National Chung Hsing University, Taichung 40227, Taiwan

14 Ph.D. Program in Translational Medicine and Rong Hsing Research Center for Translational Medicine, National Chung Hsing University, Taichung 40227, Taiwan

* Correspondence: hlin@dragon.nchu.edu.tw; Tel.: +886-4-22840-416 (ext. 311); Fax: +886-4-22874-740

† These authors contributed equally to this paper.

Received: 16 March 2019; Accepted: 26 March 2019; Published: 28 March 2019

Abstract: Arecoline is the primary alkaloid in betel nuts, which are known as a risk factor for oral submucosal fibrosis and oral cancer. Lung cancer is a severe type of carcinoma with high cell motility that is difficult to treat. However, the detailed mechanisms of the correlation between Arecoline and lung cancer are not fully understood. Here, we investigated the effect of Arecoline on migration in lung cancer cell lines and its potential mechanism through the muscarinic acetylcholine receptor 3 (mAChR3)-triggered EGFR/Src/FAK pathway. Our results indicate that different concentrations of Arecoline treatment (10 μM, 20 μM, and 40 μM) significantly increased the cell migration ability in A549 and CL1-0 cells and promoted the formation of the filamentous actin (F-actin) cytoskeleton, which is a crucial element for cell migration. However, migration of H460, CL1-5, and H520 cell lines, which have a higher migration ability, was not affected by Arecoline treatment. The EGFR/c-Src/Fak pathway, which is responsible for cell migration, was activated by Arecoline treatment, and a decreased expression level of E-cadherin, which is an epithelial marker, was observed in Arecoline-treated cell lines. Blockade of the EGFR/c-Src/Fak pathway with the inhibitors of EGFR (Gefitinib) or c-Src (Dasatinib) significantly prevented Arecoline-promoted migration in A549 cells. Gefitinib or Dasatinib treatment significantly disrupted the Arecoline-induced localization of phospho-Y576-Fak during focal adhesion in A549 cells. Interestingly, Arecoline-promoted migration in A549 cells was blocked by a specific mAChR3 inhibitor (4-DAMP) or a neutralizing antibody of matrix metalloproteinase (MMP7 or Matrilysin). Taken together, our findings suggest that mAChR3 might play an essential role in Arecoline-promoted EGFR/c-Src/Fak activation and migration in an A549 lung cancer cell line.

Keywords: Arecoline; lung cancer cells; mAchR3; EGFR; SRC; FAK

Key Contribution: Arecoline is a carcinogen that affects people who chew areca nuts. This study elucidates how Arecoline promotes migration of lung cancer cells and reminds cancer patients to avoid Arecoline exposure.

---

1. Introduction

More than a million people in Taiwan consume betel nuts, and betel nut chewing is known as a risk factor for oral cancer and oral submucosal fibrosis, as described by the International Agency for Research on Cancer (IACR) [1–3]. Feeding Swiss mice and C17 mice a betel nut extraction induces gastrointestinal tumors in 58% and 17%, respectively. Studies have shown that betel nut extraction with DMSO, given for 21 weeks, increased tumor formation in the oral mucosa and caused leukoplakia in hamsters [4,5]. A tumorigenicity study in Swiss mice showed that betel nuts and betel quid induced lung cancer formation (47% and 26%, respectively) [6]. This data suggests that the habit of chewing betel nuts not only increases the risk of head and neck cancer, but also elevates the systemic risk of carcinogenesis. Arecoline is a primary alkaline in betel nut extract, and is also detected in the saliva of people who chew betel nuts, at a concentration between 5.66 and 97.39 μg/mL (approximately 0.036 mM to 0.63 mM) [5,7], and about 7 ng/mL (approximately 0.044 μM) in peripheral blood plasma [8]. Cytological studies indicate that a higher Arecoline level (more than 0.4 mM) can cause cytotoxicity in keratinocytes, and DNA damage when used long-term [9].

Arecoline is an acetylcholine agonist that acts on the neuronal nicotinic acetylcholine receptor (nAChR) or the muscarinic acetylcholine receptor (mAChR), which are G protein-coupled receptors in the cell membrane. It has been considered that Arecoline, as an acetylcholine agonist, might be useful in neurodegenerative diseases, such as Alzheimer's disease [10]. In the previous study, betel quid extract was found to promote cell migration in an esophageal squamous carcinoma cell line CE81T/VGH, and purified Arecoline treatment also had a similar effect [11], while matrix metalloproteinase (MMP) expression, specifically MMP-1 and MMP-8, increased with Arecoline treatment. Increased cell motility was inhibited by MMP-1 [12] or MMP-8 [11] neutralizing antibody, suggesting that activation of MMPs may be involved in Arecoline-stimulated migration. Studies on human colon cancer cells have shown that after ligand binding to muscarinic acetylcholine receptor 3 (mAChR3; M3R), MMP7 was activated, with subsequent release of the EGFR ligand [13,14]. mAchR3-transactivated EGFR downstream signaling pathways, including Erk1/2 [14], PI3K/Akt [13,15] and FAK/c-Src [16,17], are involved in cell survival and migration. The activation of mAChR3 is known as a multiplying factor in non-small cell lung cancer. In a previous clinical study, 85 patients, from 148 studied patients (57.4%), had high expression of mAChR3 in paraffin-embedded non-small cell lung cancer (NSCLC) tissue samples, which correlated well with tumor metastasis and poor survival [18,19]. Down-regulation of mAChR3, by siRNA, inhibited the migration and invasion ability of the lung cancer cell lines A549 and L78 [18]. Activation of mAChR3 in A549 cells also induced epithelial–mesenchymal transition

(EMT) marker expression [19]. These studies showed that over-expression or activation of mAChR3 promotes the progression of lung cancer.

According to previous studies, acetylcholine promoted lung cancer cell proliferation and migration through the mAChR3-activated EGFR/PI3K/Akt pathway accompanied by partial MMP activation [13–15]. Arecoline, an agonist of acetylcholine, also activated MMPs and was involved in esophageal carcinoma cell migration [11,12]. Thus, we suggest that Arecoline-promoted lung cancer cell migration is mediated by MMP activation; however, the detailed mechanism has not been determined. We hypothesize that administration of Arecoline contributes to lung cancer cell migration through the EGFR/c-Src/FAK pathway via mAChR3 transactivation. Cell motility was measured by the cell migration assay, which is a standard in vitro technique for detecting cell migration, and the activation of EGFR, c-Src and FAK were analyzed. We further used inhibitors or antagonists of mAChR3, EGFR, c-Src and neutralizing antibodies of MMP7 to verify the pathway. In this study, we suggest that Arecoline consumption and activation of mAChR3 is positively correlated with lung cancer cell migration, and interrupting this potential pathway may lead to decreased lung cancer cell migration.

2. Results

2.1. Arecoline Promotes Migration in the A549 Lung Cancer Cell Line

In this study, we suggest that Arecoline activates the EGFR and c-Src/FAK signaling cascade to stimulate cell migration through regulation of mAChR3. The activation of the signal cascade was analyzed by immunoblotting of active EGFR, c-Src and FAK, and the distributions of activated proteins were observed by fluorescent confocal microscopy. We also used inhibitors of EGFR, c-Src and FAK to confirm the pathway. In the first step, different concentrations of Arecoline (Are) were administrated to several lung cancer cell lines and cell motility was measured. The covered area was measured and quantified at 0, 6, 12 and 24 h after Arecoline treatment. The results indicate that administration of Arecoline stimulated A549 and CL1-0 cell migration in a dose-dependent manner, but not in H520, H460 and CL1-5 cells (Figure 1A). After 24 h of Arecoline treatment, the coverage of A549 and CL1-0 cells significantly increased to around 50% with 10 μM, 60% with 20 μM and 70% with 40 μM Arecoline compared with the control groups. However, the coverage of H520, H460, and CL1-5 cells did not change compared with control groups in all treatments. The MTT assay showed that there was no increase in cell proliferation in Arecoline-treated A549 cells (Figure 1B). The viability results of CL1-0, CL1-5, H520 and H460 after Arecoline treatment are also provided in Figure S1. This result suggests that Arecoline did not affect cell proliferation, but it did cause an increased coverage of A549 cells. The immunofluorescent study showed that stress fiber formation increased in Arecoline-treated A549 cells (Figure 1C). This means that Arecoline stimulates a morphological change and leads to the migration of lung cancer cells. Regarding the response of non-cancerous cells, MRC-5, no apparent differences in viability and migration were observed after Arecoline treatment (Figure S2), which indicates that the action of Arecoline is specific to cancer cells.

2.2. Arecoline Treatment Activates the EGFR/c-Src/FAK Signaling Pathway in the A549 Cell Line

After Arecoline treatment at different concentrations, A549 cells were collected, and the total proteins were extracted and separated by SDS-PAGE and the activation of EGFR, c-Src, and FAK was measured by immunoblotting. Their phosphorylation antibodies determined the activation of EGFR, c-Src, and FAK at pY1068, pY416 and pY576, respectively (Figure 2). The result showed that the phosphorylation levels of pY1068-EGFR, pY416-c-Src and pY576-FAK increased in a dose-dependent manner with Arecoline treatment. The highest activation levels of EGFR, c-Src and FAK were observed with 40 μM Arecoline treatment for 24 h. The effects of Arecoline on Src and FAK in CL1-0, H520 and H460 are also provided in Figure S3.

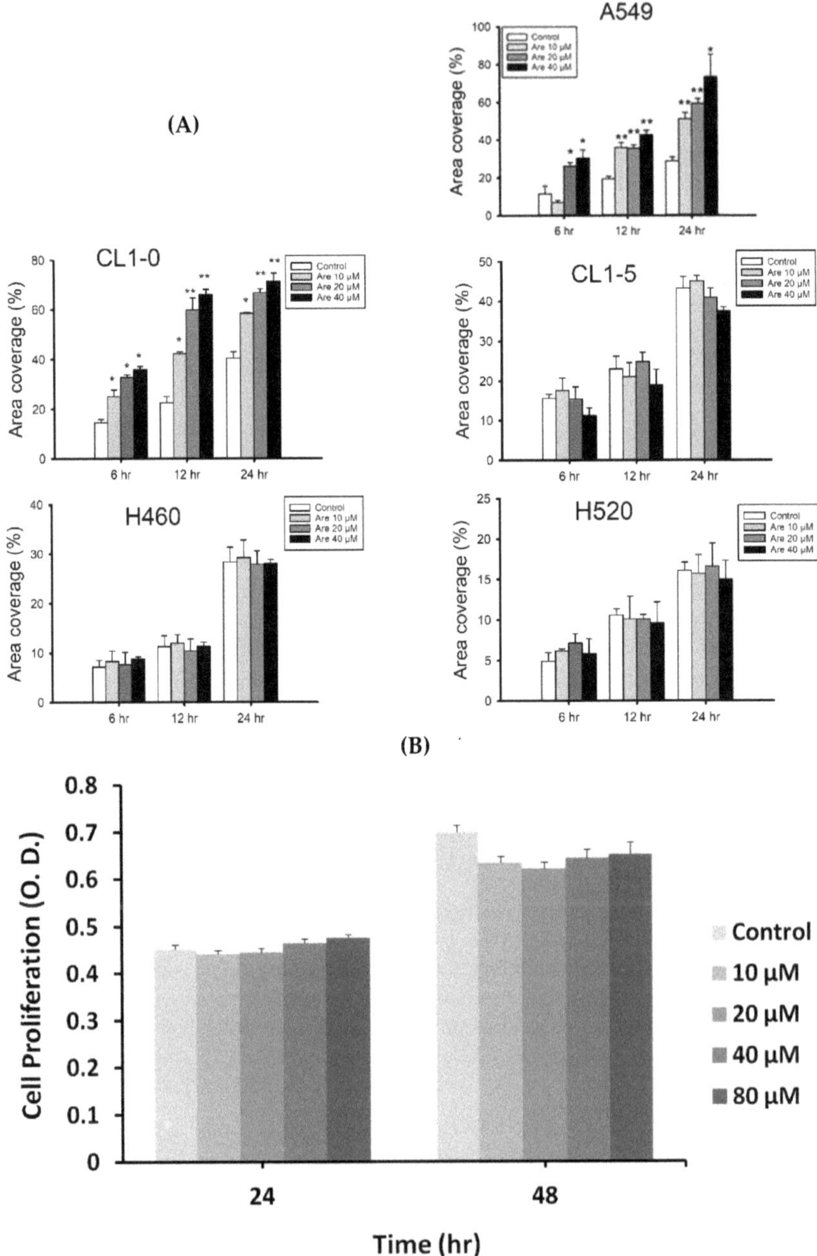

Figure 1. *Cont.*

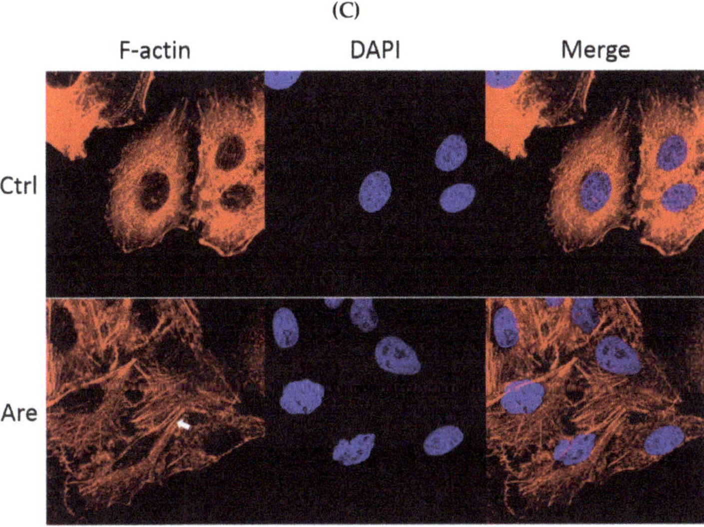

Figure 1. Arecoline promotes A549 and CL1-0 lung cancer cell migration. (**A**) Cell motility was measured by the cell migration assay. Lung cancer cells were seeded in an ibidi Culture-Insert with a cell density of 5×10^4 cells/well. Cell images were taken at indicated time points after treatment with a variable concentration of Arecoline. The results show that Arecoline promoted lung cancer cell migration in A549 and CL1-0 cells, but not in H520, H460, and CL1-5 cells. (*: $p< 0.05$; **: $p < 0.01$, compared with control groups). (**B**) The MTT assay was performed to detect cell viability. Different concentrations of Arecoline were administrated to A549 cells, and cell proliferation was measured at 24 and 48 h. (**C**) A549 cells treated with 40 μM Arecoline for 24 h and then fixed for immunostaining. F-actin was stained with phalloidin and DAPI for nuclear staining.

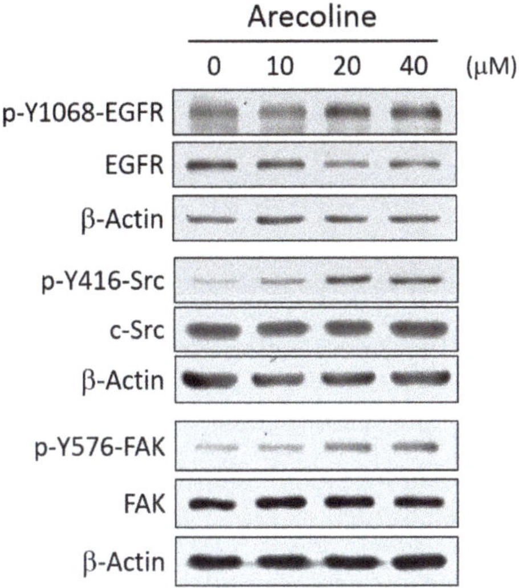

Figure 2. Arecoline activates EGFR/c-Src/FAK in a dose-dependent manner. After treatment with 10, 20 or 40 μM Arecoline for 24 h, A549 cells were collected, and proteins were analyzed by immunoblotting. The expression and phosphorylation of EGFR (pY1068-EGFR), c-Src (pY416-Src) and FAK (pY397-FAK) were measured. Actin served as an internal control.

2.3. Arecoline Stimulates Epithelial–Mesenchymal Transition (EMT) Markers in the A549 Cell Line

A549 cells were treated with different concentrations of Arecoline for 24 h and the expression of E-cadherin, N-cadherin, and vimentin were analyzed. The results indicated that the expression of E-cadherin decreased in Arecoline-treated A549 cells in a dose-dependent manner, while N-cadherin and vimentin were not affected (Figure 3).

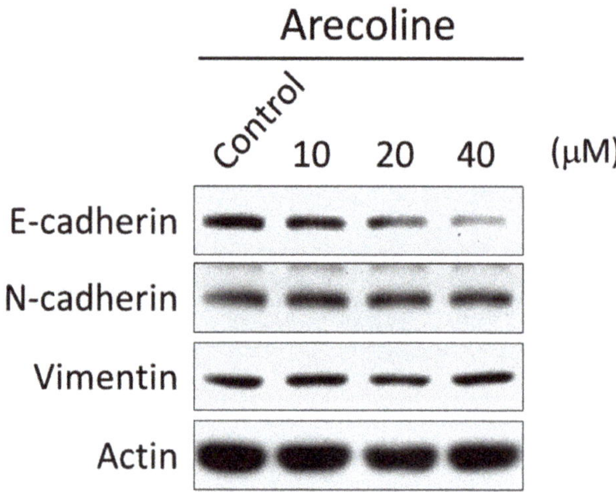

Figure 3. Arecoline administration induced epidermal–mesenchymal transition (EMT) marker expression in A549 cells. A549 cells were treated with 10, 20 or 40 μM Arecoline for 24 h and the protein expression of E-cadherin, N-cadherin, and vimentin was analyzed.

2.4. The Inhibition of EGFR or c-Src Activation Reversed Arecoline-Stimulated Migration in the A549 Cell Line

Immunoblotting and the cell migration assay showed that Arecoline stimulates A549 lung cancer cell migration. The activation of the EGFR/c-Src/FAK signaling pathway was investigated. Arecoline-stimulated migration reversed through inhibition of EGFR or c-Src activity by co-administrating 50 μM Gefitinib (Gef) or 50 nM Dasatinib (Das). The cell migration assay showed that co-administration of Gefitinib (Gef) or Dasatinib reversed A549 cell migration (Figure 4A). Furthermore, the subsequent decrease in c-Src and FAK by Gef or Das was detected by immunoblotting. After co-treatment with Arecoline and Gef or Das for 24 h, proteins were collected and analyzed by immunoblotting. The expression levels of EGFR, c-Src or FAK reduced in the Gef or Das co-treated groups (Figure 4B). Arecoline-stimulated activation of EGFR, c-Src and FAK was reversed by Gef or Das. Furthermore, the distribution of phosphorylated FAK (pY576-FAK) was also investigated by immunofluorescent staining. Accumulation of pY576-FAK showed that focal adhesion formation, as part of cell migration, was induced by Arecoline. The results indicated that pY576-FAK accumulated after Arecoline treatment for 24 h (Figure 4C). Accumulated pY576-FAK signals in Arecoline-treated cells were counted and increased 3 to 8 fold (11.06/3.64 to 17.71/1.90). Accumulation of pY576-FAK was induced by Gef and Das treatments (Figure 4C).

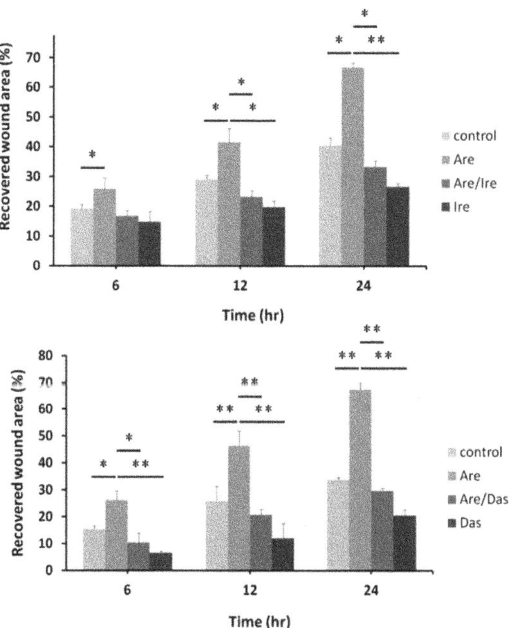

(B)

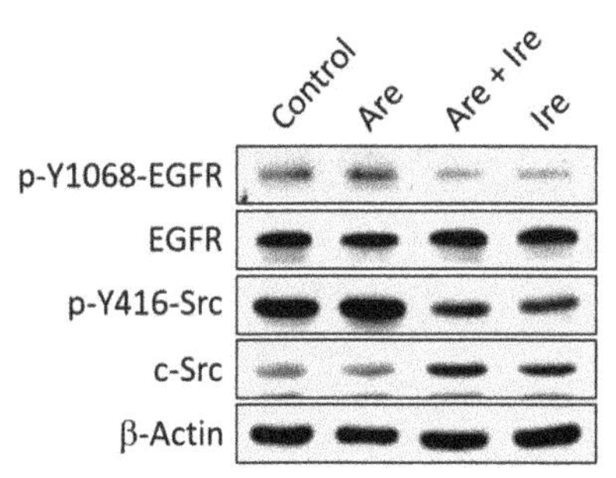

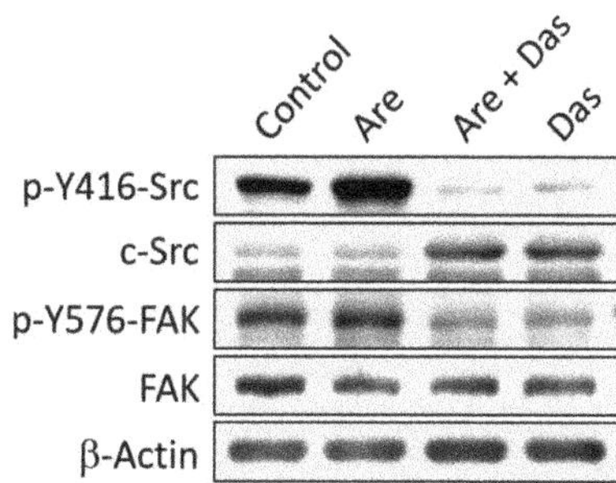

Figure 4. *Cont.*

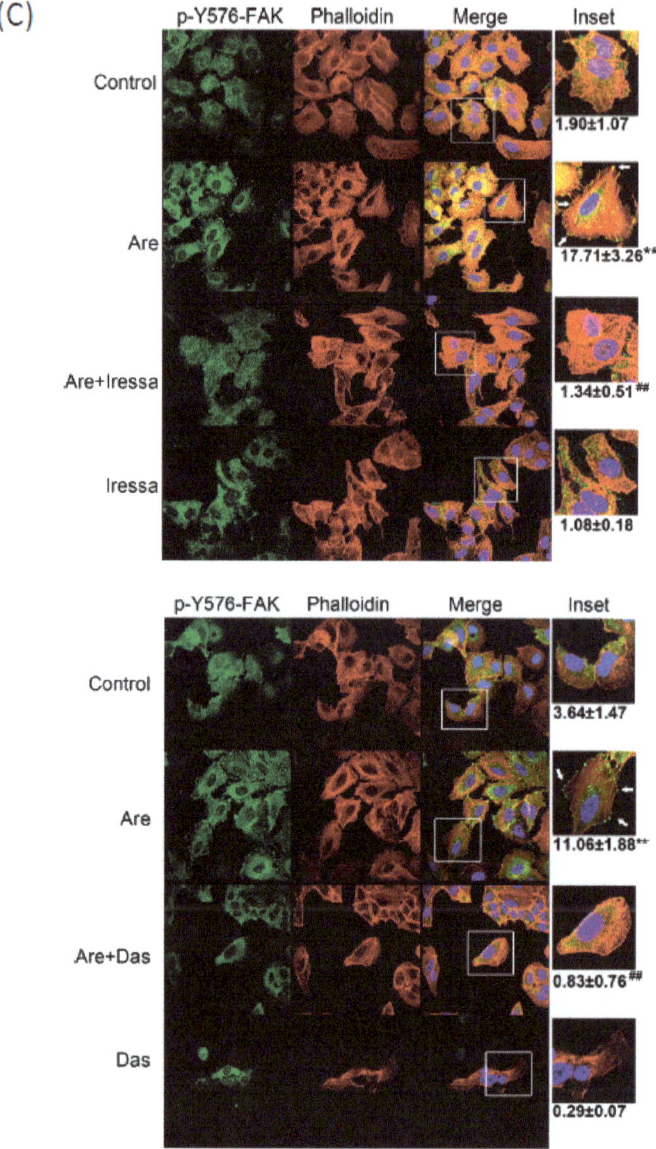

Figure 4. Gefitinib and Dasatinib reversed Arecoline-induced A549 cell migration and signaling activation. A549 cells were co-treated with 40 μM Arecoline (Are) and 50 μM Gefitinib (Gef) for 24 h. (**A**) Cell motility was recorded at indicated time points, and the result showed that Arecoline-stimulated cell migration was reversed by co-administration of Gefitinib or Dasatinib. (**B**) Protein expression and phosphorylation were measured by immunoblotting. The results showed that Arecoline-stimulated EGFR and c-Src activation reduced after Gefitinib or Dasatinib treatments. (*: $p < 0.05$; **: $p < 0.01$, compared with control groups). (**C**) Phosphorylated pY576-FAK (green), F-actin (red) and cell nucleus (blue) were detected by immunofluorescence. Activated FAK was observed in focal adhesion after Arecoline treatment, and counted. Numbers show an average activated FAK per cell (**: $p < 0.01$, compared with control groups; ##: $p < 0.01$, compared with Arecoline groups); however, the accumulation of activated FAK was reduced by Gef or Das treatments.

2.5. Arecoline Stimulates Lung Cancer Cell Migration through the Muscarinic Acetylcholine Receptor 3 (mAChR3) Transactivating EGFR Pathway in the A549 Cell line

The hypothesis is that Arecoline stimulates A549 lung cancer cell migration by mAChR3 transactivating EGFR and the following c-Src/FAK signaling pathway. Increased coverage area, stimulated by Arecoline administration, was observed, and then inhibitors that block the hypothesized pathway were further applied. After co-administration of 10 μM 4-DAMP (mAChR3 antagonist), 50 μM Gefitinib, 50 μM AG-1478 (EGFR inhibitors), 50 nM Dasatinib (c-Src inhibitor) or 2 μg/mL anti-MMP7 neutralizing antibody, Arecoline-stimulated migration recovered significantly in all treatment groups (Figure 5).

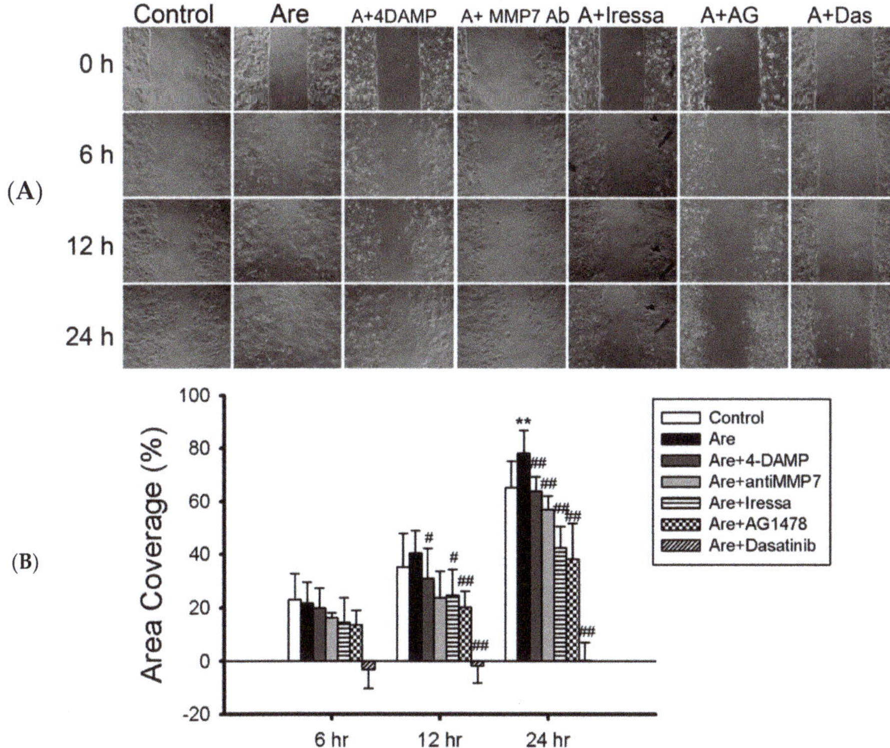

Figure 5. Muscarinic acetylcholine receptor 3-dependent A549 cell migration by Arecoline stimulation. (**A**) Cell migration assays were performed and A549 cells treated with Arecoline 40 μM (Are) were co-administrated with one of the following reagents: mAChR3 inhibitor (4-DAMP) 10 μM (Are+4-DAMP); MMP7 neutralizing antibody (MMP7 Ab) 2 μg/mL (Are+MMP7 Ab); EGFR inhibitor: Gefitinib 50 μM (Are+Gefitinib)/ AG1478 (AG) 50 μM (Are+AG); c-Src inhibitor Dasatinib (Das) 50 nM (Are+Das). (**B**) The results of cell motilities were quantified and compared with control or Arecoline treatments. (**: compared with control group, $p < 0.01$; #: compared with Arecoline-treated (Are) group, $p < 0.05$; ##: compared with Arecoline-treated (Are) group, $p < 0.01$).

3. Discussion

In Taiwan, more than a million people chew betel nuts. Betel chewing has been considered a risk factor for oral cancer and oral submucosal fibrosis; furthermore, animal studies have also shown that betel nuts can induce gastrointestinal tumors or lung cancer in Swiss mice [4,6]. These data suggest that betel-chewing habits increase not only carcinogenic risk in the oral cavity but also in the digestive tract and lungs. Furthermore, previous studies have shown that activation of muscarinic acetylcholine receptor 3 (mAChR3) promoted non-small cell lung cancer metastasis and invasion [15,18]. In this study, lung cancer cells were used, and Arecoline stimulated lung cancer cell migration. Arecoline stimulated cell migration in A549 and CL1-0 cells, but not in H460, H520, and CL1-5 cells. H520 and H460 cell lines have lower EGFR expression compared with other lung cancer cell lines, such as A549 or CL1-0 [20,21], which may be the reason why Arecoline administration did not stimulate their migration. Corresponding to the migration results, Arecoline induced phosphorylation of Src and FAK in CL1-0 cells but not in H520 and H460 cells (Supplementary Figure S3). On the other hand, CL1-5 derived from CL1-0 with higher EGFR expression and invasive ability [22]; meanwhile, CL1-0 and CL1-5 cells showed different downstream responses after EGF-stimulation [23]. The protein levels of phospho-EGFR/EGFR in CL1-5 were relatively high compared to those in CL1-0. Therefore, Arecoline might not further stimulate EGFR activation in CL1-5 when it is already at a high level.

According to the wound healing results, we found that Arecoline-stimulated cell motility can be reversed by co-administrating a mAChR3, EGFR, c-Src inhibitor or MMP-7 neutralizing antibody (Figure 5). Investigation of signaling cascade activation also presented similar results; the introduction of Arecoline elevated the phosphorylation level of EGFR, c-Src and FAK (Figure 2). The formation of focal adhesion and activated FAK accumulation was also observed (Figure 4C). The activated signaling cascade decreased by co-administration of EGFR or a c-Src inhibitor. Based on these data, Arecoline-stimulated A549 lung cancer cell migration through mAChR3 transactivating the EGFR/c-Src/FAK pathway was confirmed. Previous studies showed that mAChR3 activation stimulated cell proliferation, migration, and invasion in colon cancer and small cell lung cancer [13–15,24], and also other studies showed that the overexpression of mAChR3 in non-small cell lung cancer cells promoted NSCLC progress and poor prognosis [18,25]. Muscarinic acetylcholine receptor 3 is a member of the G-protein coupled receptor (GPCR) family; the activation of mAChR3 can transactivate EGFR by the MMP-cleaved EGF-like ligand [13–15,26–28]. Several studies showed that post-EGFR pathways are involved in promoting cell migration, including the MEK/ERK1/2 pathway in colon cancer cells [13,14], or the PI3K/AKT signaling pathway in NSCLC [15]. Receptor tyrosine kinase- (RTK), integrin- or GPCR-activated c-Src contributes to many cellular functions, including cell migration [16,29,30]. In this study, we showed that Arecoline promoted migration through mAChR3 to transactivate the EGFR/c-Src/FAK signaling pathway (Figure 6).

Furthermore, we showed that cell motility increased in Arecoline-treated lung cancer cells in a dose-dependent manner (Figure 1A). This suggests that Arecoline provides a driving force in lung cancer cell migration. Administration of 4-DAMP or anti-MMP7 neutralizing antibody reversed cell motility. This confirms that the transactivating effects were reversed in A549 lung cancer cells by a mAChR3 inhibitor and MMP7 neutralizing antibody [13,14,28,31]. Although Puthenedam's report showed that MMP7-cleaved galectin-3 inhibited the motility of intestinal epithelial cells [32], there was no further evidence indicating the phenomena could be applied to lung cancer cells. Besides MMP7, activated muscarinic acetylcholine receptors induced other MMPs, such as MMP1 [14,33], MMP3 [26], MMP9 [34] and MMP10 [14] expression, which are involved in cell invasion and migration. Several studies showed that MMP1 [12] and MMP8 [11] expressions were also elevated in esophageal carcinoma cells; however, these studies did not mention the effect of Arecoline on the muscarinic acetylcholine receptor. In this study, we elucidate that Arecoline stimulated cell migration by transactivating EGFR through the mAChR3/MMP7-cleaved EGF-like ligand and then activating the c-Src/FAK signaling pathway (Figure 6). Together, our results establish that Arecoline might be considered as a novel risk factor in NSCLC metastases.

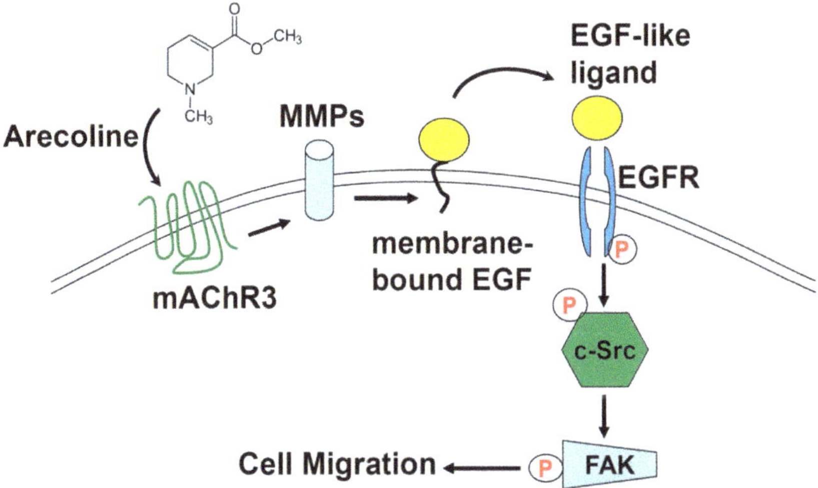

Figure 6. The scheme of the Arecoline-transactivated EGFR/c-Src/FAK pathway through mAChR3 for stimulating lung cancer cell migration. Arecoline stimulates MMP activity via mAChR3 to cleave EGF-like ligand and then subsequently trigger the EGFR/c-Src/FAK signaling cascade that stimulates lung cancer cell migration.

4. Materials and Methods

4.1. Cell Lines and Cell Culture

Non-small lung cancer cell lines A549 (ATCC number: CCL-185; BCRC number: 60074), and H520 (ATCC number: HTB-182; BCRC number: 60124) were purchased from Food Industry Research and Development Institute, Hsinchu, Taiwan. CL1-0 and CL1-5 cell lines were kindly provided by Professor Jeremy J. W. Chen, Institute of Molecular Biology, National Chung Hsing University. A549 cells were cultured in F-12K medium containing 10% FBS, 1.5 g/L $NaHCO_3$, 20 mM L-glutamine and 1% penicillin-streptomycin (P/S); H520, CL1-0, and CL1-5 cells were cultured in RPMI 1640 culture medium supplemented with 1.5 g/liter $NaHCO_3$, 10% FBS, 10 mM HEPES, 1 mM sodium pyruvate, and P/S. All cell lines were cultured at 37 °C in a humidified atmosphere with 5% CO_2.

4.2. Cell Viability Assay

Cells were seeded in the 96-well plate and incubated for 24 h after attachment. Cells were treated with different concentrations of Arecoline (Sigma-Aldrich, St. Louis, MO, USA) for the indicated number of days, and then the 3-(4,5-dimethylthiazol-2-yl)-2, 5-diphenyl tetrazolium bromide (MTT, Sigma-Aldrich, St. Louis, MO, USA) assay was used to quantify cell proliferation. The MTT stock solution (5 mg/mL) was diluted to 0.5 mg/mL with complete culture medium, and 0.1 mL MTT working solution was added to each well. Yellow MTT was converted to blue formazan by living cells, a reaction that is dependent on mitochondrial enzyme activity. After using DMSO (J. T. Baker, Center Valley, PA, USA) to dissolve the blue formazan, the absorbance of converted MTT could be measured at 570 nm.

4.3. Cell Migration Assay

The day before treatment, 5×10^4 cells/well were seeded into the culture insert (ibidi GmbH, Planegg, Germany) for attachment. After starvation with serum-free medium for 24 h, inserts were removed, and cells were treated with Arecoline and Arecoline co-treated with 10 μM 4-DAMP, 2 μg/mL MMP7 neutralizing antibody, 50 μM Gefitinib, 50 μM AG1478 or 50 nM Dasatinib. Pictures were obtained at the indicated time by optical microscopy (Olympus, CKX41, Tokyo, Japan) and analyzed by TScratch software (1.0, Computational Science & Engineering Laboratory, Swiss Federal Institute of Technology, Zurich, Switzerland, 2010) [35].

4.4. Immunoblotting

Treated cells were collected and washed with PBS then the cells were lysed by RIPA buffer (50 mM Tris, 150 mM NaCl, 2mM EDTA, 50 mM NaF, 0.1% SDS, 0.5% sodium deoxycholate and 1% NP-40). Obtained cell lysates were quantified by Bradford reagent (Sigma-Aldrich, St. Louis, MO, USA) and separated by SDS-PAGE (25 μg/lane). After being transferred, PVDF membranes (PerkinElmer Life Sciences, Shelton, CT, USA) were blocked with 5% skim milk and then incubated with primary antibodies overnight at 4 °C. After washing with PBST, horseradish peroxidase (HRP)-conjugated secondary antibodies (Jackson Immuno Research Laboratory, West Grove, PA, USA) were incubated at room temperature. The Enhanced Chemiluminescence (PerkinElmer Life Sciences, Shelton, CT, USA) reaction was performed, and the membranes were exposed to X-ray films (Fujifilm, Tokyo, Japan). Antibodies directed against the following proteins were used in this study: anti-Actin (Millipore, MAB1501, Temecula, CA, USA), anti-Tubulin (Millipore, 05829), anti-GADPH (GeneTex, GTX100118, Irvine, CA, USA), anti-cSrc (Santa Cruz, sc-8056, Dallas, TX, USA), anti-pY416-Src (Cell signaling, 2101), anti-EGFR (Santa Cruz, sc-03), anti-pY1068-EGFR (Cell signaling, 2036, Danvers, MA, USA), anti-E-cadherin (#610181, BD, Franklin Lakes, NJ, USA), anti-Vimentin (Santa Cruz, sc-32322), anti-pY397-FAK (BD, 611723), anti-pY576-FAK (Santa Cruz, sc-16563-R).

4.5. Immunostaining

Cells were seeded on a cover-slip the day before treatment; treated cells were washed with PBS and fixed in 3.7% paraformaldehyde (Sigma-Aldrich, St. Louis, MO, USA) for 30 min at room temperature. Followed by washing three times with PBS and blocking with 3% BSA in PBS, cells were incubated with anti-pY576-FAK for 1 h. After washing with PBST, cells were incubated with FITC-conjugated anti-Rabbit secondary antibody for 1 h. For F-actin and nuclear staining, 5 μg/mL phalloidin-TRITC (Sigma-Aldrich, St. Louis, MO, USA) and 1 μg/mL DAPI (Sigma-Aldrich, St. Louis, MO, USA) were applied. After further washing with PBS, slides were mounted for observation by fluorescence microscopy (Olympus, BX51, Japan) and laser confocal microscopy (Olympus, FV1000, Japan).

4.6. Statistical Analysis

The data were presented as means ± standard error of the mean (SEM). Student's t-test was used in the cell viability assay, cell migration assay, and immunoblotting. A significant difference between treatments was considered when $p < 0.05$.

Supplementary Materials: The following are available online at http://www.mdpi.com/2072-6651/11/4/185/s1, Figure S1: The effects of Arecoline on the viability of different lung cancer cell lines. Figure S2: Arecoline does not affect viability and migration of MRC-5 cells. Figure S3: The effects of Arecoline on Src and FAK in CL1-0, H520 (A), and H460 cells (B).

Author Contributions: Conceptualization, H.L. and T.-H.C.; Methodology, C.-H.C.; Validation, C.-T.R.Y. and C.-C.W.; Investigation, C.-H.C., T.-H.C., Y.-H.L., W.-C.Y., and W.-L.L.; Writing—Original Draft Preparation, C.-H.C., M.-C.C., and M.O.; Writing-Review & Editing, C.-H.L., J.-T.H., and H.L.; Visualization, C.-H.Y.; Supervision, M.-C.C.; Project Administration, H.-Y.W.; Funding Acquisition, T.-Y.Y., C.-L.J.T., K.-Y.C., and K.-C.C.

Funding: This research received no external funding.

Acknowledgments: This study was supported by Ministry of Science and Technology, Taiwan (106-2320-B-005-002-MY3) and Taichung Veterans General Hospital/National Chung Hsing University Joint Research Program (TCVGH-NCHU-1077612).

Conflicts of Interest: The authors declare no conflict of interest.

Abbreviations

EGF	Epithelial growth factor
EGFR	Epithelial growth factor receptor
EMT	Epithelial-mesenchymal transition
GPCR	G-protein coupled receptor
mAChR3	Muscarinic acetylcholine receptor 3
MMPs	Matrix metalloproteinases
RTK	Receptor tyrosine kinase

References

1. Chen, Y.J.; Chang, J.T.; Liao, C.T.; Wang, H.M.; Yen, T.C.; Chiu, C.C.; Lu, Y.C.; Li, H.F.; Cheng, A.J. Head and neck cancer in the betel quid chewing area: Recent advances in molecular carcinogenesis. *Cancer Sci.* **2008**, *99*, 1507–1514. [CrossRef]
2. Garg, A.; Chaturvedi, P.; Gupta, P.C. A review of the systemic adverse effects of areca nut or betel nut. *Indian J. Med. Paediatr. Oncol.* **2014**, *35*, 3–9. [CrossRef]
3. IARC Working Group on the Evaluation of Carcinogenic Risks to Humans. Betel-quid and areca-nut chewing and some areca-nut derived nitrosamines. *IARC Monogr. Eval Carcinog Risks Hum.* **2004**, *85*, 1–334.
4. Bhide, S.V.; Shivapurkar, N.M.; Gothoskar, S.V.; Ranadive, K.J. Carcinogenicity of betel quid ingredients: Feeding mice with aqueous extract and the polyphenol fraction of betel nut. *Br. J. Cancer* **1979**, *40*, 922–926. [CrossRef] [PubMed]
5. Suri, K.; Goldman, H.M.; Wells, H. Carcinogenic effect of a dimethyl sulphoxide extract of betel nut on the mucosa of the hamster buccal pouch. *Nature* **1971**, *230*, 383–384. [CrossRef]
6. Shirname, L.P.; Menon, M.M.; Nair, J.; Bhide, S.V. Correlation of mutagenicity and tumorigenicity of betel quid and its ingredients. *Nutr. Cancer* **1983**, *5*, 87–91. [CrossRef]
7. Ullah, M.; Cox, S.; Kelly, E.; Boadle, R.; Zoellner, H. Arecoline is cytotoxic for human endothelial cells. *J. Oral Pathol. Med.* **2014**, *43*, 761–769. [CrossRef]
8. Wu, I.C.; Chen, P.H.; Wang, C.J.; Wu, D.C.; Tsai, S.M.; Chao, M.R.; Chen, B.H.; Lee, H.H.; Lee, C.H.; Ko, Y.C. Quantification of blood betel quid alkaloids and urinary 8-hydroxydeoxyguanosine in humans and their association with betel chewing habits. *J. Anal. Toxicol.* **2010**, *34*, 325–331. [CrossRef] [PubMed]
9. Jeng, J.H.; Hahn, L.J.; Lin, B.R.; Hsieh, C.C.; Chan, C.P.; Chang, M.C. Effects of areca nut, inflorescence piper betle extracts and arecoline on cytotoxicity, total and unscheduled DNA synthesis in cultured gingival keratinocytes. *J. Oral Pathol. Med.* **1999**, *28*, 64–71. [CrossRef]
10. Chandra, J.N.; Malviya, M.; Sadashiva, C.T.; Subhash, M.N.; Rangappa, K.S. Effect of novel arecoline thiazolidinones as muscarinic receptor 1 agonist in Alzheimer's dementia models. *Neurochem. Int.* **2008**, *52*, 376–383. [CrossRef] [PubMed]
11. Liu, S.Y.; Liu, Y.C.; Huang, W.T.; Huang, G.C.; Chen, T.C.; Lin, M.H. Up-regulation of matrix metalloproteinase-8 by betel quid extract and arecoline and its role in 2d motility. *Oral Oncol.* **2007**, *43*, 1026–1033. [CrossRef] [PubMed]
12. Lee, C.H.; Liu, S.Y.; Lin, M.H.; Chiang, W.F.; Chen, T.C.; Huang, W.T.; Chou, D.S.; Chiu, C.T.; Liu, Y.C. Upregulation of matrix metalloproteinase-1 (mmp-1) expression in oral carcinomas of betel quid (bq) users: Roles of bq ingredients in the acceleration of tumor cell motility through mmp-1. *Arch. Oral Biol.* **2008**, *53*, 810–818. [CrossRef]
13. Belo, A.; Cheng, K.; Chahdi, A.; Shant, J.; Xie, G.; Khurana, S.; Raufman, J.P. Muscarinic receptor agonists stimulate human colon cancer cell migration and invasion. *Am. J. Physiol. Gastrointest Liver Physiol.* **2011**, *300*, G749–G760. [CrossRef] [PubMed]
14. Xie, G.; Cheng, K.; Shant, J.; Raufman, J.P. Acetylcholine-induced activation of m3 muscarinic receptors stimulates robust matrix metalloproteinase gene expression in human colon cancer cells. *Am. J. Physiol. Gastrointest Liver Physiol.* **2009**, *296*, G755–G763. [CrossRef]
15. Xu, R.; Shang, C.; Zhao, J.; Han, Y.; Liu, J.; Chen, K.; Shi, W. Activation of m3 muscarinic receptor by acetylcholine promotes non-small cell lung cancer cell proliferation and invasion via egfr/pi3k/akt pathway. *Tumour Biol.* **2015**, *36*, 4091–4100. [CrossRef]

16. Finn, R.S. Targeting src in breast cancer. *Ann. Oncol.* **2008**, *19*, 1379–1386. [CrossRef] [PubMed]
17. Ishizawar, R.; Parsons, S.J. C-src and cooperating partners in human cancer. *Cancer Cell* **2004**, *6*, 209–214. [CrossRef] [PubMed]
18. Lin, G.; Sun, L.; Wang, R.; Guo, Y.; Xie, C. Overexpression of muscarinic receptor 3 promotes metastasis and predicts poor prognosis in non-small-cell lung cancer. *J. Thorac. Oncol.* **2014**, *9*, 170–178. [CrossRef]
19. Yang, K.; Song, Y.; Tang, Y.B.; Xu, Z.P.; Zhou, W.; Hou, L.N.; Zhu, L.; Yu, Z.H.; Chen, H.Z.; Cui, Y.Y. Machrs activation induces epithelial-mesenchymal transition on lung epithelial cells. *BMC Pulm Med.* **2014**, *14*, 53. [CrossRef]
20. Tang, Z.; Du, R.; Jiang, S.; Wu, C.; Barkauskas, D.S.; Richey, J.; Molter, J.; Lam, M.; Flask, C.; Gerson, S.; et al. Dual met-egfr combinatorial inhibition against t790m-egfr-mediated erlotinib-resistant lung cancer. *Br. J. Cancer* **2008**, *99*, 911–922. [CrossRef]
21. Shen, H.; Yuan, Y.; Sun, J.; Gao, W.; Shu, Y.Q. Combined tamoxifen and gefitinib in non-small cell lung cancer shows antiproliferative effects. *Biomed. Pharm.* **2010**, *64*, 88–92. [CrossRef] [PubMed]
22. Tian, T.; Hao, J.; Xu, A.; Luo, C.; Liu, C.; Huang, L.; Xiao, X.; He, D. Determination of metastasis-associated proteins in non-small cell lung cancer by comparative proteomic analysis. *Cancer Sci.* **2007**, *98*, 1265–1274. [CrossRef] [PubMed]
23. Tsai, H.F.; Huang, C.W.; Chang, H.F.; Chen, J.J.; Lee, C.H.; Cheng, J.Y. Evaluation of egfr and rtk signaling in the electrotaxis of lung adenocarcinoma cells under dGefct-current electric field stimulation. *PLoS ONE* **2013**, *8*, e73418. [CrossRef] [PubMed]
24. Zhang, S.; Togo, S.; Minakata, K.; Gu, T.; Ohashi, R.; Tajima, K.; Murakami, A.; Iwakami, S.; Zhang, J.; Xie, C.; et al. Distinct roles of cholinergic receptors in small cell lung cancer cells. *Anticancer Res.* **2010**, *30*, 97–106.
25. Wu, J.; Zhou, J.; Yao, L.; Lang, Y.; Liang, Y.; Chen, L.; Zhang, J.; Wang, F.; Wang, Y.; Chen, H.; et al. High expression of m3 muscarinic acetylcholine receptor is a novel biomarker of poor prognostic in patients with non-small cell lung cancer. *Tumour Biol.* **2013**, *34*, 3939–3944. [CrossRef]
26. Kuhne, S.; Ockenga, W.; Banning, A.; Tikkanen, R. Cholinergic transactivation of the egfr in hacat keratinocytes stimulates a flotillin-1 dependent mapk-mediated transcriptional response. *Int. J. Mol. Sci.* **2015**, *16*, 6447–6463. [CrossRef]
27. Hao, L.; Du, M.; Lopez-Campistrous, A.; Fernandez-Patron, C. Agonist-induced activation of matrix metalloproteinase-7 promotes vasoconstriction through the epidermal growth factor-receptor pathway. *Circ. Res.* **2004**, *94*, 68–76. [CrossRef]
28. Cheng, K.; Zimniak, P.; Raufman, J.P. Transactivation of the epidermal growth factor receptor mediates cholinergic agonist-induced proliferation of h508 human colon cancer cells. *Cancer Res.* **2003**, *63*, 6744–6750. [CrossRef]
29. Laurent-Puig, P.; Lievre, A.; Blons, H. Mutations and response to epidermal growth factor receptor inhibitors. *Clin. Cancer Res.* **2009**, *15*, 1133–1139. [CrossRef] [PubMed]
30. Sieg, D.J.; Hauck, C.R.; Ilic, D.; Klingbeil, C.K.; Schaefer, E.; Damsky, C.H.; Schlaepfer, D.D. Fak integrates growth-factor and integrin signals to promote cell migration. *Nat. Cell Biol.* **2000**, *2*, 249–256. [CrossRef] [PubMed]
31. Cheng, K.; Xie, G.; Raufman, J.P. Matrix metalloproteinase-7-catalyzed release of hb-egf mediates deoxycholyltaurine-induced proliferation of a human colon cancer cell line. *Biochem. Pharmacol.* **2007**, *73*, 1001–1012. [CrossRef] [PubMed]
32. Puthenedam, M.; Wu, F.; Shetye, A.; Michaels, A.; Rhee, K.J.; Kwon, J.H. Matrilysin-1 (mmp7) cleaves galectin-3 and inhibits wound healing in intestinal epithelial cells. *Inflamm. Bowel Dis.* **2011**, *17*, 260–267. [CrossRef] [PubMed]
33. Raufman, J.P.; Cheng, K.; Saxena, N.; Chahdi, A.; Belo, A.; Khurana, S.; Xie, G. Muscarinic receptor agonists stimulate matrix metalloproteinase 1-dependent invasion of human colon cancer cells. *Biochem. Biophys. Res. Commun.* **2011**, *415*, 319–324. [CrossRef] [PubMed]

34. Pelegrina, L.T.; Lombardi, M.G.; Fiszman, G.L.; Azar, M.E.; Morgado, C.C.; Sales, M.E. Immunoglobulin g from breast cancer patients regulates mcf-7 cells migration and mmp-9 activity by stimulating muscarinic acetylcholine receptors. *J. Clin. Immunol.* **2013**, *33*, 427–435. [CrossRef] [PubMed]
35. Geback, T.; Schulz, M.M.; Koumoutsakos, P.; Detmar, M. Tscratch: A novel and simple software tool for automated analysis of monolayer cell migration assays. *Biotechniques* **2009**, *46*, 265–274. [CrossRef]

Article

Nigritanine as a New Potential Antimicrobial Alkaloid for the Treatment of *Staphylococcus aureus*-Induced Infections

Bruno Casciaro [1], Andrea Calcaterra [2], Floriana Cappiello [3], Mattia Mori [4], Maria Rosa Loffredo [3], Francesca Ghirga [1,*], Maria Luisa Mangoni [3,*], Bruno Botta [2] and Deborah Quaglio [2]

1 Center For Life Nano Science@Sapienza, Istituto Italiano di Tecnologia, Viale Regina Elena 291, 00161 Rome, Italy

2 Department of Chemistry and Technology of Drugs, "Department of Excellence 2018–2022", Sapienza University of Rome, P.le Aldo Moro 5, 00185 Rome, Italy

3 Laboratory affiliated to Pasteur Italia-Fondazione Cenci Bolognetti, Department of Biochemical Sciences, Sapienza University of Rome, P.le Aldo Moro 5, 00185 Rome, Italy

4 Department of Biotechnology, Chemistry and Pharmacy, "Department of Excellence 2018–2022", University of Siena, via Aldo Moro 2, 53100 Siena, Italy

* Correspondence: francesca.ghirga@iit.it (F.G.); marialuisa.mangoni@uniroma1.it (M.L.M.)

Received: 24 July 2019; Accepted: 30 August 2019; Published: 1 September 2019

Abstract: *Staphylococcus aureus* is a major human pathogen causing a wide range of nosocomial infections including pulmonary, urinary, and skin infections. Notably, the emergence of bacterial strains resistant to conventional antibiotics has prompted researchers to find new compounds capable of killing these pathogens. Nature is undoubtedly an invaluable source of bioactive molecules characterized by an ample chemical diversity. They can act as unique platform providing new scaffolds for further chemical modifications in order to obtain compounds with optimized biological activity. A class of natural compounds with a variety of biological activities is represented by alkaloids, important secondary metabolites produced by a large number of organisms including bacteria, fungi, plants, and animals. In this work, starting from the screening of 39 alkaloids retrieved from a unique *in-house* library, we identified a heterodimer β-carboline alkaloid, nigritanine, with a potent anti-*Staphylococcus* action. Nigritanine, isolated from *Strychnos nigritana*, was characterized for its antimicrobial activity against a reference and three clinical isolates of *S. aureus*. Its potential cytotoxicity was also evaluated at short and long term against mammalian red blood cells and human keratinocytes, respectively. Nigritanine showed a remarkable antimicrobial activity (minimum inhibitory concentration of 128 μM) without being toxic in vitro to both tested cells. The analysis of the antibacterial activity related to the nigritanine scaffold furnished new insights in the structure–activity relationships (SARs) of β-carboline, confirming that dimerization improves its antibacterial activity. Taking into account these interesting results, nigritanine can be considered as a promising candidate for the development of new antimicrobial molecules for the treatment of *S. aureus*-induced infections.

Keywords: natural products; alkaloids; plant secondary metabolites; β-carboline; *Staphylococcus aureus*; antimicrobial activity; cytotoxicity

Key Contribution: Starting from the screening of 39 alkaloids, we biologically characterized a heterodimer β-carboline alkaloid, named nigritanine, with potent anti-*Staphylococcus* activity and non-toxic to mammalian red blood cells and human keratinocytes at its bioactive concentration.

This manuscript is dedicated to the memory of Professor Maurizio Botta (University of Siena, Department of Biotechnology, Chemistry and Pharmacy) who prematurely passed away on 2 August 2019. During his successfully scientific career, he synthesized a huge number of small bioactive molecules for the development of new pharmaceutical agents for cancer therapy and/or treatment of microbial infections, thus providing an invaluable contribution in the field of medicinal chemistry and drug discovery, worldwide.

1. Introduction

The discovery of antibiotics in the 1900s led to a medical revolution in the fight against bacterial infections. However, during the years, bacteria have developed different mechanisms to resist the killing activity of antibiotics [1]. The human pathogen *Staphylococcus aureus* is a microorganism with high adaptability and tenacity, as highlighted by its abundance in the environment and in the normal flora, the variety of virulence factors that it produces, and the capability to colonize various human organs such as nose, pharynx, and skin [2–4]. Furthermore, multidrug-resistant *S. aureus* is one of the major microorganisms causing bloodstream infections associated with high levels of morbidity and mortality worldwide [5]. Considering that *S. aureus* has successfully evolved numerous strategies to resist the activity of practically all antibiotics, new alternative compounds able to defeat *S. aureus*-induced infections are urgently needed [6]. Notably, a significant portion of the commercial drugs occurs in nature or is derived from natural products by means of chemical transformations or de novo synthesis [7]. Alkaloids are a group of important secondary metabolites which are produced by a wide variety of organisms including bacteria, fungi, plants, and animals. Chemically, alkaloids are a large and structurally diverse group of nitrogen-containing compounds (one or more nitrogen atoms within a heterocycle ring) [8]. Alkaloids can occur as monomers, dimers (bisalkaloids), trimers, or tetramers. According to their chemical structure, alkaloids are classified in heterocyclic alkaloids (also known as typical alkaloids), containing nitrogen in the heterocycle and originating from amino acids, and nonheterocyclic alkaloids (also known as atypical or proto-alkaloids), containing a nitrogen atom derived from an amino acid which is not a part of the heterocyclic ring [9]. Heterocyclic alkaloids are divided according to their ring structure in several classes of monomeric alkaloids (e.g., pyrrole, pyrrolidine, pyridine, piperidine, indole, quinoline, isoquinoline alkaloids). Since 1940, large-scale efforts have been made to evaluate the antibacterial effects of naturally occurring alkaloids. Several potent monomer and dimer alkaloids were identified, and synthetic modifications were investigated to improve their biological activity [8–11]. However, a tremendously wide discrepancy between their historical significance and their occurrence in modern drug development exists, and no alkaloids are available in the market as antibacterial drugs [12]. In this work, an in-house library of about 1000 natural products and their derivatives was used as a unique source of lead compounds to identify new potential antibacterial alkaloids. From the screening of all the alkaloids present in this library, the rare β-carboline heterodimer nigritanine was identified and showed a potent antistaphylococcal activity. Therefore, it was thoroughly characterized for its antimicrobial and cytotoxic activities.

2. Results and Discussion

2.1. Alkaloids Collection

Natural products remain the most productive source of leads in antibacterial drug discovery, often providing novel mechanism(s) and chemical structures as useful platforms for the development of drugs. A unique in-house library of about 1000 natural compounds, mostly isolated from several plants used in traditional medicine of South America and collected over the years, is available at the Organic Chemistry Laboratory of the Department of Chemistry and Technology of Drugs (Sapienza University of Rome, Italy). This library consists of natural products belonging to different classes of organic compounds which were previously published and fully characterized [13,14]. It was then enlarged by the addition of other natural small molecules from commercially available sources and synthetic or semi-synthetic derivatives. Currently, all components of our collection are incorporated into a virtual library, and their chemical and physicochemical features are analyzed by means of cheminformatics tools, showing a satisfactory chemical diversity. Therefore, our *in-house* library is a valid source of chemotypes for the modulation of biomolecular targets, and it was successfully screened in silico and in vitro for the identification of hit and lead compounds in previous early-stage drug discovery projects [15]. One of the largest and most intriguing classes of natural occurring compounds within the library are the alkaloids, which consist of isoquinoline (**1–11**), quinoline (**12–15**), and indole (**16–39**) alkaloids (Table 1).

Table 1. List of alkaloids tested in this study.

Mol.	Common Name	Chemical Structure	M.W.	Molecular Formula	Source	Ref.
		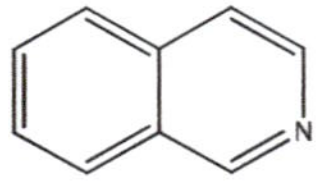 Isoquinoline Alkaloids				
1	Dihydroberberine·HCl	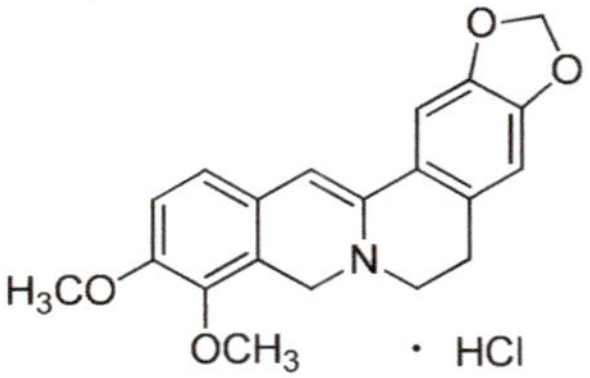	337.37 373.83 (+ HCl)	$C_{20}H_{19}NO_4 \cdot HCl$	*Berberis* species: *Berberis aristata, Berberis lyceum, Berberis petiolaris, Berberis tinctoria* (Berberidaceae family)	[16]
2	Bulbocapnine·HCl	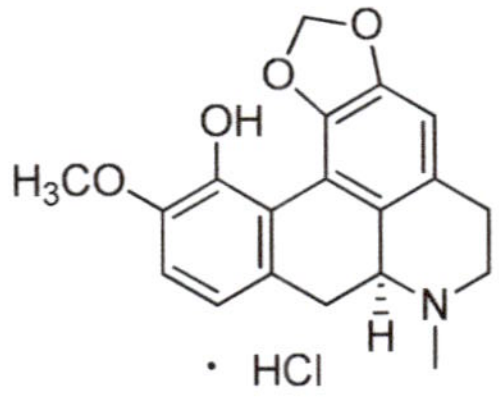	325.36 361.82 (+ HCl)	$C_{19}H_{19}NO_4 \cdot HCl$	Species: *Corydalis cava* (Papaveraceae family)	[17]
3	Boldine	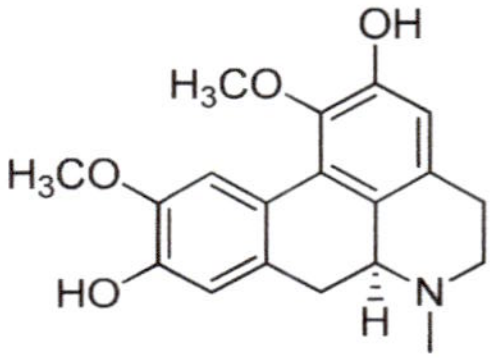	327.37	$C_{19}H_{21}NO_4$	Species: *Peumus boldus* (Monimiaceae family)	[18]

Table 1. *Cont.*

Mol.	Common Name	Chemical Structure	M.W.	Molecular Formula	Source	Ref.
4	Cotarmine·HCl	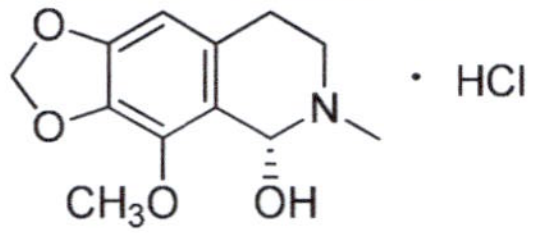	237.25 273.71	$C_{12}H_{15}NO_4 \cdot HCl$	Synthetic	[19]
5	Chelidonine	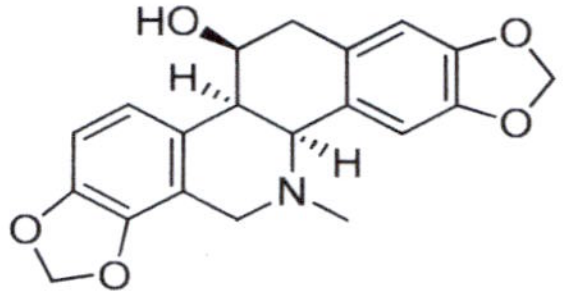	353.37	$C_{20}H_{19}NO_5$	Species: *Chelidonium majus* L. (Papaveraceae family)	[20]
6	Emetine·HCl	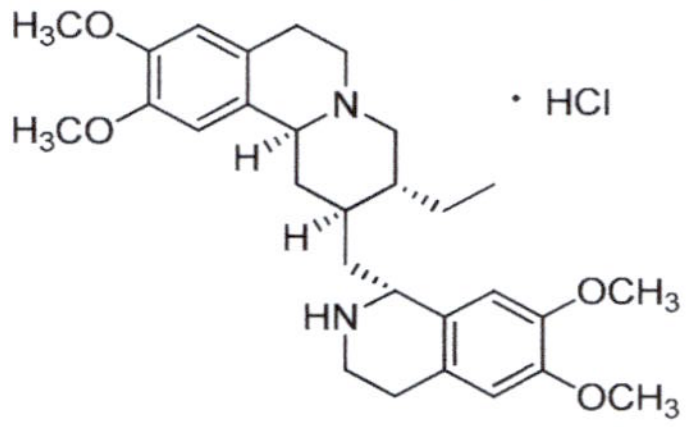	480.64 517.10 (+HCl)	$C_{29}H_{40}N_2O_4$	Species: *Psychotria ipecacuanha* Stokes (Rubiaceae family)	[21]
7	(*S*)-Glaucine	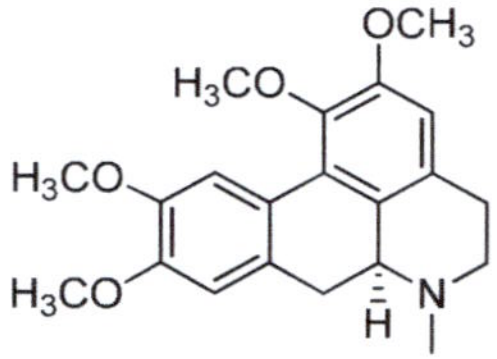	355.43	$C_{21}H_{25}NO_4$	Species: *Glaucium luteum* L. (Papaveraceae family)	[22]

Table 1. *Cont.*

Mol.	Common Name	Chemical Structure	M.W.	Molecular Formula	Source	Ref.
8	Hydrastine	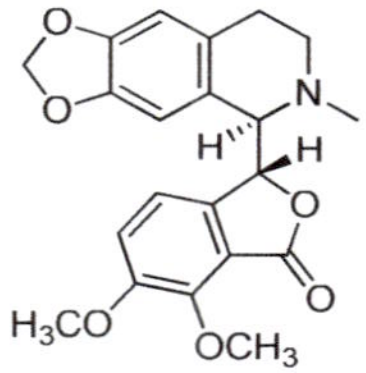	383.39	$C_{21}H_{21}NO_6$	Species: *Hydrastis canadensis* L. (Ranunculaceae family)	[23]
9	Noscapine (Narcotine)	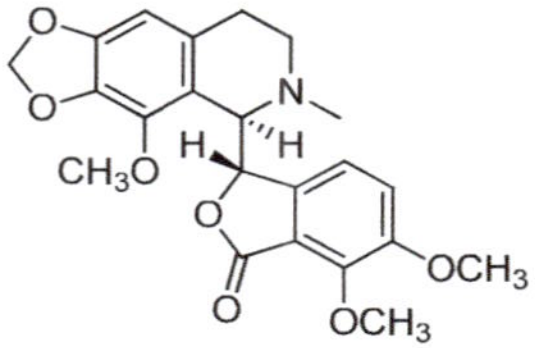	413.42	$C_{22}H_{23}NO_7$	Species: *Papaver somniferum* (Papaveraceae family)	[24]
10	Papaverine	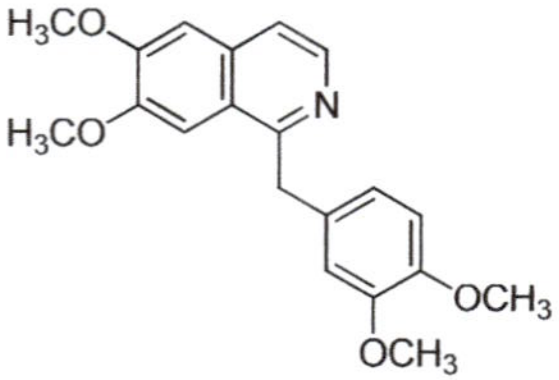	339.39	$C_{20}H_{21}NO_4$	Species: *P. somniferum* (Papaveraceae family)	[24]
11	Tubocurarine Chloride·HCl	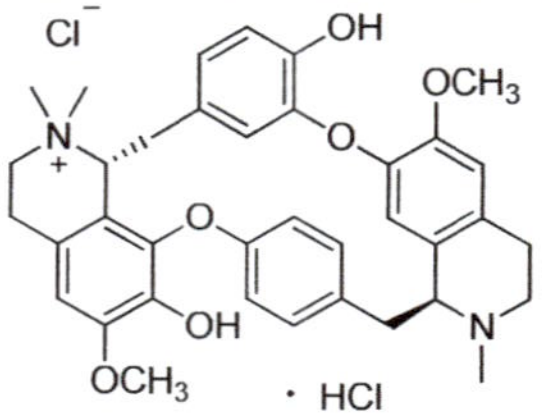	609.73 681.65 (+Cl⁻+HCl)	$C_{37}H_{41}N_2O_6 \cdot HCl$ + Cl^-	Species: *Liana Chondrodendron* (Menispermaceae family)	[25]

Table 1. *Cont.*

Mol.	Common Name	Chemical Structure	M.W.	Molecular Formula	Source	Ref.
12	Cinchonine	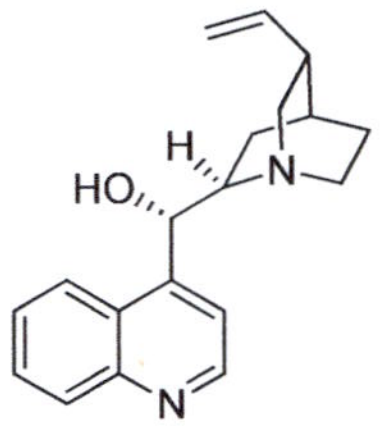 HO, H, N, N	294.39	$C_{19}H_{22}N_2O$	Species: *Cinchona ledgeriana, Remijia peruviana* (Rubiaceae family)	[26]
13	Kokusaginine	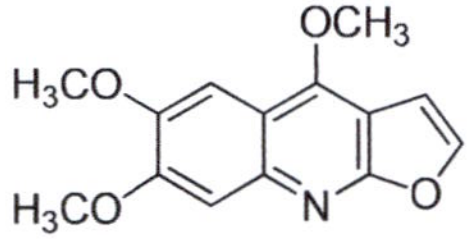 OCH_3, H_3CO, H_3CO, N, O	259.26	$C_{14}H_{13}NO_4$	Species: *Esenbeckia leiocarpa* (Rutaceae family)	[27]
14	Maculine	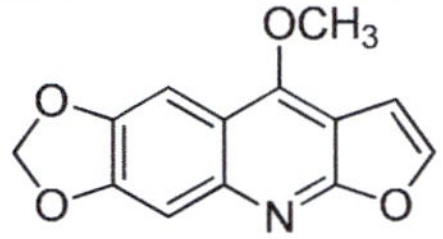 OCH_3, O, O, N, O	243.21	$C_{13}H_9NO_4$	Species: *E. leiocarpa* (Rutaceae family)	[27]
15	4-methoxy-2-(1-ethylpropyl)-quinoline	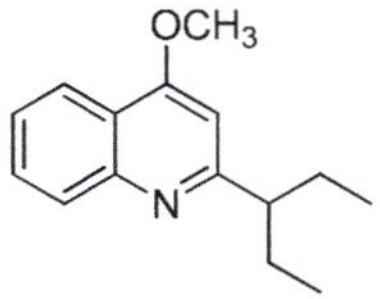 OCH_3, N	229.32	$C_{15}H_{19}NO$	Species: *E. leiocarpa* (Rutaceae family)	[27]
		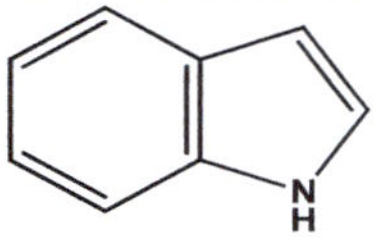 N H Indole Alkaloids				

Table 1. *Cont.*

Mol.	Common Name	Chemical Structure	M.W.	Molecular Formula	Source	Ref.
16	Aspidospermine	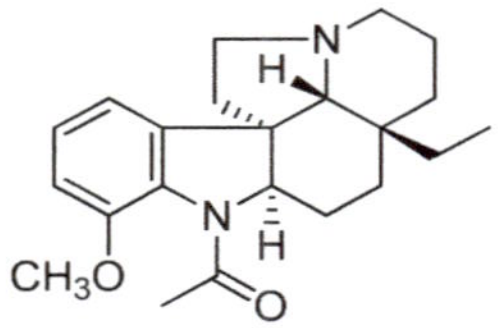 CH$_3$O, N, H, H, O	354.49	$C_{22}H_{30}N_2O_2$	*Aspidosperma* species: *Aspidosperma album*, *Aspidosperma australe*, *Aspidosperma exalatum*, *Aspidosperma peroba*, *Aspidosperma polyneuron*, *Aspidosperma pyricollum*, *Aspidosperma pyrifolium*, *Aspidosperma quebracho-blanco*, *Aspidosperma quirandy*, *Aspidosperma sessiflorum*, *Aspidosperma rhombeosignatum* (Apocynaceae family)	[28]
17	Brucine	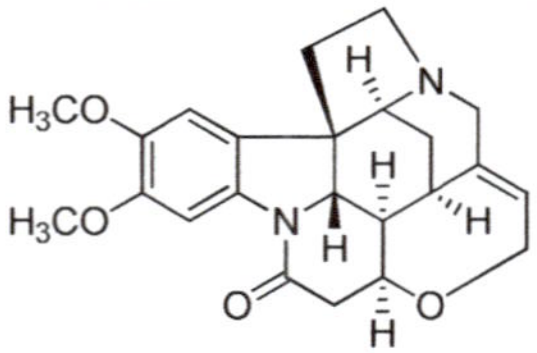 H$_3$CO, H$_3$CO, N, N, H, H, H, H, H, O, O	394.47	$C_{23}H_{26}N_2O_4$	Species: *Strychnos nux-vomica* (Apocynaceae family)	[29]
18	Diaboline	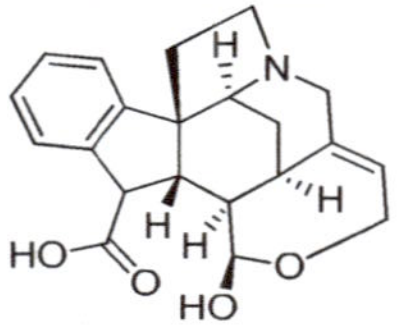 HO, O, HO, N, H, H, H, H, O	353.41	$C_{21}H_{23}NO_4$	Species: *Strychnos castelneana* (Loganiaceae family)	[27]
19	Physostigmine (Eserine)	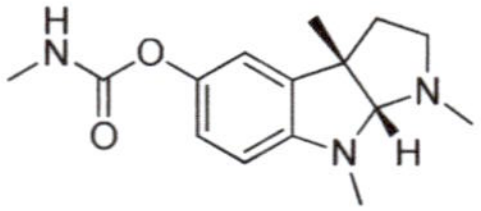 H, N, O, O, N, N, H	275.35	$C_{15}H_{21}N_3O_2$	*Physostigma venenosum* (Fabaceae family)	[30]

Table 1. *Cont.*

Mol.	Common Name	Chemical Structure	M.W.	Molecular Formula	Source	Ref.
20	Holstiine	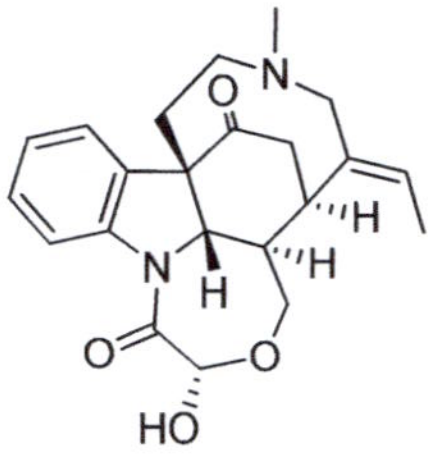	382.45	$C_{22}H_{26}N_2O_4$	Species: *Strychnos henningsii* Gilg (Loganiaceae family)	[31]
21	Pseudobrucine	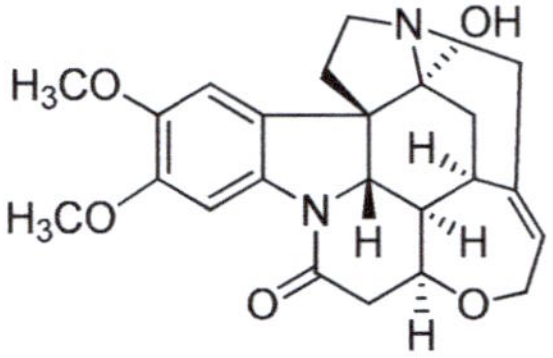	410.46	$C_{23}H_{26}N_2O_5$	Species: *S. nux-vomica* (Loganiaceae family)	[29]
22	Retuline	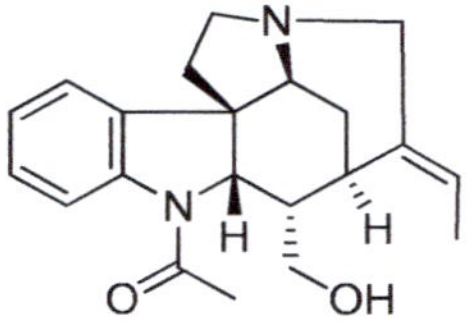	338.44	$C_{21}H_{26}N_2O_2$	*Strychnos* species: *Strychnos camptoneura,* *S. henningsii* (Loganiaceae family)	[31]
23	Serotonin	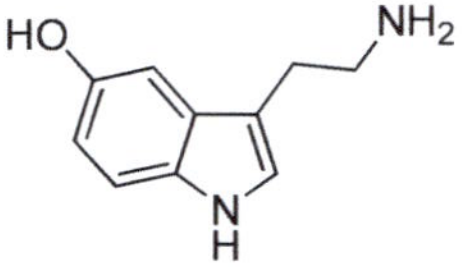	176.22	$C_{10}H_{12}N_2O$	Species: *Laphophora williamsii* (Cactaceae family)	[32]

Table 1. *Cont.*

Mol.	Common Name	Chemical Structure	M.W.	Molecular Formula	Source	Ref.
24	Triptamine·HCl	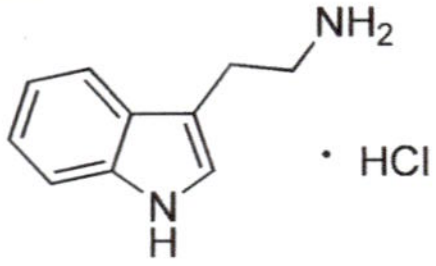	160.22 196.68 (+HCl)	$C_{10}H_{12}N_2$ HCl	*Acacia* species (Fabacee family)	[33]
25	Vomicine·HC1	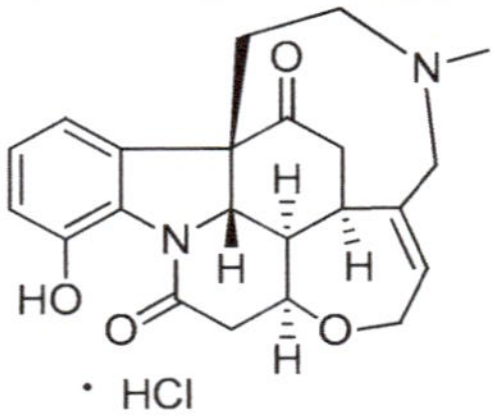	380.44 416.90 (+HCl)	$C_{22}H_{24}N_2O_4$ HCl	*Strychnos icaja* (Loganiaceae family)	[27, 31]
26	Vindoline	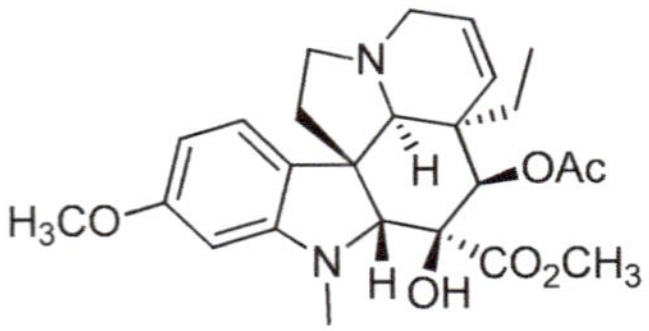	456.53	$C_{25}H_{32}N_2O_6$	*Catharanthus roseus* (Apocynaceae family)	[34]
		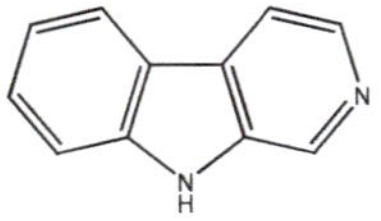 Carboline Alkaloids (Indole Subclass)				

Table 1. *Cont.*

Mol.	Common Name	Chemical Structure	M.W.	Molecular Formula	Source	Ref.
27	Akagerine	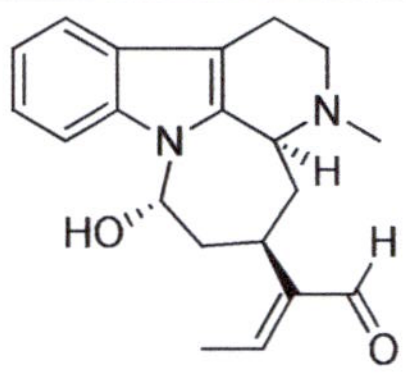	324.42	$C_{20}H_{24}N_2O_2$	*Strychnos* species: *Strychnos barteri* Solered, *S. camptoneurine*, *Strychnos nigritana* Bak (Loganiaceae family)	[31]
28	Canthin-6-one	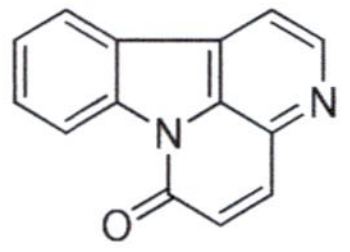	220.23	$C_{14}H_8N_2O$	Species: *Simaba ferruginea* (Simaroubaceae family)	[35]
29	α-Carboline	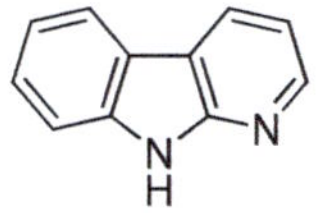	168.19	$C_{11}H_8N_2$	Synthetic	[36]
30	Harmane	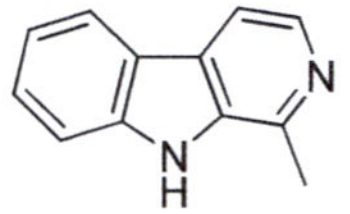	182.22	$C_{12}H_{10}N_2$	Species: *Chimarrhis turbinata*, *Ophiorrhiza communis*, *Ophiorrhiza liukiuensis*, *Ophiorrhiza tomentosa*, *Psychotria barbiflora* (Rubiaceae family)	[26]
31	Norharmane	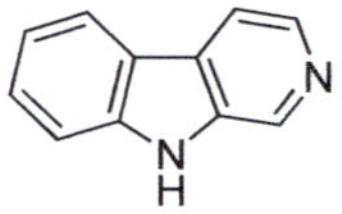	168.19	$C_{11}H_8N_2$	Species: *Hygrophorus eburneus* (Tricholomataceae family)	[37]

Table 1. *Cont.*

Mol.	Common Name	Chemical Structure	M.W.	Molecular Formula	Source	Ref.
32	Harmine	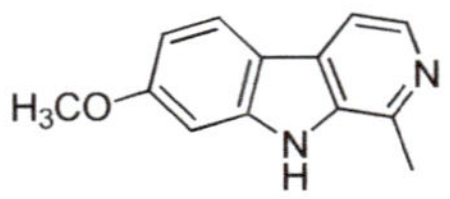 H_3CO, N, NH	212.25	$C_{13}H_{12}N_2O$	Species: *Banisteriopsis caapi* (Malpighiaceae family), *Grewia bicolor* (Malvaceae family), *Passiflora edulis* f. *flavicarpa* O. Deg., *Passiflora incarnata* L. (Passifloraceae family), *Tribulus terrestris* L., *Peganum harmala* L. (Zygophyllaceae family)	[37]
33	Ibogaine	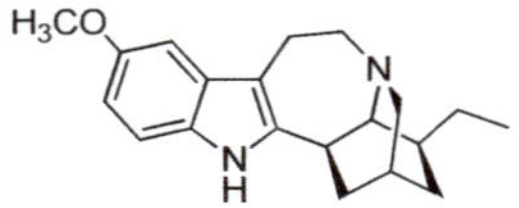 H_3CO, N, NH	310.43	$C_{20}H_{26}N_2O$	Species: *Tabernanthe iboga* (Apocynaceae family)	[38]
34	Mitragynine	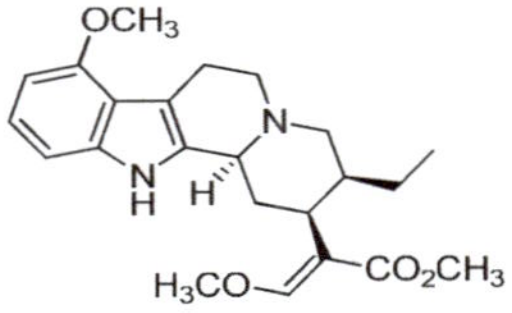 OCH_3, N, NH, H, H_3CO, CO_2CH_3	398.50	$C_{23}H_{30}N_2O_4$	Species: *Mitragyna speciosa* (Rubiaceae family)	[39]
35	Nigritanine	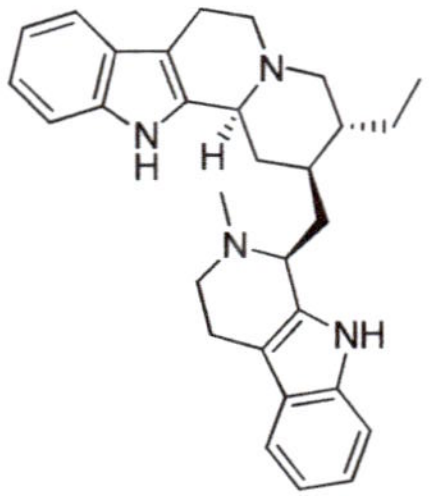 N, NH, H, N, NH	452.63	$C_{30}H_{36}N_4$	*Strychnos* species: *Strychnos borteri*, *S. nigritana* Bak. (Loganiaceae family)	[31]

Table 1. *Cont.*

Mol.	Common Name	Chemical Structure	M.W.	Molecular Formula	Source	Ref.
36	Paynantheine	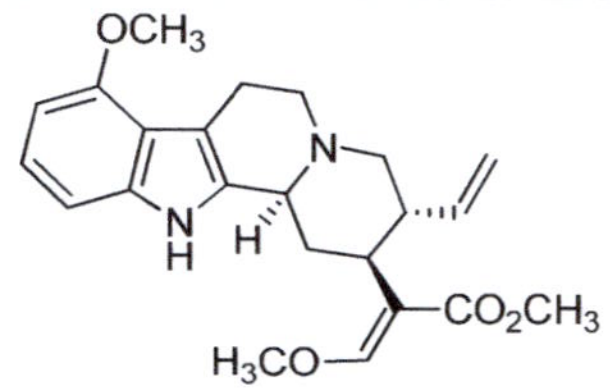	396.48	$C_{23}H_{28}N_2O_4$	Species: *M. speciosa* (Rubiaceae family)	[40]
37	Rhynchophylline	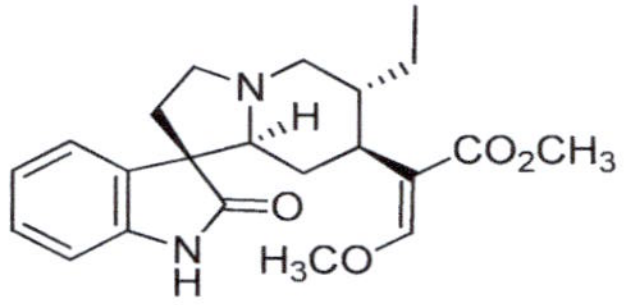	384.47	$C_{22}H_{28}N_2O_4$	Species: *M. speciosa*, *Uncaria rhynchophylla* (Rubiaceae family)	[40]
38	Speciociliatine	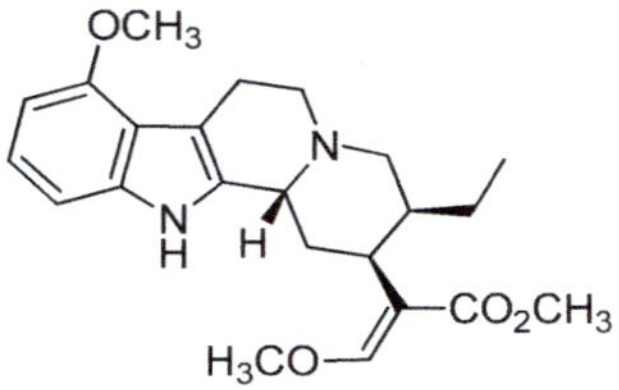	398.50	$C_{23}H_{30}N_2O_4$	Species: *M. speciosa* (Rubiaceae family)	[41]
39	Yohimbine·HCl	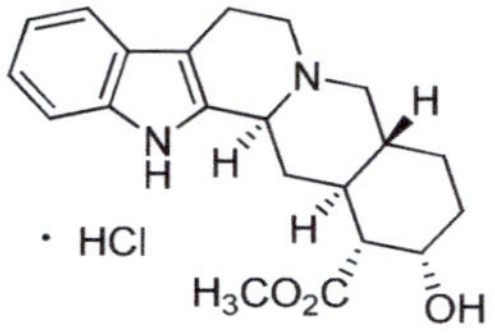	354.44 390.90 (+HCl)	$C_{21}H_{26}N_2O_3$ HCl	*Apocynaceae* species: *Aspidosperma discolor* A. DC., *Aspidosperma excelsum* Benth, *Aspidosperma eburneum* F. Allem, *Aspidosperma marcgravianum* Woodson, *Aspidosperma oblongum* A. DC.	[37]

Mol.: molecule number; M.W.: molecular weight; Ref.: references.

Considering the chemical structures of isoquinoline alkaloids, this group can be divided in two major categories: simple isoquinolines, which are composed of a benzene ring fused to a pyridine ring, and benzylisoquinolines, which contain a second aromatic ring [42]. In contrast, indole alkaloids are bicyclic structures consisting of a six-membered benzene ring fused to a five-membered nitrogen-containing pyrrole ring and are among the most numerous (at least 4100 known molecules) and complex alkaloids. In our library, the group of indole alkaloids covers several subclasses, the largest of which is represented by β-carbolines featuring a common tricyclic pyrido[3,4-b]indole ring structure (Table 1). According to the saturation of the N-containing six-membered ring, β-carbolines are categorized in fully aromatic (βCs), dihydro- (DHβCs), and tetrahydro- (THβCs) β-carbolines [43]. Alkaloids from the in-house library belonging to more than 10 plant families (e.g., Apocynaceae, Loganiaceae, Berberidaceae, Papaveraceae, Rubiaceae) are known to occur in several species (Table 1). With the aim to identify new potential antistaphylococcal agents, the in-house library of alkaloids was initially screened towards Gram(+) and Gram(-) reference bacterial strains. For the biological characterization, all the compounds were dissolved in dimethyl sulfoxide (DMSO).

2.2. Antimicrobial Activity

2.2.1. Inhibition Zone Assay

The antimicrobial activity of all collected compounds was initially tested on a reference strain of the Gram(+) *S. aureus* (*S. aureus* ATCC 25923) by the inhibition zone assay. The Gram(-) bacterial strain *Escherichia coli* ATCC 25922 was also included for comparison (Table 2). Most of the compounds resulted to be inactive against both classes of bacteria (data not shown). The only exception was the already characterized methylated derivative of β-carboline, i.e., harmane, which was able to inhibit the growth of both *S. aureus* and *E. coli* strains in an agar diffusion assay [44–46], with diameters of the inhibition zone of 4.36 and 8.46 mm, respectively (Table 2).

Table 2. Diameters of the inhibition zone of all the active tested compounds against the reference Gram(+) and Gram(-) bacterial strains.

Inhibition Zone Assay	**Diameter of Inhibition Zone (mm)** [1]	
Compound	**Gram-Positive** ***Staphylococcus aureus*** **ATCC 25923**	**Gram-Negative** ***Escherichia coli*** **ATCC 25922**
Dihydroberberine·HCl (**1**)	7.800	n.a.
(*S*)-Glaucine (**7**)	7.600	n.a.
Canthin-6-one (**28**)	6.100	n.a.
Harmane (**30**)	4.360	8.640
Harmine (**32**)	n.a.	6.250
Mytragine (**34**)	5.420	n.a.
Nigritanine (**35**)	10.39	n.a.
Paynantheine (**36**)	8.440	n.a.
Speciociliatine (**38**)	8.240	n.a.

[1] Data represent the mean of three independent experiments with standard deviation (SD) not exceeding 0.2; n.a.: not active.

Interestingly, a greater selectivity towards the human pathogen *S. aureus* was noted, especially for the β-carboline alkaloids. In fact, the rare heterodimer alkaloid nigritanine (compound **35**) as well as some of its analogues (i.e., speciociliatine, mytragine, and paynantheine) showed a powerful activity against the reference strain of *S. aureus* ATCC 25923, with a diameter of growth inhibition zone ranging from 8.24 to 10.39 mm. Mytragine was previously characterized for its selective anti-Gram(+) efficacy [47]; in contrast, no microbiological data have been provided so far for speciociliatine, paynantheine, and rhyncophylline. Because of these reasons, we decided to examine the activity of nigritanine, mytragine, and the other three abovementioned molecules against three multidrug-resistant clinical isolates of *S. aureus*. As reported in Table 3, nigritanine was the sole compound that retained a potent activity against the clinical isolates of *S. aureus* (i.e., *S. aureus* 1a, 1b, 1c). This was indicated by the similar diameters of inhibition zones. In contrast, all the others molecules completely or almost completely lost their activity towards *S. aureus* 1a, 1b, and 1c strains.

Table 3. Diameters of the inhibition zone of some active alkaloids against the reference and clinical isolates of *S. aureus* strains.

Inhibition Zone Assay	**Diameter of Inhibition Zone (mm)** [1]			
Compound	***S. aureus* ATCC 25923**	***S. aureus* 1a**	***S. aureus* 1b**	***S. aureus* 1c**
Mytragine (**34**)	5.420	4.000	n.a.	n.a.
Nigritanine (**35**)	10.39	11.20	8.440	9.100
Paynantheine (**36**)	8.440	3.800	n.a.	n.a.
Speciociliatine (**38**)	8.240	4.520	4.340	4.580

[1] Data represent the mean of three independent experiments with SD not exceeding 0.2.

A representative image of the antibacterial activity of these compounds is shown in Figure 1. The growth inhibition zone of nigritanine (zone #1) is clearly evident compared to that of the other alkaloids tested. These results are in line with other published data of alkaloids extracted from *Anabasis articulata*, showing a potent anti-Gram(+) activity when evaluated by the inhibition zone assay [48].

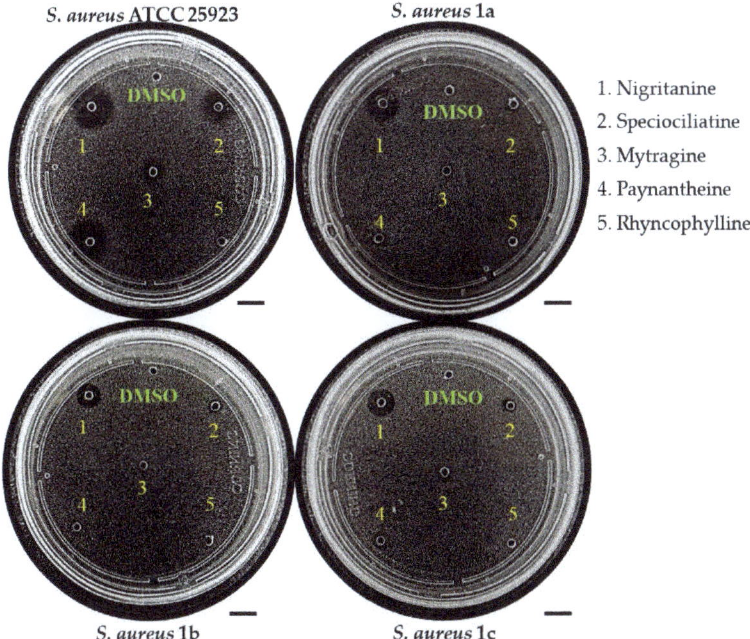

Figure 1. Representative image of the inhibition zone assays of nigritanine (**35**) and some other alkaloids against the reference strain and the three clinical isolates of *S. aureus*. Scale bars represent 1 cm.

2.2.2. Determination of the Minimum Inhibitory Concentration

The antibacterial activity of nigritanine, speciociliatine, mytragine, paynantheine, and rhyncophylline was also evaluated by the microdilution broth assay to determine the minimum inhibitory concentration (MIC) against the reference strain of *S. aureus* and the three clinical isolates after 16 hours of treatment

(Table 4). Remarkably, although the MIC of nigritanine against the reference strain of *S. aureus* was higher than the MIC of other alkaloids reported in the literature (i.e., 128 µM, 56.5 µg/mL versus 2–16 µg/mL for vincamine, atropine, allantoin, or trigonelline [49]), the ability to inhibit the growth of the clinical isolates was maintained at the same concentration of 128 µM. This is in contrast with what observed for the aforementioned compounds which completely lost activity when tested against clinical isolates [49]. Similar MIC values were also obtained with alkaloids from leaves of *Eclipta alba* [50].

Table 4. Minimum inhibitory concentration (MIC) (µM) of nigritanine, speciociliatine, mytragine, paynantheine, and rhyncophylline against the reference and clinical isolates of *S. aureus* strains. MICs are the values obtained from three identical readings out of four independent experiments.

Strains	Nigritanine	Speciociliatine	Mytragine	Paynantheine	Rhyncophylline
S. aureus ATCC 25923	128 µM	> 256 µM	> 256 µM	> 256 µM	> 256 µM
S. aureus 1a	128 µM	> 256 µM	> 256 µM	> 256 µM	> 256 µM
S. aureus 1b	128 µM	> 256 µM	> 256 µM	> 256 µM	> 256 µM
S. aureus 1c	128 µM	> 256 µM	> 256 µM	> 256 µM	> 256 µM

Notably, the MICs of nigritanine were found to correspond to the minimum bactericidal concentration (MBC) which is defined as the minimum concentration of drug causing ≥3 log killing of bacteria after 16 hours of incubation. Indeed, about five log reduction in the number of viable cells of the reference and clinical isolates of *S. aureus* were detected after treatment with nigritanine at its MIC (i.e., 128 µM) (Figure 2). Note that other plant alkaloid extracts had similar or even higher MIC and MBC values against *S. aureus* and other Gram(+) bacterial strains. For example, alkaloid extracts from the aerial part of *Sida acuta* gave MIC and MBC values ranging from 80 to >400 µg/mL against *Staphylococcus* strains [51], while the MIC of alkaloid extracts of *Mahonia aquifolium* ranged from 100 to 500 µg/mL against *Staphylococcus epidermidis* and *Staphylococcus hominis* strains [52]. Very high MIC values (>500 µg/mL) were obtained with alkaloids isolated from aerial parts of *Hypecoum erectum* L. (i.e., protopine and allocryptopine) against *S. aureus*, *Bacillus cereus*, and *Bacillus subtilis* strains [53]. Since alkaloids extracts with MICs ranging from 100 to 1000 µg/mL are considered to be compounds endowed with antimicrobial activity [54,55], nigritanine (MIC = 56.5 µg/mL) would represent a highly potential antimicrobial molecule.

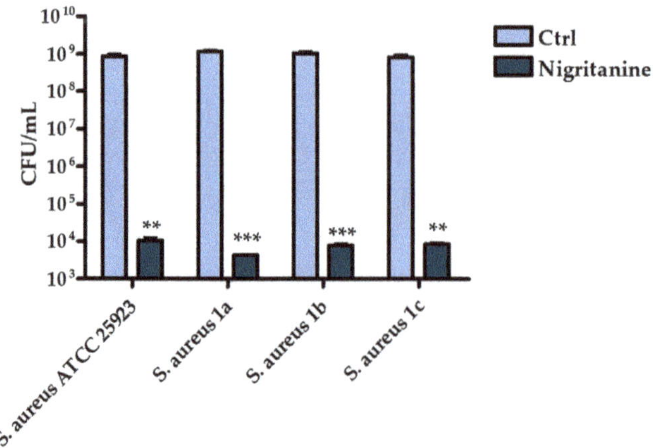

Figure 2. Reduction in the number of viable bacterial cells (evaluated by colony forming unit (CFU) counting) of the reference and clinical isolates of *S. aureus* strains after 16 hours treatment with nigritanine at the MIC (128 µM) compared to control (Ctrl) samples consisting in vehicle-treated bacterial cells. The data represent the mean of three independent experiments ± SD. The levels of statistical significance versus the Ctrl samples were $p < 0.01$ (**); $p < 0.001$ (***).

2.3. Structure–Activity Relationships (SARs) of Nigritanine for Its Antibacterial Activity

Taking into account all these microbiological data, the new β-carboline alkaloid nigritanine (**35**) emerged as a promising antibacterial agent against *S. aureus*. Nigritanine is a rare β-carboline heterodimer from different African *Strichnos* species. In particular, the interest for the *Strichnos* genus, due to the large variety of alkaloids and their use in traditional medicine, led Nicoletti et al. to isolate compound **35**, along with other alkaloids, from the leaves of *Strichnos nigritana* Bak and from the stem bark of *Strichnos barteri* Solered, two species rather common in West Africa. From a chemical standpoint, nigritanine is a heterodimer alkaloid formed by the union of a corynane (Figure 3a) and a tryptamine unit. Interestingly, this corynane heterodimer displays a substantially higher antibacterial activity than the monomeric analogs, confirming the trend observed for the β-carboline homodimer [43,56]. The structure–activity relationships (SARs) were investigated for nigritanine (**35**) and its monomeric analogs (Figure 3b). Accordingly, the analysis of the antibacterial activity related to the corynane scaffolds indicated that: (1) the tetrahydro-β-carboline scaffold exhibits good activity; (2) the methoxyl group at C9 position, the double bond at C19–C18, and the stereochemistry of C3 and C20 do not affect the activity; (3) the corynane heterodimer shows a substantially higher activity than the monomeric analogs, highlighting that the presence of the tryptamine unit is essential. Notably, for β-carboline indoles, dimerization improves the antibacterial activity possibly because the larger molecule is less susceptible to bacterial efflux [43].

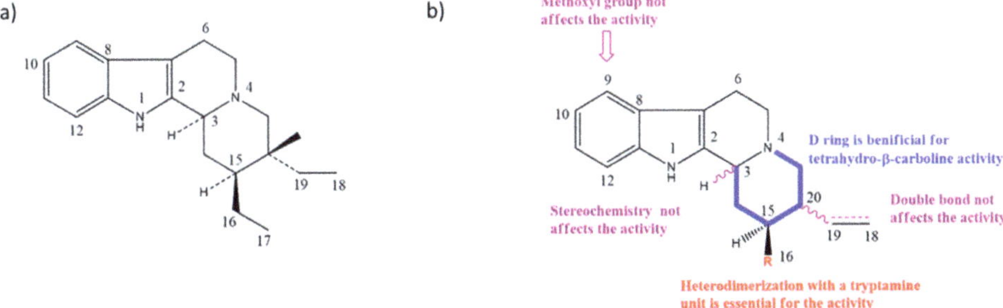

Figure 3. (**a**) Chemical structure of corynane. (**b**) Structure–activity relationship (SARs) analysis of tetrahydro-β-carboline alkaloids with respect to antibacterial activity.

2.4. Cytotoxicity

2.4.1. Hemolytic Assay

The short-term cytotoxic effect of nigritanine was tested against mammalian red blood cells after 40 minutes treatment at its MIC (128 μM), 2 × MIC (256 μM), and 4 × MIC (512 μM). The least active compound speciociliatine was tested at the same concentrations for comparison. As reported in Figure 4, both compounds displayed a weak toxic effect, causing about 30% hemolysis at the highest concentrations, while nigritanine gave rise to about 20% lysis of erythrocytes at its active concentration (128 μM). These results confirmed the potential safety of nigritanine in mammalian cells at a short term.

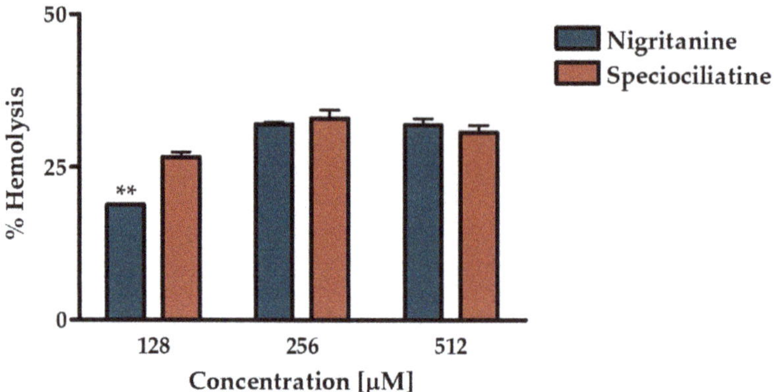

Figure 4. Hemolytic activity of nigritanine at 128 μM (MIC), 256 μM (2 × MIC), and 512 μM (4 × MIC) after 40 minutes of treatment compared to the least active speciociliatine. The data represent the mean ± standard error of the mean (SEM) of three independent experiments. The level of statistical significance between the two compounds was $p < 0.01$ (**).

2.4.2. Cytotoxic Effect on HaCaT Cells

Since the most common site of *S. aureus* infection is the skin, and keratinocytes represent the major cell type in the epidermis [57], the long-term cytotoxic effect of nigritanine was evaluated by the 3-(4,5-dimethylthiazol-2-yl)-2,5-diphenyltetrazolium bromide (MTT) assay on human immortalized keratinocytes (HaCaT) after 24 h treatment. As indicated in Figure 5, nigritanine did not induce any marked reduction in the percentage of viable keratinocytes after 24 h incubation at a concentration range between 2 μM and 200 μM. Note that other natural alkaloids were cytotoxic at much lower concentrations than 2 μM [49,58,59], which contrasts with the maximum non-toxic concentration tested for nigritanine (200 μM = 88.3 μg/mL). Moreover, the MICs of nigritanine (**35**) against reference and clinical isolates of *S. aureus* were equal to 128 μM. These results support the use of this compound as an antibacterial agent harmless to mammalian cells at its active antibacterial concentration.

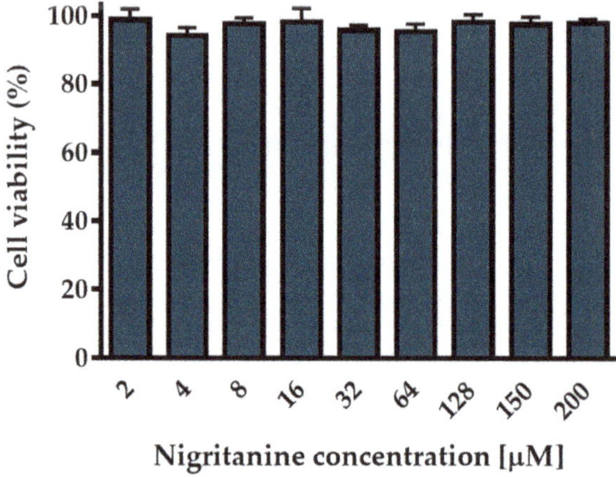

Figure 5. Effect of nigritanine on the viability of HaCaT cells determined by the MTT assay. Cell viability is expressed as a percentage with respect to the control. All data are the means of three independent experiments ± SEM.

3. Conclusions

With the increasing occurrence of multidrug-resistant bacterial infections and the low number of new antimicrobial agents on the market, the discovery of new natural compounds with antibiotic action is extremely necessary. In this work, we characterized the antibacterial profile of the heterodimer alkaloid nigritanine (isolated in the 1980s from *Strychnos* species), against *S. aureus* strains including clinical isolates. Interestingly, nigritanine resulted to have a potent anti-staphylococcal activity without being toxic to mammalian red blood cells and human keratinocytes at its active concentration. More importantly, it retained its antibacterial activity against three multidrug-resistant clinical isolates of *S. aureus*, a feature that was not observed for the other tested carboline alkaloids. Thus, the heterodimer alkaloid nigritanine has emerged as a promising scaffold for the design and development of potent and selective antibacterial compounds with low cytotoxicity.

4. Material and Methods

4.1. Chemistry

All the tested compounds (namely, **1–39**) are known structures belonging to our in-house library of natural products. The chemical identity of compounds was assessed by re-running Nuclear magnetic resonance spectroscopy (NMR) experiments and proved to be in agreement with the literature data reported below for each compound. The purity of all compounds, checked by reversed-phase High Performance Liquid Chromatography (HPLC), was always higher than 95%.

Compound **1** (dihydroberberine hydrochloride or 9,10 dimethoxy-6,8-dihydro-5*H*-[1,3]dioxolo[4,5-*g*]isoquinolino[3,2-*a*]isoquinoline hydrochloride) was purchased from Fluka (CAS: 483-15-8, St. Louis, MO, USA) and used without further purification.

Compound **2** (bulbocapnine hydrochloride or (*S*)-11-methoxy-7-methyl-6,7,7a,8-tetrahydro-5*H*-[1,3]dioxolo[4′,5′:4,5]benzo[1,2,3-*de*]benzo[*g*]quinolin-12-ol hydrochloride) was purchased from Sigma-Aldrich (CAS: 632-47-3, St. Louis, MO, USA) and used without further purification.

Compound **3** (boldine or (6a*S*)-1,10-dimethoxy-6-methyl-5,6,6a,7-tetrahydro-4*H*-dibenzo[*de,g*] quinoline-2,9-diol) was purchased from Sigma-Aldrich (CAS: 476-70-0, St. Louis, MO, USA) and used without further purification.

Compound **4** (cotarmine hydrochloride or (*R*)-4-methoxy-6-methyl-5,6,7,8-tetrahydro-[1,3]dioxolo [4,5-*g*]isoquinolin-5-ol) was purchased from MolPort (CAS: 82-54-2, Beacon, NY, USA) and used without further purification.

Compound **5** (chelidonine or (5b*R*,6*S*,12b*S*)-13-Methyl-5b,6,7,12b,13,14-hexahydro[1,3]dioxolo[4′,5′:4,5]benzo[1,2-*c*][1,3]dioxolo[4,5-*i*]phenanthridin-6-ol) was purchased from Sigma-Aldrich (CAS: 476-32-4, St. Louis, MO, USA) and used without further purification.

Compound **6** (emetine hydrochloride or (2*S*, 3 *R*, 11b *S*)-2-(((*R*)-6, 7-dimethoxy-1, 2, 3, 4-tetrahydroisoquinolin-1-yl)methyl)-3-ethyl-9,10-dimethoxy-2,3,4,6,7,11b-hexahydro-1*H*-pyrido[2,1-*a*] isoquinoline hydrochloride) was purchased from MolPort (CAS: 14198-59-5, Beacon, NY, USA) and used without further purification.

Compound **7** ((*S*)-Glaucine or (6a*S*)-1,2,9,10-tetramethoxy-6-methyl-5,6,6a,7-tetrahydro-4*H*-dibenzo [*de,g*]quinoline) was purchased from MolPort (CAS: 475-81-0, Beacon, NY, USA) and used without further purification.

Compound **8** (hydrastine or (*R*)-6,7-dimethoxy-3-((*R*)-6-methyl-5,6,7,8-tetrahydro-[1,3]dioxolo[4,5-g]isoquinolin-5-yl)isobenzofuran-1(3*H*)-one) was purchased from Sigma-Aldrich (CAS: 118-08-1, St. Louis, Mo., USA) and used without further purification.

Compound **9** (noscapine or narcotine or (3*S*)-6,7-dimethoxy-3-((5*R*)-4-methoxy-6-methyl-5,6,7,8,9,9a-hexahydro-[1,3]dioxolo[4,5-*g*]isoquinolin-5-yl)isobenzofuran-1(3*H*)-one) was purchased from Sigma-Aldrich (CAS: 128-62-1, St. Louis, MO, USA) and used without further purification.

Compound **10** (papaverine or (6,7-dimethoxyisoquinolin-1-yl)(3,4-dimethoxyphenyl)methanone) was purchased from MolPort (CAS: 58-74-2, Beacon, NY, USA) and used without further purification.

Compound **11** (tubocurarine chloride hydrochloride or (1*S*,16*R*)-10,25-dimethoxy-15,15,30-trimethyl-7,23-dioxa-30-aza-15-azoniaheptacyclo[22.6.2.2^{3,6}.1^{8,12}.1^{18,22}.0^{27,31}.0^{16,34}]hexatriaconta-3(36),4,6(35),8(34),9,11,18(33),19,21,24,26,31-dodecaene-9,21-diol chloride hydrochloride) showed NMR spectra identical to those reported in the literature [25].

Compound **12** (cinchonine or (*S*)-quinolin-4-yl((1S,2R,4S,5R)-5-vinylquinuclidin-2-yl)methanol) was purchased from Sigma-Aldrich (CAS: 118-10-5, St. Louis, MO, USA) and used without further purification.

Compound **13** (kokusaginine or 4,6,7-trimethoxyfuro[2,3-*b*]quinoline) showed NMR spectra identical to those reported in the literature [27].

Compound **14** (maculine or 9-methoxy-[1,3]dioxolo[4,5-*g*]furo[2,3-*b*]quinoline) showed NMR spectra identical to those reported in the literature [27].

Compound **15** (4-methoxy-2-(1-ethylpropyl)-quinoline) showed NMR spectra identical to those reported in the literature [27].

Compound **16** (aspidospermine or 1-((3a*R*,5a*R*,10b*R*,12b*R*)-3a-Ethyl-7-methoxy-2,3,3a,5,5a,11,12,12b-octahydro-1*H*,4*H*-6,12a-diaza-indeno[7,1-*cd*]fluoren-6-yl)-ethanone) showed NMR spectra identical to those reported in the literature [28].

Compound **17** (brucine or (4a*R*,5a*S*,8a*R*,15b*R*)-10,11-dimethoxy-4a,5,5a,7,8,13a,15,15a,15b,16-decahydro-2*H*-4,6-methanoindolo[3,2,1-*ij*]oxepino[2,3,4-*de*]pyrrolo[2,3-*h*]quinolin-14-one) was purchased from Sigma-Aldrich (CAS: 357-57-3, St. Louis, MO, USA) and used without further purification.

Compound **18** (diaboline or (4b*R*,7a*S*,8a*R*,13*R*,13a*R*,13b*R*)-13-hydroxy-5,6,7a,8,8a,11,13,13a,13b,14-decahydro-7,9-methanoindeno[1,2-*d*]oxepino[3,4-*f*]indole-14-carboxylic acid) showed NMR spectra identical to those reported in the literature [60].

Compound **19** (physostigmine or eserine or ((3a*R*,8b*S*)-3,4,8b-trimethyl-2,3a-dihydro-1*H*-pyrrolo[2,3-*b*]indol-7-yl) *N*-methylcarbamate) was purchased from MolPort (CAS: 57-47-6, Beacon, NY, USA) and used without further purification.

Compound **20** (holstiine or (15*Z*)-15-Ethylidene-10-hydroxy-17-methyl-11-oxa-8,17-diazapentacyclo[12.5.2.11,8.02,7.013,22]docosa-2,4,6-triene-9,20-dione) showed NMR spectra identical to those reported in the literature [61].

Compound **21** (pseudobrucine or (4a*R*,4a^1*R*,5a*R*,8a*S*,8a^1*S*,15a*S*)-5a-hydroxy-10,11-dimethoxy-4a^1,5,5a,7,8,8a^1,15,15a-octahydro-2*H*-4,6-methanoindolo[3,2,1-*ij*]oxepino[2,3,4-*de*]pyrrolo[2,3-*h*]quinolin-14(4a*H*)-one) was purchased from Fisherpharma (CAS: 560-30-5, Beacon, NY, USA) and used without further purification.

Compound **22** (retuline or 1-((3a*S*,5*R*,6*S*,6a*S*,11b*R*,*E*)-12-ethylidene-6-(hydroxymethyl)-1,2,4,5,6,6a-hexahydro-3,5-ethanopyrrolo[2,3-*d*]carbazol-7(3a*H*)-yl)ethanone) showed NMR spectra identical to those reported in the literature [62].

Compound **23** (serotonin or 3-(2-aminoethyl)-1H-indol-5-ol) was purchased from Sigma-Aldrich (CAS: 50-67-9, St. Louis, MO, USA) and used without further purification.

Compound **24** (triptamine hydrochloride or 2-(1*H*-indol-3-yl)ethan-1-amine hydrochloride) was purchased from Sigma-Aldrich (CAS: 343-94-2, St. Louis, MO, USA) and used without further purification.

Compound **25** (vomicine hydrochloride or (4a*R*,4a^1*R*,6a*S*,6a^1*S*,13a*S*)-10-hydroxy-16-methyl-4a,5,13,13a-tetrahydro-2*H*-6a,4-(ethanoiminomethano)indolo[3,2,1-*ij*]oxepino[2,3,4-*de*]quinoline-6,12(4a^1*H*,6a^1*H*)-dione hydrochloride) was purchased from MolPort (5969-84-6, Beacon, NY, USA) and used without further purification.

Compound **26** (vindoline or (3a*R*,3a^1*R*,4*R*,5*S*,5a*R*,10b*R*)-methyl 4-acetoxy-3a-ethyl-5-hydroxy-8-methoxy-6-methyl-3a,3a^1,4,5,5a,6,11,12-octahydro-1*H*-indolizino[8,1-*cd*]carbazole-5-carboxylate) was purchased from MolPort (CAS: 2182-14-1, Beacon, NY, USA) and used without further purification.

Compound **27** (akagerine or (*E*)-2-((3a*S*,5*R*,7*S*)-7-hydroxy-3-methyl-1,2,3,3a,4,5,6,7-octahydro-3,7a-diazacyclohepta[*jk*]fluoren-5-yl)but-2-enal) showed NMR spectra identical to those reported in the literature [63].

Compound **28** (canthin-6-one or 6*H*-indolo[3,2,1-*de*][1,5]naphthyridin-6-one) was purchased from MolPort (CAS: 479-43-6, Beacon, NY, USA) and used without further purification.

Compound **29** (α-carboline or 9*H*-pyrido[2,3-*b*]indole) was purchased from MolPort (CAS: 244-76-8, Beacon, NY, USA) and used without further purification.

Compound **30** (harmane or 1-methyl-9*H*-pyrido[3,4-*b*]indole) was purchased from Sigma-Aldrich (CAS: 486-84-0, St. Louis, MO, USA) and used without further purification.

Compound **31** (norharmane or 9*H*-pyrido[3,4-*b*]indole) was purchased from Sigma-Aldrich (CAS Number: 244-63-3, St. Louis, MO, USA) and used without further purification.

Compound **32** (harmine or 7-methoxy-1-methyl-9*H*-pyrido[3,4-*b*]indole) was purchased from Sigma-Aldrich (CAS: 442-51-3, St. Louis, MO, USA) and used without further purification.

Compound **33** (ibogaine or (6*R*,7*S*,11*S*)-7-ethyl-2-methoxy-6,6a,7,8,9,10,12,13-octahydro-5*H*-6,9-methanopyrido[1′,2′:1,2]azepino[4,5-*b*]indole) showed NMR spectra identical to those reported in the literature [38].

Compound **34** (mitragynine or (*E*)-methyl 2-((2*S*,3*S*,12b*S*)-3-ethyl-8-methoxy-1,2,3,4,6,7,12,12b-octahydroindolo[2,3-*a*]quinolizin-2-yl)-3-methoxyacrylate) was purchased from BOC Sciences (CAS: 4098-40-2, Shirley, NY, USA) and used without further purification.

Compound **35** (nigritanine or (2*S*,3*R*,12b*S*)-3-ethyl-2-(((*S*)-2-methyl-2,3,4,9-tetrahydro-1*H*-pyrido [3,4-*b*]indol-1-yl)methyl)-1,2,3,4,6,7,12,12b-octahydroindolo[2,3-*a*]quinolizine) showed NMR spectra identical to those reported in the literature [64]. The chemical characterization of compound **35** is reported below: Brown solid, m.p. 202-204°C; ^{1}H-NMR ($CDCl_3$, 400 MHz): 0.93 (t, J = 7 Hz, 3H, Me-C), 2.46 (s, 3H, Me-N), 3.56 (m, 1H, H-3), 6.40 (s, 1H, H-1 or H-1′), 7.80 (s, 1H, H-1 or H-1′); ^{13}C-NMR ($CDCl_3$, 101 MHz): 11.2 (C18), 20.7 (C6′), 21.5 (C6), 23.7 (C19), 41.9 (C20), 42.7 (N-CH_3), 34.9 (C14), 35.0 (C16), 35.8 (C15), 51.3 (C5′), 53.0 (C5), 58.6 (C17), 59.1 (C3), 60.2 (C21), 107.1 (C7), 108.8 (C7′), 110.8 (C12), 110.9 (C12′), 117.5 (C9 or C9′), 117.7 (C9 or C9′), 118.8 (C10 or C10′), 119.3 (C10 or C10′), 120.4 (C11 or C11′), 121.4 (C11 or C11′), 126.7 (C8), 127.0 (C8′), 134.7 (C2), 135.5 (C2′), 135.6 (C13 or C13′), 135.7 (C13 or C13′); m/z (ESI+) 452 (M+, 82%), 437 (6), 408 (10), 267 (8), 253 (12), 199 (27), 185 (100).

Compound **36** (paynantheine or (*E*)-methyl 3-methoxy-2-((2*S*,3*R*,12b*S*)-8-methoxy-3-vinyl-1,2,3,4, 6,7,12,12b-octahydroindolo[2,3-*a*]quinolizin-2-yl)acrylate) was purchased from BOC Sciences (CAS: 4697-66-9, Shirley, NY, USA) and used without further purification.

Compound **37** (rhynchophylline or (*E*)-methyl 2-((1′*R*,6′*R*,7′*S*,8a′*S*)-6′-ethyl-2-oxo-3′,5′,6′,7′,8′,8a′-hexahydro-2′*H*-spiro[indoline-3,1′-indolizin]-7′-yl)-3-methoxyacrylate) was purchased from BOC Sciences (CAS: 76-66-4, Shirley, NY, USA) and used without further purification.

Compound **38** (speciociliatine or (*E*)-methyl 2-((2*S*,3*S*,12b*R*)-3-ethyl-8-methoxy-1,2,3,4,6,7,12,12b-octahydroindolo[2,3-*a*]quinolizin-2-yl)-3-methoxyacrylate) was purchased from BOC Sciences (CAS: 14382-79-7, Shirley, NY, USA) and used without further purification.

Compound **39** (yohimbine hydrochloride or (1*R*,2*S*,4a*R*,13b*S*,14a*S*)-methyl 2-hydroxy-1,2,3,4,4a, 5,7,8,13,13b,14,14a-dodecahydroindolo[2′,3′:3,4]pyrido[1,2-*b*]isoquinoline-1-carboxylate hydrochloride) was purchased from Sigma-Aldrich (CAS: 65-19-0, St. Louis, MO, USA) and used without further purification.

4.2. Microorganisms and Cell Line

The reference Gram(+) and Gram(-) strains used for the antimicrobial tests were *S. aureus* ATCC 25923 and *E. coli* ATCC 25922, respectively. The multidrug-resistant clinical isolates of *S. aureus* (1a, 1b, and 1c) were kindly provided by Professor Giammarco Raponi (Sapienza, University of Rome).

HaCaT cells (AddexBio, San Diego, CA, USA) were cultured in Dulbecco's modified Eagle's medium supplemented with 4 mM glutamine (DMEMg), 10% heat-inactivated fetal bovine serum (FBS), and 0.1 mg/mL of penicillin and streptomycin at 37 °C and 5% CO_2, in 25 cm^2 or 75 cm^2 flasks.

4.3. Antimicrobial Assays

A bacterial culture inoculum was incubated at 37 °C in Luria–Bertani (LB) broth until reaching an optical density (O.D.) of 0.8 at 590 nm. For the inhibition zone assay, the bacterial culture was diluted 1:2000 and plated in LB-agarose plates. An aliquot (3 µL) of each compound at 5 mM was loaded into holes previously made in the agarose plates [65,66]. Afterwards, the plates were incubated overnight at 30 °C. Afterwards, the diameters of the inhibition zone were measured and reported in Tables 2 and 3.

For the determination of the MICs, 50 µL of bacterial suspension (2×10^5 cells/mL) in Mueller–Hinton broth (MH) was added to 50 µL MH containing serial dilutions of the compounds (from 256 µM to 2 µM) previously prepared in a 96-well plate. Controls consisted of vehicle-treated bacterial cells [67]. The plate was then incubated for 16 hours at 37 °C, and MIC was defined as the lowest concentration causing 100% visible inhibition of microbial growth. Afterwards, aliquots from the wells corresponding to the MIC of nigritanine (**35**) and to the control were withdrawn and plated onto LB agar plates for colony forming unit (CFU) counting and MBC determination.

4.4. Cytotoxicity Assays

To evaluate short-term cytotoxicity, selected alkaloids were tested on sheep red blood cells (OXOID, SR0051D). Aliquots of erythrocyte suspension (O.D. of 0.5 at 500 nm) in 0.9 % (*w*/*v*) NaCl were incubated for 40 min at 37 °C with three different concentrations (MIC, 2 × MIC, and 4 × MIC) of nigritanine (**35**) or the same concentrations of speciociliatine (**38**). The treated samples were then centrifuged for 5 min at 900× g. The amount of hemoglobin released in the supernatant by lysed red blood cells was measured at 415 nm using a microplate reader (Infinite M200; Tecan, Salzburg, Austria). The complete lysis was obtained by suspending erythrocytes in distilled water according to [68–70].

To evaluate the in vitro long-term cytotoxicity of nigritanine (**35**), a colorimetric method was employed. This assay is based on the intracellular reduction of the yellow tetrazolium salt MTT (Sigma-Aldrich, St. Luis, MO, USA) to purple formazan crystals by mitochondrial dehydrogenases of metabolically active cells. Therefore, the amount of purple color is directly proportional to the number of viable cells. About 4×10^4 HaCaT cells resuspended in DMEMg supplemented with 2% FBS, without antibiotics, were plated in each well of a 96-well plate. After overnight incubation in a humidified atmosphere containing 5% CO_2 at 37 °C, the medium was removed, and fresh serum-free DMEMg containing the compound at different concentration was added in each well. For controls, cells were treated with vehicle. The plate was incubated for 24 h at 37 °C and 5% CO_2. Afterwards, the medium was discarded, and 0.5 mg/mL of MTT in Hank's buffer (136 mM NaCl, 4.2 mM Na_2HPO_4, 4.4 mM KH_2PO_4, 5.4 mM KCl, 4.1 mM $NaHCO_3$, pH 7.2, supplemented with 20 mM D-glucose) was added to each well. After 4 h incubation at 37 °C and 5% CO_2, 100 µL of acidified isopropanol was added to each well, in order to dissolve the formazan crystals [71,72]. Absorbance was measured by a microplate reader (Infinite M200; Tecan, Salzburg, Austria) at 570 nm, and cell viability was calculated with respect to the control (cells in medium supplemented with vehicle).

4.5. Statistical Analysis

All data are expressed as the mean ± SD or SEM. Statistical analyses were performed using Student's t-test with the Prism software package (GraphPad, 6.0, San Diego, CA, USA), and the differences were considered to be statistically significant for $p < 0.05$.

Author Contributions: B.C. performed the microbiological assays; F.C. performed the cytotoxicity assays, M.R.L. partially contributed to the biological experiments. D.Q. and A.C. performed the acquisition and interpretation of the magnetic resonance spectroscopy (NMR) spectra of bioactive compounds and the determination of purity by High Performance Liquid Chromatography (HPLC). M.M. revised and edited the manuscript. B.B. provided overall guidance. F.G. and M.L.M. conceived the project and wrote the manuscript.

Funding: This work was supported by Italian Ministry of Education, University and Research—Dipartimenti di Eccellenza—L.232/2016 and Italian Institute of Technology. This work was also supported by grants from Sapienza University of Rome (project RM11816436113D8A).

Acknowledgments: We are grateful to Giammarco Raponi, Department of Public Health and Infectious Diseases, Sapienza University of Rome, for providing the clinical isolates.

Conflicts of Interest: The authors declare no conflict of interest.

References

1. Reygaert, W.C. An overview of the antimicrobial resistance mechanisms of bacteria. *AIMS Microbiol.* **2018**, *4*, 482–501. [CrossRef] [PubMed]
2. Ansari, S.; Jha, R.K.; Mishra, S.K.; Tiwari, B.R.; Asaad, A.M. Recent advances in *Staphylococcus aureus* infection: Focus on vaccine development. *Infect. Drug Resist.* **2019**, *12*, 1243–1255. [CrossRef] [PubMed]
3. Gajdacs, M. The continuing threat of methicillin-resistant *Staphylococcus aureus*. *Antibiotics* **2019**, *8*. [CrossRef] [PubMed]
4. Mermel, L.A.; Cartony, J.M.; Covington, P.; Maxey, G.; Morse, D. Methicillin-resistant *Staphylococcus aureus* colonization at different body sites: A prospective, quantitative analysis. *J. Clin. Microbiol.* **2011**, *49*, 1119–1121. [CrossRef] [PubMed]
5. Keihanian, F.; Saeidinia, A.; Abbasi, K. Epidemiology of antibiotic resistance of blood culture in educational hospitals in Rasht, North of Iran. *Infect. Drug Resist.* **2018**, *11*, 1723–1728. [CrossRef]
6. Chung, P.Y. Novel targets of pentacyclic triterpenoids in *Staphylococcus aureus*: A systematic review. *Phytomedicine Int. J. Phytother. Phytopharm.* **2019**. [CrossRef] [PubMed]
7. Calcaterra, A.; D'Acquarica, I. The market of chiral drugs: Chiral switches versus de novo enantiomerically pure compounds. *J. Pharm. Biomed. Anal.* **2018**, *147*, 323–340. [CrossRef]
8. Debnath, B.; Singh, W.S.; Das, M.; Goswami, S.; Singh, M.K.; Maiti, D.; Manna, K. Role of plant alkaloids on human health: A review of biological activities. *Mater. Today Chem.* **2018**, *9*, 56–72. [CrossRef]
9. Cushnie, T.P.T.; Cushnie, B.; Lamb, A.J. Alkaloids: An overview of their antibacterial, antibiotic-enhancing and antivirulence activities. *Int. J. Antimicrob. Agents* **2014**, *44*, 377–386. [CrossRef]
10. Ingallina, C.; D'Acquarica, I.; Delle Monache, G.; Ghirga, F.; Quaglio, D.; Ghirga, P.; Berardozzi, S.; Markovic, V.; Botta, B. The pictet-spengler reaction still on stage. *Curr. Pharm. Des.* **2016**, *22*, 1808–1850. [CrossRef]
11. Ghirga, F.; Bonamore, A.; Calisti, L.; D'Acquarica, I.; Mori, M.; Botta, B.; Boffi, A.; Macone, A. Green routes for the production of enantiopure benzylisoquinoline alkaloids. *Int. J. Mol. Sci.* **2017**, *18*. [CrossRef] [PubMed]
12. Amirkia, V.; Heinrich, M. Alkaloids as drug leads—A predictive structural and biodiversity-based analysis. *Phytochem. Lett.* **2014**, *10*. [CrossRef]
13. Infante, P.; Alfonsi, R.; Ingallina, C.; Quaglio, D.; Ghirga, F.; D'Acquarica, I.; Bernardi, F.; Di Magno, L.; Canettieri, G.; Screpanti, I.; et al. Inhibition of hedgehog-dependent tumors and cancer stem cells by a newly identified naturally occurring chemotype. *Cell Death Dis.* **2016**, *7*, E2376. [CrossRef] [PubMed]
14. Mori, M.; Tottone, L.; Quaglio, D.; Zhdanovskaya, N.; Ingallina, C.; Fusto, M.; Ghirga, F.; Peruzzi, G.; Crestoni, M.E.; Simeoni, F.; et al. Identification of a novel chalcone derivative that inhibits Notch signaling in T-cell acute lymphoblastic leukemia. *Sci. Rep.* **2017**, *7*, 2213. [CrossRef] [PubMed]
15. Infante, P.; Mori, M.; Alfonsi, R.; Ghirga, F.; Aiello, F.; Toscano, S.; Ingallina, C.; Siler, M.; Cucchi, D.; Po, A.; et al. Gli1/DNA interaction is a druggable target for Hedgehog-dependent tumors. *EMBO. J.* **2015**, *34*, 200–217. [CrossRef] [PubMed]
16. Srivastava, S.; Srivastava, M.; Misra, A.; Pandey, G.; Rawat, A. A review on biological and chemical diversity in Berberis (*Berberidaceae*). *Excli J.* **2015**, *14*, 247–267. [CrossRef]
17. Meyer, A.; Imming, P. Benzylisoquinoline alkaloids from the *Papaveraceae*: The heritage of Johannes Gadamer (1867–1928). *J. Nat. Prod.* **2011**, *74*, 2482–2487. [CrossRef]
18. Urzúa, A.P.A. Alkaloids from the bark of *Peumus boldus*. *Fitoterapia* **1983**, *54*, 175–177.
19. Imaki, N.M.Y.; Shimpuku, T.; Shirasaka, T. *A Process for Preparing Cotarnine*; Technical Report No. 4,963,684; U.S. Patent and Trademark Office: Washington, DC, USA, 1986.
20. Colombo, M.L.; Bosisio, E. Pharmacological activities of *Chelidonium majus* L. (*Papaveraceae*). *Pharmacol. Res.* **1996**, *33*, 127–134. [CrossRef]
21. Akinboye, E.; Bakare, O. Biological activities of emetine. *Open Nat. Prod. J.* **2011**, *411*, 8–15. [CrossRef]
22. Lapa, G.P.; Sheichenko, O.; Serezhechkin, A.G.N.; Tolkachev, O. HPLC determination of glaucine in yellow horn poppy grass (*Glaucium flavum* Crantz). *Pharm. Chem. J.* **2004**, *38*, 441–442. [CrossRef]

23. Brown, P.N.; Roman, M.C. Determination of hydrastine and berberine in goldenseal raw materials, extracts, and dietary supplements by high-performance liquid chromatography with UV: Collaborative study. *J. AOAC Int.* **2008**, *91*, 694–701. [PubMed]
24. Ramanathan, V.S.; Chandra, P. Recovery, separation and purification of narcotine and papaverine from Indian opium. *Bull. Narc.* **1981**, *33*, 55–64.
25. Lee, M.R. Curare: The South American arrow poison. *J. R. Coll. Physicians Edinb.* **2005**, *35*, 83–92. [PubMed]
26. Martins, D.; Nunez, C.V. Secondary metabolites from *Rubiaceae* species. *Molecules* **2015**, *20*, 13422–13495. [CrossRef] [PubMed]
27. Dellemonache, F.; Dellemonache, G.; Souza, M.A.D.; Cavalcanti, M.D.; Chiappeta, A. Isopentenylindole derivatives and other components of *Esenbeckia leiocarpa*. *Gazz. Chim. Ital.* **1989**, *119*, 435–439.
28. Guimaraes, H.A.; Braz-Filho, R.; Vieira, I.J.C. H-1 and C-13-NMR data of the simplest plumeran indole alkaloids isolated from *Aspidosperma* species. *Molecules* **2012**, *17*, 3025–3043. [CrossRef]
29. Galeffi, C.; Ciasca-Rendina, M.A.; Miranda Delle Monache, E.; Villar Del Fresno, A.; Marini Bettòlo, G.B. Gradient method for the counter-current separation of alkaloids using a heavy organic phase. *J. Chromatogr.* **1969**, *45*, 407–414. [CrossRef]
30. Zhao, B.; Moochhala, S.M.; Tham, S.Y. Biologically active components of *Physostigma venenosum*. *J. Chromatogr. B* **2004**, *812*, 183–192. [CrossRef]
31. Ohiri, F.C.; Verpoorte, R.; Baerheim Svendsen, A. The African *Strychnos* species and their alkaloids: A review. *J. Ethnopharmacol.* **1983**, *9*, 167–223. [CrossRef]
32. Mohammad-Zadeh, L.F.; Moses, L.; Gwaltney-Brant, S.M. Serotonin: A review. *J. Vet. Pharm.* **2008**, *31*, 187–199. [CrossRef] [PubMed]
33. Kousar, S.; Noreen Anjuma, S.; Jaleel, F.; Khana, J.; Naseema, S. Biomedical significance of tryptamine: A review. *J. Pharmacovigil.* **2017**, *5*. [CrossRef]
34. Song, K.M.; Park, S.W.; Hong, W.H.; Lee, H.; Kwak, S.S.; Liu, J.R. Isolation of vindoline from *Catharanthus roseus* by supercritical fluid extraction. *Biotechnol. Prog.* **1992**, *8*, 583–586. [CrossRef] [PubMed]
35. Noldin, V.F.; de Oliveira Martins, D.T.; Marcello, C.M.; da Silva Lima, J.C.; Delle Monache, F.; Cechinel Filho, V. Phytochemical and antiulcerogenic properties of rhizomes from *Simaba ferruginea* St. Hill. (*Simaroubaceae*). *Zeitschrift für Naturforschung C* **2005**, *60*, 701–706. [CrossRef]
36. Kumar, A.S.; Nagarajan, R. Synthesis of alpha-carbolines via Pd-catalyzed amidation and Vilsmeier-Haack reaction of 3-acetyl-2-chloroindoles. *Org. Lett.* **2011**, *13*, 1398–1401. [CrossRef]
37. Claudia, F.; Amaral, A.; Ramos, A.D.S.; Ferreira, J.; Santos, A.D.S.; Cruz, J.D.S.; De Luna, A.V.M.; Nery, V.V.C.; de Lima, I.C.; Chaves, M.H.d.C.; et al. LC-HRMS for the Identification of β-carboline and canthinone alkaloids isolated from natural sources. *Mass Spectrometry* **2017**, *187*. [CrossRef]
38. Alper, K.R. Ibogaine: A review. *Alkaloids. Chem. Biol.* **2001**, *56*, 1–38.
39. Takayama, H. Chemistry and pharmacology of analgesic indole alkaloids from the rubiaceous plant, *Mitragyna speciosa*. *Chem. Pharm. Bull.* **2004**, *52*, 916–928. [CrossRef]
40. Leon, F.; Habib, E.; Adkins, J.E.; Furr, E.B.; McCurdy, C.R.; Cutler, S.J. Phytochemical characterization of the leaves of *Mitragyna speciosa* grown in USA. *Nat. Prod. Commun.* **2009**, *4*, 907–910. [CrossRef]
41. Cinosi, E.; Martinotti, G.; Simonato, P.; Singh, D.; Demetrovics, Z.; Roman-Urrestarazu, A.; Bersani, F.S.; Vicknasingam, B.; Piazzon, G.; Li, J.H.; et al. Following the roots of kratom (*Mitragyna speciosa*): The evolution of an enhancer from a traditional use to increase work and productivity in Southeast Asia to a recreational psychoactive drug in western countries. *Biomed. Res. Int.* **2015**, 968786. [CrossRef]
42. Evans, W.C. *Pharmacopoeial and Related Drugs of Biological Origin*, 16th ed.; Saunders—Elsevier: Amsterdam, The Netherlands, 2009; pp. 353–416.
43. Dai, J.K.; Dan, W.J.; Schneider, U.; Wang, J.R. β-Carboline alkaloid monomers and dimers: Occurrence, structural diversity, and biological activities. *Eur. J. Med. Chem.* **2018**, *157*, 622–656. [CrossRef] [PubMed]
44. Dai, J.K.; Dan, W.J.; Ren, S.Y.; Shang, C.G.; Wang, J.R. Design, synthesis and biological evaluations of quaternization harman analogues as potential antibacterial agents. *Eur. J. Med. Chem.* **2018**, *160*, 23–36. [CrossRef] [PubMed]
45. Nenaah, G. Antibacterial and antifungal activities of (β)-carboline alkaloids of *Peganum harmala* (L) seeds and their combination effects. *Fitoterapia* **2010**, *81*, 779–782. [CrossRef] [PubMed]
46. Mohammad Reza, V.R.; Hadjiakhoondi, A. Cytotoxicity and antimicrobial activity of harman alkaloids. *J. Pharmacol. Toxicol.* **2007**, *2*, 677–680. [CrossRef]

47. Parthasarathy, S.; Bin Azizi, J.; Ramanathan, S.; Ismail, S.; Sasidharan, S.; Said, M.I.; Mansor, S.M. Evaluation of antioxidant and antibacterial activities of aqueous, methanolic and alkaloid extracts from *Mitragyna speciosa* (*Rubiaceae* family) leaves. *Molecules* **2009**, *14*, 3964–3974. [CrossRef]
48. Benziane Maatalah, M.; Kambuche Bouzidi, N.; Bellahouel, S.; Merah, B.; Fortas, Z.; Soulimani, R.; Saidi, S.; Derdour, A. Antimicrobial activity of the alkaloids and saponin extracts of *Anabasis articulate*. *J. Biotechnol. Pharm. Res.* **2012**, *3*, 54.
49. Ozcelik, B.; Kartal, M.; Orhan, I. Cytotoxicity, antiviral and antimicrobial activities of alkaloids, flavonoids, and phenolic acids. *Pharm. Biol.* **2011**, *49*, 396–402. [CrossRef]
50. Gurrapu, S.; Estari, M. In vitro antibacterial activity of alkaloids isolated from leaves of *Eclipta alba* against human pathogenic bacteria. *Pharmacogn. J.* **2017**, *9*, 573–577. [CrossRef]
51. Karou, D.; Savadogo, A.; Canini, A.; Yameogo, S.; Montesano, C.; Simpore, J.; Colizzi, V.; Traore, A.S. Antibacterial activity of alkaloids from *Sida acuta*. *Afr. J. Biotechnol.* **2006**, *5*, 195–200.
52. Slobodnikova, L.; Kost'alova, D.; Labudova, D.; Kotulova, D.; Kettman, V. Antimicrobial activity of *Mahonia aquifolium* crude extract and its major isolated alkaloids. *Phytother. Res.* **2004**, *18*, 674–676. [CrossRef]
53. Su, Y.F.; Li, S.K.; Li, N.; Chen, L.L.; Zhang, J.W.; Wang, J.R. Seven alkaloids and their antibacterial activity from *Hypecoum erectum* L. *J. Med. Plants Res.* **2011**, *5*, 5428–5432.
54. Manosalva, L.; Mutis, A.; Urzua, A.; Fajardo, V.; Quiroz, A. Antibacterial activity of alkaloid fractions from *Berberis microphylla* G. Forst and study of synergism with ampicillin and cephalothin. *Molecules* **2016**, *21*, 76. [CrossRef] [PubMed]
55. Tegos, G.; Stermitz, F.R.; Lomovskaya, O.; Lewis, K. Multidrug pump inhibitors uncover remarkable activity of plant antimicrobials. *Antimicrob. Agents Chemother.* **2002**, *46*, 3133–3141. [CrossRef] [PubMed]
56. Locher, H.H.; Ritz, D.; Pfaff, P.; Gaertner, M.; Knezevic, A.; Sabato, D.; Schroeder, S.; Barbaras, D.; Gademann, K. Dimers of nostocarboline with potent antibacterial activity. *Chemotherapy* **2010**, *56*, 318–324. [CrossRef] [PubMed]
57. Soong, G.; Paulino, F.; Wachtel, S.; Parker, D.; Wickersham, M.; Zhang, D.; Brown, A.; Lauren, C.; Dowd, M.; West, E.; et al. Methicillin-resistant Staphylococcus aureus adaptation to human keratinocytes. *mBio* **2015**, *6*, e00289-15. [CrossRef] [PubMed]
58. Malikova, J.; Zdarilova, A.; Hlobilkova, A.; Ulrichova, J. The effect of chelerythrine on cell growth, apoptosis, and cell cycle in human normal and cancer cells in comparison with sanguinarine. *Cell Biol. Toxicol.* **2006**, *22*, 439–453. [CrossRef] [PubMed]
59. Zhou, Z.S.; Li, M.; Gao, F.; Peng, J.Y.; Xiao, H.B.; Dai, L.X.; Lin, S.R.; Zhang, R.; Jin, L.Y. Arecoline suppresses HaCaT cell proliferation through cell cycle regulatory molecules. *Oncol. Rep.* **2013**, *29*, 2438–2444. [CrossRef] [PubMed]
60. Delle Monache, F.C.; Rossi, E.; Cartoni, C.; Carpi, A.; Marini Bettolo, G.B. The alkaloids of *Strychnos castelneana*. *Strychnos Alkaloids* **1970**, *33*, 279–283.
61. Martin, G.E. Configuration and total assignment of the ^{1}H-and ^{13}C-NMR spectra of the alkaloid holstiine. *J. Nat. Prod.* **1990**, *53*, 793–800.
62. Tavernier, D.; Anteunis, M.; Tits, M.; Angenot, L. The 1H NMR spectra of the strychnos alkaloids retuline isoretuline, and their N-deacetyl compounds. *Bull. Des Sociétés Chim. Belg.* **2010**, *87*, 595–607. [CrossRef]
63. Cao, M.; Muganga, R.; Nistor, I.; Tits, M.; Angenot, L.; Frederich, M. LC–SPE–NMR–MS analysis of *Strychnos usambarensis* fruits from Rwanda. *Phytochem. Lett.* **2012**, *5*, 170–173. [CrossRef]
64. Nicoletti, M.; Oguakwa, J.U.; Messana, I. On the alkaloids of two African *Strychnos*: *Strychnos nigritana* bak and *Strychnos barteri* Sol. carbon-13 NMR spectroscopy of nigritanins. *Fitoterapia* **1980**, *87*, 595–607.
65. Islas-Rodriguez, A.E.; Marcellini, L.; Orioni, B.; Barra, D.; Stella, L.; Mangoni, M.L. Esculentin 1-21: A linear antimicrobial peptide from frog skin with inhibitory effect on bovine mastitis-causing bacteria. *J. Pept. Sci. Off. Publ. Eur. Pept. Soc.* **2009**, *15*, 607–614. [CrossRef] [PubMed]
66. Mangoni, M.L.; Fiocco, D.; Mignogna, G.; Barra, D.; Simmaco, M. Functional characterisation of the 1–18 fragment of esculentin-1b, an antimicrobial peptide from *Rana esculenta*. *Peptides* **2003**, *24*, 1771–1777. [CrossRef] [PubMed]
67. Falciani, C.; Lozzi, L.; Pollini, S.; Luca, V.; Carnicelli, V.; Brunetti, J.; Lelli, B.; Bindi, S.; Scali, S.; Di Giulio, A.; et al. Isomerization of an antimicrobial peptide broadens antimicrobial spectrum to gram-positive bacterial pathogens. *PLoS ONE* **2012**, *7*, e46259. [CrossRef] [PubMed]

68. Buommino, E.; Carotenuto, A.; Antignano, I.; Bellavita, R.; Casciaro, B.; Loffredo, M.R.; Merlino, F.; Novellino, E.; Mangoni, M.L.; Nocera, F.P.; et al. The outcomes of decorated prolines in the discovery of antimicrobial peptides from Temporin-L. *ChemMedChem* **2019**, *14*, 1283–1290. [CrossRef] [PubMed]
69. Merlino, F.; Carotenuto, A.; Casciaro, B.; Martora, F.; Loffredo, M.R.; Di Grazia, A.; Yousif, A.M.; Brancaccio, D.; Palomba, L.; Novellino, E.; et al. Glycine-replaced derivatives of [Pro(3),DLeu(9)]TL, a temporin L analogue: Evaluation of antimicrobial, cytotoxic and hemolytic activities. *Eur. J. Med. Chem.* **2017**, *139*, 750–761. [CrossRef]
70. Grieco, P.; Carotenuto, A.; Auriemma, L.; Saviello, M.R.; Campiglia, P.; Gomez-Monterrey, I.M.; Marcellini, L.; Luca, V.; Barra, D.; Novellino, E.; et al. The effect of d-amino acid substitution on the selectivity of temporin L towards target cells: Identification of a potent anti-Candida peptide. *Biochimica et Biophysica Acta* **2013**, *1828*, 652–660. [CrossRef]
71. Cappiello, F.; Di Grazia, A.; Segev-Zarko, L.A.; Scali, S.; Ferrera, L.; Galietta, L.; Pini, A.; Shai, Y.; Di, Y.P.; Mangoni, M.L. Esculentin-1a-derived peptides promote clearance of *Pseudomonas aeruginosa* internalized in bronchial cells of cystic fibrosis patients and lung cell migration: Biochemical properties and a plausible mode of action. *Antimicrob. Agents Chemother.* **2016**, *60*, 7252–7262. [CrossRef]
72. Di Grazia, A.; Luca, V.; Segev-Zarko, L.A.; Shai, Y.; Mangoni, M.L. Temporins A and B stimulate migration of HaCaT keratinocytes and kill intracellular *Staphylococcus aureus*. *Antimicrob. Agents Chemother.* **2014**, *58*, 2520–2527. [CrossRef]

Article

The Activity of Isoquinoline Alkaloids and Extracts from *Chelidonium majus* against Pathogenic Bacteria and *Candida* sp.

Sylwia Zielińska [1,*], Magdalena Wójciak-Kosior [2], Magdalena Dziągwa-Becker [3], Michał Gleńsk [4], Ireneusz Sowa [2], Karol Fijałkowski [5], Danuta Rurańska-Smutnicka [6], Adam Matkowski [1,7] and Adam Junka [6]

1 Department of Pharmaceutical Biology, Wroclaw Medical University, Borowska 211, 50-556 Wroclaw, Poland
2 Department of Analytical Chemistry, Medical University of Lublin, Chodźki 4a, 20-093 Lublin, Poland
3 Departament of Weed Science and Tillage Systems, Institute of Soil Science and Plant Cultivation, Orzechowa 61, 50-540 Wrocław, Poland
4 Department of Pharmacognosy, Wroclaw Medical University, Borowska 211a, 50-556 Wroclaw, Poland
5 West Pomeranian University of Technology in Szczecin, Faculty of Biotechnology and Animal Husbandry, Department of Immunology, Microbiology and Physiological Chemistry, Piastów 45, 70-311 Szczecin, Poland
6 Pharmaceutical Microbiology and Parasitology, Wroclaw Medical University, Borowska 211a, 50-556 Wroclaw, Poland
7 Laboratory of Experimental Cultivation, Botanical Garden of Medicinal Plants, Wroclaw Medical University, Al. Jana Kochanowskiego 14, 50-556 Wroclaw, Poland
* Correspondence: sylwia.zielinska@umed.wroc.pl

Received: 13 June 2019; Accepted: 11 July 2019; Published: 12 July 2019

Abstract: *Chelidonium majus* (*Papaveraceae*) extracts exhibit antimicrobial activity due to the complex alkaloid composition. The aim of the research was to evaluate the antimicrobial potential of extracts from wild plants and in vitro cultures, as well as seven major individual alkaloids. Plant material derived from different natural habitats and in vitro cultures was used for the phytochemical analysis and antimicrobial tests. The composition of alkaloids was analyzed using chromatographic techniques (HPLC with DAD detection). The results have shown that roots contained higher number and amounts of alkaloids in comparison to aerial parts. All tested plant extracts manifested antimicrobial activity, related to different chemical structures of the alkaloids. Root extract used at 31.25–62.5 mg/L strongly reduced bacterial biomass. From the seven individually tested alkaloids, chelerythrine was the most effective against *P. aeruginosa* (MIC at 1.9 mg/L), while sanguinarine against *S. aureus* (MIC at 1.9 mg/L). Strong antifungal activity was observed against *C. albicans* when chelerythrine, chelidonine, and aerial parts extract were used. The experiments with plant extracts, individually tested alkaloids, and variable combinations of the latter allowed for a deeper insight into the potential mechanisms affecting the activity of this group of compounds.

Keywords: isoquinoline alkaloids; antimicrobial activity; *Chelidonium majus*

Key Contribution: Major isoquinoline alkaloids from *Chelidonium majus* are active against several pathogenic bacteria and yeast. Sanguinarine and chelerythrine are most efficient against Gram positive and Gram negative strains; respectively.

1. Introduction

Phytochemical characterization of the raw material enables one to recognize the production patterns of the plant compounds accumulated in response to the environmental factors. The composition and proportions of individual components are of great importance from the herbal drug usefulness point of view. Such studies should be conducted on a large number of samples due to the multitude of factors that may affect the phytochemical profile of plants. This can be done, *inter alia*, by comparing metabolite compositions in plants collected from natural habitats and in vitro cultures.

For our research, we have chosen a well-known medicinal plant *Chelidonium majus* L. (greater celandine). The species has a long tradition of being used mainly in folk medicine. It has been used in herbal medicine since Dioscorides and Pliny the Elder times, in the 1st century AD [1]. So far, anticancer, antimicrobial (antibacterial, antiviral, antifungal), antiprotozoal, anti-inflammatory, antispasmodic, spasmolytic, cholekinetic, muscle relaxant activities of *C. majus* extracts and several separated compounds have been reported [1]. The most abundant specialized metabolites produced in aerial and underground (roots and root collar, hereafter referred to as "roots") parts of the plant are isoquinoline alkaloids, mostly derivatives of benzophenantridine (chelidonine, chelerythrine, sanguinarine) protoberberine (berberine, coptisine, stylopine), and protopine (protopine, allocryptopine). The bioactivity of individual alkaloids has been linked, among others, to the presence of a methoxy substitutions, the iminium bond, a charge due to a quaternary nitrogen atom, a methyl group bonded to the quaternary nitrogen atom or lack of this substitution [2–8]. The published results do not always coincide or complement each other, or they are contradictory. Despite many multidirectional studies, the complex alkaloid composition of the species is still unexplored, and so is the bioactive potential of the plant [9–15].

In this study, we evaluated antimicrobial activity of seven alkaloids and *C. majus* extracts from plants derived from natural habitats and in vitro cultures. A comparison of the alkaloid profile of extracts obtained from aerial parts and roots of plants collected from different habitats was also performed using chromatographic techniques. Moreover, antimicrobial activity of seven major alkaloids was tested and the results were correlated with alkaloid content.

2. Results

2.1. Comparison of Extraction Effectiveness

HPLC analysis showed the presence of seven major alkaloids of the three main classes (phenanthridine: chelidonine, chelerythrine, sanguinarine; protoberberine: berberine, coptisine; protopine: allocryptopine, protopine). Chemical structures of the compounds are presented in Figure 1. The content of these alkaloids in extracts obtained with the use of various solvents expressed as μg on g of dried plant material is shown in Figures 2–4.

The aerial and underground parts differed significantly in the content of alkaloids (Figures 2 and 3). The blooming herb after separating the fruits, was rich in chelidonine, while in the separated fruits coptisine was the most abundant alkaloid (Figure 5). Roots were generally a much alkaloid richer raw material. The content of all alkaloids was much higher in roots than in aerial parts. There was a particularly high content of sanguinarine, chelidonine, chelerythrine, and allocryptopine in the samples collected from all five habitats. The most abundant compound in roots was sanguinarine (1986.43 μg/g d.w.), and its highest content was recorded in the plants from habitat E (Szukalice, Lipowa). In aerial parts, coptisine was a predominant compound (857.29 and 4979.12 μg/g d.w). The highest content of coptisine was found in plants from habitat A (Wrocław, Kochanowskiego).

The analysis across multiple test attempts using different concentrations of extraction solvent showed no tied ranks. The yield of all seven alkaloids was highest at the methanol concentration of 80%, compared to 70% and 90%. There were large differences between 70 and 80%, and much smaller between 80 and 90%.

Methanol extraction of 80% was found to be the most effective in terms of alkaloids recovery (Figure 4).

The representative chromatograms are presented in Figure 5.

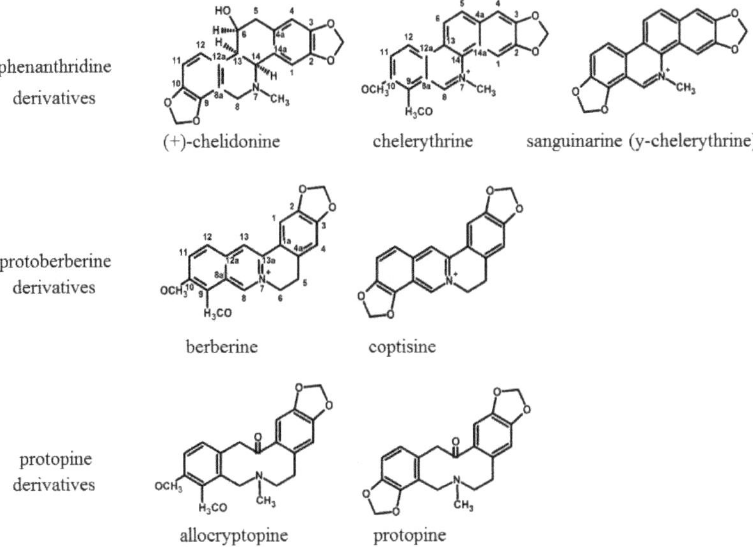

Figure 1. Structures of isoquinoline alkaloids present in *C. majus*.

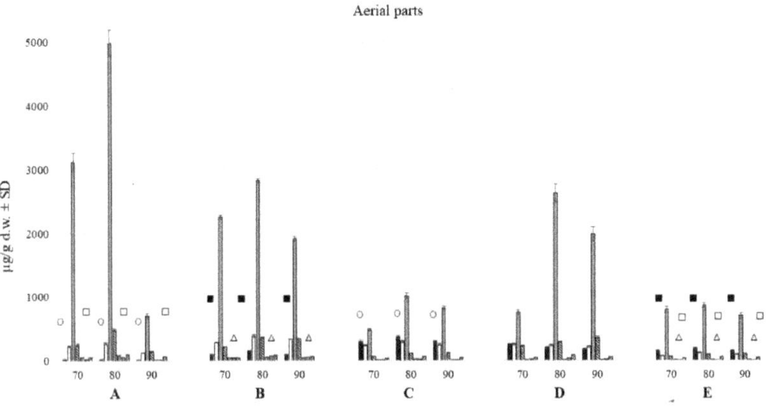

Figure 2. Isoquinoline alkaloids content in aerial parts with fruits of *C. majus* collected from different habitat (70, 80, 90% of methanol acidified with 50 mM HCl; **A**–**E**—different habitats). Statistically significant differences in the content of each compound between plants harvested from five different habitats are presented as marks of the same shape and color; protopine—white circle, chelidonine—black squares, chelerythrine—white triangles, allocryptopine—white squares, no marks—no statistically significant differences.

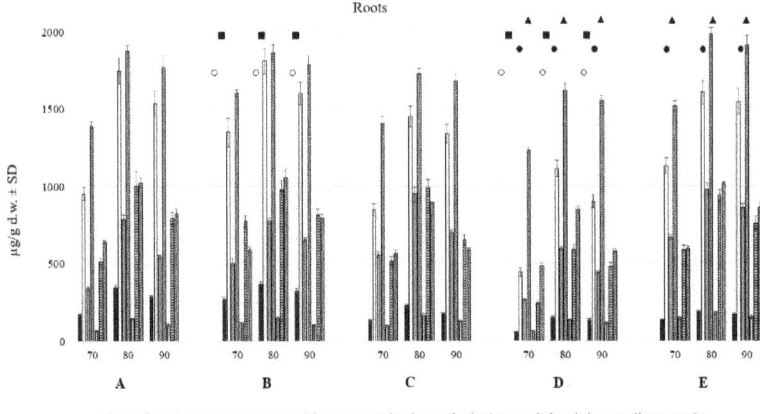

Figure 3. Isoquinoline alkaloids content in roots of *C. majus* collected from different habitat (70, 80, 90% of methanol acidified with 50 mM HCl; **A**–**E**—different habitats). Statistically significant differences in the content of each compound between plants harvested from five different habitats are presented as marks of the same shape and color; protopine—white circle, chelidonine—black squares, coptisine—black circles, sanguinarine—black triangles, chelerythrine—white triangles, no marks—no statistically significant differences.

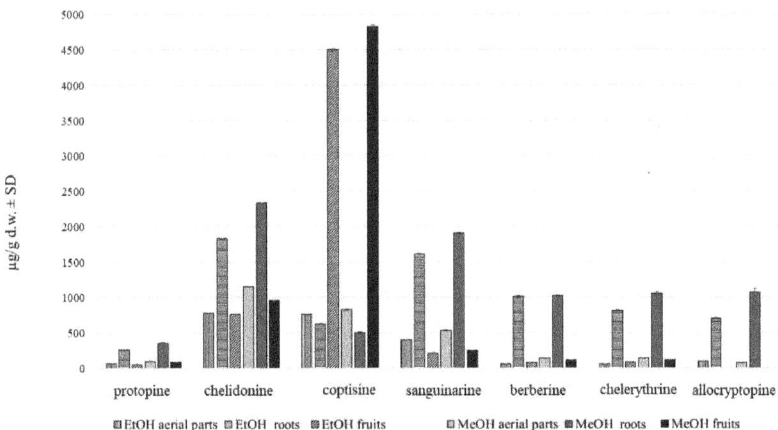

Figure 4. Alkaloid content in *C. majus* methanol and ethanol extracts of aerial parts and roots (μg/g d.w. ± SD) collected from place A.

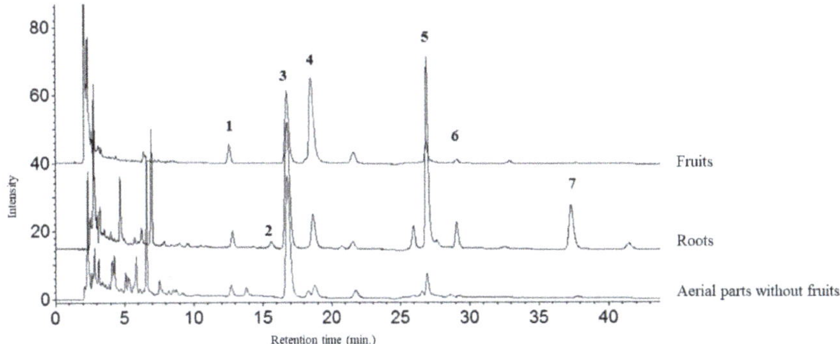

Figure 5. HPLC-DAD chromatogram of alkaloids (acquired at 290 nm) in extracts of *C. majus* fruits, aerial parts without fruits, and roots.

Differences in the alkaloids content were also found among five different habitats (A–E). The content of individual alkaloids depends on the locality of the plant harvest, and it is different for individually analyzed alkaloids from roots and aerial parts evaluated separately.

2.2. MIC Evaluation

Extracts of aerial parts and roots of *C. majus* derived from natural habitats and in vitro cultures were used for the antimicrobial assays.

None of the applied extracts was active against *K. pneumoniae*, *P. aeruginosa* and *E. coli* strains, which are Gram-negative bacteria.

In the case of Gram-positive pathogen *S. aureus*, MIC of roots MeOH extract was 62.5mg/L (Table 1). Also, a strong reduction of bacterial biomass (49%) in comparison to untreated bacteria was observed when 31.25 mg/L of this extract was used.

Individually tested alkaloids displayed MIC against *S. aureus* and *P. aeruginosa* in the range between 1.9 and 125 mg/L, depending on the compound, and between 31.25–62.5 mg/L against *C. albicans*, depending on the compound (Figure 6). Chelerythrine was the most effective, of all tested compounds against *P. aeruginosa* (MIC at 1.9 mg/L), whereas sanguinarine was the most effective against *S. aureus* (MIC at 1.9 mg/L).

Berberine and allocryptopine were the least effective—reduction of 100% of *S. aureus* bacterial biomass was observed when 125 mg/L of compounds were tested.

Strong fungal biomass reduction was observed for *C. albicans* when aerial parts extract was used, however no MIC was observed in analyzed range of concentrations.

Only chelerythrine and chelidonine were active enough to allow calculation of MIC (31.25 and 62.5 mg/L, respectively) against *C. albicans*.

The MIC of combined sanguinarine-chelerythrine-chelidonine against *S. aureus* and *P. aeruginosa* was 3.12 and 6.25 mg/L, respectively, of mixture was used (Figure 7). Sanguinearine-chelerythrine and sanguinarine-chelerythrine-berberine mixtures were much more effective against *S. sureus* than against *P. aeruginosa* and *C. albicans* (Figure 7).

Table 1. Antimicrobial activity of methanolic extracts from intact plants and in vitro plant material. Asterisk indicates concentration which was able to inhibit 100% of microbial inoculum (i.e., Minimum Inhibitory Concentration—MIC (mg/L).); no asterisk next to value represents situation when at least 70% of microbial cell was reached; "minus" sign stands for very weak ability of extract to reduce colony-forming unit (cfu) number.

Solvent/Plant Material	***Candida albicans***	***Staphylococcus aureus***
MeOH aerial	–	–
MeOH roots	–	62.5 *
Culture medium/plant material	*Candida albicans*	*Staphylococcus aureus*
MS roots	125	500 *
MS shoots	500 *	500 *
MS+N roots	3.9	500 *
MS+N shoots	3.9	500 *
$\frac{1}{2}$ MS roots	–	–
$\frac{1}{2}$ MS shoots	–	–
1.5%suc. + N roots	–	–
1.5%suc.+ N shoots	500	250
1.5%suc. roots	250	125
1.5%suc. shoots	500	500 *
B5 roots	–	250 *
B5 shoots	–	500 *
B5 + N roots	–	–
B5 + N shoots	–	250 *

MS–standard culture medium [16]; $\frac{1}{2}$ MS–medium with reduced macro-, and microelements concentration; B5–standard culture medium [17]; +N–supplementation with double amount of $NH4^{+}$ ions and simultaneous depletion of an equivalent amount of nitrate ions; 1.5%suc–MS medium supplemented with 1.5% of sucrose.

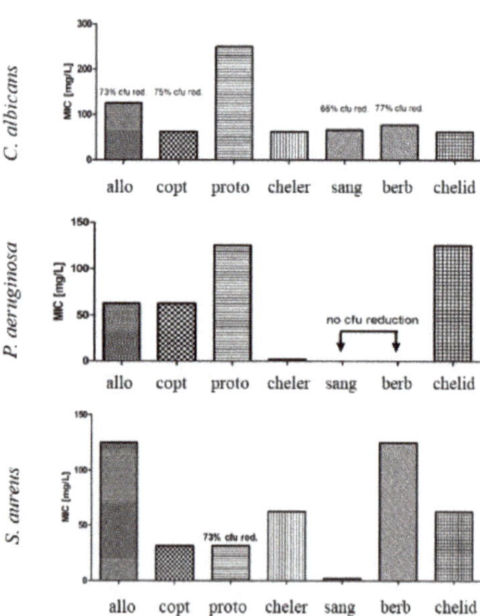

Figure 6. A comparison of Minimum Inhibitory Concentration (MIC) of the following alkaloids: allo = allocryptopine, copt = coptisine, proto = protoberberine, cheler = chelerythrine, sang = sanguinarine, berb = berberine, chelid = chelidonine.

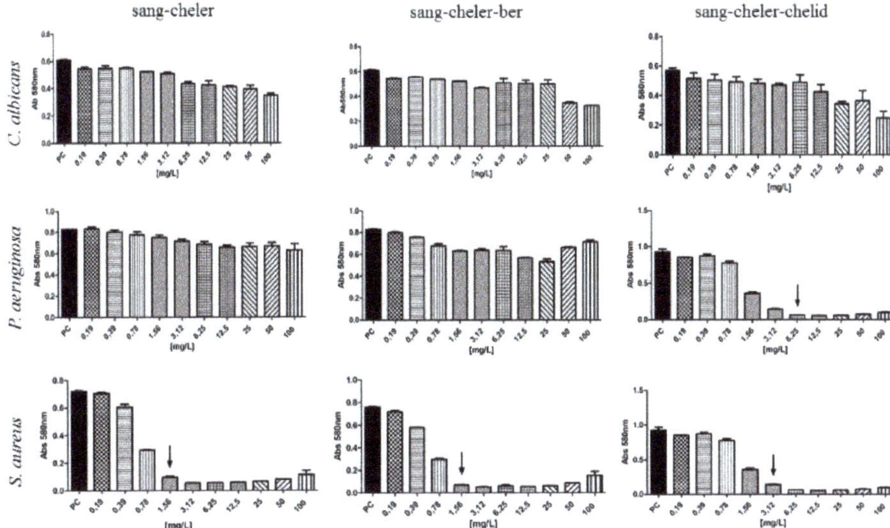

Figure 7. Antimicrobial activity of alkaloid mixtures. Arrows indicate MIC; sang = sanguinarine, cheler = chelerythrine, berb = berberine, chelid = chelidonine.

2.3. Cytotoxicity

Berberine, protopine and allocryptopine displayed no cytotoxic effect against L929 fibroblast cell lines in concentration of 15.6 mg/L; lack of cytotoxicity was observed for coptisine when 7.8 mg/L of this compound was introduced to L929 cell line; chelidonine and chelerythrinine displayed no cytotoxic effect in concentration of 3.9 mg/L. Cytotoxic properties were exhibited by sanguinarine which displayed adverse effects towards L929 cell line at concentrations of 500 μg/L or higher.

3. Discussion

The phytochemical profile that determines the biological activity of the raw material is difficult to characterize due to a large number of compounds produced in the plant and the multitude of factors determining their formation. For this reason, we compared the results of the bioactivity assays performed using plants harvested from different natural habitats, the in vitro cultures, as well as seven individual alkaloids and their mixtures.

So far, both herb and C. *majus* root and root collars as underground parts were considered to be rich in alkaloids, and hence they were used, mainly in folk medicine [1]. Our research has shown that roots of the species form a much richer source of these compounds in terms of their quality and quantity. It has been noted for both the in vitro and in vivo plant material. In turn, a kind of variation within the resulting composition of each plant parts was observed among samples collected from different habitats (Figures 2 and 3) as well as from different in vitro culture treatment [9].

MIC of methanol extracts of aerial parts and roots, as well as in vitro cultures of shoots and roots was observed against Gram-positive bacteria when 31.25–62.5 mg/L, and 125–250 mg/L were used, respectively. Sanguinarine was found to be the most abundant constituent in plant extracts (2–4 mg/g d.w. in intact plants and in vitro cultures. The amounts of other alkaloids were much lower in the extracts except coptisine (almost 5 mg/g d.w.). In individual tests, all seven alkaloids exhibited, to various extent, activity against Gram-positive strain. Sanguinarine was the most potent with MIC at 1.9 mg/L, followed by coptisine, protopine, chelerythrine, chelidonine, berberine, and allocryptopine. This kind of activity was previously reported in the earlier research on celandine alkaloids [18–21]. These experiments showed that chelerythrine and sanguinarine were more effective against these bacteria than chelidonine and berberine. In our study, sanguinarine exhibited several to several dozen times stronger activity than coptisine (MIC 31.25 mg/L), chelerythrine and chelidonine (MIC 62.5 mg/L), or allocryptopine and berberine (MIC 125 mg/L). Sanguinarine activity was also stronger than roots methanolic extract rich in the compound (1617.91–1986.43 μg/g d.w.). The extract was effective when the concentration of 62.5 mg/L was used. On the other hand, sanguinarine alone was slightly less active (MIC 1.9 mg/L) against Gram-positive bacteria than mixtures of sanguinarine with chelerythrine or sanguinarine with chelerythrine and berberine (MIC 1.56 mg/L).

In Kokoška et al. (2002) [22] experiments with multidrug-resistant bacteria existing in surgical wounds and infections of critically ill patients, *C. majus* root ethanol extract was also found to be effective against Gram-positive bacteria (*S. aureus*, *Bacillus cereus*) (MIC 15.63 and 62.5 mg of dry plant material/ml, respectively), but was inactive against Gram-negative (*P. aeruginosa*). Also, the concentration of 62.5 mg of dry plant material/ml effectively inhibited *C. candida* in these experiments. The aerial parts of *C. majus* used in the study of Kokoška et al. (2002) [22] were inactive against any of the test microbes.

Other studies in which methanolic extracts of *C. majus* were used are also consistent with the results from our experiments. Methanol extracts from leaves and petioles of *C. majus* plants grown in nature, as well as in vitro cultures [23] were potent against Gram-positive, rather than Gram-negative strains. In these studies, methanolic extracts were examined against *Bacillus subtilis, Micrococcus luteus, Sarcinia lutea,* and *S. aureus, E. coli, Proteus mirabilis, Salmonella enteritidis,* and clinically isolated *C. albicans.* Both, in vivo and in vitro plant material extracts exhibited similar bioactivity. Only some extracts were comparable against *E. coli*, *S. enteritidis*, and *C. albicans* to reference antibacterial and antifungal drugs (streptomycin, bifonazole, respectively), whereas the rest of them showed low or no activity [23].

None of the extracts, as well as sanguinarine and berberine alone were active against a Gram-negative bacteria (*P. aeruginosa*). However, other individually tested alkaloids displayed various degrees of activity against *P. aeruginosa* with MIC ranging between 1.9 and 125 mg/L (Figure 6). Chelerythrine was found to be the most potent (1.9 mg/L), followed by allocryptopine, coptisine (62.5 mg/L), and protopine, chelidonine (125 mg/L). In other studies, chelerythrine was also able to eradicate Gram-negative bacteria (*P. aeruginosa, E. coli, Klebsiella pneumoniae, Salmonella gallinarum, S. typhi, S. paratyphi, Proteus vulgaris, Shigella flexneri, Vibrio cholerae*) [18–21,24,25]. Nevertheless, sanguinarine and berberine were also listed as effective against Gram negative strains in the mentioned studies, which was not corroborated by our results.

The combined sanguinarine-chelerythrine-chelidonine were able to inhibit 100% of microbial inoculum when 6.25 mg/L of mixture was used. These results indicated that chelidonine may have an additional effect when combined with other alkaloids. It may be due to the chemical structure of chelidonine, which is different than sanguinarine and chelerythrine. Chelidonine is a benzoisoquinoline alkaloid with a tertiary nitrogen in the molecule, unlike sanguinarine and chelerythrine, which both contain a quaternary nitrogen atom (Figure 1) whose charge can depend on the pH also in the microenvironment of the bacterial cells. Various types of alkaloid bioactivity at the cellular level due to the differences in their chemical structure were well documented in the study of Barreto et al. (2003) [6]. In experiments with several groups of isoquinoline alkaloids on oxygen uptake in mouse liver mitochondria, these three alkaloids have shown a different scheme of action due to their chemical structure [6]. Generally, phenanthrene skeleton had a very low effect on oxygen uptake, while other building elements of individual alkaloids seemed to be important. Chelerythrine and sanguinarine, strongly inhibited succinate-dependent respiration and, to a lesser extent, malate–glutamate respiration, whereas chelidonine had no apparent effect. Allocryptopine, an uncharged molecule with a C=O group, similarly to chelidonine, was not effective. In Barreto et al. (2003) [6] study, the manner of compounds action was linked to the degree of substitution of a nitrogen atom in a molecule. In turn, berberine and coptisine, both with an unsubstituted quaternary nitrogen atom, have shown a marked inhibitory effect on malate-glutamate respiration and a smaller, although significant, effect on succinate respiration. The presence of a methyl group seems to be of less importance for the direction of biological activity.

Chelerythrine, sanguinarine, and chelidonine contain methyl group but they present different activity pattern against microorganisms [26].

In case of antifungal assays, among seven individually tested alkaloids, only chelerythrine and chelidonine, as well as extracts from in vitro-derived shoots, were able to inhibit 100% of C *albicans* colony forming units (31.25, 62.5, and 500 mg/L, respectively). Five other alkaloids, their mixtures, as well as in vivo and in vitro plant extracts showed no MIC in the analyzed range of concentrations. However, a strong fungal biomass reduction was observed (Figure 6, Table 1). The observation of other researchers has shown similar results, only with chelerythrine, sanguinarine, and their derivatives being up to several times more effective against pathogenic fungi than chelidonine [26–28]. Extracts from roots and shoots cultured in vitro on two different standard media, namely MS and B5, exhibited varied effect against *C. albicans*. It was probably related to the differences in the composition of micro and macro elements, sucrose and vitamins between the two culture media.

Microbes, during their the long evolutionary journey developed a plethora of systems aiming to evade or neutralize antimicrobial agents. It concerns not only human or animal pathogens but also environmental strains which co-exist in complex, multi-species habitat in water or soil having contact with plants [29–31].

In turn, it was proven, that bacterial pathogens in nosocomial environment are able to actively pump biocides out from their cytoplasm using "efflux pump" systems. It is also known that these mechanisms may be of un-specific nature, i.e., efflux pump activation as result of presence of specific biocide may activate bacterial organism to pump out wide spectrum of other biocides. An increased resistance to chlorhexidine antiseptic and cross-resistance to colistin antibiotic following exposure to chlorhexidine in *Klebsiella pneumoniae* is one of the most studied examples of such a phenomenon [32]. Such a mechanism may explain results obtained by us and other researchers showing various levels of antimicrobial activity of alkaloids provided alone or in mixture.

4. Conclusions

The complex composition of *C. majus* alkaloids contained in plant extracts can manifest a wide spectrum of antimicrobial activity, arising from structural diversity of the compounds. Alkaloids from all parts of *C. majus* and from in vitro biomass may find an application in eradication of both Gram-positive and Gram-negative cocci. They are also promising against *Candida* pathogens. Further detailed studies are necessary to fully understand the mechanisms of activity associated with the chemical structures of isoquinoline alkaloids. The rational design of the composition of alkaloids should be envisaged to combat various microbes depending on the activity of individual components and for this to happen, the natural proportions of the alkaloids in plant matricies can be manipulated both by means of plant treatment and post-harvest processing of the crude biomass and extracts obtained thereof. In further studies on *C. majus*, the factors at work that orchestrate the alkaloid profile and their strain-specific activity should be elucidated.

5. Materials and Methods

5.1. Plant Material

5.1.1. Wild Growing Plants

The studied plants were collected from five different locations in Poland: A: (Wrocław, Kochanowskiego Street 51°07′01.6″N 17°04′26.7″E51.117121, 17.074088—28.05.2017), B: (Wrocław, Kosciuszki Street 51°06′08.0″N 17°02′08.9″E51.102220, 17.035811—30.05.2017), C: (Turawa 50°44′45.9″N 18°02′29.1″E50.746071, 18.041428—11.06.2017), D: (Wrocław, Borowska Street 51°04′42.0″N 17°01′51.9″E51.078325, 17.031080—25.05.2017), E: (Szukalice, Lipowa Street 51°00′07.4″N 17°00′40.4″E51.002062, 17.011213—22.05.2017). The habitats of plants were overshadowed roadsides, as well as forest edges and shrubbery. The maximum distance between the localities was up to 150 km and minimum was 3 km.

5.1.2. In Vitro Plant Material

Plant material from in vitro cultures was used for the analysis and bioactivity assays. The in vitro cultures establishment, as well as extraction procedures for the phytochemical analysis, were presented previously [9].

5.2. Phytochemistry

5.2.1. Reagents and Standards

Alkaloid standards such as protopine (purity ≥ 95), berberine (purity ≥ 95%), chelidonine (purity ≥ 95%), chelerythrine (purity ≥ 90%), and sanguinarine (purity ≥ 90%) were purchased from Extrasynthese (France) and allocryptopine, (purity ≥ 95), coptisine (purity ≥ 98%) from Sigma (St. Louis, MO, USA). Ammonium acetate, acetic acid, HPLC grade methanol (MeOH), and acetonitrile (ACN) were from Merck (Darmstadt, Germany). Water was deionized and purified by ULTRAPURE Millipore Direct-QVR 3UV-R (Merck, Darmstadt, Germany).

Alkaloid mixtures used for bioactivity assays were obtained from the following sources:

sanguinarine (Sang), chelerythrine (Cheler) were isolated as a mixture from *Coptis chinensis* rhizoma (19g/100g yield). Three mixures were used: Sang-Cheler (0.2:1 *w*/*w*); Sang-Cheler-Chelid (0.2:1:1, *w*/*w*); Sang-Cheler-Berb (0.2:1:1, *w*/*w*).

5.2.2. Sample Preparation

Dried raw material was divided into aerial parts (stems with leaves, flowers and fruits) and underground (roots and root collars) parts and powdered with mortar and pestle. The extraction of intact plants was performed in round-bottom flasks with a solvent to solid ratio of 1:20 (*v*:*w*) in ultrasonic bath (3 × 15 min). Samples were extracted with methanol or ethanol and 50 mM hydrochloric acid according to the procedure by Kulp et al. [33]. Additionally, extraction was conducted with acidified (50 mM HCl) aqueous methanol in three different proportions (90:10, 80:20 and 70:30 *v*/*v*). The extracts were combined, evaporated and dissolved in 20 ml of methanol.

5.2.3. HPLC Analysis

Chromatography was carried out using a VWR Hitachi Chromaster 600 chromatograph (Merck, Darmstadt, Germany) with a spectrophotometric detector (DAD) and EZChrom Elite software (Merck). The samples were analyzed on an XB-C18 reversed phase core-shell column (Kinetex, Phenomenex, Aschaffenburg, Germany) (25 cm × 4.6mm i.d., 5 μm particle size), kept at 25 °C. Mobile phase consisted of acetonitrile (A) and 10 mM water solution of ammonium acetate adjusted to pH 4 with acetic acid (B). Gradient elution program was as follows: from 0 to 20 min: 20% A; from 20.5 to 27 min 25% A and from 27.5–60 min. 30% A at the flow rate of 1 mL/min. Chromatograms were recorded in the range of wavelength from 220 to 400 nm. The identity of compounds in plant extracts was confirmed by comparison of retention times and spectra with corresponding standards. Peak homogeneity was established comparing the spectrum recorded at the three peak sections upslope, apex, and downslope with the reference spectrum. Additionally, the chromatographic fractions eluted at the retention time characteristic for the investigated alkaloids were collected using a Foxy R1 fraction collector (Teledyne Isco, Lincoln, NE, USA), and their identity was confirmed by direct injection mass spectrometry (micrOTOF-Q II, Bruker Daltonics, Bremen, Germany) using Compass DataAnalysis software version 4.1. The operating conditions were as follows: positive ionization mode, ion spray voltage: 4500 V; fragmentator voltage: −500 V, corona discharge: 4000 nA, flow rate of nitrogen: 3.5 L/min, temperature of nitrogen: 200 °C and evaporator temperature: 350 °C. The generated ions were analyzed in the range of 50–500 *m*/*z*.

Quantitative analyses were performed at following wavelengths: 290 nm for protopine, allocryptopine and chelidonine, 359 nm for coptisine, 329 nm for sanguinarine, 346 nm for berberine, 318 nm for chelerythrine. Validation of the method was performed in our previous study [10].

5.3. *Statistical Analysis*

To explore differences in treatments (the content of protopine, allocryptopine, chelidonine, coptisine, chelerythrine, sanguinarine, and berberine) across multiple test attempts (different concentrations of extraction solvent, different habitats of intact plants) the Friedman ANOVA analysis followed by the Dunn test and Bonferroni correction was performed. The statistical significance of differences between treatments was considered significant at $p < 0.05$. All statistical processing was performed using Microsoft Excel (Office 365, Microsoft, Redmont, WA, USA).

5.4. *Experimental Design for Bioactivity Assays*

To conduct the bioactive potential of *C. majus* extracts bioactivity assays against Gram-positive (*Staphylococcus aureus*), Gram-negative (*Pseudomonas aeruginosa, Klebsiella pneumonia, Escherichia coli*), and pathogenic yeast (*Candida albicans*).

5.4.1. Strains

The following microbial strains from ATCC collection were used for experimental purposes: *Staphylococcus aureus* 6538; *Pseudomonas aeruginosa* 14452; *Klebsiella pneumoniae* 700603; *Escherichia coli* 25922. Clinical fungal strain: *Candida albicans* 10231.

5.4.2. MIC Evaluation

Standard microdilution technique according to EUCAST guidelines was used to assess antimicrobial potential of the methanol extracts (from underground and aerial plant parts). Survival of cells subjected to extract's activity was assessed by the TTC assay (based on ability of colorless triphenyl tetrazolium chloride (TTC) compound to change into red formazan in the presence of living microbes; qualitative technique) and using spectrophotometry ($\lambda = 600$; semi-quantitative technique).

Author Contributions: Conceptualization, S.Z. and A.J.; Methodology, S.Z., A.J., M.W.-K., I.S., K.F., and M.G.; Software, M.W.-K. and M.D.-B.; Validation, M.W.-K. and M.D.-B.; Formal Analysis, S.Z., M.W.-K., D.R.-S., A.J.; Investigation, S.Z. and A.J.; Resources, S.Z., A.M., A.J.; Data Curation, S.Z.; Writing—Original Draft Preparation, S.Z.; Writing—Review & Editing, S.Z., M.W.-K., A.M., A.J.; Visualization, S.Z.; Supervision, A.J. and A.M.; Project Administration, A.J., A.M.; Funding Acquisition, A.J., A.M.

Funding: This research was funded by National Research Center grant number 2018/02/X/NZ4/01169 NCN MINIATURA-2" and the Wroclaw Medical University project SUB.D033.19.009.

Acknowledgments: This study was supported by National Research Center grant no. 2018/02/X/NZ4/01169 NCN MINIATURA-2 and the Wroclaw Medical University project SUB.D033.19.009. The authors thank Andrzej Dryś from the Department of Physical Chemistry, Wroclaw Medical University for his valued assistance in statistical analysis.

Conflicts of Interest: The authors declare no conflicts of interest.

References

1. Zielinska, S.; Jezierska-Domaradzka, A.; Wójciak-Kosior, M.; Sowa, I.; Junka, A.; Matkowski, A.M. Greater Celandine's ups and downs—21 centuries of medicinal uses of Chelidonium majus from the viewpoint of today's pharmacology. *Front. Pharmacol.* **2018**, *9*, 1–29. [CrossRef]
2. Miao, F.; Yanga, X.-J.; Zhou, L.; Hu, H.-J.; Zheng, F.; Ding, X.-D.; Sun, D.-M.; Zhou, C.-D.; Sun, W. Structural modification of sanguinarine and chelerythrine and their antibacterial activity. *Nat. Prod. Res.* **2011**, *25*, 863–875. [CrossRef]
3. Hossain, M.; Khan, A.Y.; Kumar, G.S. Interaction of the anticancer plant alkaloid sanguinarine with bovine serum albumin. *PLoS ONE* **2011**, *6*, e18333. [CrossRef]

4. Giri, P.; Kumar, G.S. Specific binding and self-structure induction to poly(A) by the cytotoxic plant alkaloid sanguinarine. *Biochim. Biophys. Acta* **2007**, *1770*, 1419–1426. [CrossRef]
5. Bai, L.-P.; Zhao, Z.-Z.; Cai, Z.; Jiang, Z.-H. DNA-binding affinities and sequence selectivity of quaternary benzophenanthridine alkaloids sanguinarine, chelerythrine, and nitidine. *Bioorgan. Med. Chem.* **2006**, *14*, 5439–5445. [CrossRef]
6. Barreto, M.C.; Pinto, R.E.; Arrabaça, J.D.; Pavão, M.L. Inhibition of mouse liver respiration by Chelidonium majus isoquinoline alkaloids. *Toxicol. Lett.* **2003**, *146*, 37–47. [CrossRef]
7. Sedo, A.; Vlasicova, K.; Bartak, P.; Vespalec, R.; Vicar, J.; Simanek, V.; Ulrichova, J. Quaternary Benzo[c]phenanthridine Alkaloids as Inhibitors of Aminopeptidase N and Dipeptidyl Peptidase IV. *Phytother. Res.* **2002**, *16*, 84–87. [CrossRef]
8. Ulrichova, J.; Dvorak, Z.; Vicar, J.; Lata, J.; Smrzova, J.; Sedo, A.; Simanek, V. Cytotoxicity of natural compounds in hepatocyte cell culture models. The case of quaternary benzo[c]phenanthridine alkaloids. *Toxicol. Lett.* **2001**, *125*, 125–132. [CrossRef]
9. Zielinska, S.; Wójciak-Kosior, M.; Płachno, B.J.; Sowa, I.; Włodarczyk, M.; Matkowski, A. Quaternary alkaloids in Chelidonium majus in vitro cultures. *Ind. Crop. Prod.* **2018**, *123*, 17–24. [CrossRef]
10. Sowa, I.; Zielinska, S.; Sawicki, J.; Bogucka-Kocka, A.; Staniak, M.; Bartusiak-Szczesniak, E.; Podolska-Fajks, M.; Kocjan, R.; Wojciak-Kosior, M. Systematic evaluation of chromatographic parameters for isoquinoline alkaloids on XB-C18 core-shell column using different mobile phase compositions. *J. Anal. Methods Chem.* **2018**, *3*, 1–8. [CrossRef]
11. Borghini, A.; Pietra, D.; di Trapani, C.; Madau, P.; Lubinu, G.; Bianucci, A.M. Data mining as a predictive model for *Chelidonium majus* extracts production. *Ind. Crop Prod.* **2015**, *64*, 25–32. [CrossRef]
12. Gilca, M.; Gaman, L.; Panait, E.; Stoian, I.; Atanasiu, V. Chelidonium majus—An integrative review: Traditional knowledge versus modern findings. *Complement. Med. Res.* **2010**, *17*, 241–248. [CrossRef]
13. Sarközi, A.; Janicsak, G.; Kursinszki, L.; Kery, A. Alkaloid composition of Chelidonium majus L. studied by different chromatographic techniques. *Chromatographia* **2006**, *63*, S81–S86. [CrossRef]
14. Osbaldeston, T.A.; Wood, R.P.A. *Dioscorides de Materia Medica, Being a Herbal with Many other Materials Written in Greek in the First Century of the Common Era*; An Indexed Version in Modern English; Ibidis Press: Johannesburg, South Africa, 2000; pp. 352–355.
15. Jones, W.H.S. *Pliny Natural History with an English Translation in Ten Volumes*; Harvard University Press: London, UK, 1966.
16. Murashige, T.; Skoog, F. A revised medium for rapid growth and bioassays with tobacco tissue cultures. *Physiol. Plant.* **1962**, *15*, 473–497. [CrossRef]
17. Gamborg, O.L.; Miller, R.A.; Ojima, K. Nutrient requirements of suspension culture of soybean root cells. *Exp. Cell Res.* **1968**, *50*, 151–158. [CrossRef]
18. Kędzia, B.; Hołderna-Kędzia, E.; Goździcka-Józefiak, A.; Buchwald, W. The antimicrobial activity of *Chelidonium majus* L. *Postępy Fitoterapii* **2013**, *4*, 236–243.
19. Artini, M.; Papa, R.; Barbato, G.; Scoarughi, G.L.; Cellini, A.; Morazzoni, P.; Bombardelli, E.; Selan, L. Bacterial biofilm formation inhibitory activity revealed for plant derived natural compounds. *Bioorg. Med. Chem.* **2012**, *20*, 920–926. [CrossRef]
20. Mitscher, L.A.; Leu, R.P.; Bathala, M.S.; Wu, W.N.; Beal, J.L. Antimicrobial agents from higher plants. I. Introduction, rationale, and methodology. *Lloydia* **1972**, *35*, 157–166.
21. Mitscher, L.A.; Park, Y.H.; Clark, D.; Clark, G.W.; Hammesfahr, P.D.; Wu, W.N.; Beal, J.L. Antimicrobial agents from higher plants. An investigation of Hunnemannia fumariaefolia pseudoalcoholates of sanguinarine and chelerythrine. *Lloydia* **1978**, *41*, 145–149.
22. Kokoška, L.; Polesny, Z.; Rada, V.; Nepovim, A.; Vanek, T. Screening of some Siberian medicinal plants for antimicrobial activity. *J. Ethnopharmacol.* **2002**, *82*, 51–53. [CrossRef]
23. Ciric, A.; Vinterhalter, B.; Šavikin-Fodulovi'c, K.; Sokovic, M.; Vinterhalter, D. Chemical analysis and antimicrobial activity of methanol extracts of celandine (*Chelidonium majus* L.) plants growing in nature and cultured in vitro. *Arch. Biol. Sci.* **2008**, *60*, 7–8. [CrossRef]
24. Wongbutdee, J. Physiological effects of berberine—Review article. *Thai Pharm. Health Sci. J.* **2009**, *4*, 78–83.
25. Imanshahidi, M.; Hosseinzadeh, H. Pharmacological and therapeutic effects of *Berberis vulgaris* and its active constituent, Berberine—Review article. *Phytother. Res.* **2008**, *22*, 999–1012. [CrossRef]

26. Kedzia, B.; Hołderna-Kedzia, E. The effect of alkaloids and other groups of plant compounds on bacteria and fungi. *Postęp. Fitoter.* **2013**, *1*, 8–16.
27. Ma, W.G.; Fukushi, Y.; Tahava, S.; Osawa, T. Fungitoxic alkaloids from Hokkaido Papaveraceae. *Fitoterapia* **2000**, *71*, 527–534. [CrossRef]
28. Meng, F.; Zuo, G.; Hao, X.; Wang, G.; Xiao, H.; Zhang, J.; Xu, G. Antifungal activity of the benzo [c] phenanthridine alkaloids from *Chelidonium majus* Linn against resistant clinical yeast isolates. *J. Ethnopharmacol.* **2009**, *125*, 494–496. [CrossRef]
29. Young, I.M.; Crawford, J.W. Interactions and Self-Organization in the Soil-Microbe Complex. *Science* **2004**, *304*, 1634–1637. [CrossRef]
30. Djenane, Z.; Nateche, F.; Amziane, M.; Gomis-Cebolla, J.; El-Aichar, F.; Khorf, H.; Ferre, J. Assessment of the Antimicrobial Activity and the Entomocidal Potential of Bacillus thuringiensis Isolates from Algeria. *Toxins* **2017**, *9*, 139. [CrossRef]
31. Schirawski, J.; Perlin, H. Plant-Microbe Interaction 2017—The Good, the Bad and the Diverse. *Int. J. Mol. Sci.* **2018**, *19*, 1374. [CrossRef]
32. Wand, M.E.; Bock, L.J.; Bonney, L.C.; Sutton, J.M. Mechanisms of Increased Resistance to Chlorhexidine and Cross-Resistance to Colistin following Exposure of Klebsiella pneumoniae Clinical Isolates to Chlorhexidine. *Antimicrob. Agents Chemother.* **2017**, *61*, 01162-16. [CrossRef]
33. Kulp, M.; Bragina, O.; Kogerman, P.; Kaljurand, M. Capillary electrophoresis with LED-induced native fluorescence detection for determination of isoquinoline alkaloids and their cytotoxicity in extracts of *Chelidonium majus* L. *J. Chromatogr. A* **2011**, *1218*, 5298–5304. [CrossRef]

Review

The Biological Activity of Natural Alkaloids against Herbivores, Cancerous Cells and Pathogens

Amin Thawabteh [1], Salma Juma [2], Mariam Bader [2], Donia Karaman [3], Laura Scrano [4], Sabino A. Bufo [3] and Rafik Karaman [2,*]

1 Samih Darwazah Institute for Pharmaceutical Industries, Faculty of Pharmacy Nursing and Health Professions, Birzeit University, Bir Zeit 71939, Palestine; athawabtah@birzeit.edu
2 Pharmaceutical Sciences Department, Faculty of Pharmacy, Al-Quds University, Jerusalem 20002, Palestine; salmajuma_1989@yahoo.com (S.J.); mariam407@hotmail.com (M.B.)
3 Department of Science, University of Basilicata, 85100 Potenza, Italy; kdonia65@yahoo.com (D.K.); sabino.bufo@unibas.it (S.A.B.)
4 Department of European Cultures (DICEM), University of Basilicata, 75100 Matera, Italy; laura.scrano@unibas.it
* Correspondence: dr_karaman@yahoo.com; Tel.: +972-5-9875-5052

Received: 25 September 2019; Accepted: 5 November 2019; Published: 11 November 2019

Abstract: The growing incidence of microorganisms that resist antimicrobials is a constant concern for the scientific community, while the development of new antimicrobials from new chemical entities has become more and more expensive, time-consuming, and exacerbated by emerging drug-resistant strains. In this regard, many scientists are conducting research on plants aiming to discover possible antimicrobial compounds. The secondary metabolites contained in plants are a source of chemical entities having pharmacological activities and intended to be used for the treatment of different diseases. These chemical entities have the potential to be used as an effective antioxidant, antimutagenic, anticarcinogenic and antimicrobial agents. Among these pharmacologically active entities are the alkaloids which are classified into a number of classes, including pyrrolizidines, pyrrolidines, quinolizidines, indoles, tropanes, piperidines, purines, imidazoles, and isoquinolines. Alkaloids that have antioxidant properties are capable of preventing a variety of degenerative diseases through capturing free radicals, or through binding to catalysts involved indifferent oxidation processes occurring within the human body. Furthermore, these entities are capable of inhibiting the activity of bacteria, fungi, protozoan and etc. The unique properties of these secondary metabolites are the main reason for their utilization by the pharmaceutical companies for the treatment of different diseases. Generally, these alkaloids are extracted from plants, animals and fungi. Penicillin is the most famous natural drug discovery deriving from fungus. Similarly, marines have been used as a source for thousands of bioactive marine natural products. In this review, we cover the medical use of natural alkaloids isolated from a variety of plants and utilized by humans as antibacterial, antiviral, antifungal and anticancer agents. An example for such alkaloids is berberine, an isoquinoline alkaloid, found in roots and stem-bark of *Berberis asculin* P. Renault plant and used to kill a variety of microorganisms.

Keywords: alkaloids; natural sources; anticancer; antibacterial; antiviral; antifungal

Key Contribution: Alkaloids are secondary plant metabolites that have been shown to possess potent pharmacological activities. These activities are mainly beneficial, except for pyrrolizidine alkaloids, which are known to be toxic.

---

1. Introduction

Ample research has been conducted on natural products in order to obtain new antimicrobial agents to compensate for the increasingly microbial resistance. In fact, the problem with the currently used antiviral drugs is the development of resistance by the microorganism.

Many traditionally used plants for viral infections have been studied. Extracted compounds, including terpenes (e.g., mono-, di- and tri-), flavonoids, phenols and polyphenols have been found to be active against HSV virus [1,2]. Flavonoids extracted from plants are utilized by people for their health benefits, and have been shown to have viral activity against HCMV [3]. Fourteen new alkaloids were isolated from *Cladosporium* species (spp.) PJX-41 fungi and showed inhibitory activity against influenza *virus A (H1N1)* [4].

The Biological Activities of Alkaloids

Alkaloids are grouped into several classes. This classification is based on their heterocyclic ring system and biosynthetic precursor. They include tropanes, pyrrolidines, isoquinoline purines, imidazoles, quinolizidines, indoles, piperidines and pyrrolizidines. There is a great interest in the chemical nature of these alkaloids and their biosynthetic precursors. Alkaloids have been extensively researched because of their biological activity and medicinal uses. Serotonin and other related compounds are belonging to the commonly used insole alkaloids. It is estimated that about 2000 compounds are classified as indole alkaloids. They include vinblastine, strychnine, ajmaline, vincamine, vincristine and ajmalicine, which are among the most researched members, due to their pharmacological activities. For example, vincristine and vinblastine, named spindle poison, are generally utilized as anticancer agents [5]. Convolvulaceae, Erythroxylaceae and Solanaceae families include the pharmacologically active tropane alkaloids which have an 8-azabicyclo octane moiety derived from ornithine [5,6]. Hyoscyamine, cocaine, scopolamine and atropine alkaloids are the most known members of this group and possess a variety of pharmacological effects. Quinoline and isoquinoline known as benzopyridines are heterocyclic entities containing fused benzene, and pyridine rings have many medical uses [6]. Quinine, a quinoline alkaloid isolated from *Cinchona ledgeriana* (Howard) and *Calendula officinalis* L. was proved to be poisonous to *Plasmodium vivax* and organisms with single cell or *Protozoans* that cause malaria. Other members of the quinine alkaloids include cinchonidin, folipdine, camptothecin, chinidin, dihydroquinine, echinopsine and homocamptothecin [5,7]. These chemical entities have demonstrated significant pharmacological effects, such as anticonvulsant, analgesic, antifungal, anthelmintic, anti-inflammatory, antimalarial, anti-bacterial andcardiotonic [7]. Other important alkaloids are those derived from isoquinoline, a quinoline isomer, which are classified into various classes, based on the addition of certain groups: Phthalide isoquinolines, simple isoquinolines and benzylisoquinolines. Among the well-known alkaloids belong to this category are morphine (analgesic and narcotic drug), codeine (cough suppressant), narcotines, protopines, and thebaine [8]. In addition, this class of alkaloids has demonstrated various pharmacological activities, such as antitumor, antihyperglycemic and antibacterial [6]. Among the most important alkaloids from the purine class (xanthenes) are theophylline, aminophyline and caffeine. This class of alkaloids possesses a variety of pharmacological activities, including anti-inflammatory, antioxidant, antidiabetic, anti-obesity and anti-hyperlipidemic [9]. On the other hand, the alkaloids derived from piperidine are generally obtained from *Piper nigrum* L. and *Conium maculatum* L. plants. It is estimated that 700 members of this class have been researched. These alkaloids possess a saturated heterocyclic ring (piperidine nucleus) and are familiar with their toxicity. They have many pharmacological activities which include anticancer, antibacterial, antidepressant, herbicidal, anti-histaminic, central nervous system stimulant, insecticidal and fungicidal [10,11]. The famous poison of hemlock known as *Conium maculatum* presents in the piperidine alkaloids. Members of the piperidine alkaloids include lobeline, coniine and cynapine. The pyridine alkaloids have a quite similar chemical structure to that of piperidine alkaloids except the unsaturated bonds exist in their heterocyclic nucleus. Anatabin, anatabine, anabasin, epibatidine and nicotine are some members of the pyridine alkaloids [12]. Imidazole alkaloids are compounds containing an imidazole ring in their chemical structure and are derived from L-histidine.

The most known member of this class is pilocarpine which is obtained from *Pilocarpus cearensis* Rizziniand is used as a drug in ophthalmic preparations to treat glaucoma [13]. The pyrrolizidine alkaloids, containing a necine base, are present only in plants, such as Leguminosae, Convolvulaceae, Boraginaceae, Compositae, Poaceae and Orchidaceae. Among the most known members of this class are heliotrine, echinatine, senecionine and clivorine which are biosynthesized by the plants for protection from herbivores. These are hepatotoxic causing several diseases, such as liver cancer. Due to their glycosidase inhibition activity, they are used to treat diabetes and cancer [14]. Pyrrolidine alkaloids are compounds composed of aza five membered rings that are derived from ornithine and lysine. Hygrine, cuscohygrine and putrescine are some members of this class. Biological studies conducted on this class have revealed significant antifungal, antitubercular and antibacterial activities among a large number of these compounds [15].

Quinolizidine alkaloids contain two fused 6-membered rings that share nitrogen and derived from the genus Lupinus and are known as lupine alkaloids. Among the members of this class are lupinine and lupanine cytisine and sparteine. The last two members are the most distributed quinolizidine alkaloids and are characterized by their antimicrobial activities [16].

Bacterial infections are considered as a major health problem worldwide. Moreover, they are increasing, due to multidrug resistance, which subsequently causes mortality and morbidity. Therefore, new antibacterial remedies are needed, and the plants represent a wide source for novel natural compounds [17]. Three alkaloids solanine, solasodine and *B*-solamarine have been extracted from *Solanum dulcamara* L. (Solanaceae), commonly known as bittersweet plant, and have demonstrated significant antibacterial activity against *Staphycoccus aureus* [17]. Bis-indole alkaloids were obtained from *marine invertebrates* and showed antibacterial activity against *S. aureus*, including MRSA (methicillin resistance *Staphycoccusaureus*) [18]. Berberine and hydrastine alkaloids were extracted from Goldenseal (*Hydrastis canadensis* L., *Ranunculaceae*) and have demonstrated a potent antibacterial activity mostly against *Streptococcus pyogenes* and *Staphycoccus aureus* [19]. Cocsoline alkaloid was isolated from *Epinetrumvillosum* (Exell), it has a wide antibacterial activity; inhibits *Shigella strains*, *Campylobacter jejuni* and *Campylobacter coli* [20].

Antifungal agents to treat fungal infections have serious side effects and developed fungal resistance, hence, there is a pressing need to look for new and novel antifungal agents. Alkaloids extracted from the leaves of *Ruta graveolens* L. were shown to possess fungitoxic activity [21]. Tomadini Glycoalkaloids have been extracted from tomato and proved to have antifungal activity [22]. Quinoline alkaloids and flavonoids extracted from *WaltheriaIndica* L. Roots were approved to have antifungal activity against *Candida albicans* [23].

Cancer is second in the list of diseases causing death worldwide. Phytochemicals represent a source for anticancer agents, due to their low toxicity and high effectiveness. Hersutin alkaloid is a major alkaloid found in *Uncaria* genus; hersutin was found to cause apoptosis in HER2-positive and the p53-mutated breast cancer cells [24]. Oxymatrine, a natural alkaloid extracted from *Sophora chrysophylla* (Salisb.) roots, was found to have anticancer activity in human cervical cancer Hela cells, due to its cytotoxic effects and apoptosis [25].

Herein, we report a comprehensive review on the medical use of some natural alkaloids, such as antibacterial, antiviral, antifungal and anticancer agents.

2. Natural Alkaloid Used to Control Agricultural Pests (Herbivores)

Glycoalkaloids extracted from Potato leaves were demonstrated to exert negative effects on the hatching success of *Spodoptera exigua* eggs, and on the heart contractile activity of three beetle species *Zophobas atratus*, *Tenebrio molitor*, and *Leptinotarsa decemlineata* [26].

Similar effects were shown on *Zophobas atratus* F. and *Tenebrio molitor* L. by commercial glycoalkaloids (solamargine, solasonine, α-chaconine, α-solanine, α-tomatine) and by aqueous extracts from *Solanum etuberosum* L., *Lycopersicon esculentum* Mill., and *Solanum nigrum* L. [27]. Furthermore, Potato leaf extracts and commercial α-solanine were proved to influence the life history parameters and antioxidative enzyme activities in the midgut and fat body of *Galleria mellonella* L. [28]. Additionally, *Solanum tuberosum* L, *Lycopersicon esculentum* Mill.,and *Solanum nigrum* L. Leaf extracts and single pure glycoalkaloids have been demonstrated to affect the development and reproduction of *Drosophila melanogaster* [29,30].

3. Natural Alkaloid Used as Anticancer Agents

Alkaloids are the most biologically active compounds found in natural herbs and the source of some important drugs currently marketed. These include some anticancer agents, such as camptothecin (CPT) and vinblastine. The cytotoxicity and mechanisms of action for the following derived alkaloids, berberine, evodiamine, matrine, piperine, piplartine, sanguinarine, tetrandrine, aporphine, harmine, harmaline, harmalacidine and vasicinone, (**1–12**, respectively, in Figure 1), is our main focus in this section, since they are believed to have fewer side effects and lower resistance compared to other chemotherapeutic agents.

Berberine (**1** in Figure 1) is an isoquinoline derivative extract from *Coptis chinensis* Franch, Berberidaceae.This secondary metabolite possesses a variety of pharmacological effects, which include antibacterial, antidiabetes, anti-inflammatory and antiulcer ones. In addition, Also, it had been found to be beneficial for the cardiovascular system [19,20,22,23]. It has been demonstrated that *Coptis chinensis* Franch, the plant in which berberine present, has an inhibitory effect on proliferation of breast and liver cancer cells. The anticarcinogenic activity of berberine was studied in FaDu cells, which are human pharyngeal squamous carcinoma cells; berberine was found to have a cytotoxic effect and has decreased the viability of these cells in a concentration-dependent manner [31]. In vitro and in vivo experiments on berberine have demonstrated anticancer activity by causing cell cycle arrest at the G1 or G2/M phases and tumor cell apoptosis [24,32]. In addition, berberine was found to cause endoplasmic reticulum stress and autophagy, which had resulted in inhibition of tumor cell metastasis and invasion [2,33–36].

Besides its apoptotic effects, berberine was found to reduce angiogenesis by reducing VEGF expression. Also, it resulted in decreased cancer cell migration [31]. The anticancer activity of berberine was studied in the human promonocytic U937 and murine melanoma B16 cell line; cytotoxic activity was found to be concentration-dependent. Intraperitoneal administration of berberine in mice had caused a reduction of 5 to 10 kg of tumor weight after a treatment of 16 days. This reduction in tumor weight was found to be time and concentration-dependent [33].

Berberine binds to DNA or RNA to form the corresponding complexes [37,38]. Berberine also inhibits a number of enzymes, including cyclooxygenase-2 (COX-2), *N*-acetyltransferase (NAT) and tolemerase [24]. It has many effects on tumor cells, which include cyclin-dependent kinase (CDK) regulation [24,37] and expression regulation of *B*-cell lymphoma 2 (Bcl2) proteins (Bax, Bcl-2, Bcl-xL) [24,33,37] and caspases [33,37]. Further, berberine has an inhibition activity on nuclear factor κ-light-chain enhancer of activated B cells (NF-κB) and activation on the synthesis of intracellular ROS [24,33]. It is worth noting that berberine activity is selective for cancer cells [24]. Studies have shown that berberine affects tumor cell progression by the inhibition of focal adhesion kinase (FAK), urokinase, matrix metalloproteinase 9 (MMP-9), NFκB and matrix metalloproteinase2 (MMP-2) [2,39]. In addition, it reduced Rho kinase-mediated Ezrin phosphorylation [36], inhibited COX-2 synthesis, prostaglandin receptors and prostaglandin E [40]. Other activities of berberine include inhibition of the synthesis of hypoxia-inducible factor 1 (HIF-1), proinflammatory mediators and vascular endothelial growth factor (VEGF) [41,42]. Berberine also activates P53 gene, which results in apoptosis and cell cycle arrest. It has been demonstrated that berberine also caused apoptosis by mitochondrial-dependent pathway and interactions with DNA, as shown in Figure 2 [32].

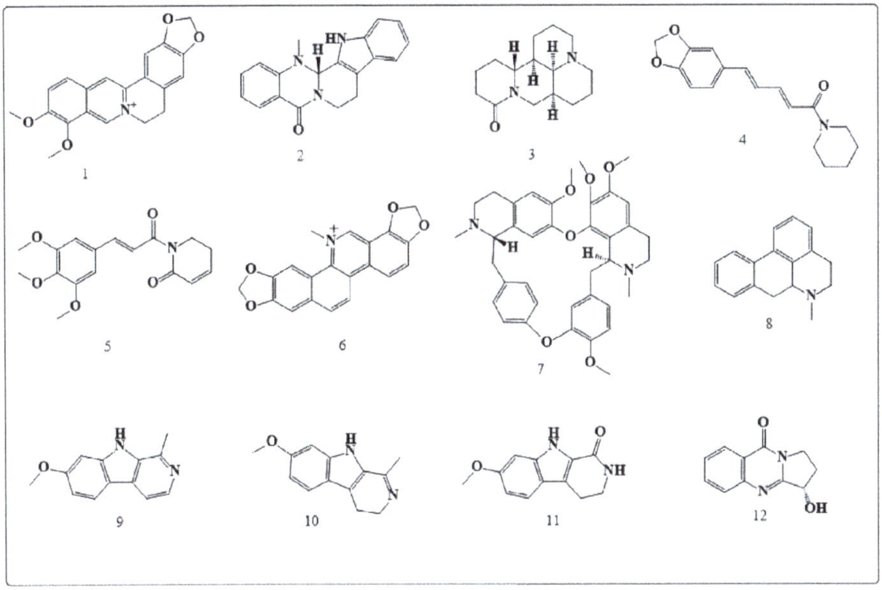

Figure 1. Chemical structures ofanticancer natural alkaloids: Berberine (**1**), evodiamine (**2**), matrine (**3**), piperine (**4**), piplartine (**5**), sanguinarine (**6**), tetrandrine (**7**), aporphine (**8**), harmine (**9**), harmaline (**10**), harmalacidine (**11**) and vasicinone (**12**).

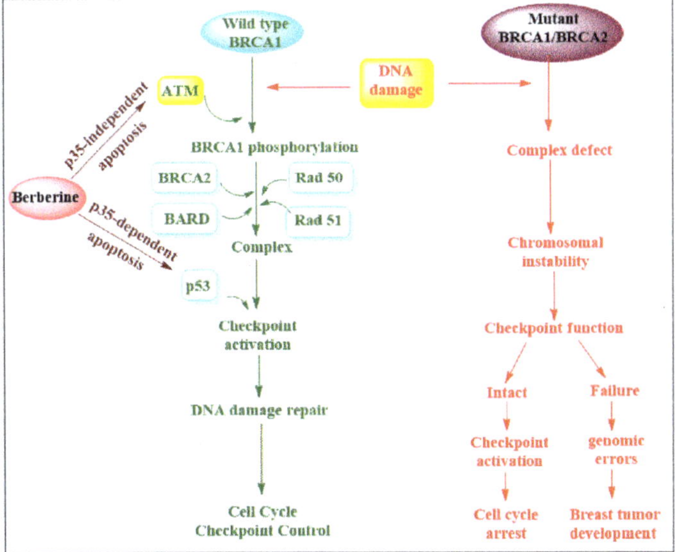

Figure 2. Anticancer mechanism of action.

Evodiamine (**2** in Figure 1), a quinolone alkaloid, isolated from the Chinese plant *Evodia rutaecarpa* has a variety of pharmacological activities against obesity, allergy, inflammation, anxiety, nociception, cancer, thermoregulation. In addition, it is a vessel-relaxing activator and an excellent protector of myocardial ischemia-reperfusion injury [43–46]. It causes cell cycle arrest, apoptosis, inhibits the angiogenesis, invasion, and metastasis in different cancer cells at G2/M phase in most cancer cell lines [47–51]. In vitro studies have shown evodiamine to be quite active against cancer cell progression at micromolar-nanomolar concentrations [52,53]. Additionally, evodiamine stimulates autophagy, which prolongs survival [54]. Evodiamine is selective for tumor cells and less toxic to normal human cells, such as human peripheral blood mononuclear cells. Studies have shown that it inhibits the proliferation of Adriamycin resistant human breast cancer NCI/ADR-RES cells, both in vitro and in vivo, and was found to be active when administered orally [49]. Moreover, it was found that the administration of 10 mg/kg evodiamine from the 6th day after tumor injection into mice reduced lung metastasis without affecting the mice weight [47]. Studies showed that evodiamine inhibits TopI enzyme, forms a DNA covalent complex with a close concentration, 2.4 μM and 4.8 μM, to that of CPT, and induces DNA damage [55–57]. It exhibited G2/M phase arrest [47,49,58] and not S phase arrest, which is not consistent with the mechanism of TopI inhibitors, such as CPT, which indicates that evodiamine has targeted other than TopI, such as tubulin polymerization [58]. Evodiamine was found to induce intracellular ROS production and cause mitochondrial depolarization [59]. Mitochondria apoptosis is caused by the generation of ROS and nitric oxide [54]. Evodiamine was also found to trigger caspases dependent and caspase-independent apoptosis, downregulates Bcl-2 expression, and upregulates Bax expression in some cancer cells [50,52]. The phosphatidylinositol 3kinase/Akt/caspase and Fas ligand (Fas-L)/NF-κB and ubiquitin-proteasome pathway are considered the route in which evodiamine causes induction to cells death [53].

Matrine (**3** in Figure 1) is an alkaloid isolated from *Sophora flavescens* Aitonplants [60]. It possesses a variety of pharmacological activities which include diuretic, choleretic, hepatoprotective, nephroprotective, antibacterial, antiviral, anti-inflammatory, antiasthmatic, antiarrhythmic, antiobesity, cardioprotective and anticancer [61–67]. In the treatment of cancer matrine is used in high doses, but it has not shown any major effects on normal cells viability [61–71]. G1 cell cycle arrests mediation and apoptosis is the mechanism, by which matrine exerts its inhibition effects on the proliferation of different cancer cells [68,69,71–73]. Matrine triggers apoptosis and autophagy in cancer cells, such as hepatoma G2 cells and SGC7901 cells. Matrine also induces the differentiation of K562 tumor cells and has antiangiogenesis activity [74]. The in vivo anticancer effects of matrine has been tested in H22 cells, MNNG/HOS cells, 4T1 cells and BxPC-3 cells in BALB/c mice [70,71,74,75]. It was found that matrine at 50 mg/kg or 100 mg/kg inhibits MNNG/HOS *Xenograft* growth, and reduces the pancreatic tumor size compared to controls at similar doses [71]. Targets of matrine are still under study until now, but it was found that it affects many proteins involved in cell proliferation and apoptosis, such as E2F-1, Bax, Bcl-2, Fas, and Fas-L [68,70–73,76]. It also inhibits cancer cell progression by the inhibition of MMP-2 and MMP-9 expression and modulation of the NF-κB signaling pathway [77–79].

Piperine (**4** in Figure 1) is a piperidine alkaloid found in *Piper nigrum* and *Piper longum* [80]. It exhibits antioxidant, antidiarrheal, anti-inflammatory, anticonvulsant, anticancer and antihyperlipidemic properties in addition to being a trigger for bile secretion production [65,81] and suppressant to the CNS system [82,83]. An administration of 50 mg/kg or 100 mg/kg of piperine daily for a week significantly reduced the size of solid tumor in mice transplanted with sarcoma 180 cells. Recently, a study has revealed that piperine has managed to inhibit the progression of breast cancer in a selective manner [84]. It has been shown that this secondary metabolite triggered cell cycle arrest in G2/M phase and apoptosis in 4T1 cells [85,86]. Piperine in a concentration of 200 μM/kg is also active against lung cancer metastasis induced by B16F-10 melanoma cells in mice [87] and causes suppression of phorbol-12-myristate-13acetate (PMA), which induce tumor cell invasion [88]. Piperine inhibits NF-κB, c-Fos, cAMP response element-binding (CREB) and activated transcription factor 2 (ATF-2) [89]. It suppresses PMA-induced MMP-9 expression through the inhibition of PKCα/extracellular signal-regulated kinase (ERK) $\frac{1}{2}$ and reduction of NF-κB/AP-1activation [88]. Piperine also inhibits P-glycoprotein (P-gp) and CYP3A4 activity, which affects drug metabolism and also re-sensitizes multidrug resistant (MDR) cancer cells [90,91]. Piperine increases the effect of docetaxel without inducing more side effects on the treated mice by inhibiting CYP3A4 which is the main metabolizing enzymes of docetaxel [92].

Piplartine, the less common name for piperlongumine, (**5** in Figure 1) is an amide-alkaloid obtained from *Piper longum* L.It has resulted in tumor growth inhibition in *Sarcoma180* cells transplanted in mice. This antitumor activity was due to its antiproliferative effect [93].

Sanguinarine (**6** in Figure 1) is a benzophenanthridine obtained from *Sanguinaria canadensis L.* and *Chelidoniummajus L.* [94,95]. It is active against bacterial, fungal and schistosomal infections, antiplatelet, and anti-inflammatory properties [96–98], it is also utilized for schistosomiasis control and cancer treatment. In vitro studies showed that it presents anticancer effects at concentrations less than ten micromoles. Sanguinarine triggers cell cycle arrest at different phases of apoptosis in many tumor cells [99–104]. It also is active against angiogenesis [105–107]. The administration of COX-2 inhibitors and sanguinarine has been recommended for prostate cancer treatment. Sanguinarine can also be used for the treatment of conditions caused by ultraviolet exposure, such as skin cancer [108]. The mechanism of its anticancer activity could be its interactions with glutathione (GSH). These interactions decrease cellular GSH and increase ROS generation [102,109]. Sanguinarine-induced ROS production and cytotoxicity can be blocked by pretreatment with N-acetyl cysteine or catalase. This mode of action is similar to that of TopII inhibitor salvicine [110,111]. Sanguinarine is considered a potent inhibitor of MKP-1 (mitogen-activated protein kinase phosphatase 1) [112]. It also interferes with microtubule assembling [113] and the nucleocytoplasmic trafficking of cyclin D1and TopII, and causes DNA damage leading to anticancer activity. Sanguinarine suppresses NF-κB activation triggered by TNF, interleukin-1, okadaic acid, and phorbol ester, but not that induced by hydrogen peroxide or ceramide [114]. It also inhibits the signal transducer and activator of transcription 3 activation (STAT-3) [115]; downrgulates CDKs, cyclins, MMP-9 and MMP-2 [106,109]; upregulates p21, p27 [104,107] and the phosphorylation of p53 [99]; alters the members of the Bcl-2 family, including Bax, Bid, Bak Bcl-2, and Bcl-xL; activates caspases [101–103]; and upregulates death receptor 5 (DR-5) [101].

Tetrandrine (**7** in Figure 1) is a bisbenzylisoquinoline alkaloid extracted from *Stephania tetrandra* S.MooreIt is used as an immune modulator, antihepatofibrogenetic, antiarrhythmic, anti-inflammatory, antiportal hypertension, anticancer and neuroprotective agent [116]. It is a potent anticancer agent. Tetrandrine triggers different phases of cell cycle arrest depending on the cancer cell type, and also induces apoptosis in many human cancer cells, including leukemia, colon, bladder, lung and hepatoma [117,118]. In vivo experiments demonstrated that the survival of mice subcutaneously injected with CT-26 cells was prolonged after daily treatment with tetrandrine [119]. Tetrandrine suppresses the expression of VEGF in glioma cells, has an anticancer effect on ECV304 human umbilical vein endothelial cells, and aniangiogenesis effects [120]. Tetrandrine had no acute toxicity were noticed [121]. Hence, tetrandrine has a great promise as an MDR modulator for the treatment of P-gp-mediated MDR cancers. Tetrandrine could be used in combination with 5-fluorouracil and cisplatin drugs [122,123]. If combined with tamoxifen it increases the efficacy by inhibiting phosphoinositide-dependent kinase 1 [124]. It also increases the radio-sensitivity of many cancer cells by affecting the radiation-induced cell cycle arrest and interfering with the cell cycle. So tetrandrine can be used in combination with cancer chemotherapy or radiotherapy. Activation of glycogen synthase kinase 3β (GSK-3β), generation of ROS, activation of p38 mitogen-activated protein kinase (p38 MAPK), and inhibition of Wnt/betacaten in signaling might cause the anticancer activity of tetrandrine [119,125]. Tetrandrine up-regulates p53, p21, p27, and Fas; down-regulates Akt phosphorylation, cyclins, and CDKs and activates caspases [120,125–127].

Aporphine alkaloids (**8** in Figure 1) are extracted from the aerial part of *Pseuduvariasetosa*. These alkaloids demonstrated moderate antitumor activity against lung and breast cancer cells, in addition to their high antitumor activity against *epidermoid carcinoma* (KB) and breast cancer [36]. Mohamed et al. have studied *Magnolia grandiflora* L., characterized the isolated aporphine alkaloids and researched their cytotoxic activity. They found that the magnoflorine and lanuginosine alkaloids have cytotoxic activity against liver carcinoma cell lines [2].

Apomorphine alkaloids were isolated from *Nelumbo nucifera* Gaertn leaves and were demonstrated to have antioxidant and antipoliferative activities [41].

Peganum. Harmala has been used traditionally for cancer therapy, harmine (**9** in Figure 1) and harmaline (**10** in Figure 1) are the major alkaloids found in this plant in ratios of 4.3% and 5.6%, respectively [128]. Four alkaloids; harmine, (**9** in Figure 1), harmaline (**10** in Figure 1), harmalacidine (**11** in Figure 1) and vasicinone (**12** in Figure 1) were isolated from plant seeds and have shown a significant cytotoxic activity [128]. The mechanism of action of P. *harmala* seeds has been investigated by Sobhani et al.; it was found that these alkaloids inhibit topoisomerase **1**, resulting in antiproliferation of cancer cells [129].

A concise summary of the anticancer activities of alkaloids **1–12** is depicted in Table 1.

Table 1. The anticancer compounds discussed in this section and their activities.

Alkaloids	Anticancer Activity
Berberine	Inhibits the proliferation of breast, lung, colon and liver cancer cell lines Inducing the cell cycle arrest or apoptosis in cancer cell
Matrine	Inhibits the proliferation of cancer cell by G1 cell cycle arrest or apoptosis
Piplartine	Antitumor related to its antiproliferative effect
Piperine	Antitumor and immunomodulatory
Sanguinarine	Induces cell cycle arrest at different phases or apoptosis in a variety of cancer cell
Tetrandrine	Induces different phases of cell cycle arrest depends on cancer cell types
Aporphine	Antitumor activity against small cell, lung cancer and breast cancer cells, in addition to a high antitumor activity against epidermoid carcinoma
Apomorphine	Antioxidant and antipoliferative activities
Harmine, harmaline harmalacidine, vasicinone	Inhibit topoisomerase 1 resulting in antiproliferation of cancer cells
Evodiamine Sanguinarine	Induce cell cycle arrest or apoptosis, inhibiting the angiogenesis, invasion, and metastasis in a variety of cancer cell lines
Matrine	Inhibits the proliferation of various types of cancer cells mainly through the mediation of G1 cell cycle arrest or apoptosis
Tetrandrine	Induces apoptosis in many human cancer cells, including leukemia, bladder, colon, hepatoma, and lung

4. Antibacterial Activities

Anti-bacterial agent is a chemical entity that has the potential to kill or inhibit the production of bacteria. Since antiquity several nations have been using different plants for healing and treating several kinds of diseases, like using, *Ayurveda,* in providing many medicines from the *Neem tree, Azadirachta indica* A.Juss.in India, and *Valerian (Valeriana officinalis),* the medicinal plant endogenous to Europe and Asia and widely introduced in North America [37,39,130]. However, at the beginning of the 1980s this trend has declined, and the sights of researchers were turned into synthesizing new compounds. More recently the interest in using natural products has been renewed, due to the huge increase in antibiotics resistance, the limited availability of new synthetic antibacterials [37], and the discovery of many new natural products [38]. Finding new anti-infective agents is a necessity. Even with the careful use of antibiotics, each antibiotic has a limited life span [37]. Nothing can rule out the need for antibiotics, although biologics can sometimes be used in several cases, but they are usually accompanied by many limitations [39,40].

Several studies have demonstrated that much plant extracts containingalkaloids, flavonoids, phenolics and other compounds have significant antibacterial activity. However, in this section, we are specifically concerned with alkaloids as antibacterial agents.

Alkaloids have a reputation of being a natural curse and blessing [131]. They are a wide-range and diverse group of natural compounds that exist in plants, animals, bacteria, and fungi. The only thing they have in common is the occurrence of a basic nitrogen [132], which can bea primary, secondary or tertiary amine. Currently, there are more than 18,000 discovered alkaloids [133]. The unique bioactivity of alkaloids is attributed to the presence of nitrogen, that capable of accepting a proton, and one or more amine donating hydrogen atoms, which is usually accompanied by proton-accepting and -donating functional [127].

Alkaloids have played a very important role in developing new antibacterial agents. Many interesting examples include the synthesis of quinolones from quinine, derivatization of azomycin to afford metronidazole, and alteration of the quinoline scaffold to furnish bedaquiline. In other antibacterial agents, alkaloids are utilized as scaffold substructures as seen with linezolid and trimethoprim [134–136].

4.1. Antibacterial Indole Alkaloids

Clausenine (**13** in Figure 3) extracted from the stem bark of *Clausena anisate* (Willd.) were shown to have antibacterial and antifungal activities [137]. *I*-Mahanine indole alkaloid (**14** in Figure 3) obtained from *Micromelum minutum* Wight and Arn. has demonstrated antimicrobial activity against *Bacillus cereus* (MIC100 values of 6.25 μg/mL) and *Staphylococcus aureus* (MIC100 values of 12.5 μg/mL) [138].

Hapalindole alkaloids, hapalindole X (**15** in Figure 3), deschlorohapalindole I (**16** in Figure 3), and 13-hydroxy dechlorofontonamide (**17** in Figure 3) and hapalindoles A, C, G, H, I, J, and U, hapalonamide H, anhydrohapaloxindole A, and fischerindole L were obtained from *cyanobacteria sp.* Results demonstrated that **15** and **18** possess a very strong activity against both *Mycobacterium tuberculosis* and *Candida albicans* with MIC values in the range of 0.6 to 2.5 μM [139].

The indolizdine alkaloid 2, 3-dihydro-1H-indolizinium chloride (**19** in Figure 3) isolated from *Prosopisglandulosa* Torr. *var. glandulosa* was found to be a strong antifungal and antibacterial agent against *Cryptococcus neoformans, Aspergillus fumigatus* with IC50 values of 0.4 and 3.0 μg/mL, respectively, and antibacterial activity against methicillin-resistant *Staphylococcus aureus* and *Mycobacterium* with IC50 values of 0.35 and 0.9 μg/mL respectively [140].

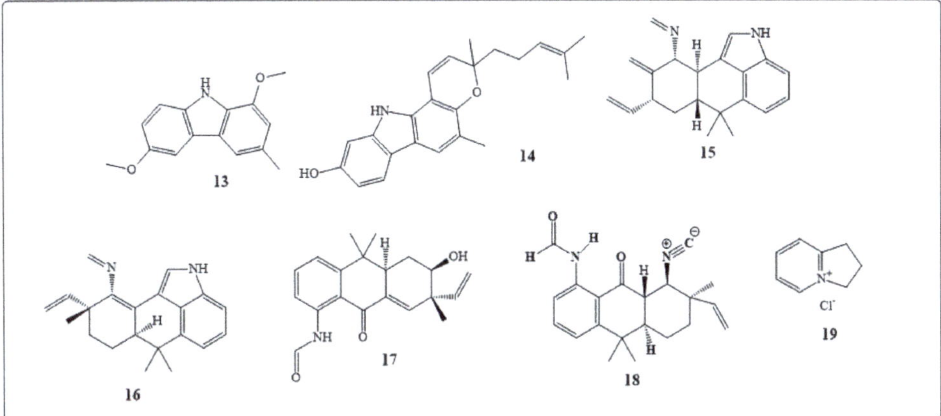

Figure 3. Chemical structures of natural antibacterial alkaloids: Clausenine (**13**), *(R)*-Mahanine (**14**), hapalindole X (**15**), deschlorohapalindole I (**16**), 13-hydroxy dechlorofontonamide (**17**), hapalonamide H (**18**) and 2,3-dihydro-1 H-indolizinium chloride (**19**).

The antimicrobial, antimalarial, cytotoxic, and anti-HIV activities of 26 isoquinolines were studied. The results showed that compound **20** (Figure 4) has antimicrobial effects, compounds **21, 22** and **23** (Figure 4) have antimalarial effects, compounds **20** and **21** have cytotoxic effects and compounds **24** and **25** (Figure 4) have anti-HIV effects. It is expected that these compounds have the potential to be used as lead compounds for further research and investigation [141].

Berberine alkaloid (**1** in Figure 1) was tested against the oral pathogens *Fusobacterium nucleatum*, *Enterococcus faecalis*, and *Prevotella intermedia*. The MIC values of berberine against these pathogens were 31.25 μg/mL, 3.8 μg/mL and 500 μg/mL, respectively [142].

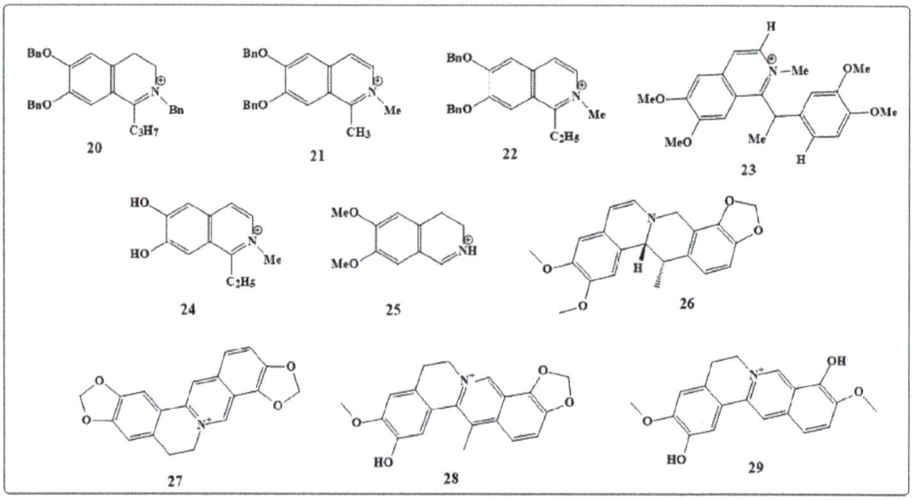

Figure 4. Chemical structures of isoquinolines alkaloids (antimicrobial, antimalarial, cytotoxic, and anti-HIV agents, **20–25**), 2-benzyl-6,7-bis(benzyloxy)-1-propyl-3,4-dihydroisoquinolin-2-ium (**20**), 6,7-bis(benzyloxy)-1,2-dimethylisoquinolin-2-ium (**21**), 6,7-bis(benzyloxy)-1-ethyl-2-methylisoquinolin-2-ium (**22**), 1-(1-(3,4-dimethoxyphenyl)ethyl)-6,7-dimethoxy-2-methylisoquinolin-2-ium (23), 1-ethyl-6,7-dihydroxy-2-methylisoquinolin-2-ium (**24**), 6,7-dimethoxy-3,4-dihydro- isoquinolin-2-ium (**25**), dehydrocavidine (**26**), coptisine (**27**), dehydroapocavidine (**28**) and tetradehydroscoulerine (**29**).

Sanguinarine (**6** in Figure 1), a benzophenanthridine alkaloid obtained from the root of *Sanguinaria canadensis* L., has shown inhibition activity against methicillin-resistant *Staphylococcus aureus* (MRSA) bacteria, an organism known for its resistance to almost all antibacterial agents. MRSA is responsible for a large number of life-threatening infections. MRSA infections that reach the bloodstream are responsible for numerous complications and fatalities, killing 10–30% of patients. An important predictor of morbidity and mortality in adults is the blood concentrations of vancomycin, the antibiotic of choice to treat this condition [18]. Many of them are skin-related conditions: Skin glands, and mucous membranes. Sanguinarine activity against MRSA strains is in the range of 3.12 to 6.25 μg/mL. Whereas, its MIC values against the two MRSA strains are 3.12 μg/mL and 1.56 μg/mL. Sanguinarine causes lysis of the cell by induction the release of autolytic enzymes [143]. Further, sanguinarine has shown to kill cells and destroy tissues when applied to the skin. The biosynthesis of sanguinarine in plants is via the action of dihydrobenzophe-anthridineoxidase on dihydrosanguinarine.

Dehydrocavidine, coptisine, dehydroapocavidine and tetradehydroscoulerine (**26–29,** respectively, in Figure 4) represent the active compounds of Yanhuanglian that extracted from *Corydalis saxicola* Bunting and used in traditional Chinese medicine, and they have shown antibacterial, antiviral and anticancer activities in *in-vivo* studies [144].

Five new quinolone alkaloids, euocarpines A–E (**30–35** in Figure 5), were isolated from the fruits of *Evodia ruticarpa* var. *officinalis* (Dode) *C.C.Huang* and were found to exhibit moderate antibacterial activities MIC values of 4–128 μg/mL Compounds with thirteen carbons on side chain showed the highest antibacterial activity with favorable low cytotoxicity [145].

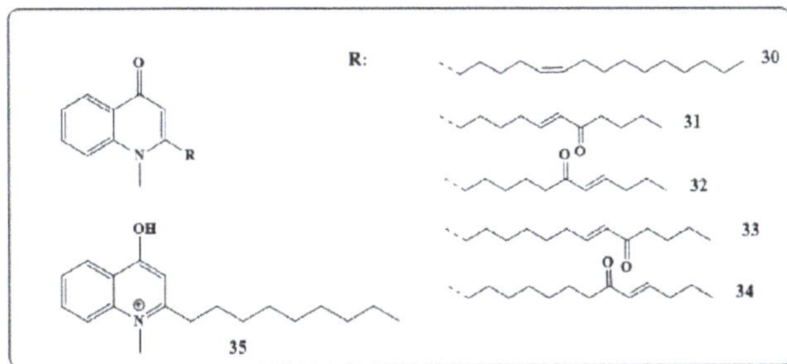

Figure 5. Chemical structures of alkaloids; euocarpines A–E, **30–35**, respectively.

The three quinolone alkaloids kokusaginin, maculine, kolbisine (**36–38** in Figure 6) were obtained from *Teclea afzelii* Engl.stem bark, and their antimicrobial and antifungal activities were studied. The results revealed that kokusaginine, **36**, was active against Gram-positive and negativebacteria, fungi and *Mycobacterium smegmatis*. While the crude extracts maculine, **37**, and kolbisine, **38**, have resulted in inhibition of 87.5% of the microbes with an MIC value of 19.53 μg/mL. While maculine, **37**, demonstrated a moderate activity against *M. smegmatis* with greater MIC value (156.25 μg/mL) [146].

Eight new quinolone alkaloids were extracted from the *actinomycete Pseudonocardia* sp. CL38489 were found to be potent and selective anti *Helicobacter pylori* agents [147]. Diterpene alkaloids, such as agelasines O–U (**39–45** in Figure 6), were obtained from *Okinawan marine sponge Agelas sp*. Studies have shown that agelasines O–R (**39–42** in Figure 6) and T (**44** in Figure 6) had antimicrobial activities against several bacteria and fungi [148].

The polyamine alkaloid squalamine (**46** in Figure 6) which was extracted from tissues of the dogfish shark *Squalus acanthiasis* is now considered as a broad-spectrum steroidal antibiotic with potent bactericidal properties against both *Gram-negative* and *Gram-positive* bacteria [149].

Aaptamine (**47** in Figure 6) extracted from the Indonesian marine sponge of the genus Xestospongia has shown significant antibacterial activity against *gram-negative bacteria* [150]. Table 2 illustrates the antibacterial activity of the compounds discussed in this section.

Table 2. Antibacterial compounds, discussed in this section, and their antibacterial activity.

Alkaloids	Antibacterial Activity
Clausenol, Kokusaginine, Maculine, Kolbisine, squalamine, Aaptamine	Active against *Gram-positive* and *negative* bacteria and fungi
R- Mahanine	Antimicrobial activity against *Bacillus cereus* and *Staphylococcus aureus*
hapalindole X, deschlorohapalin-dole I, 13-hydroxy dechlorofonto-namide, hapalonamide H	Potent activity against both *Mycobacterium tuberculosis* and *Candida albicans*
Indolizdine	Antibacterial activities against *Cryptococcus neoformans*, *Aspergillus fumigatus*, methicillin-resistant *Staphylococcus aureus* and *Mycobacterium* intracellular
Isoquinolines	Antimalarial, cytotoxic, and anti-HIV effects.
Sanguinarine,	Antibacterial activity against methicillin-resistant *Staphylococcus aureus*

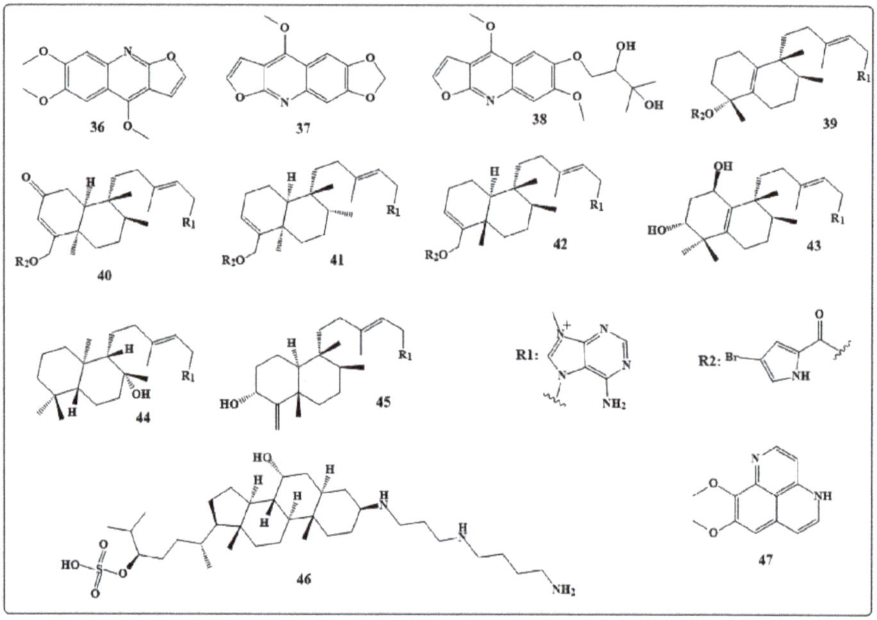

Figure 6. Chemical structures of natural antibacterial alkaloids: Kokusaginine (**36**), maculine (**37**), kolbisine (**38**), agelasines O–U (**39–45**), squalamine (**46**) and aptamine (**47**).

4.2. Antibacterial Mechanism of Action of Alkaloids

Most alkaloids are found to be bactericidal rather than being bacteriostatic. For example, squalamine, **42**, was found to be with potent bactericidal properties killing *Gram-positive* and *Gram-negative* pathogens by ≥99.99% in about 1–2 h [151]. The mechanism of action of squalamine, **42**, is illustrated in Figure 7.

The alkaloids mechanism of action as antibacterial agents has been found to be different between each class. In the alkaloids pergularinine and tylophorinidine from the indolizine class, the antibacterial action is due to inhibition of the enzyme dihydrofolate reductase resulting in the inhibition of nucleic acid synthesis [152].

Within the isoquinolone class two mechanisms of bacterial inhibition were shown to occur; Ungereminea phenanthridine isoquinoline exerts its effect through the inhibition of nucleic acid synthesis, while from the studies with benzophenanthridine and protoberberine isoquinolines it was suggested that those agents act by the perturbation of the Z-ring and cell division inhibition, and this was further proved by many studies [97,153–155].

On the other hand, synthetic quinolones inhibit the type II topoisomerase enzymes, while the natural quinolone alkaloids lacking the 3-carboxyl function act as respiratory inhibitors by reducing oxygen consumption in the treated bacteria [156,157].

Agelasines alkaloids exert their antibacterial activity by the inhibition of the dioxygenase enzyme BCG 3185c, causing a disturbance in the bacterial hemostasis. This result was revealed from overexpression and binding affinity experiments on the anti-mycobacterial alkaloid agelasine D [158]. Saqualamine from the polyamine alkaloid class acts by disturbing bacterial membrane integrity [158,159].

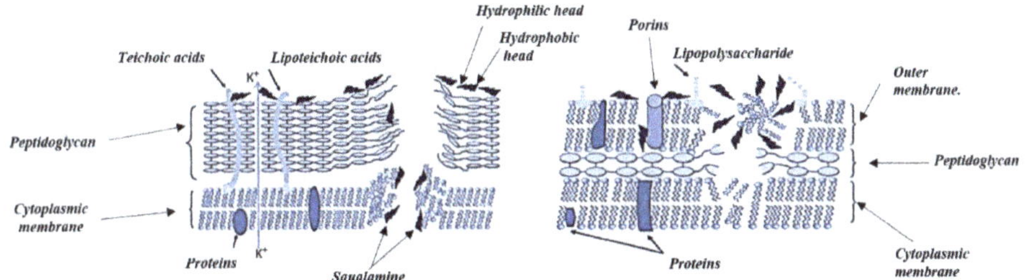

Figure 7. A representation of the mode of action of squalamine antibacterial agent; intracellular ion efflux in *Gram-positive* bacteria caused by the depolarization of the membranes, and membrane disruption in *Gram-negative* bacteria. PG is peptidoglycan, CM is cytoplasmic membrane, and OM is outer membrane.

5. Antiviral Activity

Plants and animals, host a vast number of viruses, often transmitted by insects, such as aphids and bugs. Viral infection resistance can be caused either by biochemical mechanisms that prevent the development and multiplication of the virus or by preventing vectors, such as aphids. The evaluation of the antiviral activity is relatively difficult. Several researchers have studied the effect of alkaloids on viral reproduction. The studies revealed that about 40 alkaloids possess antiviral properties.

Leurocristine, periformyline, perivine and vincaleucoblastine (**48–51** in Figure 8) are natural alkaloids obtained from *Catharanthus roseus* (L.) and *lanceusPich* (Apocycaceae). Leurocristine, 48, is an active against *mengovirus* extracellular virucidal, poliovirus, *vaccinia,* and *influenza* viruses [160]. Periformyline, **49**, inhibits poliovirus type viruses. Whereas, perivine, **50**, exhibits polio extracellular virucidal activity against *vaccinia,* and vincaleucoblastine, **51**, possesses extracellular virucidal activity against *poliovirus vaccinia and influenza* virus.

The michellamines D and F (**52** and **53** in Figure 8), naphthyl-isoquinoline alkaloids, obtained from the tropical *liana Ancistrocladuskorupensis* have demonstrated HIV-inhibitory [161]. A series of isoquinoline alkaloids as lycorine, lycoricidine (**54** in Figure 8), narciclasine, and cis-dihydronarciclasine, obtained from *Narcissus poeticus* (Amaryllidaceae), have shown significant in vitro activity against *flaviviruses and bunyaviruses.* Poliomyelitis virus inhibition by the above-mentioned compounds occurred at 1 mg/mL [162].

The alkaloids homonojirimycin (**55** in Figure 9) and 1-deoxymanojirimycin (**56** in Figure 9) were obtained from *Omphaleadiandra* (Euphorbiaceae). Homonojirimycin, **55**, is an inhibitor of several *α*-glucosidases, glucosidase I and glucosidase II. Deoxymanojirimycin, **56**, is an inhibitor of glycoprocessing mannosidase [163]. Castanospermine (**57** in Figure 9) and australine (**58** in Figure 9) are alkaloids present in seeds of *Castanospermum austral* A. Cunn. and C. Fraser (Leguminosae) and reduce the ability of HIV to infect cultured cells and have the potential for treating AIDS [164]. Sesquiterpene alkaloids obtained from *Tripterygium hypoglaucum* (H. Lév.) and *Tripterygium wilfordii* Hook. f. (Celastraceae) were found to possess anti-HIV activity [165].

The acridone alkaloids, 5-hydroxynoracronycine—same compound of 11-hydroxynorac- ronycine (**59** in Figure 9) and Acrimarine F (**60** in Figure 9) were extracted from *Citrus alata* (Tanaka) plants and were found to be very effectiveagainst Epstein-Barr virus [166–168]. Columbamine, palmitine(**61, 62** in Figure 9), and berberine are alkaloids with potent activity against HIV-1 and can be found in many plants, includingAnnonaceae (Coelocline), *Berberis vulgaris* (Berberidaceae), Menispermaceae and Papaveraceae [169–171]. A concise summary of the antiviral activities of alkaloids 48–62 is depicted in Table 3.

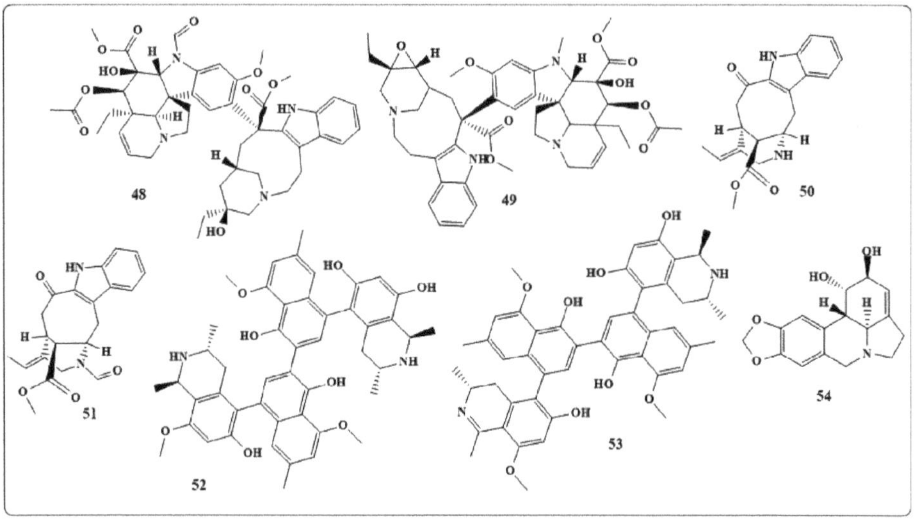

Figure 8. Chemical structures of the antiviral natural alkaloids: Leurocristine (**48**), periformyline (**49**), perivine (**50**), vincaleucoblastine (**51**), michellamines D and F (**52** and **53**) and lycoricidine (**54**).

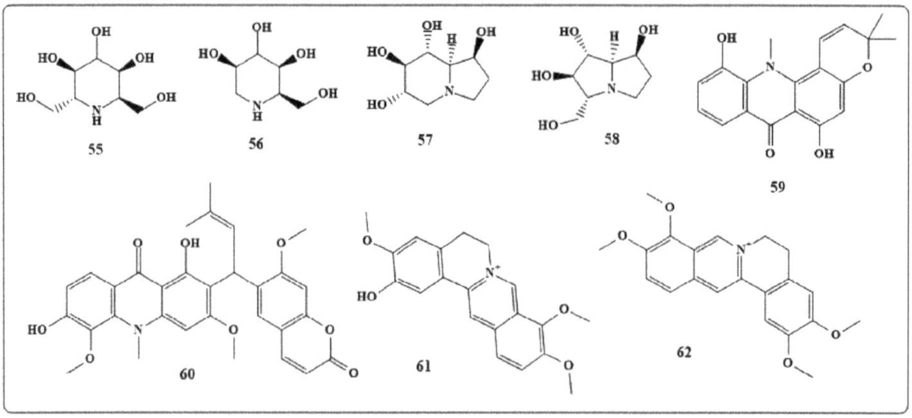

Figure 9. Chemical structures of antiviral natural alkaloid: Homonojirimycin (**55**), 1-deoxymanojirimycin (**56**), castanospermine (**57**), australine (**58**), 5-hydroxynoracronycine (**59**), acrimarine F (**60**), columbamine (**61**), palmitine (**62**).

Table 3. Antiviral compounds, discussed in this section, and their antiviral activity.

Alkaloids	Antiviral Activity
Leurocristine	Active against *mengovirus* extracellular virucidal, poliovirus, *vaccinia*, and *influenza* viruses
Periformyline	Inhibits poliovirus type viruses
Perivine	Exhibits polio extracellular virucidal activity against *vaccinia*
Vincaleucoblastine	Possesses extracellular virucidal activity against poliovirus *vaccinia* and *influenza* virus.

Table 3. *Cont.*

Michellamines D and F	HIV-inhibitory
Homonojirimycin	Inhibitor of several a-glucosidases
Deoxymanojirimycin	Inhibitor of glycoprocessing mannosidase
Castanospermine, australine	Reduce the ability of the human immunodeficiency virus (HIV) to infect cultured cells, and have potential for treating AIDS
Sesquiterpene	Anti-HIV activity
5-hydroxynoracronycine, Acrimarine F	Remarkable inhibitory effects on *Epstein-Barr* virus activation
Columbamine, Berberine, Palmitine	Inhibitors alkaloids against HIV-1

6. Antifungal Activity

The spread of resistance to antifungal agents has accelerated in the recent few years. The rate of morbidity and mortality has enhanced, due to resistance to antifungal agents. Since the molecular processes in humans and fungi are similar, there is always a risk that the fungal cytotoxic substance is toxic to the host cells. Consequently, patients with a reduced immune system, such as those with transplants, cancer and diabetics who do not respond appropriately to the current medications have ignited the need for new antifungal medicines [93]. The current antifungal drugs suffer from many side effects as irritation, diarrhea, and vomiting, furthermore, it is less effective, due to the development of resistance to those drugs by the diverse available fungi. The potency and effectiveness of the antifungal medicines developed during 1980–1995, such as the imidazoles and triazoles, which inhibit the fungal cell processes have gradually decreased, due to the development of resistance by the microorganism [172]. Therefore, researchers are looking for new potent entities to replace the currently used antifungal agents. A respected number of biologically active compounds were isolated from medicinal plants and are widely used as mixtures or pure compounds to cure a variety of diseases. It is estimated that about 250,000 to 500,000 species of plants are growing on our planet. However, humans are using only 1 to 10% of those plants [173]. This treasure should be utilized to isolate and develop new antifungal agents by using new methods and techniques [174]. For example, chemical entities isolated from plants, such as the indole derivatives, dimethyl pyrrole and hydroxydihydrocornin-aglycones have shown promising antifungal activities [174]. However, pure medicines derived from these chemical entities still to be developed.

It should be emphasized that the annual mortality rate, due to fungi was constant through several decades, and the resistance to antifungal drugs has emerged only in the recent two to three decades [175–177]. Thus, a combined effort from chemists, biologists, pharmacologists and etc. is crucially needed to combat the issue of microorganisms' resistance to drugs. Consequently, great emphasis has been placed on developing a more detailed understanding of antimicrobial resistance mechanisms, improved methods for detecting resistance when they occur, new antimicrobial options for treating infections caused by resistant organisms, and ways to prevent the emergence and spread of resistance in the first place [174].

In this section of the review, we attempted to report on some important antifungal compounds obtained from plants.

Antifungal Activity of Alkaloids

Extraction of the opium poppy *Papaver somniferum* L. has resulted in the isolation of morphine which is considered the first member [156]. Recently, a new alkaloid, 2-(3,4-dimethyl-2,5-dihydro-1H-pyrrol-2-yl)- 1-methylethyl pentanoate (**63** in Figure 10) was isolated from the plant *Datura metel* var. *fastuosa* (L.) *Saff.*, and has demonstrated activity against *Aspergillus* and *Candida* species when tested both in vitro and in vivo [178,179].

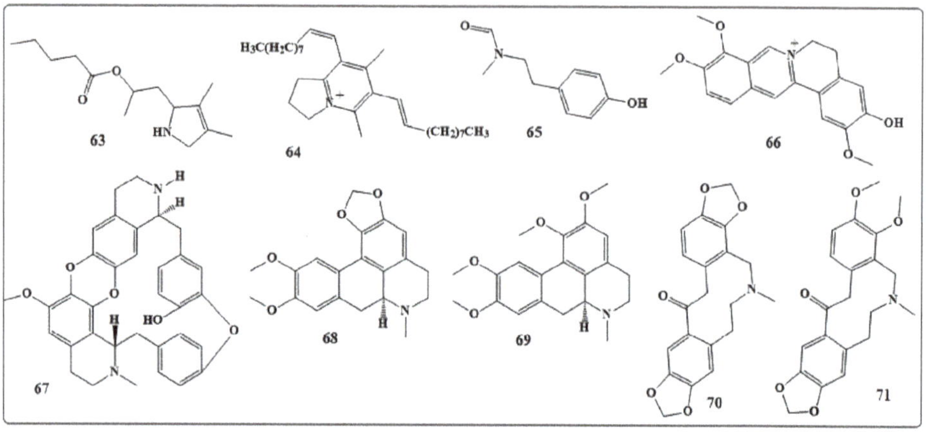

Figure 10. Chemical structures of antifungal natural alkaloids: 2-(3,4-dimethyl-2,5-dihydro-1H-pyrrol-2-yl)-1-methylethyl pentanoate (**63**), 6,8-didec-(1Z)-enyl-5,7-dimethyl-2,3-dihydro-1H-indolizinium (**64**), N-methyl-N-formyl-4-hydroxy-beta-phenylethylamine (**65**), Jatrorrhizine (**66**), (+)-Cocsoline (**67**), dicentrine (**68**), glaucine (**69**), protopine (**70**) and alpha-allocryptopin (**71**).

Another alkaloid, 6,8-didec-(1Z)-enyl-5,7-dimethyl-2,3-dihydro-1H-indolizinium (**64** in Figure 10) was isolated from *Anibapanurensis* and demonstrated excellent activity against the strain of *C. albicans* [180].

β-Carboline, a tryptamine and twophenylethylamine-derived alkaloids along with N-methyl-N-formyl-4-hydroxy-beta-phenylethylamine (**65** in Figure 10) from *Cyathobasisfruticulosa* [20] and *haloxylines A* and *B*, new piperidine from *Haloxylon schmittianum* Pomel have shown a potential antifungal activity [181].

Jatrorrhizine (**66** in Figure 10) isolated from *Mahonia aquifolium (Pursh)* Nutt. has shown the most potent antifungal inhibitory in all studied fungi with MIC between 62.5 to 125 μg/mL, while the crude extract, berberine, and palmatine showed reduced inhibitory activity with MIC of 500 to >/= 1000 μg/mL [182].

(+)-Cocsoline (**67** in Figure 10) is a bisbenzylisoquinoline alkaloid isolated from *Epinetrumvillosum* has demonstrated a good antifungal activity [183]. The alkaloids N-methylhydrasteinehydroxylactam and 1-methoxyberberine chloride isolated from *Corydalis longipes* D. Don demonstrated great inhibitory activity [184]. Four alkaloids, dicentrine (**68** in Figure 10), glaucine (**69** in Figure 10), protopine (**70** in Figure 10), and alpha-allocryptopin (**71** in Figure 10) were isolated from *Glaucium oxylobum* Boiss and Buhse have exhibited good activity against *Microsporumgypseum, Microsporumcanis, T. mentagrophytes* and *Epidermophytonfloccosum* [185].

Flindersine (**72** in Figure 11) and haplopine (**73** in Figure 11) obtained from *Haplophyllum sieversii* Fisch. were growth-inhibitory compounds against various fungi [186]. Canthin-6-one (**74** in Figure 11) and 5-methoxy-canthin-6-one (**75** in Figure 11) of *Zanthoxylumchiloperone* var. *angustifolium* exhibited antifungal activity against *C. albicans,* A. *fumigatus* and T. *mentagrophytes* [187]. Frangulanine (**76** in Figure 11), a cyclic peptide alkaloid and waltherione A (**77** in Figure 11), quinolinone alkaloids from leaves of *Melochiaodorata* were reported to exhibit antifungal activities against a broad spectrum of pathogenic fungi [188].

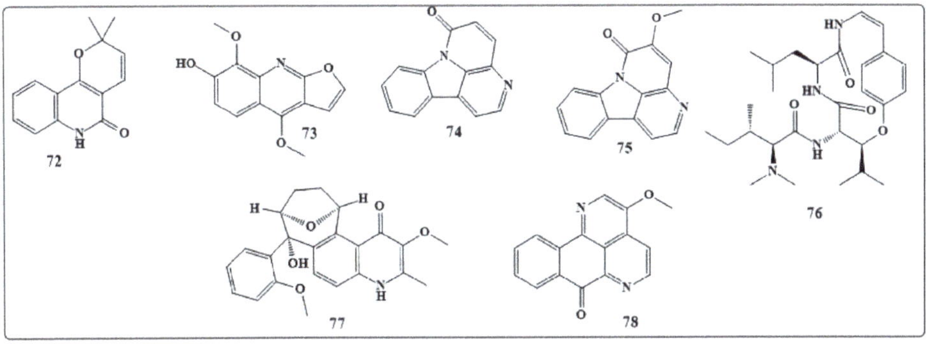

Figure 11. Chemical structures of antifungal natural alkaloids: Flindersine (**72**), haplopine (**73**), canthin-6-one (**74**), 5-methoxy-canthin-6-one (**75**), frangulanine (**76**), waltherione A (**77**) and 3-methoxisampangin (**78**).

Furthermore, anodic alkali aninolinate has demonstrated inhibitory activity against all 10 fungi tested and demonstrated a particularly high sensitivity to this compound, showing germination levels of less than 10% [189]. From the root of *Dictamnusdasycarpus* two antifungal fructoxin alkaloids were isolated. 3-Methoxisampangin (**78** in Figure 11) of *cleistopholispatens* showed vast inhibitory activity against each of *C. albicans, A. fumigatus,* and *C. neoformans* [190]. Table 4 lists the antifungal activity of the compounds discussed in this section.

Table 4. Antifungal compounds, discussed in this section, and their antifungal activity.

Alkaloids	Antifungal Activity
2-(3,4-dimethyl-2,5-dihydro-1H-pyrrol-2-yl)-1-methylethyl pentanoate	Activity against *Aspergillus* and *Candida* species
6,8-didec-(1Z)-enyl-5,7-dimethyl-2,3-dihydro-1H-indolizinium	Good activity against a drug-resistant strain of *C. albicans*
β-carboline, Cocsoline	Effective against all fungal species and marginal activity
Dicentrine, glaucine, Protopine, alpha-allocryptopin	Good activity against *Microsporumgypseum, Microsporumcanis, T. mentagrophytes* and *Epidermophytonfloccosum*
Canthin-6-one, 5-methoxy-canthin-6-one	Antifungal activity against *C. albicans, A. fumigatus* and *T. mentagrophytes*
Frangulanine, Waltherione, quinolinone alkaloids	Exhibit antifungal activities against a broad spectrum of pathogenic fungi
3-methoxisampangin	Significant antifungal activity against *C. albicans, A. fumigatus*, and *C. neoformans*

7. Toxicity of Plant Secondary Metabolites

Through hundreds of years, people have been used plant extracts for the treatment of a variety of diseases, such as snakebite, fever and insanity. However, a number of plants containing alkaloids are classified as main plant toxins because of their vast structural diversity and different mechanism of actions. Inhalation or swallowing toxic alkaloids by humans or animals might result in a certain mechanism involving transporters, enzymes and receptors at certain cells and tissues, and therefore, causing musculoskeletal deformities and hepatotoxic effects. These toxic alkaloids include tropane, piperidine, pyrrolizidine and indolizidine. The most adverse effects of these entities are vomiting, mild gastrointestinal perturbation, teratogenicity, arrhythmias, itching, nausea, psychosis, paralysis, and death [188].

8. Conclusions

Secondary metabolites isolated from different plants have proven to be useful for humans and animals alike. Studies on a number of these alkaloids have demonstrated that a vast number of them possess many pharmacological activities. These activities include anticancer, antibacterial, antiviral, antioxidant and antifungal. However, when high doses of the secondary metabolites are consumed, toxic effects may be observed and in sometimes fatalities are imminent. For instance, some alkaloids have proven to lead to paralysis, asphyxia, and in sometimes to death. During the past few years researches have been conducted to make new derivatives of the isolated secondary metabolites aiming to obtain medicines with more potency, less toxicity, and more resistant to different microorganisms. It is hoped that the state-of-the-art methods and sophisticated techniques along with computational methods, will make the path for generating novel potent drugs, based on natural products, be shorter and beneficial.

Several plants contain secondary metabolites which are toxic and may cause danger to humans when administered. However, the cases in which fatal plant poisonings occur are negligible. The most common incidents of toxicity are those involving the abuse of plants for hallucinogenic purposes. Utilizing toxicological analysis of such secondary metabolites may help in the diagnosis of poisoning or abuse cases. Among the toxic secondary metabolites are harmaline, ibogaine, kawain, cytisine, dimethyltryptamine, harmine, aconitine, atropine, coniine, colchicine, taxine, mescaline, and scopolamine, which are often involved in fatal poisonings [190].

Author Contributions: All authors contributed to writing and editing the manuscript.

Funding: This research received no external funding.

Acknowledgments: Authors are thankful to Basilicata University for supporting the present study.

Conflicts of Interest: The authors declare no conflict of interest.

References

1. Khan, M.T.H.; Ather, A.; Thompson, K.D.; Gambari, R. Extracts and molecules from medicinal plants against herpes simplex viruses. *Antivir. Res.* **2005**, *67*, 107–119. [CrossRef] [PubMed]
2. Mohamed, S.; Hassan, E.; Ibrahim, N. Cytotoxic and antiviral activities of aporphine alkaloids of *Magnolia Grandiflora* L. *Nat. Prod. Res.* **2010**, *24*, 1395–1402. [CrossRef] [PubMed]
3. Evers, D.L.; Chao, C.F.; Wang, X.; Zhang, Z.; Huong, S.M.; Huang, E.S. Human cytomegalovirus-inhibitory flavonoids: Studies on antiviral activity and mechanism of action. *Antivir. Res.* **2005**, *68*, 124–134. [CrossRef] [PubMed]
4. Peng, J.; Lin, T.; Wang, W.; Xin, Z.; Zhu, T.; Gu, Q.; Li, D. Antiviral alkaloids produced by the mangrove-derived fungus *Cladosporium* sp. PJX-41. *J. Nat. Prod.* **2013**, *76*, 1133–1140. [CrossRef] [PubMed]
5. Kainsa, S.; Kumar, P.; Rani, P. Medicinal plants of Asian origin having anticancer potential: Short review. *Asian J. Biomed. Pharm. Sci.* **2012**, *2*, 1–7.
6. Ziegler, J.; Facchini, P.J. Alkaloid biosynthesis: Metabolism and trafficking. *Annu. Rev. Plant Biol.* **2008**, *59*, 735–769. [CrossRef] [PubMed]
7. Marella, A.; Tanwar, O.P.; Saha, R.; Ali, M.R.; Srivastava, S.; Akhter, M.; Shaquiquzzaman, M.; Alam, M.M. Quinoline: A versatile heterocyclic. *Saudi Pharm. J.* **2013**, *21*, 1–12. [CrossRef] [PubMed]
8. Frick, S.; Kramell, R.; Schmidt, J.; Fist, A.J.; Kutchan, T.M. Comparative qualitative and quantitative determination of alkaloids in narcotic and condiment Papaver s omniferum cultivars. *J. Nat. Prod.* **2005**, *68*, 666–673. [CrossRef] [PubMed]
9. Nassiri, M. Simple, one-pot, and three-component coupling reactions of azaarenes (phenanthridine, isoquinoline, and quinoline), with acetylenic esters involving methyl propiolate or ethyl propiolate in the presence of nh-heterocyclic or 1, 3-dicarbonyl compounds. *Synth. Commun.* **2013**, *43*, 157–168. [CrossRef]
10. Herman, A.; Herman, A. Caffeine's mechanisms of action and its cosmetic use. *Skin Pharmacol. Physiol.* **2013**, *26*, 8–14. [CrossRef] [PubMed]

11. Li, S.; Lo, C.Y.; Pan, M.H.; Lai, C.S.; Ho, C.T. Black tea: Chemical analysis and stability. *Food Funct.* **2013**, *4*, 10–18. [CrossRef] [PubMed]
12. Singh, A.K.; Chawla, R.; Rai, A.; Yadav, L.D.S. NHC-catalysed diastereoselective synthesis of multifunctionalised piperidines via cascade reaction of enals with azalactones. *Chem. Commun.* **2012**, *48*, 3766–3768. [CrossRef] [PubMed]
13. Machado, P.A.; Hilário, F.F.; Carvalho, L.O.; Silveira, M.L.; Alves, R.B.; Freitas, R.P.; Coimbra, E.S. Effect of 3-Alkylpyridine Marine Alkaloid Analogues in Leishmania Species Related to American Cutaneous Leishmaniasis. *Chem. Biol. Drug Des.* **2012**, *80*, 745–751. [CrossRef] [PubMed]
14. Cronemberger, S.; Calixto, N.; Moraes, M.N.; Castro, I.D.; Lana, P.C.; Loredo, A.F. Efficacy of one drop of 2% pilocarpine to reverse the intraocular pressure peak at 6: 00 am in early glaucoma. *Vision Pan-Am. Pan-Am. J. Ophthalmol.* **2012**, *11*, 14–16.
15. Parmar, N.J.; Pansuriya, B.R.; Barad, H.A.; Kant, R.; Gupta, V.K. An improved microwave assisted one-pot synthesis, and biological investigations of some novel aryldiazenyl chromeno fused pyrrolidines. *Bioorg. Med. Chem. Lett.* **2012**, *22*, 4075–4079. [CrossRef] [PubMed]
16. Singh, K.S.; Das, B.; Naik, C.G. Quinolizidines alkaloids: Petrosin and xestospongins from the sponge Oceanapia sp. *J. Chem. Sci.* **2011**, *123*, 601–607. [CrossRef]
17. Kumar, P.; Sharma, B.; Bakshi, N. Biological activity of alkaloids from Solanum dulcamara L. *Nat. Prod. Res.* **2009**, *23*, 719–723. [CrossRef] [PubMed]
18. Zoraghi, R.; Worrall, L.; See, R.H.; Strangman, W.; Popplewell, W.L.; Gong, H.; Samaai, T.; Swayze, R.D.; Kaur, S.; Vuckovic, M. Methicillin-resistant Staphylococcus aureus (MRSA) pyruvate kinase as a target for bis-indole alkaloids with antibacterial activities. *J. Biol. Chem.* **2011**, *286*, 44716–44725. [CrossRef] [PubMed]
19. Villinski, J.; Dumas, E.; Chai, H.B.; Pezzuto, J.; Angerhofer, C.; Gafner, S. Antibacterial activity and alkaloid content of Berberis thunbergii, Berberis vulgaris and Hydrastis canadensis. *Pharm. Biol.* **2003**, *41*, 551–557. [CrossRef]
20. Otshudi, A.L.; Apers, S.; Pieters, L.; Claeys, M.; Pannecouque, C.; De Clercq, E.; Van Zeebroeck, A.; Lauwers, S.; Frederich, M.; Foriers, A. Biologically active bisbenzylisoquinoline alkaloids from the root bark of Epinetrum villosum. *J. Ethnopharmacol.* **2005**, *102*, 89–94. [CrossRef] [PubMed]
21. Oliva, A.; Meepagala, K.M.; Wedge, D.E.; Harries, D.; Hale, A.L.; Aliotta, G.; Duke, S.O. Natural fungicides from Ruta graveolens L. leaves, including a new quinolone alkaloid. *J. Agric. Food Chem.* **2003**, *51*, 890–896. [CrossRef] [PubMed]
22. Simons, V.; Morrissey, J.P.; Latijnhouwers, M.; Csukai, M.; Cleaver, A.; Yarrow, C.; Osbourn, A. Dual effects of plant steroidal alkaloids on Saccharomyces cerevisiae. *Antimicrob. Agents Chemother.* **2006**, *50*, 2732–2740. [CrossRef] [PubMed]
23. Cretton, S.; Dorsaz, S.; Azzollini, A.; Favre-Godal, Q.; Marcourt, L.; Ebrahimi, S.N.; Voinesco, F.; Michellod, E.; Sanglard, D.; Gindro, K. Antifungal quinoline alkaloids from Waltheria indica. *J. Nat. Prod.* **2016**, *79*, 300–307. [CrossRef] [PubMed]
24. Lou, C.; Yokoyama, S.; Saiki, I.; Hayakawa, Y. Selective anticancer activity of hirsutine against HER2-positive breast cancer cells by inducing DNA damage. *Oncol. Rep.* **2015**, *33*, 2072–2076. [CrossRef] [PubMed]
25. Li, M.; Su, B.S.; Chang, L.H.; Gao, Q.; Chen, K.L.; An, P.; Huang, C.; Yang, J.; Li, Z.F. Oxymatrine induces apoptosis in human cervical cancer cells through guanine nucleotide depletion. *Anti-Cancer Drugs* **2014**, *25*, 161–173. [CrossRef] [PubMed]
26. Marciniak, P.; Adamski, Z.; Bednarz, P.; Slocinska, M.; Ziemnicki, K.; Lelario, F.; Scrano, L.; Bufo, S.A. Cardioinhibitory properties of potato glycoalkaloids in beetles. *Bull. Environ. Contam. Toxicol.* **2010**, *84*, 153–156. [CrossRef] [PubMed]
27. Marciniak, P.; Kolińska, A.; Spochacz, M.; Chowański, S.; Adamski, Z.; Scrano, L.; Falabella, P.; Bufo, S.A.; Rosiński, G. Differentiated Effects of Secondary Metabolites from Solanaceae and Brassicaceae Plant Families on the Heartbeat of Tenebrio molitor Pupae. *Toxins* **2019**, *11*, 287. [CrossRef] [PubMed]
28. Büyükgüzel, E.; Büyükgüzel, K.; Erdem, M.; Adamski, Z.; Marciniak, P.; Ziemnicki, K.; Ventrella, E.; Scrano, L.; Bufo, S.A. The influence of dietary α-solanine on the waxmoth Galleria mellonella L. *Arch. Insect Biochem. Physiol.* **2013**, *83*, 15–24. [CrossRef] [PubMed]

29. Ventrella, E.; Adamski, Z.; Chudzińska, E.; Miądowicz-Kobielska, M.; Marciniak, P.; Büyükgüzel, E.; Büyükgüzel, K.; Erdem, M.; Falabella, P.; Scrano, L. Solanum tuberosum and Lycopersicon esculentum leaf extracts and single metabolites affect development and reproduction of Drosophila melanogaster. *PLoS ONE* **2016**, *11*, e0155958. [CrossRef] [PubMed]
30. Spochacz, M.; Chowański, S.; Szymczak, M.; Lelario, F.; Bufo, S.; Adamski, Z. Sublethal Effects of Solanum nigrum Fruit Extract and Its Pure Glycoalkaloids on the Physiology of Tenebrio molitor (Mealworm). *Toxins* **2018**, *10*, 504. [CrossRef] [PubMed]
31. Seo, Y.S.; Yim, M.J.; Kim, B.H.; Kang, K.R.; Lee, S.Y.; Oh, J.S.; You, J.S.; Kim, S.G.; Yu, S.J.; Lee, G.J. Berberine-induced anticancer activities in FaDu head and neck squamous cell carcinoma cells. *Oncol. Rep.* **2015**, *34*, 3025–3034. [CrossRef] [PubMed]
32. Kaboli, P.J.; Rahmat, A.; Ismail, P.; Ling, K.H. Targets and mechanisms of berberine, a natural drug with potential to treat cancer with special focus on breast cancer. *Eur. J Pharmacol.* **2014**, *740*, 584–595. [CrossRef] [PubMed]
33. Letašiová, S.; Jantová, S.; Čipák, L.U.; Múčková, M. Berberine—Antiproliferative activity in vitro and induction of apoptosis/necrosis of the U937 and B16 cells. *Cancer Lett.* **2006**, *239*, 254–262. [CrossRef] [PubMed]
34. Bezerra, D.; Castro, F.; Alves, A.; Pessoa, C.; Moraes, M.; Silveira, E.; Lima, M.; Elmiro, F.; Costa-Lotufo, L. In vivo growth-inhibition of Sarcoma 180 by piplartine and piperine, two alkaloid amides from Piper. *Braz. J. Med. Biol. Res.* **2006**, *39*, 801–807. [CrossRef] [PubMed]
35. Sunila, E.; Kuttan, G. Immunomodulatory and antitumor activity of Piper longum Linn. and piperine. *J. Eethnopharmacol.* **2004**, *90*, 339–346. [CrossRef] [PubMed]
36. Wirasathien, L.; Boonarkart, C.; Pengsuparp, T.; Suttisri, R. Biological Activities of Alkaloids from Pseuduvaria setosa. *Pharm. Biol.* **2006**, *44*, 274–278. [CrossRef]
37. Chopra, I. The 2012 Garrod lecture: Discovery of antibacterial drugs in the 21st century. *J. Antimicrob. Chemother.* **2012**, *68*, 496–505. [CrossRef] [PubMed]
38. Imhoff, J.F.; Labes, A.; Wiese, J. Bio-mining the microbial treasures of the ocean: New natural products. *Biotechnol. Adv.* **2011**, *29*, 468–482. [CrossRef] [PubMed]
39. Oleksiewicz, M.B.; Nagy, G.; Nagy, E. Anti-bacterial monoclonal antibodies: Back to the future? *Arch. Insect Biochem.* **2012**, *526*, 124–131. [CrossRef] [PubMed]
40. Henein, A. What are the limitations on the wider therapeutic use of phage? *Bacteriophage* **2013**, *3*, e24872. [CrossRef] [PubMed]
41. Liu, C.M.; Kao, C.L.; Wu, H.M.; Li, W.J.; Huang, C.T.; Li, H.T.; Chen, C.Y. Antioxidant and anticancer aporphine alkaloids from the leaves of Nelumbo nucifera Gaertn. cv. Rosa-plena. *Molecules* **2014**, *19*, 17829–17838. [CrossRef] [PubMed]
42. Lamchouri, F.; Zemzami, M.; Jossang, A.; Abdellatif, A.; Israili, Z.H.; Lyoussi, B. Cytotoxicity of alkaloids isolated from Peganum harmala seeds. *Pak. J. Pharm. Sci.* **2013**, *26*, 699–706. [PubMed]
43. Kobayashi, Y.; Nakano, Y.; Kizaki, M.; Hoshikuma, K.; Yokoo, Y.; Kamiya, T. Capsaicin-like anti-obese activities of evodiamine from fruits of Evodia rutaecarpa, a vanilloid receptor agonist. *Planta Med.* **2001**, *67*, 628–633. [CrossRef] [PubMed]
44. Kobayashi, Y. The nociceptive and anti-nociceptive effects of evodiamine from fruits of Evodia rutaecarpa in mice. *Planta. Med.* **2003**, *69*, 425–428. [PubMed]
45. Shin, Y.W.; Bae, E.A.; Cai, X.F.; Lee, J.J.; Kim, D.H. In vitro and in vivo antiallergic effect of the fructus of Evodia rutaecarpa and its constituents. *Biol. Pharm. Bull.* **2007**, *30*, 197–199. [CrossRef] [PubMed]
46. Ko, H.C.; Wang, Y.H.; Liou, K.T.; Chen, C.M.; Chen, C.H.; Wang, W.Y.; Chang, S.; Hou, Y.C.; Chen, K.T.; Chen, C.F. Anti-inflammatory effects and mechanisms of the ethanol extract of Evodia rutaecarpa and its bioactive components on neutrophils and microglial cells. *Eur. J. Pjarmacol.* **2007**, *555*, 211–217. [CrossRef] [PubMed]
47. Du, J.; Wang, X.F.; Zhou, Q.M.; Zhang, T.L.; Lu, Y.Y.; Zhang, H.; Su, S.B. Evodiamine induces apoptosis and inhibits metastasis in MDA-MB-231 human breast cancer cells in vitro and in vivo. *Oncol. Rep.* **2013**, *30*, 685–694. [CrossRef] [PubMed]
48. Ogasawara, M.; Matsunaga, T.; Takahashi, S.; Saiki, I.; Suzuki, H. Anti-invasive and metastatic activities of evodiamine. *Biol. Pharm. Bull.* **2002**, *25*, 1491–1493. [CrossRef] [PubMed]

49. Fei, X.F.; Wang, B.X.; Li, T.J.; Tashiro, S. i.; Minami, M.; Xing, D.J.; Ikejima, T. Evodiamine, a constituent of Evodiae Fructus, induces anti-proliferating effects in tumor cells. *Cancer Sci.* **2003**, *94*, 92–98. [CrossRef] [PubMed]
50. Zhang, Y.; Wu, L.J.; Tashiro, S.I.; Onodera, S.; Ikejima, T. Intracellular regulation of evodiamine-induced A375-S2 cell death. *Biol. Pharm. Bull.* **2003**, *26*, 1543–1547. [CrossRef] [PubMed]
51. Shyu, K.G.; Lin, S.; Lee, C.C.; Chen, E.; Lin, L.C.; Wang, B.W.; Tsai, S.C. Evodiamine inhibits in vitro angiogenesis: Implication for antitumorgenicity. *Life Sci.* **2006**, *78*, 2234–2243. [CrossRef] [PubMed]
52. Lee, T.J.; Kim, E.J.; Kim, S.; Jung, E.M.; Park, J.W.; Jeong, S.H.; Park, S.E.; Yoo, Y.H.; Kwon, T.K. Caspase-dependent and caspase-independent apoptosis induced by evodiamine in human leukemic U937 cells. *Mol. Cancer Ther.* **2006**, *5*, 2398–2407. [CrossRef] [PubMed]
53. Wang, C.; Li, S.; Wang, M.W. Evodiamine-induced human melanoma A375-S2 cell death was mediated by PI3K/Akt/caspase and Fas-L/NF-κB signaling pathways and augmented by ubiquitin—Proteasome inhibition. *Toxicol. Vitr.* **2010**, *24*, 898–904. [CrossRef] [PubMed]
54. Yang, J.; Wu, L.J.; Tashino, S.I.; Onodera, S.; Ikejima, T. Reactive oxygen species and nitric oxide regulate mitochondria-dependent apoptosis and autophagy in evodiamine-treated human cervix carcinoma HeLa cells. *Free Radic. Res.* **2008**, *42*, 492–504. [CrossRef] [PubMed]
55. Tsai, H.P.; Lin, L.W.; Lai, Z.Y.; Wu, J.Y.; Chen, C.E.; Hwang, J.; Chen, C.S.; Lin, C.M. Immobilizing topoisomerase I on a surface plasmon resonance biosensor chip to screen for inhibitors. *J. Biomed. Sci.* **2010**, *17*, 49. [CrossRef] [PubMed]
56. Chan, A.; Chang, W.S.; Chen, L.M.; Lee, C.M.; Chen, C.E.; Lin, C.M.; Hwang, J.L. Evodiamine stabilizes topoisomerase I-DNA cleavable complex to inhibit topoisomerase I activity. *Molecules* **2009**, *14*, 1342–1352. [CrossRef] [PubMed]
57. Dong, G.; Sheng, C.; Wang, S.; Miao, Z.; Yao, J.; Zhang, W. Selection of evodiamine as a novel topoisomerase I inhibitor by structure-based virtual screening and hit optimization of evodiamine derivatives as antitumor agents. *J. Med. Chem.* **2010**, *53*, 7521–7531. [CrossRef] [PubMed]
58. Huang, Y.C.; Guh, J.H.; Teng, C.M. Induction of mitotic arrest and apoptosis by evodiamine in human leukemic T-lymphocytes. *Life Sci.* **2004**, *75*, 35–49. [CrossRef] [PubMed]
59. Yang, J.; Wu, L.J.; Tashino, S.I.; Onodera, S.; Ikejima, T. Critical roles of reactive oxygen species in mitochondrial permeability transition in mediating evodiamine-induced human melanoma A375-S2 cell apoptosis. *Free Radic. Res.* **2007**, *41*, 1099–1108. [CrossRef] [PubMed]
60. Lai, J.P.; He, X.W.; Jiang, Y.; Chen, F. Preparative separation and determination of matrine from the Chinese medicinal plant Sophora flavescens Ait by molecularly imprinted solid-phase extraction. *Anal. Bioanal. Chem.* **2003**, *375*, 264–269. [CrossRef] [PubMed]
61. Li, X.; Zhou, R.; Zheng, P.; Yan, L.; Wu, Y.; Xiao, X.; Dai, G. Cardioprotective effect of matrine on isoproterenol-induced cardiotoxicity in rats. *J. Pharm. Pharmacol.* **2010**, *62*, 514–520. [CrossRef] [PubMed]
62. Xing, Y.; Yan, F.; Liu, Y.; Liu, Y.; Zhao, Y. Matrine inhibits 3T3-L1 preadipocyte differentiation associated with suppression of ERK1/2 phosphorylation. *Bio. Chem. Bioph Res. Commun.* **2010**, *396*, 691–695. [CrossRef] [PubMed]
63. Zheng, J.; Zheng, P.; Zhou, X.; Yan, L.; Zhou, R.; Fu, X.Y.; Dai, G.D. Relaxant effects of matrine on aortic smooth muscles of guinea pigs. *Biomed. Environ. Sci.* **2009**, *22*, 327–332. [CrossRef]
64. Long, Y.; Lin, X.; Zeng, K.; Zhang, L. Efficacy of intramuscular matrine in the treatment of chronic hepatitis B. *Hepatobiliary Pancreat. Dis. Int.* **2004**, *3*, 69–72. [PubMed]
65. Yadav, V.; Krishnan, A.; Vohora, D. A systematic review on Piper longum L.: Bridging traditional knowledge and pharmacological evidence for future translational research. *J. Ethnopharmacol.* **2019**. [CrossRef] [PubMed]
66. Yixiang, H.; Shenghui, Z.; Jianbo, W.; Kang, Y.; Yu, Z.; Lihui, Y.; Laixi, B. Matrine induces apoptosis of human multiple myeloma cells via activation of the mitochondrial pathway. *Leuk. Lymphoma* **2010**, *51*, 1337–1346. [CrossRef] [PubMed]
67. Zhang, L.; Wang, T.; Wen, X.; Wei, Y.; Peng, X.; Li, H.; Wei, L. Effect of matrine on HeLa cell adhesion and migration. *Eur. J. Pharmacol.* **2007**, *563*, 69–76. [CrossRef] [PubMed]
68. Dai, Z.J.; Gao, J.; Ji, Z.Z.; Wang, X.J.; Ren, H.T.; Liu, X.X.; Wu, W.Y.; Kang, H.F.; Guan, H.T. Matrine induces apoptosis in gastric carcinoma cells via alteration of Fas/FasL and activation of caspase-3. *J. Ethnopharmacol.* **2009**, *123*, 91–96. [CrossRef] [PubMed]

69. Zhang, P.; Wang, Z.; Chong, T.; Ji, Z. Matrine inhibits proliferation and induces apoptosis of the androgen-independent prostate cancer cell line PC-3. *Mol. Med. Rep.* **2012**, *5*, 783–787. [CrossRef] [PubMed]
70. Liang, C.Z.; Zhang, J.K.; Shi, Z.; Liu, B.; Shen, C.Q.; Tao, H.M. Matrine induces caspase-dependent apoptosis in human osteosarcoma cells in vitro and in vivo through the upregulation of Bax and Fas/FasL and downregulation of Bcl-2. *Cancer Chemother. Pharmacol.* **2012**, *69*, 317–331. [CrossRef] [PubMed]
71. Liu, T.; Song, Y.; Chen, H.; Pan, S.; Sun, X. Matrine inhibits proliferation and induces apoptosis of pancreatic cancer cells in vitro and in vivo. *Biol. Pharm. Bull.* **2010**, *33*, 1740–1745. [CrossRef] [PubMed]
72. Zhang, Z.; Wang, X.; Wu, W.; Wang, J.; Wang, Y.; Wu, X.; Fei, X.; Li, S.; Zhang, J.; Dong, P. Effects of Matrine on Proliferation and Apoptosis in Gallbladder Carcinoma Cells (GBC-SD). *Phytother. Res.* **2012**, *26*, 932–937. [CrossRef] [PubMed]
73. Jiang, H.; Hou, C.; Zhang, S.; Xie, H.; Zhou, W.; Jin, Q.; Cheng, X.; Qian, R.; Zhang, X. Matrine upregulates the cell cycle protein E2F-1 and triggers apoptosis via the mitochondrial pathway in K562 cells. *Eur. J. Pharmacol.* **2007**, *559*, 98–108. [CrossRef] [PubMed]
74. Li, H.; Tan, G.; Jiang, X.; Qiao, H.; Pan, S.; Jiang, H.; Kanwar, J.R.; Sun, X. Therapeutic effects of matrine on primary and metastatic breast cancer. *Am. J. Chin. Med.* **2010**, *38*, 1115–1130. [CrossRef] [PubMed]
75. Ma, L.; Wen, S.; Zhan, Y.; He, Y.; Liu, X.; Jiang, J. Anticancer effects of the Chinese medicine matrine on murine hepatocellular carcinoma cells. *Planta Med.* **2008**, *74*, 245–251. [CrossRef] [PubMed]
76. Zhang, W.; Dai, B.T.; Xu, Y.H. Effects of matrine on invasion and metastasis of leukemia cell line Jurkat. *Chin. J. Integr. Tradit. West. Med.* **2008**, *28*, 907–911. [CrossRef] [PubMed]
77. Yu, P.; Liu, Q.; Liu, K.; Yagasaki, K.; Wu, E.; Zhang, G. Matrine suppresses breast cancer cell proliferation and invasion via VEGF-Akt-NF-κB signaling. *Cytotechnology* **2009**, *59*, 219–229. [CrossRef] [PubMed]
78. Yu, H.B.; Zhang, H.F.; Li, D.Y.; Zhang, X.; Xue, H.Z.; Zhao, S.H. Matrine inhibits matrix metalloproteinase-9 expression and invasion of human hepatocellular carcinoma cells. *J. Asian Nat. Prod. Res.* **2011**, *13*, 242–250. [CrossRef] [PubMed]
79. Luo, C.; Zhong, H.J.; Zhu, L.M.; Wu, X.G.; Ying, J.E.; Wang, X.H.; Lü, W.X.; Xu, Q.; Zhu, Y.L.; Huang, J. Inhibition of matrine against gastric cancer cell line MNK45 growth and its anti-tumor mechanism. *Mol. Biol. Rep.* **2012**, *39*, 5459–5464. [CrossRef] [PubMed]
80. Szallasi, A. Piperine: Researchers discover new flavor in an ancient spice. *Trends Pharmacol. Sci.* **2005**, *26*, 437–439. [CrossRef] [PubMed]
81. Srinivasan, K. Black pepper and its pungent principle-piperine: A review of diverse physiological effects. *Crit. Rev. Food Sci.* **2007**, *47*, 735–748. [CrossRef] [PubMed]
82. Lee, S.A.; Hong, S.S.; Han, X.H.; Hwang, J.S.; Oh, G.J.; Lee, K.S.; Lee, M.K.; Hwang, B.Y.; Ro, J.S. Piperine from the fruits of Piper longum with inhibitory effect on monoamine oxidase and antidepressant-like activity. *Chem. Pharm. Bull.* **2005**, *53*, 832–835. [CrossRef] [PubMed]
83. Li, S.; Wang, C.; Wang, M.; Li, W.; Matsumoto, K.; Tang, Y. Antidepressant like effects of piperine in chronic mild stress treated mice and its possible mechanisms. *Life Sci.* **2007**, *80*, 1373–1381. [CrossRef] [PubMed]
84. Kakarala, M.; Brenner, D.E.; Korkaya, H.; Cheng, C.; Tazi, K.; Ginestier, C.; Liu, S.; Dontu, G.; Wicha, M.S. Targeting breast stem cells with the cancer preventive compounds curcumin and piperine. *Breast Cancer Res. Treat.* **2010**, *122*, 777–785. [CrossRef] [PubMed]
85. Lai, L.H.; Fu, Q.H.; Liu, Y.; Jiang, K.; Guo, Q.M.; Chen, Q.Y.; Yan, B.; Wang, Q.Q.; Shen, J.G. Piperine suppresses tumor growth and metastasis in vitro and in vivo in a 4T1 murine breast cancer model. *Acta Pharmacol. Sin.* **2012**, *33*, 523. [CrossRef] [PubMed]
86. Song, Q.; Qu, Y.; Zheng, H.; Zhang, G.; Lin, H.; Yang, J. Differentiation of erythroleukemia K562 cells induced by piperine. *Chin. J. Cancer* **2008**, *27*, 571–574.
87. Pradeep, C.; Kuttan, G. Effect of piperine on the inhibition of lung metastasis induced B16F-10 melanoma cells in mice. *Clin. Exp. Metastasis* **2002**, *19*, 703–708. [CrossRef] [PubMed]
88. Hwang, Y.P.; Yun, H.J.; Kim, H.G.; Han, E.H.; Choi, J.H.; Chung, Y.C.; Jeong, H.G. Suppression of phorbol-12-myristate-13-acetate-induced tumor cell invasion by piperine via the inhibition of PKCα/ERK1/2-dependent matrix metalloproteinase-9 expression. *Toxicol. Lett.* **2011**, *203*, 9–19. [CrossRef] [PubMed]
89. Pradeep, C.; Kuttan, G. Piperine is a potent inhibitor of nuclear factor-κB (NF-κB), c-Fos, CREB, ATF-2 and proinflammatory cytokine gene expression in B16F-10 melanoma cells. *Int. Immunopharmacol.* **2004**, *4*, 1795–1803. [CrossRef] [PubMed]

90. Bhardwaj, R.K.; Glaeser, H.; Becquemont, L.; Klotz, U.; Gupta, S.K.; Fromm, M.F. Piperine, a major constituent of black pepper, inhibits human P-glycoprotein and CYP3A4. *J. Pharmacol. Exp. Ther.* **2002**, *302*, 645–650. [CrossRef] [PubMed]
91. Li, S.; Lei, Y.; Jia, Y.; Li, N.; Wink, M.; Ma, Y. Piperine, a piperidine alkaloid from Piper nigrum re-sensitizes P-gp, MRP1 and BCRP dependent multidrug resistant cancer cells. *Phytomedicine* **2011**, *19*, 83–87. [CrossRef] [PubMed]
92. Makhov, P.; Golovine, K.; Canter, D.; Kutikov, A.; Simhan, J.; Corlew, M.M.; Uzzo, R.G.; Kolenko, V.M. Co-administration of piperine and docetaxel results in improved anti-tumor efficacy via inhibition of CYP3A4 activity. *Prostate* **2012**, *72*, 661–667. [CrossRef] [PubMed]
93. Wang, Y.H.; Morris-Natschke, S.L.; Yang, J.; Niu, H.M.; Long, C.L.; Lee, K.H. Anticancer principles from medicinal Piper plants. *J. Tradit. Complement. Med.* **2014**, *4*, 8–16. [CrossRef] [PubMed]
94. Mahady, G.B.; Beecher, C. Quercetin-induced benzophenanthridine alkaloid production in suspension cell cultures of Sanguinaria canadensis. *Planta Med.* **1994**, *60*, 553–557. [CrossRef] [PubMed]
95. Vavrečková, C.; Gawlik, I.; Müller, K. Benzophenanthridine alkaloids of Chelidonium majus; I. Inhibition of 5-and 12-lipoxygenase by a non-redox mechanism. *Planta. Med.* **1996**, *62*, 397–401. [CrossRef] [PubMed]
96. Lenfeld, J.; Kroutil, M.; Maršálek, E.; Slavik, J.; Preininger, V.; Šimánek, V. Antiinflammatory activity of quaternary benzophenanthridine alkaloids from Chelidonoum majus. *Planta Med.* **1981**, *43*, 161–165. [CrossRef] [PubMed]
97. Beuria, T.K.; Santra, M.K.; Panda, D. Sanguinarine blocks cytokinesis in bacteria by inhibiting FtsZ assembly and bundling. *Biochemistry* **2005**, *44*, 16584–16593. [CrossRef] [PubMed]
98. Jeng, J.H.; Wu, H.L.; Lin, B.R.; Lan, W.H.; Chang, H.H.; Ho, Y.S.; Lee, P.H.; Wang, Y.J.; Wang, J.S.; Chen, Y.J. Antiplatelet effect of sanguinarine is correlated to calcium mobilization, thromboxane and cAMP production. *Atherosclerosis* **2007**, *191*, 250–258. [CrossRef] [PubMed]
99. Ahsan, H.; Reagan-Shaw, S.; Breur, J.; Ahmad, N. Sanguinarine induces apoptosis of human pancreatic carcinoma AsPC-1 and BxPC-3 cells via modulations in Bcl-2 family proteins. *Cancer Lett.* **2007**, *249*, 198–208. [CrossRef] [PubMed]
100. Chang, M.C.; Chan, C.P.; Wang, Y.J.; Lee, P.H.; Chen, L.I.; Tsai, Y.L.; Lin, B.R.; Wang, Y.L.; Jeng, J.H. Induction of necrosis and apoptosis to KB cancer cells by sanguinarine is associated with reactive oxygen species production and mitochondrial membrane depolarization. *Toxicol. Appl. Pharm.* **2007**, *218*, 143–151. [CrossRef] [PubMed]
101. Hussain, A.R.; Al-Jomah, N.A.; Siraj, A.K.; Manogaran, P.; Al-Hussein, K.; Abubaker, J.; Platanias, L.C.; Al-Kuraya, K.S.; Uddin, S. Sanguinarine-dependent induction of apoptosis in primary effusion lymphoma cells. *Cancer. Res.* **2007**, *67*, 3888–3897. [CrossRef] [PubMed]
102. Kim, S.; Lee, T.J.; Leem, J.; Choi, K.S.; Park, J.W.; Kwon, T.K. Sanguinarine-induced apoptosis: Generation of ROS, down-regulation of Bcl-2, c-FLIP, and synergy with TRAIL. *J. Cell Biochem.* **2008**, *104*, 895–907. [CrossRef] [PubMed]
103. Adhami, V.M.; Aziz, M.H.; Mukhtar, H.; Ahmad, N. Activation of prodeath Bcl-2 family proteins and mitochondrial apoptosis pathway by sanguinarine in immortalized human HaCaT keratinocytes. *Clin. Cancer Res.* **2003**, *9*, 3176–3182. [PubMed]
104. Adhami, V.M.; Aziz, M.H.; Reagan-Shaw, S.R.; Nihal, M.; Mukhtar, H.; Ahmad, N. Sanguinarine causes cell cycle blockade and apoptosis of human prostate carcinoma cells via modulation of cyclin kinase inhibitor-cyclin-cyclin-dependent kinase machinery. *Mol. Cancer Ther.* **2004**, *3*, 933–940. [PubMed]
105. Weerasinghe, P.; Hallock, S.; Tang, S.C.; Trump, B.; Liepins, A. Sanguinarine overcomes P-glycoprotein-mediated multidrug-resistance via induction of apoptosis and oncosis in CEM-VLB 1000 cells. *Exp. Toxicol. Pathol.* **2006**, *58*, 21–30. [CrossRef] [PubMed]
106. De Stefano, I.; Raspaglio, G.; Zannoni, G.F.; Travaglia, D.; Prisco, M.G.; Mosca, M.; Ferlini, C.; Scambia, G.; Gallo, D. Antiproliferative and antiangiogenic effects of the benzophenanthridine alkaloid sanguinarine in melanoma. *Biochem. Pharmacol.* **2009**, *78*, 1374–1381. [CrossRef] [PubMed]
107. Choi, Y.H.; Choi, W.Y.; Hong, S.H.; Kim, S.O.; Kim, G.Y.; Lee, W.H.; Yoo, Y.H. Anti-invasive activity of sanguinarine through modulation of tight junctions and matrix metalloproteinase activities in MDA-MB-231 human breast carcinoma cells. *Chem. Biol. Interact.* **2009**, *179*, 185–191. [CrossRef] [PubMed]

108. Ahsan, H.; Reagan-Shaw, S.; Eggert, D.M.; Tan, T.C.; Afaq, F.; Mukhtar, H.; Ahmad, N. Protective effect of sanguinarine on ultraviolet B-mediated damages in SKH-1 hairless mouse skin: Implications for prevention of skin cancer. *Photochem. Photobiol.* **2007**, *83*, 986–993. [CrossRef] [PubMed]
109. Jang, B.C.; Park, J.G.; Song, D.K.; Baek, W.K.; Yoo, S.K.; Jung, K.H.; Park, G.Y.; Lee, T.Y.; Suh, S.I. Sanguinarine induces apoptosis in A549 human lung cancer cells primarily via cellular glutathione depletion. *Toxicol. Vitr.* **2009**, *23*, 281–287. [CrossRef] [PubMed]
110. Cai, Y.J.; Lu, J.J.; Zhu, H.; Xie, H.; Huang, M.; Lin, L.P.; Zhang, X.W.; Ding, J. Salvicine triggers DNA double-strand breaks and apoptosis by GSH-depletion-driven H2O2 generation and topoisomerase II inhibition. *Free Radic. Bio. Med.* **2008**, *45*, 627–635. [CrossRef] [PubMed]
111. Cai, Y.; Lu, J.; Miao, Z.; Lin, L.; Ding, J. Reactive oxygen species contribute to cell killing and P-glycoprotein downregulation by salvicine in multidrug resistant K562/A02 cells. *Cancer Biol. Ther.* **2007**, *6*, 1794–1799. [CrossRef] [PubMed]
112. Vogt, A.; Tamewitz, A.; Skoko, J.; Sikorski, R.P.; Giuliano, K.A.; Lazo, J.S. The benzo [c] phenanthridine alkaloid, sanguinarine, is a selective, cell-active inhibitor of mitogen-activated protein kinase phosphatase-1. *J. Biol. Chem.* **2005**, *280*, 19078–19086. [CrossRef] [PubMed]
113. Lopus, M.; Panda, D. The benzophenanthridine alkaloid sanguinarine perturbs microtubule assembly dynamics through tubulin binding: A possible mechanism for its antiproliferative activity. *J. FEBS* **2006**, *273*, 2139–2150. [CrossRef] [PubMed]
114. Chaturvedi, M.M.; Kumar, A.; Darnay, B.G.; Chainy, G.B.; Agarwal, S.; Aggarwal, B.B. Sanguinarine (pseudochelerythrine) is a potent inhibitor of NF-κB activation, IκBα phosphorylation, and degradation. *J. Biol. Chem.* **1997**, *272*, 30129–30134. [CrossRef] [PubMed]
115. Sun, M.; Liu, C.; Nadiminty, N.; Lou, W.; Zhu, Y.; Yang, J.; Evans, C.P.; Zhou, Q.; Gao, A.C. Inhibition of Stat3 activation by sanguinarine suppresses prostate cancer cell growth and invasion. *Prostate* **2012**, *72*, 82–89. [CrossRef] [PubMed]
116. Li, D.G.; Wang, Z.R.; Lu, H.M. Pharmacology of tetrandrine and its therapeutic use in digestive diseases. *World J. Gastroenterol.* **2001**, *7*, 627. [CrossRef] [PubMed]
117. Ng, L.T.; Chiang, L.C.; Lin, Y.T.; Lin, C.C. Antiproliferative and apoptotic effects of tetrandrine on different human hepatoma cell lines. *Am. J. Chin. Med.* **2006**, *34*, 125–135. [CrossRef] [PubMed]
118. Meng, L.H.; Zhang, H.; Hayward, L.; Takemura, H.; Shao, R.G.; Pommier, Y. Tetrandrine induces early G1 arrest in human colon carcinoma cells by down-regulating the activity and inducing the degradation of G1-S–specific cyclin-dependent kinases and by inducing p53 and p21Cip1. *Cancer Res.* **2004**, *64*, 9086–9092. [CrossRef] [PubMed]
119. Wu, J.M.; Chen, Y.; Chen, J.C.; Lin, T.Y.; Tseng, S.H. Tetrandrine induces apoptosis and growth suppression of colon cancer cells in mice. *Cancer Lett.* **2010**, *287*, 187–195. [CrossRef] [PubMed]
120. Chen, Y.; Chen, J.C.; Tseng, S.H. Tetrandrine suppresses tumor growth and angiogenesis of gliomas in rats. *Int. J. Cancer* **2009**, *124*, 2260–2269. [CrossRef] [PubMed]
121. Chang, K.H.; Liao, H.F.; Chang, H.H.; Chen, Y.Y.; Yu, M.C.; Chou, C.J.; Chen, Y.J. Inhibitory effect of tetrandrine on pulmonary metastases in CT26 colorectal adenocarcinoma-bearing BALB/c mice. *Am. J. Chin. Med.* **2004**, *32*, 863–872. [CrossRef] [PubMed]
122. Zhang, Y.; Wang, C.; Wang, H.; Wang, K.; Du, Y.; Zhang, J. Combination of Tetrandrine with cisplatin enhances cytotoxicity through growth suppression and apoptosis in ovarian cancer in vitro and in vivo. *Cancer Lett.* **2011**, *304*, 21–32. [CrossRef] [PubMed]
123. Wei, J.; Liu, B.; Wang, L.; Qian, X.; Ding, Y.; Yu, L. Synergistic interaction between tetrandrine and chemotherapeutic agents and influence of tetrandrine on chemotherapeutic agent-associated genes in human gastric cancer cell lines. *Cancer Chemother. Pharmacol.* **2007**, *60*, 703–711. [CrossRef] [PubMed]
124. Iorns, E.; Lord, C.J.; Ashworth, A. Parallel RNAi and compound screens identify the PDK1 pathway as a target for tamoxifen sensitization. *Biochem. J.* **2009**, *417*, 361–371. [CrossRef] [PubMed]
125. Li, C.X.; Huan Ren, K.; He, H.W.; Shao, R.G. Involvement of PI3K/AKT/GSK3ÃŸ pathway in tetrandrine-induced G1 arrest and apoptosis. *Cancer Biol. Ther.* **2008**, *7*, 1073–1078.
126. Oh, S.H.; Lee, B.H. Induction of apoptosis in human hepatoblastoma cells by tetrandrine via caspase-dependent Bid cleavage and cytochrome c release. *Biochem. Pharmacol.* **2003**, *66*, 725–731. [CrossRef]

127. Cho, H.S.; Chang, S.H.; Chung, Y.S.; Shin, J.Y.; Park, S.J.; Lee, E.S.; Hwang, S.K.; Kwon, J.T.; Tehrani, A.M.; Woo, M. Synergistic effect of ERK inhibition on tetrandrine-induced apoptosis in A549 human lung carcinoma cells. *J. Vet. Sci.* **2009**, *10*, 23–28. [CrossRef] [PubMed]
128. Kittakoop, P.; Mahidol, C.; Ruchirawat, S. Alkaloids as important scaffolds in therapeutic drugs for the treatments of cancer, tuberculosis, and smoking cessation. *Curr. Top. Med. Chem.* **2014**, *14*, 239–252. [CrossRef] [PubMed]
129. Sobhani, A.M.; Ebrahimi, S.A.; Mahmoudian, M. An in vitro evaluation of human DNA topoisomerase I inhibition by *Peganum harmala* L. seeds extract and its beta-carboline alkaloids. *J. Pharm. Pharm. Sci.* **2002**, *5*, 19–23. [PubMed]
130. Heinrich, M.; Barnes, J.; Gibbons, S.; Williamson, E. *A Text Book of Fundamentals of Pharmacognosy and Phytotherapy*; Elsevier: New York, NY, USA, 2004; Volume 8, pp. 60–105.
131. Hesse, M. *Alkaloids: Nature's Curse or Blessing*; John Wiley & Sons: zürich, Switzerland, 2002.
132. Evans, W.C. *Trease and Evans' Pharmacognosy E-Book*; Elsevier Health Sciences: Edinburgh, UK, 2009.
133. Robbers, J.E.; Speedie, M.K.; Tyler, V.E. *Pharmacognosy and Pharmacobiotechnology*; Williams & Wilkins: Baltimore, MD, USA, 1996.
134. Hraiech, S.; Brégeon, F.; Brunel, J.M.; Rolain, J.-M.; Lepidi, H.; Andrieu, V.; Raoult, D.; Papazian, L.; Roch, A. Antibacterial efficacy of inhaled squalamine in a rat model of chronic Pseudomonas aeruginosa pneumonia. *J. Antimicrob. Chemother.* **2012**, *67*, 2452–2458. [CrossRef] [PubMed]
135. Bogatcheva, E.; Hanrahan, C.; Nikonenko, B.; De los Santos, G.; Reddy, V.; Chen, P.; Barbosa, F.; Einck, L.; Nacy, C.; Protopopova, M. Identification of SQ609 as a lead compound from a library of dipiperidines. *Bioorg. Med. Chem. Lett.* **2011**, *21*, 5353–5357. [CrossRef] [PubMed]
136. Parhi, A.; Kelley, C.; Kaul, M.; Pilch, D.S.; LaVoie, E.J. Antibacterial activity of substituted 5-methylbenzo [*c*] phenanthridinium derivatives. *Bioorg. Med. Chem. Lett.* **2012**, *22*, 7080–7083. [CrossRef] [PubMed]
137. Chakraborty, A.; Chowdhury, B.; Bhattacharyya, P. Clausenol and clausenine—Two carbazole alkaloids from Clausena anisata. *Phytochemistry* **1995**, *40*, 295–298. [CrossRef]
138. Nakahara, K.; Trakoontivakorn, G.; Alzoreky, N.S.; Ono, H.; Onishi-Kameyama, M.; Yoshida, M. Antimutagenicity of some edible Thai plants, and a bioactive carbazole alkaloid, mahanine, isolated from Micromelum minutum. *J. Agric. Food Chem.* **2002**, *50*, 4796–4802. [CrossRef] [PubMed]
139. Kim, H.; Lantvit, D.; Hwang, C.H.; Kroll, D.J.; Swanson, S.M.; Franzblau, S.G.; Orjala, J. Indole alkaloids from two cultured cyanobacteria, Westiellopsis sp. and Fischerella muscicola. *Bioorgan. Med. Chem.* **2012**, *20*, 5290–5295. [CrossRef] [PubMed]
140. Samoylenko, V.; Ashfaq, M.K.; Jacob, M.R.; Tekwani, B.L.; Khan, S.I.; Manly, S.P.; Joshi, V.C.; Walker, L.A.; Muhammad, I. Indolizidine, antiinfective and antiparasitic compounds from Prosopis glandulosa Torr. var. glandulosa. *Planta med.* **2009**, *75*, 48. [CrossRef]
141. Iwasa, K.; Moriyasu, M.; Tachibana, Y.; Kim, H.S.; Wataya, Y.; Wiegrebe, W.; Bastow, K.F.; Cosentino, L.M.; Kozuka, M.; Lee, K.H. Simple isoquinoline and benzylisoquinoline alkaloids as potential antimicrobial, antimalarial, cytotoxic, and anti-HIV agents. *Bioorgan. Med. Chem.* **2001**, *9*, 2871–2884. [CrossRef]
142. Xie, Q.; Johnson, B.R.; Wenckus, C.S.; Fayad, M.I.; Wu, C.D. Efficacy of berberine, an antimicrobial plant alkaloid, as an endodontic irrigant against a mixed-culture biofilm in an in vitro tooth model. *J. Endod.* **2012**, *38*, 1114–1117. [CrossRef] [PubMed]
143. Obiang-Obounou, B.W.; Kang, O.H.; Choi, J.G.; Keum, J.H.; Kim, S.B.; Mun, S.H.; Shin, D.W.; Kim, K.W.; Park, C.B.; Kim, Y.G. The mechanism of action of sanguinarine against methicillin-resistant Staphylococcus aureus. *J. Toxicol. Sci.* **2011**, *36*, 277–283. [CrossRef] [PubMed]
144. Li, H.L.; Zhang, W.D.; Zhang, C.; Liu, R.H.; Wang, X.W.; Wang, X.L.; Zhu, J.B.; Chen, C.L. Bioavailabilty and pharmacokinetics of four active alkaloids of traditional Chinese medicine Yanhuanglian in rats following intravenous and oral administration. *J. Pharma. Biomed.* **2006**, *41*, 1342–1346. [CrossRef] [PubMed]
145. Wang, X.X.; Zan, K.; Shi, S.P.; Zeng, K.W.; Jiang, Y.; Guan, Y.; Xiao, C.L.; Gao, H.Y.; Wu, L.J.; Tu, P.F. Quinolone alkaloids with antibacterial and cytotoxic activities from the fruits of Evodia rutaecarpa. *Fitoterapia* **2013**, *89*, 1–7. [CrossRef] [PubMed]
146. Kuete, V.; Wansi, J.; Mbaveng, A.; Sop, M.K.; Tadjong, A.T.; Beng, V.P.; Etoa, F.X.; Wandji, J.; Meyer, J.M.; Lall, N. Antimicrobial activity of the methanolic extract and compounds from Teclea afzelii (Rutaceae). *S. Afr. J. Bot.* **2008**, *74*, 572–576. [CrossRef]

147. Dekker, K.A.; Inagaki, T.; Gootz, T.D.; Huang, L.H.; Kojima, Y.; Kohlbrenner, W.E.; Matsunaga, Y.; McGuirk, P.R.; Nomura, E.; Sakakibara, T. New Quinolone Compounds from Pseudonocardia sp. with Selective and Potent Anti-Helicobacter pylori Activity. *Jpn. J. Antibiot.* **1998**, *51*, 145–152. [CrossRef] [PubMed]
148. Kubota, T.; Iwai, T.; Takahashi-Nakaguchi, A.; Fromont, J.; Gonoi, T.; Kobayashi, J.I. Agelasines O–U, new diterpene alkaloids with a 9-N-methyladenine unit from a marine sponge Agelas sp. *Tetrahedron* **2012**, *68*, 9738–9744. [CrossRef]
149. Moore, K.S.; Wehrli, S.; Roder, H.; Rogers, M.; Forrest, J.N.; McCrimmon, D.; Zasloff, M. Squalamine: An aminosterol antibiotic from the shark. *Proc. Natl. Acad. Sci. USA* **1993**, *90*, 1354–1358. [CrossRef] [PubMed]
150. Calcul, L.; Longeon, A.; Al Mourabit, A.; Guyot, M.; Bourguet-Kondracki, M.L. Novel alkaloids of the aaptamine class from an Indonesian marine sponge of the genus Xestospongia. *Tetrahedron* **2003**, *59*, 6539–6544. [CrossRef]
151. Alhanout, K.; Malesinki, S.; Vidal, N.; Peyrot, V.; Rolain, J.M.; Brunel, J.M. New insights into the antibacterial mechanism of action of squalamine. *J. Antimicrob. Chemother.* **2010**, *65*, 1688–1693. [CrossRef] [PubMed]
152. Rao, K.N.; Venkatachalam, S. Inhibition of dihydrofolate reductase and cell growth activity by the phenanthroindolizidine alkaloids pergularinine and tylophorinidine: The in vitro cytotoxicity of these plant alkaloids and their potential as antimicrobial and anticancer agents. *Toxicol. Vitr.* **2000**, *14*, 53–59. [CrossRef]
153. Casu, L.; Cottiglia, F.; Leonti, M.; De Logu, A.; Agus, E.; Tse-Dinh, Y.-C.; Lombardo, V.; Sissi, C. Ungeremine effectively targets mammalian as well as bacterial type I and type II topoisomerases. *Bioorg. Med. Chem. Lett.* **2011**, *21*, 7041–7044. [CrossRef] [PubMed]
154. Domadia, P.N.; Bhunia, A.; Sivaraman, J.; Swarup, S.; Dasgupta, D. Berberine targets assembly of Escherichia coli cell division protein FtsZ. *Biochemistry* **2008**, *47*, 3225–3234. [CrossRef] [PubMed]
155. Boberek, J.M.; Stach, J.; Good, L. Genetic evidence for inhibition of bacterial division protein FtsZ by berberine. *PLoS ONE* **2010**, *5*, e13745. [CrossRef] [PubMed]
156. Heeb, S.; Fletcher, M.P.; Chhabra, S.R.; Diggle, S.P.; Williams, P.; Cámara, M. Quinolones: From antibiotics to autoinducers. *FEMS Microbiol. Rev.* **2011**, *35*, 247–274. [CrossRef] [PubMed]
157. Tominaga, K.; Higuchi, K.; Hamasaki, N.; Hamaguchi, M.; Takashima, T.; Tanigawa, T.; Watanabe, T.; Fujiwara, Y.; Tezuka, Y.; Nagaoka, T. In vivo action of novel alkyl methyl quinolone alkaloids against Helicobacter pylori. *J. Antimicrob. Chemother.* **2002**, *50*, 547–552. [CrossRef] [PubMed]
158. Arai, M.; Yamano, Y.; Setiawan, A.; Kobayashi, M. Identification of the Target Protein of Agelasine D, a Marine Sponge Diterpene Alkaloid, as an Anti-dormant Mycobacterial Substance. *ChemBioChem* **2014**, *15*, 117–123. [CrossRef] [PubMed]
159. Salmi, C.; Loncle, C.; Vidal, N.; Letourneux, Y.; Fantini, J.; Maresca, M.; Taïeb, N.; Pagès, J.M.; Brunel, J.M. Squalamine: An appropriate strategy against the emergence of multidrug resistant gram-negative bacteria? *PLoS ONE* **2008**, *3*, e2765. [CrossRef] [PubMed]
160. Patil, P.J.; Ghosh, J.S. Antimicrobial activity of Catharanthus roseus—A detailed study. *Br. J. Pharmcol. Toxicol.* **2010**, *1*, 40–44.
161. Hallock, Y.F.; Manfredi, K.P.; Dai, J.R.; Cardellina, J.H.; Gulakowski, R.J.; McMahon, J.B.; Schäffer, M.; Stahl, M.; Gulden, K.P.; Bringmann, G. Michellamines D–F, new HIV-inhibitory dimeric naphthylisoquinoline alkaloids, and korupensamine E, a new antimalarial monomer, from Ancistrocladus korupensis. *J. Nat. Prod.* **1997**, *60*, 677–683. [CrossRef] [PubMed]
162. Song, B.; Yang, S.; Jin, L.H.; Bhadury, P.S. *Environment-Friendly Antiviral Agents for Plants*; Springer Science & Business Media: Guizhou, China, 2011.
163. Sasidharan, S.; Chen, Y.; Saravanan, D.; Sundram, K.; Latha, L.Y. Extraction, isolation and characterization of bioactive compounds from plants' extracts. *Afr. J. Tradit. Complement. Altern. Med.* **2011**, *8*, 1–10. [CrossRef] [PubMed]
164. Watson, A.A.; Fleet, G.W.; Asano, N.; Molyneux, R.J.; Nash, R.J. Polyhydroxylated alkaloids—Natural occurrence and therapeutic applications. *Phytochemistry* **2001**, *56*, 265295. [CrossRef]
165. Duan, H.; Takaishi, Y.; Imakura, Y.; Jia, Y.; Li, D.; Cosentino, L.M.; Lee, K.H. Sesquiterpene Alkaloids from Tripterygium h ypoglaucum and Tripterygium w ilfordii: A New Class of Potent Anti-HIV Agents. *J. Nat. Prod.* **2000**, *63*, 357–361. [CrossRef] [PubMed]

166. Rahman, M.M.; Gray, A.I. Antimicrobial constituents from the stem bark of Feronia limonia. *Phytochemistry* **2002**, *59*, 73–77. [CrossRef]

167. Tabarrini, O.; Manfroni, G.; Fravolini, A.; Cecchetti, V.; Sabatini, S.; De Clercq, E.; Rozenski, J.; Canard, B.; Dutartre, H.; Paeshuyse, J. Synthesis and anti-BVDV activity of acridones as new potential antiviral agents. *J. Med. Chem.* **2006**, *49*, 2621–2627. [CrossRef] [PubMed]

168. Ito, C.; Itoigawa, M.; Sato, A.; Hasan, C.M.; Rashid, M.A.; Tokuda, H.; Mukainaka, T.; Nishino, H.; Furukawa, H. Chemical Constituents of Glycosmis a rborea: Three New Carbazole Alkaloids and Their Biological Activity. *J. Nat. Prod.* **2004**, *67*, 1488–1491. [CrossRef] [PubMed]

169. Manske, R.H.F.; Holmes, H.L. *The Alkaloids: Chemistry and Physiology*; Elsevier: New York, NY, USA, 2014.

170. Ududua, U.O.; Monanu, M.O.; Chuku, L.C. Proximate Analysis and Phytochemical Profile of Brachystegia eurycoma Leaves. *Asian J. Res. Biochem.* **2019**, *4*, 1–11. [CrossRef]

171. Petrov, N.M. Antiviral Activity of Plant Extract from Tanacetum Vulgare Against Cucumber Mosaic Virus and Potato Virus Y. *J. Biomed Biotechnol.* **2016**, *5*, 189–194.

172. Rex, J.H.; Rinaldi, M.; Pfaller, M. Resistance of Candida species to fluconazole. *Antimicrob. Agents. Chemother.* **1995**, *39*, 1. [CrossRef] [PubMed]

173. Borris, R.P. Natural products research: Perspectives from a major pharmaceutical company. *J. Ethnopharmacol.* **1996**, *51*, 29–38. [CrossRef]

174. Lewis, W.H.; Elvin-Lewis, M.P. Medicinal plants as sources of new therapeutics. *Ann. Mo. Bot. Gard.* **1995**, *82*, 16–24. [CrossRef]

175. Anaissie, E.; Bodey, G.; Rinaldi, M. Emerging fungal pathogens. *Eur. J. Clin. Microbiol. Infect. Dis.* **1989**, *8*, 323–330. [CrossRef] [PubMed]

176. Wey, S.B.; Mori, M.; Pfaller, M.A.; Woolson, R.F.; Wenzel, R.P. Hospital-acquired candidemia: The attributable mortality and excess length of stay. *Arch. Intern. Med.* **1988**, *148*, 2642–2645. [CrossRef] [PubMed]

177. Beck-Sague, C.; Banerjee, S.; Jarvis, W.R. Infectious diseases and mortality among US nursing home residents. *Am. J. Public Health* **1993**, *83*, 1739–1742. [CrossRef] [PubMed]

178. Schultes, R.E. Plants and plant constituents as mind-altering agents throughout history. In *Stimulants*; Springer: Boston, MA, USA, 1978; pp. 219–241.

179. Küçükosmanoğlu Bahçeevli, A.; Kurucu, S.; Kolak, U.; Topçu, G.; Adou, E.; Kingston, D.G. Alkaloids and Aromatics of Cyathobasis f ruticulosa (Bunge) Aellen. *J. Nat. Prod.* **2005**, *68*, 956–958. [CrossRef] [PubMed]

180. Ferheen, S.; Ahmed, E.; Afza, N.; Malik, A.; Shah, M.R.; Nawaz, S.A.; Choudhary, M.I. Haloxylines A and B, antifungal and cholinesterase inhibiting piperidine alkaloids from Haloxylon salicornicum. *Chem. Pharm. Bull.* **2005**, *53*, 570–572. [CrossRef] [PubMed]

181. Singh, N.; Azmi, S.; Maurya, S.; Singh, U.; Jha, R.; Pandey, V. Two plant alkaloids isolated fromCorydalis longipes as potential antifungal agents. *Folia Microbiol.* **2003**, *48*, 605–609. [CrossRef] [PubMed]

182. Jung, H.J.; Sung, W.S.; Yeo, S.H.; Kim, H.S.; Lee, I.S.; Woo, E.R.; Lee, D.G. Antifungal effect of amentoflavone derived fromSelaginella tamariscina. *Arch. Pharm. Res.* **2006**, *29*, 746. [CrossRef] [PubMed]

183. Morteza-Semnani, K.; Amin, G.; Shidfar, M.; Hadizadeh, H.; Shafiee, A. Antifungal activity of the methanolic extract and alkaloids of Glaucium oxylobum. *Fitoterapia* **2003**, *74*, 493–496. [CrossRef]

184. Thouvenel, C.; Gantier, J.C.; Duret, P.; Fourneau, C.; Hocquemiller, R.; Ferreira, M.E.; de Arias, A.R.; Fournet, A. Antifungal compounds from Zanthoxylum chiloperone var. angustifolium. *Phytother. Res.* **2003**, *17*, 678–680. [CrossRef] [PubMed]

185. Singh, U.; Sarma, B.; Mishra, P.; Ray, A. Antifungal activity of venenatine, an indole alkaloid isolated fromAlstonia venenata. *Folia Microbiol.* **2000**, *45*, 173. [CrossRef] [PubMed]

186. Balls, A.; Hale, W.; Harris, T. A crystalline protein obtained from a lipoprotein of wheat flour. *Cereal Chem.* **1942**, *19*, 279–288.

187. Liu, S.; Oguntimein, B.; Hufford, C.; Clark, A. 3-Methoxysampangine, a novel antifungal copyrine alkaloid from Cleistopholis patens. *Antimicrob. Agents Chemother.* **1990**, *34*, 529–533. [CrossRef] [PubMed]

188. Emile, A.; Waikedre, J.; Herrenknecht, C.; Fourneau, C.; Gantier, J.C.; Hnawia, E.; Cabalion, P.; Hocquemiller, R.; Fournet, A. Bioassay-guided isolation of antifungal alkaloids from Melochia odorata. *Phytother. Res.* **2007**, *21*, 398–400. [CrossRef] [PubMed]

189. Cantrell, C.; Schrader, K.; Mamonov, L.; Sitpaeva, G.; Kustova, T.; Dunbar, C.; Wedge, D. Isolation and identification of antifungal and antialgal alkaloids from Haplophyllum sieversii. *J. Agric. Food Chem.* **2005**, *53*, 7741–7748. [CrossRef] [PubMed]
190. Salehi, B.; Sharopov, F.; Boyunegmez Tumer, T.; Ozleyen, A.; Rodríguez-Pérez, C.; M Ezzat, S.; Azzini, E.; Hosseinabadi, T.; Butnariu, M.; Sarac, I. Symphytum Species: A Comprehensive Review on Chemical Composition, Food Applications and Phytopharmacology. *Molecules* **2019**, *24*, 2272. [CrossRef] [PubMed]

Article

Differentiated Effects of Secondary Metabolites from *Solanaceae* and *Brassicaceae* Plant Families on the Heartbeat of *Tenebrio molitor* Pupae

Paweł Marciniak [1,*], Angelika Kolińska [1], Marta Spochacz [1], Szymon Chowański [1], Zbigniew Adamski [1,2], Laura Scrano [3], Patrizia Falabella [4], Sabino A. Bufo [4,5] and Grzegorz Rosiński [1]

1 Department of Animal Physiology and Development, Institute of Experimental Biology, Faculty of Biology, Adam Mickiewicz University in Poznań, 61-614 Poznań, Poland; kolinska.angelika@gmail.com (A.K.); marta.spochacz@amu.edu.pl (M.S.); szyymon@amu.edu.pl (S.C.); ed@amu.edu.pl (Z.A.); rosin@amu.edu.pl (G.R.)

2 Electron and Confocal Microscope Laboratory, Faculty of Biology, Adam Mickiewicz University in Poznań, 61-614 Poznań, Poland

3 Department of European and Mediterranean Cultures, University of Basilicata, 75100 Matera, Italy; laura.scrano@unibas.it

4 Department of Sciences, University of Basilicata, 85100 Potenza, Italy; patrizia.falabella@unibas.it (P.F.); sabino.bufo@unibas.it (S.A.B.)

5 Department of Geography, Environmental Management & Energy Studies, University of Johannesburg, Auckland Park Kingsway Campus, Johannesburg 2092, South Africa

* Correspondence: pmarcin@amu.edu.pl

Received: 15 April 2019; Accepted: 20 May 2019; Published: 22 May 2019

Abstract: The usage of insects as model organisms is becoming more and more common in toxicological, pharmacological, genetic and biomedical research. Insects, such as fruit flies (*Drosophila melanogaster*), locusts (*Locusta migratoria*), stick insects (*Baculum extradentatum*) or beetles (*Tenebrio molitor*) are used to assess the effect of different active compounds, as well as to analyse the background and course of certain diseases, including heart disorders. The goal of this study was to assess the influence of secondary metabolites extracted from *Solanaceae* and *Brassicaceae* plants: Potato (*Solanum tuberosum*), tomato (*Solanum lycopersicum*), black nightshade (*Solanum nigrum*) and horseradish (*Armoracia rusticana*), on *T. molitor* beetle heart contractility in comparison with pure alkaloids. During the in vivo bioassays, the plants glycoalkaloid extracts and pure substances were injected at the concentration 10^{-5} M into *T. molitor* pupa and evoked changes in heart activity. Pure glycoalkaloids caused mainly positive chronotropic effects, dependant on heart activity phase during a 24-h period of recording. Moreover, the substances affected the duration of the heart activity phases. Similarly, to the pure glycoalkaloids, the tested extracts also mainly accelerated the heart rhythm, however *S. tuberosum* and *S. lycopersicum* extracts slightly decreased the heart contractions frequency in the last 6 h of the recording. Cardioacceleratory activity of only *S. lycopersicum* extract was higher than single alkaloids whereas *S. tubersoum* and *S. nigrum* extracts were less active when compared to pure alkaloids. The most cardioactive substance was chaconine which strongly stimulated heart action during the whole recording after injection. *A. rusticana* extract which is composed mainly of glucosinolates did not significantly affect the heart contractions. Obtained results showed that glycoalkaloids were much more active than glucosinolates. However, the extracts depending on the plant species might be more or less active than pure substances.

Keywords: plant secondary metabolites; glycoalkaloids; insect heart; beetles; insect; *Tenebrio molitor*

Key Contribution: *Solanaceae* alkaloids and other secondary metabolites might be used to disturb heart activity in insects.

---

1. Introduction

Two groups of assimilates stand out in the processes of biosynthesis occurring in plants: basic substances and products of secondary metabolism [1]. Primary metabolites (carbohydrates, proteins, and fats) are found in all plants where they perform basic physiological functions. Secondary substances occur only in specific systematic groups, that means they do not perform the functions necessary for life. However, they often have the character of physiologically active compounds, that makes them applicable in different fields, for instance in medicine or agriculture.

A broad spectrum of physiological activity is demonstrated by alkaloids [2]. They have been proved to exhibit antioxidant, anti-inflammatory, anti-aggregation, hypocholesteric, immunostimulant or anticancer properties. Many of them belong to the group of compounds modifying or normalizing the activity of the heart. The mode of action is often connected with the ability to control calcium channels, similarly to synthetic substances/drugs [3].

Alkaloids are produced by different plants species such as *Papaveraceae, Fabaceae, Ranunculaceae* or *Solanaceae* plant families or lower plants such as forks or creaks [4]. They are usually basic nitrogen-containing organic compounds, mainly synthesized from amino acids. Substrates for about five thousand of this type of compound are three protein amino acids: phenylalanine, lysine and tryptophan.

Nowadays, emerging interest is put on *Solanaceae* alkaloids. They are mainly steroidal glycoalkaloids (GAs)-glycosidic derivatives of nitrogen-containing steroids that are produced in more than 350 plant species [5]. The major representatives of this glycoalkaloid family are α-solanine and α-chaconine in potato plants (*Solanum tuberosum* L.), solasonine and solamargine in black nightshade (*Solanum nigrum* L.) or α-tomatine and dehydrotomatine in tomato plants (*Solanum lycopersicum* L.) [6,7] (Figure 1). They have been shown to possess a wide spectrum of biological activity at the molecular, cellular and organismal levels [6,8].

Insects are suitable models for the study of various active compounds, including plant derived substances. They have a short life cycle and are easy to grow. There are also reports suggesting possible similarities in the mechanisms of action of various active substances in relation to vertebrate and invertebrate animals [9]. Increasingly, they are used in various toxicological or pharmacological researches [9,10].

Previous investigation has shown that *Solanaceae* GAs or *Solanaceae* plant extracts are able to influence the heart activity of *Zophobas atratus* Fab. beetle [11]. Chosen GAs irreversibly stopped *Z. atratus* hearts in in vitro conditions but surprisingly stimulated the activity in in vivo bioassays, alternating the duration of phases in the heart cycle. Thus, further experiments were needed to explain whether GAs are cardiotoxic or just cardioregulatory. Using a different model beetle species *Tenebrio molitor*, here, it is presented further evidence that *Solanaceae* GAs differentially affect the insect heart activity. Furthermore, a different group of plant secondary metabolites have been tested—glucosinolates (GLSs) present in *Armoria rusticana* extract [7], in order to check whether they possess cardioregulatory activity.

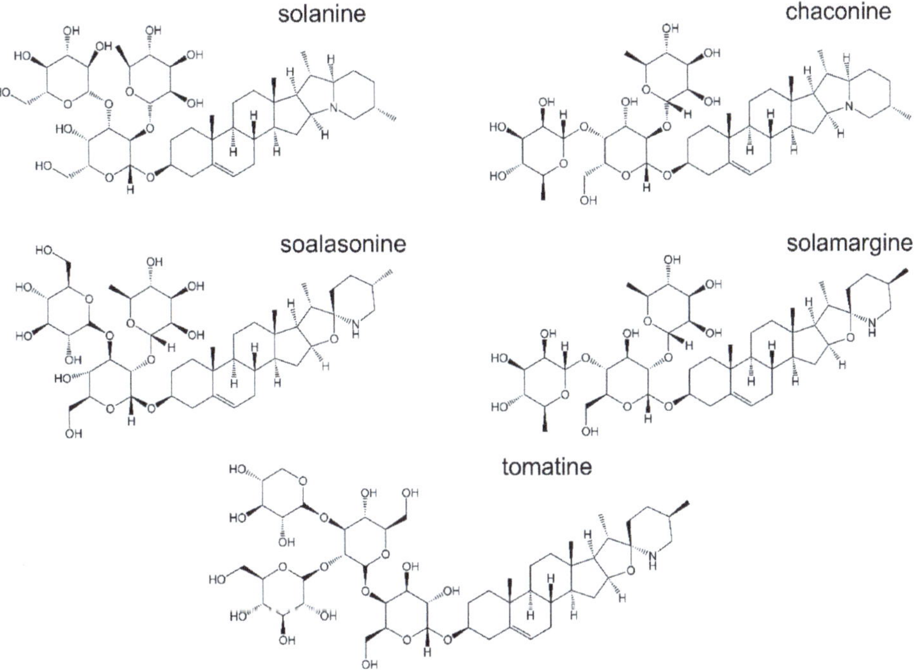

Figure 1. Structures of major potato, black nightshade and tomato glycoalkaloids.

2. Results

2.1. Heart Rhythm of the Tenebrio molitor Pupae

Cardiac contractile activity of *T. molitor* intact pupae was examined by a non-invasive optoelectronic technique. In order to detect circadian changes in the heart rhythm, 24-h recordings were conducted. The recordings indicated that the heart rhythm of one-day-old pupae is complex. A constant pattern in the work of this organ was manifested by regular alternations of: (1) forward orientated, fast—anterograde phase, (2) backward orientated, slow—retrograde phase, and (3) no contraction phase—diastasis. These phases, apart from the direction of propagation of peristalsis contractions wave, differed also in the frequency of contractions of myocardium and their duration (Figure 2). Average heart contraction frequency of intact *T. molitor* pupae was 26/min in anterograde phase and 10/min in retrograde phase (Figure 3A). The average duration of each phase was 3 min 40 s for anterograde phase, 4 min 20 s for retrograde phase and 5 min 30 s for the diastasis (Figure 3B). The length of the phases depends on the time of recording (day or night). The injection of physiological saline caused minor effects in the heartbeat (Figure 3C) and had no effects in the duration of the phases (Figure 3D). These effects were not statistically significant and thus were not taken under consideration during the experiments with tested substances.

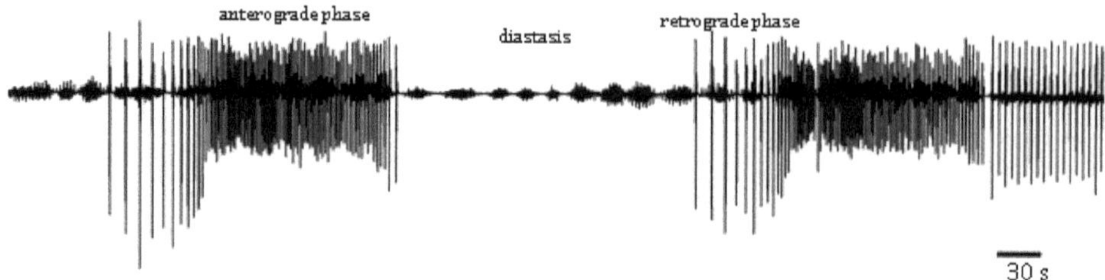

Figure 2. Myocardiograms of the *Tenebrio molitor* pupal heartbeat registered under control conditions.

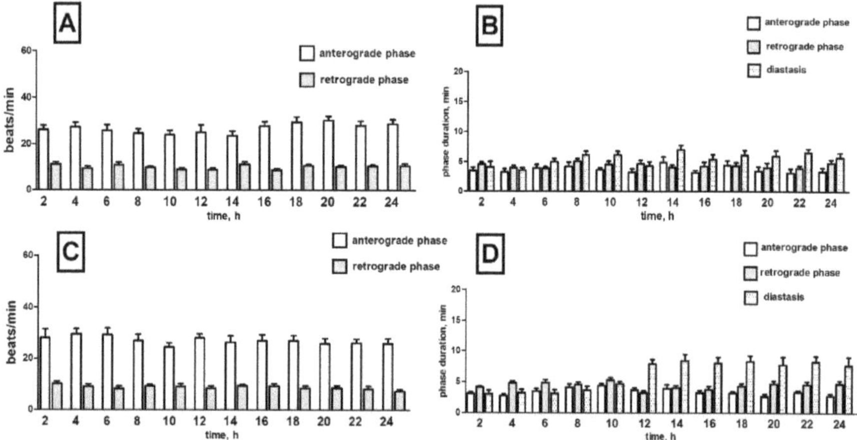

Figure 3. Changes in the heart rate (**A**,**C**) and in the duration of anterograde, retrograde phases and diastase in the cardiac cycle (**B**,**D**) of *T. molitor* pupae in 2-h intervals, registered without injection (**A**,**B**) and after injection of physiological saline (**C**,**D**). Statistically significant differences ($p \leq 0.05$, Mann-Whitney test) when compared to control are indicated with an asterisk, $n \geq 10$.

2.2. Effects of Potato Glycoalkaloids on the Heart Rhythm of T. molitor Pupae

2.2.1. Pure Alkaloids

α-Solanine

Injection of α-solanine caused ambiguous effects which strongly depended on time after application. In the first two hours, we noticed strong cardioinhibitory effects in the anterograde phase with the frequency of heart contraction decreasing by 47% when compared to the control (physiological saline injections). The effect was evident in just two first hours after injection. The negative chronotropic effect in this phase was also recorded in the 6th and 12th hour of the recording (decreasing by 20 and 16%) but the results were not statistically significant (Figure 4A). Otherwise, in the 10th and 22nd hour of recording, the frequency of the heart contractions in anterograde phase significantly increased (both by 22%) (Table S1). During the first part of recording (4th–18th h), an increase in the heart rate of 18% on average was observed in retrograde phase. A clear chronotropic-positive effect occurred in 6th and 14th hour of registration (up by 38 and 49% respectively) (Table S1). Nevertheless, in the final hours of recording (20th–24th h), cardioinhibitory effects of α-solanine were observed. The frequency of contraction 20, 22 and 24 h after injection was reduced by 38, 38 and 77%, respectively when compared to the control (Figure 4A).

Injection of α-solanine altered duration of phases of pupal heart rhythm. Shortening of fast anterograde phase by 1 min and 39 s on average was observed between the 2nd and 16th hour, later the effect silenced (Figure 4B). Changes in retrograde phase varied depending on time after compound application. Moreover, α-solanine caused reduction of diastase duration between the 2nd and 16th hour of the recording period by an average of 4 min (Table S2). In the next hours, the effect weakened and finally in the 20th and 22nd hour of registration, strong elongation of the diastase with an average of 4 min 40 s was observed (Figure 4B).

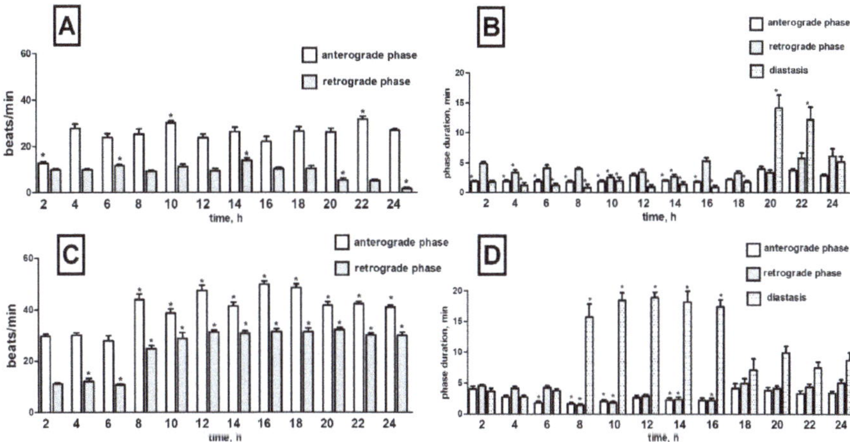

Figure 4. Changes in the heart rate (**A**,**C**) and in the duration of anterograde, retrograde phases and diastase in the cardiac cycle (**B**,**D**) of *T. molitor* pupae in 2-h intervals, registered after injection of α-solanine (1×10^{-5} M) (**A**,**B**) and α-chaconine (1×10^{-5} M) (**C**,**D**). Statistically significant differences ($p \leq 0.05$, Mann-Whitney test) when compared to control are indicated with an asterisk, $n \geq 10$.

α-Chaconine

Injection of α-chaconine induced an increase in the contraction frequency of 66% on average, in the anterograde phase, and 200% in the retrograde phase during the entire registration period (Table S1). The highest-positive chronotropic effect in fast phase was observed at 16 and 18 h after compound injection (Figure 4C). In this period, the frequency of contractions increased by 86%. In slow phase, increased contraction frequency was noticed from the 4th to the 24th hour of registration, and it was 210% higher on average than in the control (Figure 4C). Application of α-chaconine shortened the anterograde phase between the 6th and 16th hour during the recording by 1 min and 32 s on average. Retrograde phase was reduced by an average of 1 min and 46 s between the 8th to 16th hour after injection (Table S2). The opposite effect was observed during the diastase. α-Chaconine injections caused a significant prolongation of this phase between the 8th and 16th hour of recording, by 10 min and 50 s on average (Figure 4D).

2.2.2. *S. tuberosum* Extract

In order to ascertain whether potato (*S. tuberosum)* leaf extract containing known ratio of α-chaconine/α-solanine and some other minor GAs caused similar cardiotropic effects as pure substances, the pupae were injected with the potato extract prepared as described above. Application of potato extract caused, in the initial hours of registration (2nd–14th h), a weak increase in frequency of heart contractions in both phases by an average of 21% in the anterograde phase and 23% in retrograde phase. In the later hours of the registration (16th–24th h) frequency decreased by 9% in the anterograde and 3% in retrograde phases (Table S1). In the fast phase the highest positive chronotropic effect was recorded at the 10th hour after injection (by 40% compared to the control). In the slow phase the strongest increase occurred at the 12th hour of registration (by 67% compared to control) (Figure 5A). The injection of the potato extract slightly shortened the duration of the anterograde phase between 8th and 20th hours of the registration (by 44 s on average). In the retrograde phase differentiated changes were noticed depending on the examined interval in the circadian cycle (Table S2). The duration of diastase between the 8th and 24th hour was reduced by an average of 1 min and 52 s (Figure 5B).

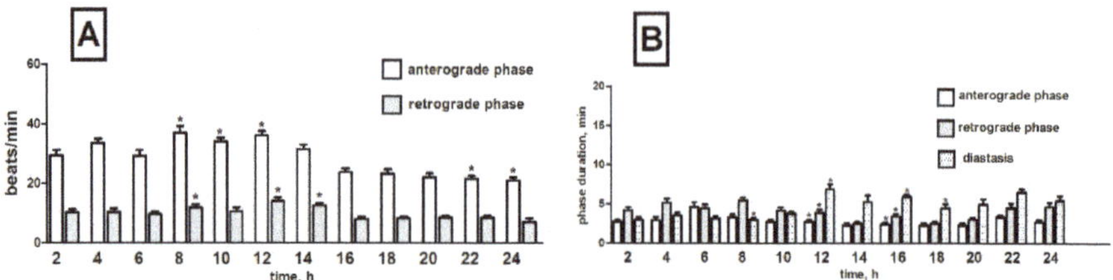

Figure 5. Changes in the heart rate (**A**) and in the duration of anterograde, retrograde phases and diastase in the cardiac cycle (**B**) of *T. molitor* pupae in 2-h intervals, registered after injection of potato extract (1×10^{-5} M). Statistically significant differences ($p \leq 0.05$, Mann-Whitney test) when compared to control are indicated with an asterisk, $n \geq 10$.

2.3. *Effects of Black Nightshade Glycoalkaloids on the Heart Rhythm of* T. molitor *Pupae*

2.3.1. Pure Alkaloids

Solamargine

Two major *S. nigrum* glycoalkaloids are solamargine and solasonine. Solamargine injection resulted in a decrease of the contractions frequency in the anterograde phase of initial part of the recording time (2nd–6th h) by 30% on average (Table S1). In the following hours (8th–24th h), an increase in contractile activity in this phase by 84% on average occurred. In the slow phase, contractions frequency increased by 56% on average throughout the whole period of registration (Figure 6A).

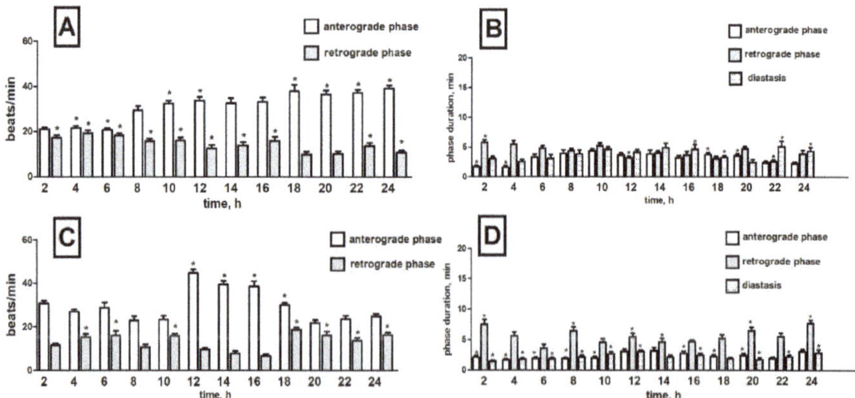

Figure 6. Changes in the heart rate (**A**,**C**) and in the duration of anterograde, retrograde phases and diastase in the cardiac cycle (**B**,**D**) of *T. molitor* pupae in 2-h intervals, registered after injection of solamargine (1×10^{-5} M) (**A**,**B**) and solasonine (1×10^{-5} M) (**C**,**D**). Statistically significant differences ($p \leq 0.05$, Mann-Whitney test) when compared to control are indicated with an asterisk, $n \geq 10$.

Injection of solamargine evoked slight shortening of the duration of the fast phase in the first 4 h of registration by 1 min and 4 s on average. Following hours of anterograde phase duration did not differ significantly from the control values. In addition, under the influence of solamargine, prolongation of the retrograde phase in the 2nd hour of recording by 1 min and 30 s was observed. In the next hours, the effect weakened and finally shortening in the 18th and 22nd hour by 1 min 10 s and 1 min and 50 s, respectively, was observed (Table S2). The length of retrograde phase in other intervals did not differ significantly from the control values. The observed diastasis duration shortened on average by 2 min 15 s throughout the whole circadian cycle (Figure 6B).

Solasonine

After solasonine application, an increase in the frequency of contractions in the fast phase between the 12th and 18th hour of registration by 43% on average, in relation to the control, was observed (Table S1). In the retrograde phase, contractile activity increased by 43% on average, throughout the period of registration. The highest positive chronotropic effect was recorded in the 18th and 24th hour when the frequency of contractions was two times higher than in the control (Figure 6C).

In the pupae treated with solasonine, we noticed a shortened duration of the anterograde phase by 1 min 20 s on average throughout the whole registration period. In the retrograde phase, varied changes in its duration were observed that depended on the analysed interval of the circadian cycle. Significant changes were noted during the phase of diastase (Table S2). Solasonine caused the shortening of this phase by 3 min 55 s on average throughout the registration period (Figure 6D).

2.3.2. *S. nigrum* Extract

Changes in the cardiac anterograde phase after injection of black nightshade (*S. nigrum*) extract were minor and showed different changes depending on the considered time interval in the circadian cycle (Table S1). During the retrograde phase, an increase in the frequency of contractions by 19% on average throughout the whole period of registration was observed (Figure 7A).

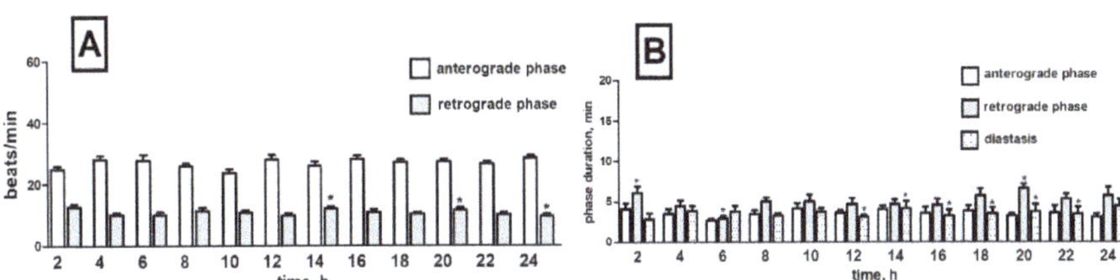

Figure 7. Changes in the heart rate (**A**) and in the duration of anterograde, retrograde phases and diastase in the cardiac cycle (**B**) of *T. molitor* pupae in 2-h intervals, registered after injection of *S. nigrum* extract (1×10^{-5} M). Statistically significant differences ($p \leq 0.05$, Mann-Whitney test) when compared to control are indicated with an asterisk, $n \geq 10$.

Similarly, injection of black nightshade extract did not cause significant changes in the duration of the fast phase when compared to control values in each recording interval. During the slow phase, a clear reduction occurred in the duration of this phase at the 6th hour by 2 min 2 s and elongation at 2nd and 20th hour by 1 min 50 s and 1 min 43 s, respectively. At other intervals, the length of the retrograde phases was comparable to control values. The extract of black nightshade significantly influenced the duration of diastase but the observed effects were delayed and appeared between the 12th and 24th hour of recording (Table S2). In that period, the resting phase was shortened by an average of 4 min and 10 s (Figure 7B).

2.4. *Effects of Tomato Glycoalkaloids on the Heart Rhythm of T. molitor Pupae*

2.4.1. Synthetic α-Tomatine

The major glycoalkaloid in *S. lycopesicum* is α-tomatine; thus, it was tested in order to evaluate the effect of this substance to the heart rhythm of the pupae. Injection of α-tomatine caused a small decrease in the contractile activity of the heart in the anterograde phase with an average of 10% over the entire registration period and a decrease in the frequency of contractions of this organ in the retrograde phase between the 4th and 22nd hour with an average of 13% (Figure 8A).

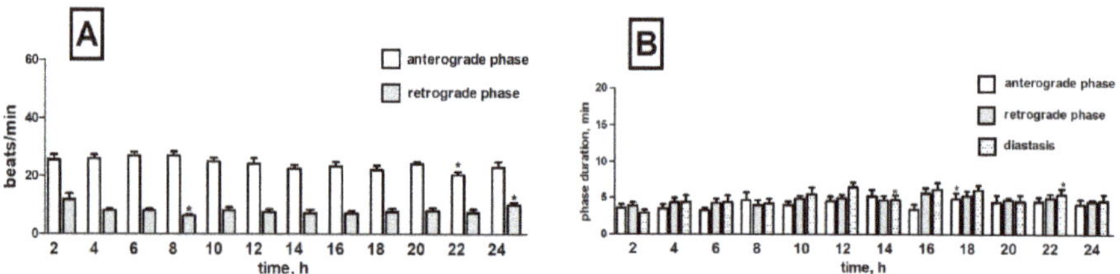

Figure 8. Changes in the heart rate (**A**) and in the duration of anterograde, retrograde phases and diastase in the cardiac cycle (**B**) of *T. molitor* pupae in 2-h intervals, registered after injection of α-tomatine (1×10^{-5} M). Statistically significant differences ($p \leq 0.05$, Mann-Whitney test) when compared to control are indicated with an asterisk, $n \geq 10$.

α-Tomatine prolonged the duration of the anterograde phase in the second half of the circadian cycle (between 12th and 24th hour) by 1 min on average for each interval compared to the control. The duration of retrograde phases throughout the registration period varied from one period to the next (Table S2). The duration of diastasis under the influence of α-tomatine, similarly to anterograde phase, was prolonged in the second half of the recording period (between 12th and 24th hour) by 2 min 33 s on average (Figure 8B).

2.4.2. *S. lycopersicum* Extract

After the injection of tomato (*S. lycopersicum*) leaf extract, an increase in the frequency of contractions in the anterograde phase between the 2nd and 10th hour of recording, by 26% on average, was observed. In subsequent periods (12th–24th h), the contractile activity of the heart decreased by 15% on average in comparison to the control (Table S1). The response of the heart to the injection of the tomato extract in the retrograde phase was slight and showed varied changes depending on the time interval in the circadian cycle of recording (Figure 9A).

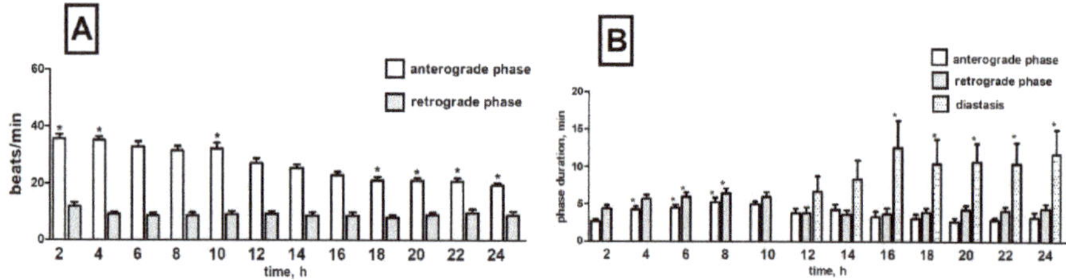

Figure 9. Changes in the heart rate (**A**) and in the duration of anterograde, retrograde phases and diastase in the cardiac cycle (**B**) of *T. molitor* pupae in 2-h intervals, registered after injection of *S. locypersicum* extract (1×10^{-5} M). Statistically significant differences ($p \leq 0.05$, Mann-Whitney test) when compared to control are indicated with an asterisk, $n \geq 10$.

The extract obtained from tomato extended the duration of the anterograde phase in time between the 4th and 8th hour of the registration period by 1 min and 10 s on average. In addition, the duration of the retrograde phase increased by 1 min and 22 s average in the period between 6th and 8th hour of registration and shortened by 20 s on average in the second half of registration (between 14th and 24th hour). The duration of diastasis in the initial hours of the registration (2nd–10th h) decreased by an average of 2 min and 48 s and was not present (Table S2), while at later intervals (16th–24th h) increased by an average of 3 min and 4 s (Figure 9B).

2.5. Effects of Horseradish Glucosinolates on the Heart Rhythm of T. molitor Pupae

When the pupae were treated with the horseradish extract, a slight decrease in the frequency of contractions over the entire registration period by 5% on average in the anterograde phase and by 10% in the retrograde phase was noticed (Figure 10A).

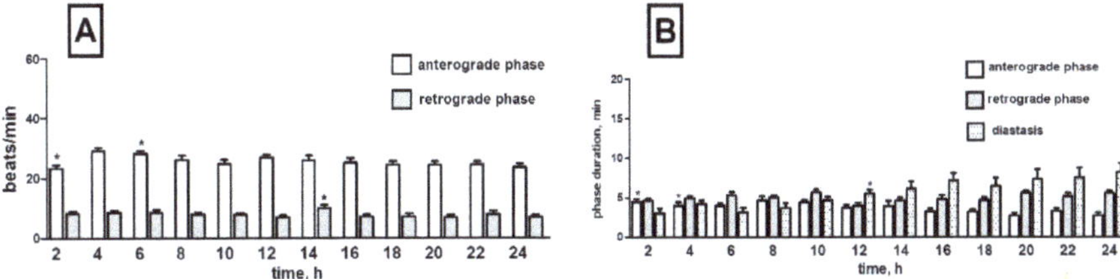

Figure 10. Changes in the heart rate of *T. molitor* pupae in 2-h intervals registered after injection of *A. rusticana* extract in concentration 1×10^{-5} M (**A**) together with changes in the duration of anterograde, retrograde phases and diastasis in the cardiac cycle of these pupae after injection of this extract (1×10^{-5} M) (**B**). Statistically significant differences ($p \leq 0.05$, Mann-Whitney test) when compared to control are indicated with an asterisk, $n \geq 10$.

Horseradish extract did not significantly affect the duration of individual phases in the heart cycle. Slightly extended duration time of the anterograde phase in the range from the 2nd to 4th hour of registration was observed (Table S2). Duration of individual retrograde phases was comparable to the corresponding values in the control. During the second half of the recording period (between the 12th and 22nd hour), the diastase was shortened by 1 min on average in each interval (Figure 10B).

3. Discussion

Insects are increasingly used as model organisms in toxicological, pharmacological, genetic or biomedical research. In addition to economic aspects, such as low farming costs and a short development cycles, the arguments for using insects in experiments are the similarities which they exhibit in relation to mammals [9,12]. Functional analogies at the molecular and cellular level between the myocardium of the insect and the mammalian heart were shown in the research on the fruit fly *D. melanogaster*, among others, by Bier and Bodmer [13] and Ocorr et al. [14]. Nowadays, other insects, such as locusts (*L. migratoria*), stick insects (*B. extradentatum*), moths (*B. mori*) or beetles (*T. molitor*) are used more and more often to evaluate the action of various bioactive substances, as well as to analyse the basis and the course of some diseases [15–17].

The research conducted so far has shown that the effects of different alkaloids on animal circulatory systems manifest, among others, through a significant increase or decrease of blood pressure, bradycardia, reduced respiratory activity, and haemolysis. It is believed that these changes are the result of inhibition of acetylcholinesterase activity, increase of cell membrane permeability, as well as disruption of steroid metabolism [18]. However, no detailed data is available on the activity of *Solanaceae* GAs on heart activity. To the best of our knowledge, there are only two published reports that study cardiotropic properties of GAs on vertebrates. Bergers and Alink [19] examined effects of α-solanine and α-tomatine in cultured beating rat heart cells and observed ceased beating within a few minutes after the application of α-solanine (80 µg/mL) or α-tomatine (20 µg/mL). What is interesting, is that at lower concentrations, both compounds increased the contraction frequency of the heart cells. That indicates the high dependence of observed effects on the dose. Different GAs have also been tested for cardiotropic activities on the isolated frog heart. This study showed cardio-stimulatory properties of these compounds which are directly related to the kind of sugars contained in the GAs [20]. Introduction of a new in vivo insect model (pupae of beetles) gave new data on the potent cardiotropic activity of different GAs and GLSs. It also proved, that myocardium of insects can be successfully used to evaluate the cardioactive properties of various active substances, including plant derived molecules, in relatively fast, cheap and ethically more acceptable tests.

Previous research showed that potato GAs applied as extract caused a decrease in the heart contraction in *Z. atratus* beetles in in vitro conditions while two other tested species (*T. molitor* and *Leptinotarsa decemlineata*) were resistant to these extracts [21]. Further detailed studies showed that potato and tomato extracts applied on *Z. atratus* semi-isolated heart caused irreversible cardiac arrests, while an extract from black nightshade caused fast but reversible arrests [11]. *S. nigrum* extract showed also reversed inhibition of heart activity when applied on the *T. molitor* adult heart in vitro. The reversible few-second heart arrest was observed after the application of extract at a 1% concentration [22]. In *Z. atratus*, pure commercial GAs (α-tomatine, α-chaconine, α-solanine, solamargine, and solasonine) caused similar but less evident effects compared with extracts, whereas in *T. molitor* no significant alterations after in vitro application of pure solasonine and solamargine were observed [22]. These results supported the hypothesis that the bioactivity of tested compounds depended on their structure and suggested the existence of synergistic interactions with the extracts [11]. Interestingly, injection of tomato and potato extracts in 1-day-old pupae (in vivo studies) of *Z. atratus*, produced completely opposite effects, inducing reversible positive chronotropic effects and decreased the duration of both phases (anterograde and retrograde) of heart contractile activity [11]. Different results of in vivo and in vitro studies suggest indirect effects of GAs, through alterations of insect physiology that are not known yet. These results are partially in tune with results obtained in the present study. Most tested extracts and single substances exerted a cardiostimulatory effects (chronotropic positive effects). Only *S. nigrum* extract and *A. rusticana* extract showed no-significant effects on the heart cycle of *T. molitor* pupae. This proves that cardiostimulatory effects of GAs are not species-specific in vivo. This was also confirmed when single synthetic substances were tested. Among six tested GAs only tomatine did not produce some effects. However, on the contrary to the results obtained on *Z. atratus*, the effects produced by *S.tuberosum*, *S. nigrum*, and *S. lycopresicum* extracts were similar or even smaller than that of single substances. This indicates that in in vivo studies synergistic interactions of major molecules are not so important for the specific effect. The reason for this is not known so far and needs further explanation. Perhaps, the detoxifying mechanisms of the organism may be activated more by the presence of a variety of minor toxic agents in extracts or they reveal an antagonistic effect on the activity to the heart. Next, some other, minor substances present in the extract, may additionally affect the effect of major substances.

All tested substances and extracts showed no-cardiotoxic activity. The most cardioactive turned out to be chaconine. The chronotropic positive effect caused by this molecule was the strongest however, it also strongly affected heart cycle phases, increasing the duration of diastasis. These results are different than those obtained on the frog heart [20]. α-Tomatine showed the most potent cardiostimulatory activity on the frog heart in comparison to α-chaconine and α-solanine. Thus, GAs may reveal species-specific effects. However, our results support the hypothesis that the cardioactivity of *Solanaceae* GAs depends on the nature of the aglycone and the number of sugar moieties rather than on the kinds of sugars or their stereochemical configuration in the molecule [20].

Lack of cardiotoxicity of tested substances is a quite promising and interesting result and gives new possible direction to use GAs as cardioregulatory factors supporting for example therapies with synthetic molecules. This is also quite surprising because research performed so far showed cytotoxic properties of *Solanaceae* GAs on the isolated heart cells [19]. Possibly, in vitro models give different results compared with in vivo bioassays due to the direct interaction between the tested substances and the cells. Also, research performed on different insect species showed that different GAs or GAs-containing plant extracts alter the health of individuals. For example, *S. nigrum* alkaloids and *A. rusticana* GLSs affected the development and caused imago malformations of model species *D. melanogaster* [23], whereas tomato and potato GAs and their extracts interrupted the functioning of the fat body cells of the moth *Spodoptera exigua* [24] and enzymes' activity in *Galleria mellonella* moths [25]. Thus far, glycoalkaloids have been considered as potent bioinsecticides and cytotoxic agents rather than therapeutic molecules. On the other hand, *Solanaceae* GAs have been shown to have anticancer activity [26,27]. However, research obtained in this study showed new pharmacological possibilities of the tested plant-derived substances in the modulation of heart muscle activity. The results also indicate that the method of application may play a crucial role in the mode of action and switching from therapeutic to toxic activity. Therefore, further, more detailed analyses of the biological activity of plant-derived substances are highly likely to bring new, valuable data and suggest their future applications. This conclusion is strongly supported by the increasing trend of scientific interest in the activity of GAs and GLs, observed worldwide [28,29].

4. Materials and Methods

4.1. Insects

T. molitor L. (Coleoptera: Tenebrionidae) 1-day-old pupae were obtained from a colony maintained at the Department of Animal Physiology and Development, Adam Mickiewicz University, Poznań, Poland, according to the procedure described previously [30].

4.2. Chemical Standards

Pure solamargine (97.5%) and solasonine (98.3%) were purchased from Glycomix (Compton, UK), pure α-chaconine (≥95%) and α-solanine (≥95%) were obtained from Lab Service Analytica (Bologna, Italy), while hydrate α-tomatine was supplied by Sigma–Aldrich (Taufkirchen, Germany). Commercial α-tomatine contained dehydrotomatine as an impurity (α-tomatine:dehydrotomatine 10:1 *w*/*w*). Standards were dissolved in physiological saline for beetles (274 mM/L of NaCl, 19 mmol/L of KCl, 9 mmol/L of $CaCl_2$) to obtain 10^{-3} M concentration. Samples were stored at −20 °C. Immediately before use, the dilutions of tested compounds were prepared to the desired concentration.

4.3. Plant Material and Extracts Preparation

The extracts used for bioassays were prepared from plants grown in Bari, southern Italy, according to the previously described procedure [31]. Potato *S. tuberosum* L. and tomato *S. lycopersicum* Mill. leaves were collected before the fruiting. Green unripe fruits were harvested from the black nightshade. Collected plant material was immediately frozen to suppress the ripening process. Before the extraction the leaves and fruits were stored at −20 °C. The extraction of alkaloids was performed with 1% aqueous acetic acid solution. Each 1.5 g of sample of the material was placed in 20 mL of extraction solution. In order to enhance the contact of plant tissues with the solution, the suspension was stirred for about 2 h, and then centrifuged at 6000 rpm for about 30 min. After extraction the supernatant was lyophilized. *A. rusticana* extract was prepared according to previously described protocol [32]. Lyophilized samples were dissolved in physiological saline for beetles containing 0.1% of acetic acid to obtain 1×10^{-3} M concentration and stored at −20 °C before using. Immediately before use, the dilutions of test compounds were prepared to the desired concentration (10^{-5} M). Stock solutions of extracts were obtained at a concentration of 1×10^{-3} M for their main component metabolite (α-chaconine, α-tomatine, and solamargine, respectively). The voucher specimens of all plants were deposited at Herbarium Lucanum (HLUC, Potenza, Italy) with the ID Codes: 5809 *S. tuberosum*, 4433 *S. lycopersicum*, 2320 *S. nigrum* and 9197 *A. rusticana*.

4.4. In Vivo Heart Bioassay

The experiments were performed according to the optoelectronic technique performed previously by Slama and Rosiński [15] and Chowański and Rosiński [33]. In brief, it uses alternating visible red light (1–5 kHz) emitted from common light emitting diodes. The light beam is carried from the diode to the dorsal pericardial region of the pupa through a thin outgoing optic fibre. An associated incoming optic fibre collects the reflected pulse light (modulated mainly by movements of the heart and slightly also by other organs) and delivers it to a phototransistor at the opposite end of the fibre. Rarely, during the registration, very high amplitude of recorded contractions was observed, which corresponded to the movement of other organs or the whole body—respiratory movements. These movements were not assessed. The injection of the pupae was performed using a Hamilton syringe, puncturing the cuticle between the 2nd and 3rd segments of the abdomen, on the dorsal side. The needle was inserted towards the head, to a depth of about 3 mm tangentially to the surface of the body. The tested solutions were applied in a volume of 2 μL. For the injection, alkaloids with a concentration of 1×10^{-5} M were used, which corresponds with the final concentration in haemolymph of 10^{-6} M. The needle of the syringe was left for a few seconds in the body of the insect, which accelerated the clotting of the hemolymph and prevented the outflow from the opening after the puncture. The implanted pupae were put aside to check for possible haemolymph loss.

4.5. Statistical Analyses

Statistical analysis was performed with GraphPad Prism 6 statistical software (GraphPad Software, Inc., San Diego, CA, USA, PM license) using Mann-Whitney test. Significant changes were considered as those with a *p*-value of $p \leq 0.05$ (*). A period of around 5 to 8 min in each 2-h period of each of the recording was assessed which correspond to adequate phase (anterograde and retrograde separately). The data are presented as average (± SD) values obtained from *n* replicates.

Supplementary Materials: The following are available online at http://www.mdpi.com/2072-6651/11/5/287/s1, Table S1: Percentage changes in the heart beat frequency of *T. molitor* pupae during 24 h recordings including anterograde (A) and retrograde (R) phases. Red marked font presents the increase of heartbeats per minute, while blue marked font, the decrease. Presented data includes only statistically significant results ($p < 0.05$). Table S2: Percentage changes in the duration anterograde (A), retrograde (R) and diastasis (D) phases of *T. molitor* pupae heart beat during 24 h recordings. Blue colored font presents the phase reduction, while red font the phase elongation. Sign "-" means lack of appearance of the phase. Presented data includes only statistically significant results ($p < 0.05$).

Author Contributions: Conceptualization, P.M. and Z.A.; methodology, G.R.; validation, M.S. and S.C.; formal analysis, S.C.; investigation, A.K. and L.S.; resources, P.M., L.S., P.F., S.A.B.; data curation, M.S.; writing—original draft preparation, P.M.; writing—review and editing, M.S., S.C., Z.A., L.S., P.F., S.A.B. and G.R.; visualization, P.M.; supervision, P.F., S.A.B. and G.R.; project administration, P.M. and A.K.; funding acquisition, S.A.B.

Funding: This research received no external funding.

Conflicts of Interest: The authors declare no conflict of interest. The funders had no role in the design of the study; in the collection, analyses, or interpretation of data; in the writing of the manuscript, or in the decision to publish the results.

References

1. Shitan, N. Secondary metabolites in plants: Transport and self-tolerance mechanisms. *Biosci. Biotechnol. Biochem.* **2016**, *80*, 1283–1293. [CrossRef]
2. Habli, Z.; Toumieh, G.; Fatfat, M.; Rahal, O.N.; Gali-Muhtasib, H. Emerging Cytotoxic Alkaloids in the Battle against Cancer: Overview of Molecular Mechanisms. *Molecules* **2017**, *22*, 250. [CrossRef]
3. Chrobot, A.; Matkowski, A. Plant natural drugs as calcium channel blockers in the cardiovascular system. *Postępy Fitoter.* **2007**, *2*, 95–108.
4. Lichota, A.; Gwozdzinski, K. Anticancer Activity of Natural Compounds from Plant and Marine Environment. *Int. J. Mol. Sci.* **2018**, *19*, 3533. [CrossRef]
5. Roddick, J.G. Steroidal glycoalkaloids: Nature and consequences of bioactivity. *Adv. Exp. Med. Biol.* **1996**, *404*, 277–295.
6. Chowanski, S.; Adamski, Z.; Marciniak, P.; Rosinski, G.; Buyukguzel, E.; Buyukguzel, K.; Falabella, P.; Scrano, L.; Ventrella, E.; Lelario, F.; et al. A Review of Bioinsecticidal Activity of Solanaceae Alkaloids. *Toxins* **2016**, *8*, 60. [CrossRef]
7. Agneta, R.; Rivelli, A.R.; Ventrella, E.; Lelario, F.; Sarli, G.; Bufo, S.A. Investigation of glucosinolate profile and qualitative aspects in sprouts and roots of horseradish (Armoracia rusticana) using LC-ESI-hybrid linear ion trap with Fourier transform ion cyclotron resonance mass spectrometry and infrared multiphoton dissociation. *J. Agric. Food Chem.* **2012**, *60*, 7474–7482. [CrossRef]

8. Spochacz, M.; Chowanski, S.; Walkowiak-Nowicka, K.; Szymczak, M.; Adamski, Z. Plant-Derived Substances Used Against Beetles-Pests of Stored Crops and Food-and Their Mode of Action: A Review. *Compr. Rev. Food Sci. Food Saf.* **2018**, *17*, 1339–1366. [CrossRef]
9. Chowanski, S.; Adamski, Z.; Lubawy, J.; Marciniak, P.; Pacholska-Bogalska, J.; Slocinska, M.; Spochacz, M.; Szymczak, M.; Urbanski, A.; Walkowiak-Nowicka, K.; et al. Insect Peptides-Perspectives in Human Diseases Treatment. *Curr. Med. Chem.* **2017**, *24*, 3116–3152. [CrossRef]
10. Adamski, Z.; Bufo, S.A.; Chowanski, S.; Falabella, P.; Lubawy, J.; Marciniak, P.; Pacholska-Bogalska, J.; Salvia, R.; Scrano, L.; Slocinska, M.; et al. Beetles as Model Organisms in Physiological, Biomedical and Environmental Studies—A Review. *Front. Physiol.* **2019**, *10*. [CrossRef]
11. Ventrella, E.; Marciniak, P.; Adamski, Z.; Rosinski, G.; Chowanski, S.; Falabella, P.; Scrano, L.; Bufo, S.A. Cardioactive properties of Solanaceae plant extracts and pure glycoalkaloids on Zophobas atratus. *Insect Sci.* **2015**, *22*, 251–262. [CrossRef] [PubMed]
12. Szymczak, M.; Marciniak, P.; Rosiński, G. Insect myocardium-model to biomedical research. *Adv. Cell Biol.* **2014**, *41*, 59–77.
13. Bier, E.; Bodmer, R. Drosophila, an emerging model for cardiac disease. *Gene* **2004**, *342*, 1–11. [CrossRef] [PubMed]
14. Ocorr, K.; Akasaka, T.; Bodmer, R. Age-related cardiac disease model of Drosophila. *Mech. Ageing Dev.* **2007**, *128*, 112–116. [CrossRef] [PubMed]
15. Slama, K.; Rosinski, G. Delayed pharmacological effects of proctolin and CCAP on heartbeat in pupae of the tobacco hornworm, Manduca sexta. *Physiol. Entomol.* **2005**, *30*, 14–28. [CrossRef]
16. Pacholska-Bogalska, J.; Szymczak, M.; Marciniak, P.; Walkowiak-Nowicka, K.; Rosinski, G. Heart mechanical and hemodynamic parameters of a beetle, *Tenebrio molitor*, at selected ages. *Arch. Insect Biochem. Physiol.* **2018**, *99*. [CrossRef] [PubMed]
17. Ejaz, A.; Lange, A.B. Peptidergic control of the heart of the stick insect, *Baculum extradentatum*. *Peptides* **2008**, *29*, 214–225. [CrossRef]
18. Al Chami, L.; Mendez, R.; Chataing, B.; O'Callaghan, J.; Usubillaga, A.; LaCruz, L. Toxicological effects of alpha-solamargine in experimental animals. *Phytother. Res.* **2003**, *17*, 254–258. [CrossRef] [PubMed]
19. Bergers, W.W.A.; Alink, G.M. Toxic Effect of the Glycoalkaloids Solanine and Tomatine on Cultured Neonatal Rat-Heart Cells. *Toxicol. Lett.* **1980**, *6*, 29–32. [CrossRef]
20. Nishie, K.; Fitzpatrick, T.J.; Swain, A.P.; Keyl, A.C. Positive Inotropic Action of Solanaceae Glycoalkaloids. *Res. Commun. Chem. Pathol. Pharmacol.* **1976**, *15*, 601–607.
21. Marciniak, P.; Adamski, Z.; Bednarz, P.; Slocinska, M.; Ziemnicki, K.; Lelario, F.; Scrano, L.; Bufo, S.A. Cardioinhibitory properties of potato glycoalkaloids in beetles. *Bull. Environ. Contam. Toxicol.* **2010**, *84*, 153–156. [CrossRef] [PubMed]
22. Spochacz, M.; Chowanski, S.; Szymczak, M.; Lelario, F.; Bufo, S.A.; Adamski, Z. Sublethal Effects of *Solanum nigrum* Fruit Extract and Its Pure Glycoalkaloids on the Physiology of *Tenebrio molitor* (Mealworm). *Toxins* **2018**, *10*, 504. [CrossRef]
23. Chowanski, S.; Chudzinska, E.; Lelario, F.; Ventrella, E.; Marciniak, P.; Miadowicz-Kobielska, M.; Spochacz, M.; Szymczak, M.; Scrano, L.; Bufo, S.A.; et al. Insecticidal properties of *Solanum nigrum* and *Armoracia rusticana* extracts on reproduction and development of *Drosophila melanogaster*. *Ecotoxicol. Environ. Saf.* **2018**, *162*, 454–463. [CrossRef]
24. Adamski, Z.; Radtke, K.; Kopiczko, A.; Chowanski, S.; Marciniak, P.; Szymczak, M.; Spochacz, M.; Falabella, P.; Lelario, F.; Scrano, L.; et al. Ultrastructural and developmental toxicity of potato and tomato leaf extracts to beet armyworm, Spodoptera exigua (lepidoptera: Noctuidae). *Microsc. Res. Tech.* **2016**, *79*, 948–958. [CrossRef] [PubMed]
25. Adamski, Z.; Adamski, Z.; Marciniak, P.; Ziemnicki, K.; Buyukguzel, E.; Erdem, M.; Buyukguzel, K.; Ventrella, E.; Falabella, P.; Cristallo, M.; et al. Potato leaf extract and its component, alpha-solanine, exert similar impacts on development and oxidative stress in *Galleria mellonella* L. *Arch. Insect Biochem. Physiol.* **2014**, *87*, 26–39. [CrossRef]
26. Wang, X.; Zou, S.; Lan, Y.L.; Xing, J.S.; Lan, X.Q.; Zhang, B. Solasonine inhibits glioma growth through anti-inflammatory pathways. *Am. J. Transl. Res.* **2017**, *9*, 3977–3989.

27. Zhang, X.H.; Yan, Z.P.; Xu, T.T.; An, Z.T.; Chen, W.Z.; Wang, X.S.; Huang, M.M.; Zhu, F.S. Solamargine derived from *Solanum nigrum* induces apoptosis of human cholangiocarcinoma QBC939 cells. *Oncol. Lett.* **2018**, *15*, 6329–6335. [CrossRef]
28. Abellán, Á.; Domínguez-Perles, R.; Moreno, D.A.; García-Viguera, C. Sorting out the value of cruciferous sprouts as sources of bioactive compounds for nutrition and health. *Nutrients* **2019**, *11*, 429. [CrossRef]
29. Chaudhary, P.; Sharma, A.; Singh, B.; Nagpal, A.K. Bioactivities of phytochemicals present in tomato. *J. Food. Sci. Technol.* **2018**, *55*, 2833–2849. [CrossRef]
30. Rosinski, G.; Pilc, L.; Obuchowicz, L. Effect of Hydrocortisone on Growth and Development of Larvae Tenebrio-Molitor. *J. Insect Physiol.* **1978**, *24*, 97–99. [CrossRef]
31. Cataldi, T.R.; Lelario, F.; Bufo, S.A. Analysis of tomato glycoalkaloids by liquid chromatography coupled with electrospray ionization tandem mass spectrometry. *Rapid Commun. Mass Spectrom.* **2005**, *19*, 3103–3110. [CrossRef]
32. Agneta, R.; Lelario, F.; De Maria, S.; Mollers, C.; Bufo, S.A.; Rivelli, A.R. Glucosinolate profile and distribution among plant tissues and phenological stages of field-grown horseradish. *Phytochemistry* **2014**, *106*, 178–187. [CrossRef]
33. Chowanski, S.; Rosinski, G. Myotropic Effects of Cholinergic Muscarinic Agonists and Antagonists in the Beetle *Tenebrio molitor* L. *Curr. Pharm. Biotechnol.* **2017**, *18*, 1088–1097. [CrossRef] [PubMed]

Article

Tissue Accumulations of Toxic Aconitum Alkaloids after Short-Term and Long-Term Oral Administrations of Clinically Used *Radix Aconiti Lateralis* Preparations in Rats

Xiaoyu Ji [1,†], Mengbi Yang [1,†], Ka Hang Or [2], Wan Sze Yim [2] and Zhong Zuo [1,*]

1 School of Pharmacy, The Chinese University of Hong Kong, Hong Kong SAR, China; SharonChi@link.cuhk.edu.hk (X.J.); yangmengbi@cuhk.edu.hk (M.Y.)

2 School of Chinese Medicine, The Chinese University of Hong Kong, Hong Kong SAR, China; orkh@cuhk.edu.hk (K.H.O.); wendy.yim@cuhk.edu.hk (W.S.Y.)

* Correspondence: joanzuo@cuhk.edu.hk

† These authors contributed equally to this work.

Received: 31 May 2019; Accepted: 14 June 2019; Published: 18 June 2019

Abstract: Although *Radix Aconiti Lateralis* (Fuzi) is an extensively used traditional Chinese medicine with promising therapeutic effects and relatively well-reported toxicities, the related toxic aconitum alkaloid concentrations in major organs after its short-term and long-term intake during clinical practice are still not known. To give a comprehensive understanding of Fuzi-induced toxicities, current study is proposed aiming to investigate the biodistribution of the six toxic alkaloids in Fuzi, namely Aconitine (AC), Hypaconitine (HA), Mesaconitine (MA), Benzoylaconine (BAC), Benzoylhypaconine (BHA) and Benzoylmesaconine (BMA), after its oral administrations at clinically relevant dosing regimen. A ultra-performance liquid chromatography-tandem mass spectrometry (UPLC–MS/MS) method was developed and validated for simultaneous quantification of six toxic alkaloids in plasma, urine and major organs of Sprague Dawley rats after oral administrations of two commonly used Fuzi preparations, namely Heishunpian and Paofupian, at their clinically relevant dose for single and 15-days. Among the studied toxic alkaloids and organs, BMA demonstrated the highest concentrations in all studied organs with liver containing the highest amount of the studied alkaloids, indicating their potential hepatotoxicity. Moreover, tissue accumulation of toxic alkaloids after multiple dose was observed, suggesting the needs for dose adjustment and more attention to the toxicities induced by chronic use of Fuzi in patients.

Keywords: *Radix Aconiti Lateralis* preparations; short-term and long-term usage; di-ester diterpenoid alkaloids; mono-ester diterpenoid alkaloids; biodistribution

Key Contribution: Both short-term and long-term use of *Radix Aconiti Lateralis* could result in distribution of toxic aconitum alkaloids mainly in liver, while multiple dosing of *Radix Aconiti Lateralis* preparations could lead to toxic aconitum alkaloids accumulation in major organs in Sprague Dawley rats.

1. Introduction

The processed lateral root of *Radix Aconiti Lateralis* is known as Fuzi, an extensively used traditional Chinese medicine in China and other Asian countries [1]. Fuzi is recognized as a treatment for cardiovascular diseases, rheumatism arthritis, bronchitis, pains and hypothyroidism, etc. [1,2]. Among the 500 well-known traditional Chinese medicine (TCM) formulae used in clinics, approximately 13.2% include Fuzi [1]. In clinical practice, Fuzi-containing prescriptions are usually given orally, with 81.8% of them used at 5 g/person/day, 9.1% of them used at 10 g/person/day and the rest used at 15 g/person/day [3].

Although Fuzi has promising therapeutic effects, aconitum poisoning has been extensively reported with 17 cases from Taiwan during 1990–99, 39 cases form Hong Kong during 2012–19, 2017 cases from mainland of China during 1989–2008, and 121 cases from Korea during 1995–2007 [4,5]. In clinical poisoning cases, the patients usually had acute onset of illness within half an hour to several hours after consumption of Fuzi with cardiotoxic and neurotoxic effects as the main outcomes. Cardiotoxicity symptoms included palpitation, arrhythmia with slow heart rate, and low blood pressure. Neurotoxicity features included numbness in the oral cavity, tongue, face, extremities and body, muscle weakness and dizziness. Arrhythmia and muscle weakness-induced breathing difficulties may lead to death in severe cases [6].

The principal toxic ingredients in Fuzi are C_{19}-diterpenoid alkaloids, including three di-ester diterpenoid alkaloids (DDAs) and three mono-ester diterpenoid alkaloids (MDAs) [1,3]. As shown in Figure 1, processing of Fuzi would make DDAs lose their acetyl group at C_8 and become MDAs. Based on their LD_{50} values on mice, the toxicities of MDAs were 1/700–1/100 of that of DDAs [7]. MDAs could subsequently lose the benzoyl ester group at C_{14} to generate non-toxic non-ester type alkaloids (NDAs) [7–11].

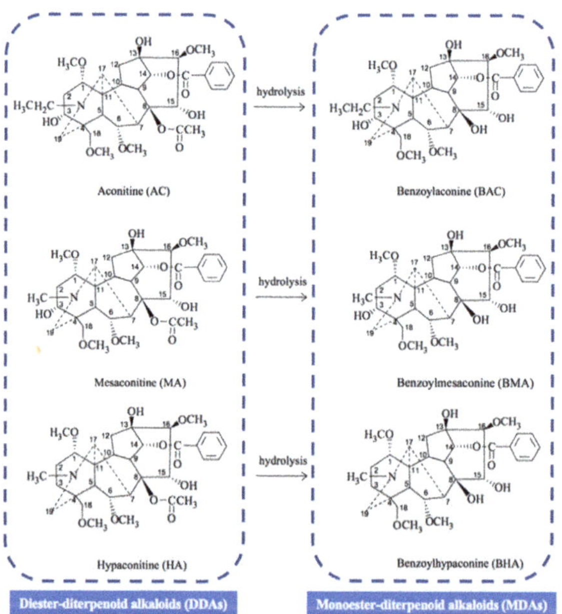

Figure 1. Relationships between the studied di-ester- and mono-ester- diterpenoid alkaloids of Fuzi.

Previous mechanistic studies on Fuzi-induced toxicity mainly focused on the cardiotoxicity of the DDAs. The DDA-induced arrhythmogenic effects, which have been frequently observed after ingestion of Fuzi, were well recognized to be caused by activating the voltage-dependent sodium channel on the cell membranes in excitable tissues such as myocardium [1,12–14]. More recently, other cardiotoxic mechanisms for toxic aconitum alkaloids were proposed. Aconitine was found to be able to perturb intracellular calcium homeostasis via Na^+-Ca^{2+} exchange system [1] and sarco/endoplasmic reticulum Ca^{2+}-ATPase [12], leading to further cell apoptosis in heart [12]. Therefore, damaged intracellular calcium homeostasis was considered as the key cause of ventricular arrhythmias of aconitine [15].

In addition to heart, aconitine induced apoptosis was found in other organs including liver [16], brain [17], and kidney [18].

Based on various processing methods used by Chinese medicine practitioners, there are ten types of Fuzi preparations [19–21], among which Heishunpian and Paofupian are the most common ones and serve as the major components in classical Chinese medicine formulae Sini Tang for acute treatment (usually single dose) and Fuzi Lizhong Tang for sub-chronic treatment (could be as long as two weeks), respectively. After excavation, crude Fuzi was boiled in hot water, roasted with salts and finally stained into black slice to make Heishunpian, which could be further roasted with sands until inflated and slightly discolored to form Paofupian [22]. It was reported that some poisoning cases occurred after consumption of Heishunpian even at doses within its therapeutic range of 3–15 g/person/day [23]. Although compared with Heishunpian, further hydrolysis processing procedure for Paofupian is expected to lower its toxic alkaloid contents (especially for DDAs), the toxicities of Paofupian should not be overlooked considering its wider indications and longer period of intake.

The concentrations of the toxic alkaloids DDAs in plasma and cardiac tissue have been found positively correlated with the cardiac toxicity after single dose of Fuzi in mice and rats [24], suggesting that plasma concentrations of DDAs could be applied to monitor cardiac toxicities of Fuzi in clinical practice. However, the relationship between plasma/tissue concentration of toxic alkaloids and the other potential organ toxicity of Fuzi is largely unknown. Previous studies depicted in vivo biodistribution of toxic alkaloids after single dose of Fuzi, while there is a lack of information on the biodistribution profiles of toxic alkaloids after multiple dose of Fuzi preparation, leading to no established toxicokinetic-toxicity correlation for long-term use of Fuzi. In addition, most of the previous biodistribution studies [25,26] on Fuzi used their own Fuzi extract, and the effect of Fuzi preparations with different processing method on biodistribution profiles of toxic alkaloids remains unclear. It is hard to generate any conclusion without standardizing the Fuzi preparations used in the studies.

To give a more complete understanding of the biodistribution profiles of toxic alkaloids to see how they relate to Fuzi-induced toxicities, the current study was conducted to establish an LC/MS/MS method for simultaneous determination of six toxic aconitum alkaloids in rat plasma, urine and major organs, followed by its application to investigate their biodistributions after single and 15-day oral administrations of Heishunpian and Paofupian at their clinical dosing regimen in rats.

2. Results

2.1. *LC/MS/MS Method Development and Validation for Simultaneous Determination of Six Toxic Aconitum Alkaloids in Different Biological Matrix*

2.1.1. Optimization of Chromatographic and Mass Conditions

The optimal parameters including precursor ion, product ion, fragmentor and collision energy are listed in Table 1. Under such optimized MS/MS condition, the response for each analyte was high enough for detection in biological matrix. Spectra of product ion scan on six targeted toxic alkaloids are shown in Figure 2.

Table 1. List of selected Multiple Reaction Monitoring (MRM) parameters, fragmentor, and collision energy for studied analytes and internal standard (IS).

Compound	Precursor Ion (m/z)	Product Ion (m/z)	Fragmentor (mV)	Collision Energy (mV)
AC	646	586	130	25
MA	632	572	150	37
HA	616	556	125	33
BAC	604	554	145	41
BMA	590	540	130	33
BHA	574	542	185	37
Berberine (IS)	336.2	320.2	130	25

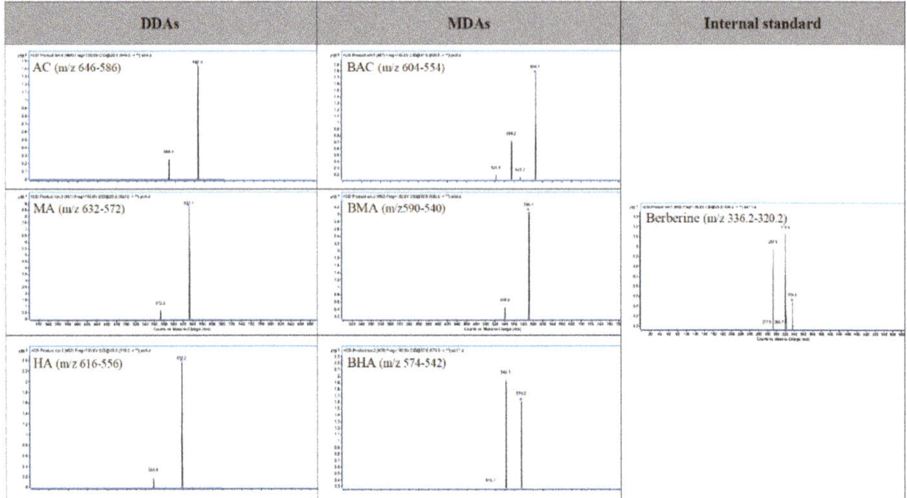

Figure 2. Product ion spectra of AC, MA, HA, BCA, BHA, BMA, and Berberine (IS).

2.1.2. Optimization of Solid Phase Extraction (SPE) Conditions for Sample Treatment

Solid phase extraction (SPE) using a mixed-mode cation-exchange (MCX) cartridge was adopted to separate the targeted toxic alkaloids from the biological matrix in brain and liver homogenates. According to previous published studies [27], 5% ammonium hydroxide in 70% methanol was used to elute compounds binding on the MCX cartridge. However, by adopting such elution method, we found the response of targeted alkaloids were not high enough for detection, due to the extremely low recoveries of Hypaconitine (HA) and Benzoylmesaconine (BHA). Under such elution condition, the recovery of Aconitine (AC), Mesaconitine (MA), HA, Benzoylaconine (BAC), BMA and Benzoylhypaconine (BHA) were 75%, 86%, 9%, 39%, 48% and 0%, respectively. Therefore, a series of eluent with different percentage of methanol ranging from 15% to 95% were tested further (data not shown). The results revealed that the best recoveries were achieved by 5% ammonium hydroxide in 95% methanol with over 93% recoveries for all six analytes. Based on such optimization, 5% ammonium hydroxide in 95% methanol was selected out as final eluent for solid phase extraction.

2.1.3. Method Validation

As demonstrated in Table A1, the optimized liquid chromatography tandem-mass spectrometry (LC-MS/MS) conditions for the six toxic alkaloids could provide satisfactory linearities ($r^2 \geq 0.99$) in plasma (0.5–100 ng/mL), urine (0.5–200 ng/mL) and different organs (range from 0.5 to 200 ng/mL). Lower Limit of Quantification (LLOQs) in plasma was 0.5 ng/mL, while no higher than 2 ng/mL in urine and different organs. The results for intra- and inter-day accuracy and precision are shown in Tables A2–A7. The accuracy and precision of the plasma and organs assays at low, medium and high concentrations of targeted alkaloids were within ±15% bias and 15% Relative Standard Deviation (RSD), which met the criteria set in the guidance issued by U.S. Food and Drug Administration (FDA) [28]. The extraction recoveries in biological matrices remained consistent.

The stability results were also shown in Tables A1–A7. According to the results, alkaloids concentration detected in stability samples were within 15% of nominal concentrations after three freeze-thaw cycles, 4 h on bench top, 12 h in the auto-sampler of LC/MS/MS system, and 30 days at −80 °C.

2.2. Content of Six Toxic Aconitum Alkaloids in Studied Radix Aconiti Lateralis Preparation

Contents of the six toxic alkaloids in Heishunpian and Paofupian are depicted in Figure 3. It was found that the content of BMA was the highest among all six toxic alkaloids and the content of HA was the highest among the studied three DDAs. It was also noted that total contents of MDAs were much higher than those of DDAs in both Fuzi preparations with the contents of MDAs 50–5000 times of DDAs in Heishunpian and 2000–30,000 folds of DDAs in Paofupian. Comparing the toxic alkaloids contents in two Fuzi preparations, higher contents of toxic alkaloids were detected in Heishunpian than that in Paofupian, especially for HA, MA, BHA and BMA. The content of AC was 12.94 ng/g in Heishunpian, while it was not detectable in Paofupian.

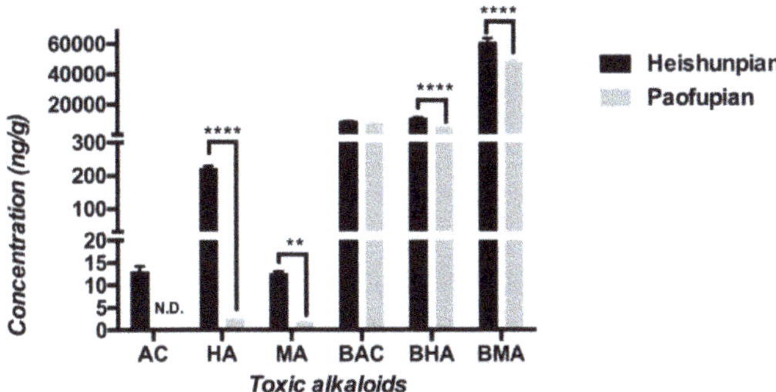

Figure 3. Concentrations of AC, HA, MA, BAC, BHA, and BMA in the Heishunpian and Paofupian crude herb ($n = 3$, ** $p < 0.01$, **** $p < 0.0001$, N.D.: not detectable, below the listed LLOQ of each rat tissue in Table A1).

Based on the contents of six toxic alkaloids in two Fuzi preparations, the toxic alkaloid dosages of 30 g/kg Heishunpian in rat were: BMA (1814 μg/kg), BHA (321 μg/kg), BAC (254 μg/kg), HA (6.61 μg/kg), AC (0.39 μg/kg) and MA (0.38 μg/kg), and the toxic alkaloid dosages of 30 g/kg Paofupian were: BMA (1430 μg/kg), BAC (227 μg/kg), BHA (150 μg/kg), HA (0.07 μg/kg), MA (0.05 μg/kg) with no detectable AC.

2.3. Biodistributions of Toxic Aconitum Alkaloids after Oral Administrations of the Studied Radix Aconiti Lateralis Preparations

Toxic alkaloid concentrations in rat tissues after oral administrations of 30 g/kg Heishunpian and 30 g/kg Paofupian both for single dose and multiple dose to rats are illustrated respectively in Figure 4. The contents of AC and MA in all the tested tissues of all treatment groups were below their detection limits and therefore are not shown in Figure 4.

Consistent with their contents in the dosed Fuzi preparations (Figure 3), the contents of DDAs in all the tested tissues are lower than those of MDAs (Figure 4a,b). Three types of MDAs could be detected in the majority of the organs regardless giving either single dose or multiple dose of both two Fuzi preparations (Figure 4a,b). The biodistribution levels of three MDAs into rat tissues followed the order of BMA >> BHA ≈ BAC (Figure 4a,b). In terms of DDAs, HA was the only detectable one after single dose of Heishunpian (Figure 4a), while no DDA can be detected after single or multiple dose of Paofupian due to the extremely low contents of DDAs in the administrated Paofupian (Figure 3).

Among all the studied organs, liver contained highest amount of all detectable toxic alkaloids. Closely following liver, kidney contained relative high concentration of toxic alkaloids. Only once exception occurred in biodistribution of BHA after multiple dose of Paofupian, where concentration of BHA in kidney went beyond that in liver.

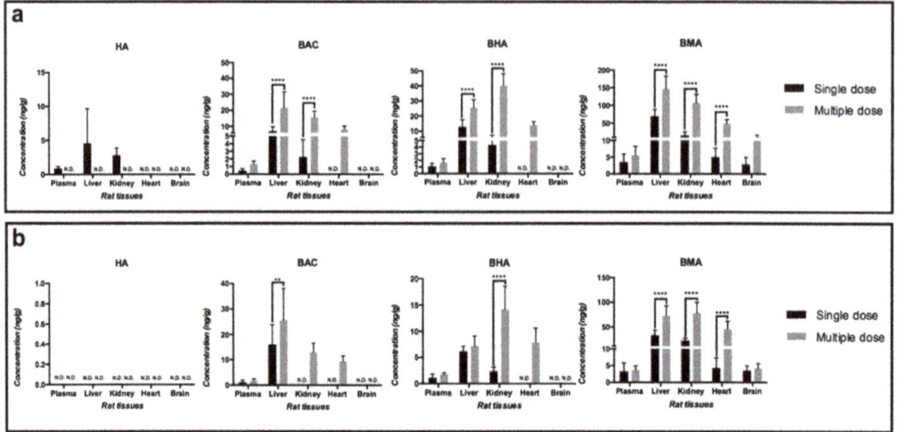

Figure 4. Profiles of HA, BAC, BHA, and BMA in rat plasma and different organs at 2 h after both bolus dose and 15-day consecutively dose of (**a**) Heishunpian and (**b**) Paofupian at 30 g/kg to rats (n = 8, ** $p < 0.01$, **** $p < 0.0001$, N.D.: not detected, below the listed LLOQ of each rat tissue in Table A1).

In addition to plasma and organs, concentration of toxic alkaloids in urine had also been tested. Althought no difference was shown after single or multiple dose of Paofupian, after oral administration of Heishunpian, more numbers of studied toxic alkaloids could be detected in urine sample (all toxic alkaloids for single dose, four toxic alkaloids for multiple dose) than that in plasma sample (four toxic alkaloids for single dose, three toxic alkaloids for multiple dose). Furthermore, concentrations of toxic alkaloids in urine were 10–80 times higher than those in plasma after dosing Heishunpian or Paofupian to rats.

Tissue accumulation of MDAs due to long-term treatment–giving Heishunpian or Paofupian preparation once daily for consecutive 15 days–was clearly observed (Figure 4a,b). Compared with single dose, multiple dose of Heishunpian resulted in higher concentrations of MDAs in liver, kidney and heart (Figure 4a). Similarly, multiple dose of Paofupian resulted in more content of BAC in liver or BHA in kidney, and BMA in four major organs except brain (Figure 4a). Considering the only variable behind such comparison was treatment duration, it was reasonable that attributing higher content of toxic alkaloids in rat tissues to longer treatment duration, or more dosing times in other words.

2.4. Comparison of Dose-Normalized Toxic Aconitum Alkaloid Contents in Rat Plasma, Urine and Major Organs

Since the doses of each toxic alkaloids in two Fuzi preparations varied, the contents of toxic alkaloids in each organ were normalized by their individual doses as shown in Table 2 (single dose) and Table 3 (multiple dose). The detected DDAs had higher dose-normalized contents than MDAs in plasma, urine, and major organs including liver, kidney and heart, which was more evident in rats after giving single dose of Heishunpian. In addition, the dose-normalized contents of all detectable toxic alkaloids in different rat organs were in the order of liver > kidney ≈ heart > brain.

Dose-normalized toxic alkaloid contents in each organ were adopted for comparison of the biodistribution of toxic alkaloids between two different Fuzi preparations, Heishunpian and Paofupian. It could be summarized from Table 2 that the dose-normalized total toxic alkaloids contents in rat tissues (plasma, urine, liver, kidney and heart) from single dose of Heishunpian treated group were significantly higher than that from single dose of Paofupian treated group. Table 3, where gave the dose-normalized content of toxic alkaloids after multiple dosing of Fuzi preparations, demonstrated that compared with Paofupian, toxic alkaloids from Heishunpian could more easily distribute to rat liver, kidney, brain as well as more possibly be detected in rat urine preparation.

Table 2. Comparison of dose-normalized toxic alkaloids content ($\times 10^{-3}$) in rat plasma, urine and different organs at 2 h post-dosing after single dose of 30 g/kg Heishunpian and single dose of 30 g/kg Paofupian to rats. ($n = 8$).

Tissue	Liver		Kidney		Heart		Brain		Plasma		Urine	
Compound	Heishunpian	Paofupian	Heishunpian	Paofupian	Heishunpian	Paofupian	Heishunpian	Paofupian	Heishunpian	Paofupian	Heishunpian	Paofupian
AC	N.D.	N.D.	N.D.	N.D.	N.D.	N.D.	N.D.	N.D.	N.D.	N.D.	N.D.	N.D.
HA	10.89 ± 11.11	N.D.	1.38 ± 0.47	N.D.	0.09 ± 0.14	N.D.	N.D.	N.D.	3.91 ± 1.49	N.D.	17.10 ± 13.51	N.D.
MA	N.D.	N.D.	N.D.	N.D.	N.D.	N.D.	N.D.	N.D.	N.D.	N.D.	9.63 ± 6.93	N.D.
Total DDAs	10.89 ± 11.11	N.D.	1.38 ± 0.47	N.D.	0.09 ± 0.14	N.D.	N.D.	N.D.	3.91 ± 1.49	N.D.	26.73 ± 19.88	N.D.
BAC	0.44 ± 0.13	2.84 ± 0.12	0.03 ± 0.03	N.D.	N.D.	N.D.	N.D.	N.D.	0.05 ± 0.03	0.04 ± 0.00	0.25 ± 0.19	0.11 ± 0.00
BHA	0.66 ± 0.25	1.37 ± 0.06	0.04 ± 0.03	0.10 ± 0.00	0.01 ± 0.00	N.D.	0.01 ± 0.00	N.D.	0.10 ± 0.05	0.06 ± 0.00	0.28 ± 0.17	0.07 ± 0.00
BMA	0.64 ± 0.21	0.82 ± 0.04	0.03 ± 0.02	0.11 ± 0.00	N.D.	0.00 ± 0.00	N.D.	0.02 ± 0.00	0.06 ± 0.04	0.02 ± 0.00	0.31 ± 0.20	0.11 ± 0.00
Total MDAs	1.74 ± 0.47	5.03 ± 0.22	0.10 ± 0.07	0.21 ± 0.01	0.01 ± 0.01	0.00 ± 0.00	**** 0.01 ± 0.01	0.02 ± 0.00	0.21 ± 0.09	0.11 ± 0.00	0.84 ± 0.53	0.29 ± 0.01
Total toxic alkaloids	** 12.63 ± 10.97	5.03 ± 0.22	**** 1.48 ± 0.51	0.21 ± 0.01	* 0.09 ± 0.14	0.00 ± 0.00	N.D.	0.02 ± 0.00	**** 4.13 ± 1.56	0.11 ± 0.00	**** 27.57 ± 20.31	0.29 ± 0.01

* $p < 0.05$, ** $p < 0.01$, **** $p < 0.0001$ compared with corresponding Paofupian group; N.D.: not detectable, below the listed LLOQ of each rat tissue in Table A1.

Table 3. Comparison of dose-normalized toxic alkaloids content ($\times 10^{-3}$) in rat plasma, urine and different organs between multiple dose of 30 g/kg Heishunpian and multiple dose of 30 g/kg Paofupian to rats. ($n = 8$).

Tissue	Liver		Kidney		Heart		Brain		Plasma		Urine	
Compound	Heishunpian	Paofupian	Heishunpian	Paofupian	Heishunpian	Paofupian	Heishunpian	Paofupian	Heishunpian	Paofupian	Heishunpian	Paofupian
AC	N.D.	N.D.	N.D.	N.D.	N.D.	N.D.	N.D.	N.D.	N.D.	N.D.	N.D.	N.D.
HA	N.D.	N.D.	N.D.	N.D.	N.D.	N.D.	N.D.	N.D.	N.D.	N.D.	102.96 ± 46.64	N.D.
MA	N.D.	N.D.	N.D.	N.D.	N.D.	N.D.	N.D.	N.D.	N.D.	N.D.	N.D.	N.D.
Total DDAs	N.D.	N.D.	N.D.	N.D.	N.D.	N.D.	N.D.	N.D.	N.D.	N.D.	102.96 ± 46.64	N.D.
BAC	3.01 ± 1.51	4.23 ± 0.28	0.51 ± 0.11	0.30 ± 0.02	0.13 ± 0.02	0.11 ± 0.01	N.D.	N.D.	0.16 ± 0.05	0.14 ± 0.01	12.18 ± 8.84	3.02 ± 0.20
BHA	** 2.82 ± 0.83	0.78 ± 0.05	**** 1.04 ± 0.20	0.54 ± 0.04	**** 0.17 ± 0.03	0.26 ± 0.02	N.D.	N.D.	0.16 ± 0.06	0.25 ± 0.02	14.38 ± 8.43	6.74 ± 0.45
BMA	*** 2.87 ± 0.97	0.81 ± 0.05	0.49 ± 0.11	0.35 ± 0.02	0.11 ± 0.03	0.09 ± 0.01	**** 0.05 ± 0.01	0.03 ± 0.00	0.09 ± 0.05	0.03 ± 0.00	8.06 ± 7.33	2.84 ± 0.19
Total MDAs	**** 8.70 ± 2.52	5.82 ± 0.39	**** 2.05 ± 0.40	1.21 ± 0.08	** 0.41 ± 0.04	0.45 ± 0.03	**** 0.05 ± 0.01	0.03 ± 0.00	0.41 ± 0.10	0.38 ± 0.15	34.62 ± 23.87	12.60 ± 0.85
Total toxic alkaloids	**** 8.70 ± 2.52	5.82 ± 0.39	**** 2.05 ± 0.40	1.21 ± 0.08	** 0.41 ± 0.04	0.45 ± 0.03	**** 0.05 ± 0.01	0.03 ± 0.00	0.41 ± 0.10	0.38 ± 0.15	**** 137.58 ± 49.70	12.60 ± 0.85

** $p < 0.01$, *** $p < 0.001$, **** $p < 0.0001$ compared with corresponding Paofupian group; N.D.: not detectable, below the listed LLOQ of each rat tissue in Table A1.

3. Discussion

Existing studies mainly focused on cardiac toxicities induced by DDAs from Fuzi. Mechanisms behind DDAs-induced arrhythmogenic effects have been well elaborated [1,12–15], and the relationship between DDAs concentrations in heart tissue and cardiac toxicity has been established [24]. However, only few studies have mentioned the hepatic and renal toxicities of Fuzi [16,18]. Furthermore, there has been a lack of information on concentrations of major toxic alkaloids in liver and kidney after both short-term and long-term intake of Fuzi preparations. Therefore, the current study for the first time investigated the biodistributions of the six DDAs and MDAs from commonly used processed Fuzi preparations after both single and multiple oral administrations in rats at the their clinically relevant doses. Such preclinical tissue distribution profiles were expected to provide us with better understanding of in vivo distribution behaviors of these alkaloids for more in-depth investigation of their toxicities and safer use of Fuzi. The current LC/MS/MS analytical method achieved simultaneously detection of all six toxic aconitum alkaloids across urine, plasma, and four major organs, providing a wider range of application with higher efficiency than previous analytical methods which were applied to detect fewer types of toxic alkaloids in urine or plasma only [29,30].

Different contents of six alkaloids in the two studied Fuzi preparations were highly related to different processing procedure of Heishunpian and Paofupian. Paofupian had one more step of roasting with sands to do from Heishunpian. Long time high temperature condition facilitated the hydrolyzation of DDAs and MDAs to MDAs and NDAs [10]. As a result, there were less DDAs and MDAs in Paofupian, explaining why traditional Chinese medicine prefer Paofupian for long-term treatment. Percentages of DDAs in Heishunpian and Paofupian were determined to be $2.46 \times 10^{-5}\%$, and $4.11 \times 10^{-7}\%$ respectively. The DDA contents in both two preparations were much less than 0.01%—the criteria listed in 2015 Chinese Pharmacopoeia [31]. Therefore, the tested Heishunpian and Paofupian met the criteria in 2015 Chinese Pharmacopoeia. In contrary to the extremely low contents of DDAs, the contents of MDAs were 300 and 15,000 folds higher in Heishunpinan and Paofupian respectively, leading to relatively high dose of the toxic MDAs in this two Fuzi preparation.

Our biodistribution data for the first time revealed the effect of long-term dose of Fuzi preparations (both Heishunpian and Paofupian) on the biodistribution profiles of toxic alkaloids. It was found that long-term exposure of Fuzi preparation could lead to significant accumulation of MDAs in liver, kidney and heart, while no accumulation was found for DDAs. Since the phase I and phase II metabolic rate of MDAs in human liver microsome are less than 10% [32], the accumulation of MDAs in the major organs may not be related to altered metabolic enzyme activities. On the other hand, altered activities of transporters [33] and impaired organ functions might be the reasons responsible for such accumulation, which warrants further investigation on the underlining mechanism. Our novel findings on the accumulation of MDAs in liver, kidney and heart after ingestion of Fuzi preparations suggested higher risk of organ injury when Fuzi was used for sub-chronic treatment over 15 days.

As shown by the contents of toxic alkaloids in the two Fuzi preparations (Figure 3), the doses of MDAs were 300–15,000 folds higher than those of DDAs. As a result of the high doses, MDAs demonstrated the remarkably higher amounts in plasma, urine and major organs regardless of the treatment duration compared with DDAs. Based on the reported LD_{50} for a single oral dose of MDAs in mice (0.81 g/kg for BMA, 1.50 g/kg for BAC, and 0.83 g/kg for BHA [7]), MDAs were also toxic and lethal. The high in vivo exposure and the marked cumulation of MDAs in organs suggested the toxicity of MDAs warranted further attention and investigation.

In addition, our findings also revealed that the biodistributions of six toxic alkaloids from the two Fuzi preparations still differed from each other even after dose normalization, (Tables 2 and 3). Such difference between Heishunpian and Paofupian was highly possibly due to composition and relative ratio changes of co-occurring components existing in Heishunpian and Paofupian. Different processing procedures and substances introduced during decoction and roasting, such as sand and salt, may give further explanation of co-occurring component's source in Heishunpian and Paofupian.

It was also noted that the dose-normalized contents of DDAs were higher than those of MDAs in majority of studied organs (Tables 2 and 3). One of the explanations for such phenomena could be the higher lipophilicities of DDAs compared with that of MDAs (calculated LogP by ALOGPS 2.1 software of studied six toxic alkaloids are 1.68 for AC vs 1.09 for BAC, 2.07 for HA vs 1.35 for BHA, and 1.37 for MA vs 0.69 for BMA,), which may potentially result in higher membrane permeability, higher tissue binding affinity and less excretion. Moreover, the toxicities of DDAs was around 1000-fold stronger than MDAs, according to their LD_{50} values [7]. Therefore, due to their higher recoveries in organs and more potent toxicity compared with MDAs, the quality control on the contents of DDAs was essential to avoid Fuzi-induced toxicity. Apart from quality control, sufficient decoction time of Fuzi crude herb which allowed effective hydrolysis of DDAs to MDAs was necessary.

Our findings on higher possibility of toxic alkaloids occurrence in urine sample when compared to plasma sample provided a reasonable explanation of a previous clinical poisoning case [1] reported that concentration of toxic alkaloids in urine was much higher than those in blood and supported urine test for Fuzi poisoning diagnosis in clinic.

Last but not the least, current findings confirmed that the toxic alkaloid levels were remarkably high in the liver and kidney, relatively low in the heart and blood with only a trace amount recovered in the cerebrum, consistent with previous biodistribution result in clinical poisoning cases [34] and preclinical studies [25,26]. However, it should be emphasized that they gave extremely high dose of Fuzi extract ranging from 0.2–10 g/kg [35,36] or pure toxic alkaloid such as AC or HA ranging from 0.1–2 mg/kg [37,38] to rats. Differently, 30 g/kg of Fuzi used in our study referred to reported maximum dose in clinics [39]. Different from others using organic solvent to extract Fuzi [40], current study mimicked patient's intake habit that suspending the granule in hot water and stirring well. Hepatotoxicity and nephrotoxicity caused by Fuzi have been previously evidenced by upregulated cell apoptosis factor [16], dose-dependent edema in liver tissue, high level expression of Alanine Aminotransferase (ALT), Aminotransferase (AST) and Lactate Dehydrogenase (LDH) in serum [41] after oral administrations of Fuzi extract or pure toxic alkaloid compounds. Current results gave a warning that hepatotoxicity and nephrotoxicity caused by Fuzi preparations—no matter if taken at one time or multiple times—are worthy of closer attention in clinical practice. Our current study revealed the remarkably high tendency of the toxic alkaloids to distribute into the liver after oral ingestion of Fuzi at a clinically relevant dosing regimen, which could partially explain the hepatotoxicity reported in previously conducted studies.

4. Conclusions

The current developed and optimized LC/MS/MS method for quantitative determination of the six toxic aconitum alkaloids was specific and sensitive for biodistribution study in rats. Among the six toxic alkaloids, BMA had the highest content in the investigated Fuzi preparation, which also led to its highest content in rat tissues.

Our current study for the first time demonstrated not only the remarkably high tendency of toxic aconitum alkaloids distributing into the liver and kidney, but also their significant accumulation in major organs after long-term oral administrations of the Heishunpian and Paofupian concentrated granules at clinically relevant dosage to rats. Therefore, clinical use of Fuzi in patients with sub-chronic and even chronic disease may need more precise adjustment on dosage and more attention to liver injury induced by Fuzi. Additionally, our study revealed that co-occurring components existing in Heishunpian increase the in vivo exposure and tissue content of six toxic aconitum alkaloids existing in Fuzi.

5. Materials and Methods

5.1. Materials, Reagents and Animals

All the reagents used in the current study were at least of analytical grade. Formic acid was supplied by BDH Laboratory Supplied Ltd. (Kampala, Ukraine). Acetonitrile and methanol were purchased from RCI Laboscan Ltd. (Bangkok, Thailand) and both of them were HPLC-grade. Deionized water was used for the preparation of all solutions. Six toxic aconitum alkaloids, including aconitine, hypaconitine, mesaconitine, benzoylaconine, benzoylhypaconine and benzoylmesaconine, and Berberine hydrochloride which was used as internal standard (IS) were all purchased from Sigma-Aldrich (St. Louis, MO, USA). Heishunpian or Paofupian concentrated granule were obtained from Purapharm Co., Ltd. (Hong Kong SAR, P. R. China, Batch Number A1601163 for Heishunpian, A1701434 for Paofupian).

Adult male Sprague Dawley rats (weighing 200–220 g) were supplied by the Laboratory Animal Services Centre, the Chinese University of Hong Kong, HKSAR, China. This study was approved on 9 February 2018 by the Animal Experimentation Ethics Committee of the university (Reference No. 17/219/HMF-5-B).

5.2. Chromatographic and Mass Spectrometric Conditions

Separation and quantification of DDAs including Aconitine (AC), Hypaconitine (HA), Mesaconitine (MA), MDAs including Benzoylaconine (BAC), Benzoylhypaconine (BHA), Benzoylmesaconine (BMA) and internal standard (Berberine) were performed on Agilent 1290 Ultrahigh performance liquid chromatograph coupled to an Agilent 6430 Triple Quad LC/MS (UPLC–MS/MS) with electrospray ionization (ESI) (Agilent Technologies Inc., Santa Clara, CA, USA). A Waters Acquity UPLC BEH C18 column (2.1 × 50 mm, 1.7 μm, Waters Corporation, Milford, MA, USA) was used for chromatographic separation. The mobile phase was composed of an aqueous solution of 0.1% formic acid (Solvent A) and acetonitrile containing 0.1% formic acid (Solvent B). The gradient profile was as follows: 0–4 min with a linear gradient of B from 10 to 45 % and 4–8 min with a linear gradient of B from 45 to 70%, for 8–9 min the composition of B is maintained at 70%, finally the column was re-equilibrated. The flow rate was 0.15 mL/min. The injection volume was 10 μL. The entire eluent was ionized via an ESI source operating in the positive mode and monitored by MS/MS detection in the multiple reaction monitoring (MRM) mode. The optimization of the MS/MS conditions were conducted by directly injecting the individual analyte solutions in methanol at a concentration of 500 ng/ml with a mobile phase composition and flow rate equivalent to those at the time the analyte would elute from the UPLC column.

5.3. Preparation of Standard Solution and Quality Control Samples

Tested alkaloid (1 mg) was dissolved in 1 mL 50% methanol in water (*v/v*, containing 2.5% formic acid) to obtain a stock solution (1 mg/mL). The stock solutions of each tested alkaloid were then mixed accordingly to obtain a stock solution containing 1 mg/mL of each toxic alkaloid.

The stock solutions with mixed alkaloids were diluted with 50 % methanol in water (*v/v*, containing 2.5% formic acid, for quality control (QC) samples in granule content determination) or deionized water (for QC samples in rat tissue content detection to a series of concentration level ranging from 0.5–1000 ng/mL). Detailed concentration levels depended on sample content in various matrix to be tested as shown in Tables A2–A7.

5.4. Sample Preparations

All the collected organs were homogenized with two volumes normal saline containing 1 mmol/L hydrochloric acid to obtain the relevant homogenate.

After centrifuging blood at 8000 rpm for 3 min, 300 μL plasma was obtained for further protein precipitation with 600 μL acetonitrile according previous description [42]. In addition, collected urine sample (300 μL) and organ homogenate from kidney (300 μL) and heart (300 μL) were also subject to precipitate protein with 600 μL acetonitrile. All the above mixtures were centrifuged at 13,000 rpm for 10 min to obtain the supernatant.

For liver and brain sample preparations, 300 μL of their homogenate was loaded to MCX cartridge (Oasis® Part Number 186000252). After washing the cartridge with 1 mL of 0.1 M HCl in H_2O followed by 1 mL of methanol, the analytes were eluted with 750 μL methanol containing 5% ammonium hydroxide.

All the supernatant or eluent resulted from above sample treatment experienced dryness under a stream of nitrogen gas. The residues were reconstituted with 100 μL 50% methanol containing 2.5% formic acid with 10 μL of which injected to LC/MS/MS for analyses of the six toxic alkaloids.

5.5. LC/MS/MS Method Validation

The method developed was validated based on the guidelines provided by U.S. Food and Drug Administration (FDA) [28], including specificity, linearity, lower limit of quantification (LLOQ), accuracy, precision, recovery, matrix effect, and stability.

5.5.1. Specificity

The specificity of the method was conducted by comparing the chromatographs of blank rat plasma, urine and organ homogenate samples with that of blank rat samples spiked with standard solutions and rat plasma/brain homogenate after oral administration of Heishunpian or Paofupian.

5.5.2. Linearity and Sensitivity

Calibration samples were prepared as described in Section 5.3. The calibration curves were obtained by plotting the ratio of peak area of six toxic alkaloids to IS against the concentration of six toxic alkaloids. The coefficient of determination (r^2) was calculated, which with a value greater than 0.99 was considered as an indicator of good linearity. The lower limit of quantification (LLOQ) was defined as the lowest concentration of the calibration curve with a signal-to-noise (S/N) peak ratio greater than 5:1.

5.5.3. Accuracy and Precision

The intra-day accuracy and precision test were conducted by analyzing QC samples at nominal concentrations (described in Tables A2–A7) with at least 5 replicates for each concentration within one day. The inter-day accuracy and precision were determined on three days separately. Accuracy and precision should be within ± 15% bias and 15% RSD respectively.

5.5.4. Recovery and Matrix Effect

The recovery of extraction was determined by the peak area of six toxic alkaloids spiked with the biological matrices followed by extraction against the peak area of six toxic alkaloids spiked to the extracted biological matrices. The matrix effects caused by different biological matrix including plasma, urine, liver, kidney, brain and heart homogenate were calculated by comparing the standard curve in above biomatrix with that in 50% methanol in water (*v/v*, containing 2.5% formic acid). Recoveries and matrix effects of six toxic alkaloids at the levels of Limit of Quantification (LOQs), Middle of Quantification (MOQs) and High of Quantification (HOQs) should be consistent and reproducible.

5.5.5. Stability

Freeze-thaw stability test was conducted by putting the QC samples (prepared in Section 5.3) to three cycles of freeze (−80 °C) thaw (25 °C) before extraction. The stability of six toxic alkaloids at bench top and in the auto-sampler were conducted after sample extraction on the bench (25 °C) for 4 h or in the auto-sampler (8 °C) for 12 h respectively. Long-term stability of six toxic alkaloids was tested after keeping sample at −80 °C for 30 days.

5.6. Application of the Developed LC/MS/MS Method for Biodistribution Study of Radix Aconiti Lateralis Preparations in Rats

5.6.1. Quality Control of the Studied Radix Aconiti Lateralis Preparations

Heishunpian or Paofupian concentrated granules are most frequently prescribed Fuzi processing products in local clinics. Based on content analyses of DDAs in Fuzi processed products in 2015 edition

Chinese Pharmacopoeia [29] and the concentrated ratio of current Heishunpian or Paofupian granules to its processed herb equals (1:5), the contents of DDAs in the granules were analyzed as follows to compare with the criteria in 2015 Chinese Pharmacopoeia.

About 5 g concentrated granules were firstly suspended in 10 mL 50% Methanol which containing 2.5% formic acid, then sonicated at 25 °C for 30 min. Filtrating through 0.45 μm polypropylene filter, filtrate was divided into two part. One of filtrate (200 μL) was mixed with 200 μL 20 ng/mL internal standard then injected into LC/MS/MS spectrometry, focusing on DDAs quantification. The other filtrate need be 20 times diluted before mixing with internal standard, focusing on MDAs detection. The content of DDAs come from per gram Fuzi processed herb will be indicated as DDAs percentage (%) and calculated by following Equations (1–2) for its comparison with those criteria for Fuzi processed product:

$$\text{DDAs percentage }(\%) = \frac{\text{Content of DDAs in per gram concentrated granule}}{\text{Concentrated ratio}} \times 100\% \quad (1)$$

$$\text{Concentrated ratio} = 5 \quad (2)$$

5.6.2. Animal Studies

Commercially available concentrated granules of Heishunpian and Paofupian were orally given to Sprague Dawley rats (n = 8 for each group) at a bolus of 6 g/kg (equivalent to 30 g/kg crude herb) for both one time and once daily lasting 15 days.

To mimic clinical use of Fuzi preparations, 5 g concentrated granule of Heishunpian and Paofupian (Batch number A1601163 and A1701434 respectively) were suspended in 10 mL boiling water followed by sonication until evenly suspense in water. Since reported cardiotoxicity usually happened at 1–2 h post-dosing [2,14], two hours after the single dose of Fuzi preparations or two hours after the oral administration of Fuzi preparations on the last day for multiple dosing, rats were sacrificed followed by collection of blood and organs including liver, kidney, heart, and brain after cardiac perfusion with 150–200 mL saline. For single dose of Heishunpian or Paofupian, urine from each rat was collected using metabolic cage during the period from dosing to sacrificing. For multiple dose of above Fuzi preparations, urine from each rat was collected using metabolic cage during the period from dosing on Day 15 to sacrificing. All collected samples were treated as described in Section 5.4 and analyzed by the developed and validated LC/MS/MS method shown in Section 5.2.

5.7. Data Analyses

Toxic alkaloid contents were expressed as the mean ± standard deviation. To give further explanation to biodistribution results, dose-normalized content of toxic alkaloid in each organ is calculated using following Equations (3)–(5). For Paofupian treated group, the dose on last day was served as dosed alkaloid content to calculate dose-normalized content of toxic alkaloid in each tissue.

Partition coefficient parameter (LogP) of studied six toxic alkaloids from Fuzi was calculated using ALOGPS 2.1 software.

$$\text{Dose normalized content of toxic alkaloid} = \frac{\text{Detected Alkaloid content}}{\text{Dosed Alkaloid content}} \quad (3)$$

$$\text{Detected Alkaloid content (ng)} = \text{detected Alkaloid conc.(ng/g)} \times \text{tissue weight (g)} \quad (4)$$

$$\text{Dosed Alkaloid content (ng)} = \text{granule alkaloid conc.(ng/g)} \times \text{dosed granule content (g)} \quad (5)$$

Author Contributions: Conceptualization, Z.Z., K.H.O., W.S.Y. and M.Y.; methodology, X.J. and M.Y.; formal analysis, X.J. and M.Y; writing—original draft preparation, X.J.; writing—review and editing, Z.Z., M.Y and X.J.

Funding: This research was funded by Health and Medical Research Fund (15161541) by the Food and Health Bureau, Hong Kong SAR, P. R. China.

Conflicts of Interest: The authors declare no conflict of interest.

Appendix A

Table A1. Lower Limit of Quantification (LLOQ) and linearity of AC, HA, MA, BAC, BHA and BMA in rat heart, liver, brain, urine, kidney and plasma by the current developed LC/MS/MS method.

Tissue	Heart				Liver				Brain				Urine				Kidney				Plasma			
Compound	LLOQ ng/mL	RSD %	RE %	Linear Range ng/mL	LLOQ ng/mL	RSD %	RE %	Linear Range ng/mL	LLOQ ng/mL	RSD %	RE %	Linear Range ng/mL	LLOQ ng/mL	RSD %	RE %	Linear Range ng/mL	LLOQ ng/mL	RSD %	RE %	Linear Range ng/mL	LLOQ ng/mL	RSD %	RE %	Linear Range ng/mL
AC	1	16.24	−2.06	1–20	2	5.12	−0.41	2–20	2	3.05	2.35	2–100	2	3.88	5.4	1–200	1	1.46	1.49	1–50	0.5	4.74	16.16	0.5–100
HA	1	12.14	−2.93	1–20	1	1.7	6.19	1–20	2	7.81	0.39	2–100	0.5	[illegible].87	7.27	0.5–200	0.5	2.66	2.44	0.5–50	0.5	7.30	5.84	0.5–100
MA	1	8.32	0.01	1–20	1	1.74	17.21	1–20	2	7.53	0.23	2–100	0.5	[illegible].48	0.47	0.5–200	0.5	12.03	−8.39	0.5–50	0.5	9.89	5.79	0.5–100
BAC	1	13.46	0.11	1–20	1	6.42	9.14	1–20	2	5.87	−7.78	2–100	0.5	[illegible].67	−2.25	0.5–200	0.5	9.32	2.67	0.5–50	0.5	5.72	5.89	0.5–100
BHA	1	2.27	7.05	1–20	1	3.89	10.02	1–20	2	10.18	−5.5	2–100	0.5	5.1	−12.24	0.5–200	0.5	17.27	1.1	0.5–50	0.5	1.18	−1.64	0.5–100
BMA	1	4.99	11.42	1–20	1	5.39	2.83	1–20	2	11.83	−1.03	2–100	2	5.13	−4.32	2–200	0.5	10.62	−0.05	0.5–50	0.5	1.22	5.52	0.5–100

Table A2. Summary of accuracy, precision, recovery, matrix effect, and stability information of AC, HA, MA, BAC, BHA and BMA in rat heart homogenate ($n = 5$).

Analytes	Concentration (ng/mL)	Intra-Day (%, RSD)	Inter-Day (%, RSD)	Accuracy (%, RE)	Absolute Recovery (%, Mean ± SD)	Matrix Effect (%, RSD)	Stability (%, Mean ± SD)			
							4 h at Room Temperature	12 h at 8 °C	3 Freeze-Thaw Cycles	30 Days at −80 °C
AC	2	11.21	5.32	2.11	48.83 ± 11.47	10.04	66.64 ± 12.85	71.12 ± 9.60	105.41 ± 12.97	69.46 ± 0.48
	5	0.04	1.55	0.03	36.19 ± 4.35	3.81	N.A.	N.A.	N.A.	N.A.
	10	0.08	3.99	0.95	35.58 ± 1.53	8.72	88.50 ± 10.76	72.77 ± 7.56	109.17 ± 4.33	129.82 ± 5.89
HA	2	11.25	1.31	−11.08	52.47 ± 4.89	11.25	80.32 ± 4.42	97.99 ± 14.73	101.72 ± 14.32	94.47 ± 6.18
	5	6.54	4.67	2.95	48.65 ± 2.96	6.54	N.A.	N.A.	N.A.	N.A.
	10	10.52	8.86	−13.62	34.06 ± 1.17	9.89	92.43 ± 9.29	75.79 ± 9.84	107.78 ± 7.09	127.00 ± 5.00
MA	2	8.71	6.65	−3.11	27.08 ± 3.41	9.26	68.43 ± 3.93	90.60 ± 13.11	98.31 ± 7.88	103.54 ± 1.54
	5	9.37	7.02	0.78	22.02 ± 0.91	8.86	N.A.	N.A.	N.A.	N.A.
	10	10.92	2.07	−1.19	15.93 ± 0.10	11.42	82.54 ± 9.02	80.64 ± 8.44	94.76 ± 1.83	125.96 ± 4.38
BAC	2	11.06	3.06	1.76	55.06 ± 9.84	11.06	104.47 ± 10.69	146.84 ± 11.79	86.81 ± 5.92	86.34 ± 6.75
	5	7.74	5.39	−6.22	36.39 ± 4.63	7.74	N.A.	N.A.	N.A.	N.A.
	10	5.17	6.85	2.49	29.75 ± 2.12	5.17	115.46 ± 6.90	157.93 ± 13.13	111.46 ± 7.11	105.17 ± 2.94
BHA	2	7.2	7.3	1.22	38.20 ± 7.50	7.2	102.36 ± 3.86	146.72 ± 5.14	94.32 ± 11.40	89.70 ± 2.61
	5	9.19	7.8	9.41	25.50 ± 2.69	9.19	N.A.	N.A.	N.A.	N.A.
	10	6.24	3.76	2.16	19.91 ± 1.68	6.24	86.60 ± 10.07	131.51 ± 3.44	97.57 ± 6.87	105.87 ± 1.33
BMA	2	5.56	7.73	−10.38	59.61 ± 4.38	5.14	92.84 ± 5.27	185.65 ± 14.98	118.67 ± 6.02	69.01 ± 4.47
	5	11	6.58	10.14	60.62 ± 1.69	13.23	N.A.	N.A.	N.A.	N.A.
	10	6.59	2.29	−0.44	51.37 ± 2.95	6.59	78.04 ± 5.01	113.08 ± 5.77	73.44 ± 4.82	139.14 ± 10.84

N.A.: No requirements for stability test on analytes' Middle of Quantification (MOQs) in U.S. FDA [28] Guidelines.

Table A3. Summary of accuracy, precision, recovery, matrix effect and stability information of AC, HA, MA, BAC, BHA and BMA in rat liver homogenate (n = 5).

Analytes	Concentration (ng/mL)	Intra-Day (%, RSD)	Inter-Day (%, RSD)	Accuracy (%, RE)	Absolute Recovery (%, Mean ± SD)	Matrix Effect (%, RSD)	Stability (%, Mean ± SD)			
							4 h at Room Temperature	12 h at 8 °C	3 Freeze-Thaw Cycles	30 Days at −80 °C
AC	5	10.83	9.21	−0.74	16.27 ± 1.76	10.83	109.37 ± 9.10	95.40 ± 4.17	119.10 ± 11.54	123.35 ± 14.20
	10	4.2	2.04	2.89	15.44 ± 0.65	4.2	N.A.	N.A.	N.A.	N.A.
	50	1.8	0.06	0.6	16.85 ± 0.30	1.8	88.31 ± 5.74	99.15 ± 2.97	91.81 ± 2.92	98.03 ± 14.28
HA	5	6.17	5.05	1.36	17.48 ± 1.08	6.17	101.74 ± 2.56	91.27 ± 6.35	108.72 ± 3.18	89.69 ± 4.36
	10	7.85	0.08	0.07	17.00 ± 1.33	7.85	N.A.	N.A.	N.A.	N.A.
	50	1.75	0.06	−3.88	18.35 ± 0.32	1.75	106.79 ± 1.71	95.52 ± 2.14	110.29 ± 4.89	95.45 ± 13.53
MA	5	6.08	3	−2.47	16.46 ± 1.00	6.08	101.76 ± 0.57	93.24 ± 4.46	120.36 ± 6.72	92.81 ± 10.90
	10	2.75	8.07	-0.01	15.78 ± 0.43	2.75	N.A.	N.A.	N.A.	N.A.
	50	3.52	0.05	0.14	16.94 ± 0.60	3.52	94.11 ± 4.29	96.41 ± 3.02	99.94 ± 4.80	92.27 ± 12.05
BAC	5	9.37	2.81	−1.07	2.64 ± 0.25	9.37	138.93 ± 7.85	85.36 ± 8.33	141.29 ± 11.86	127.23 ± 9.12
	10	8.85	0.53	2.23	3.38 ± 0.30	8.85	N.A.	N.A.	N.A.	N.A.
	50	2.65	3.04	−0.23	2.96 ± 0.08	2.65	97.68 ± 5.58	97.81 ± 4.03	100.15 ± 4.94	106.27 ± 6.59
BHA	5	5.69	6.67	9.08	2.60 ± 0.15	5.69	141.14 ± 9.36	93.21 ± 7.16	161.20 ± 10.89	137.34 ± 12.78
	10	11.98	5.63	−0.17	4.22 ± 0.51	11.98	N.A.	N.A.	N.A.	N.A.
	50	3.61	3.99	−5.59	3.12 ± 0.11	3.61	106.85 ± 3.50	95.41 ± 2.42	112.54 ± 3.40	140.18 ± 14.86
BMA	5	5.51	2.22	6.12	2.47 ± 0.14	5.51	104.19 ± 9.20	87.37 ± 3.96	113.15 ± 9.93	100.35 ± 11.19
	10	5.84	4.62	−3.09	2.71 ± 0.16	5.84	N.A.	N.A.	N.A.	N.A.
	50	2.78	5.54	−10.32	2.36 ± 0.07	2.78	106.43 ± 5.48	93.28 ± 4.63	93.12 ± 0.68	86.70 ± 6.88

N.A.: No requirements for stability test on analytes' Middle of Quantification (MOQs) in U.S. FDA [28] Guidelines.

Table A4. Summary of accuracy, precision, recovery, matrix effect and stability information of AC, HA, MA, BAC, BHA and BMA in rat brain homogenate (n = 5).

Analytes	Concentration (ng/mL)	Intra-Day (%, RSD)	Inter-Day (%, RSD)	Accuracy (%, RE)	Absolute Recovery (%, Mean ± SD)	Matrix Effect (%, RSD)	Stability (%, Mean ± SD)			
							4 h at Room Temperature	12 h at 8 °C	3 Freeze-Thaw Cycles	30 Days at −80 °C
AC	2	5.55	15.2	−1.88	27.48 ± 1.53	5.55	95.56 ± 10.84	100.24 ± 12.29	105.91 ± 5.19	114.35 ± 9.72
	5	7.87	6.76	−1.72	34.63 ± 2.73	7.87	N.A.	N.A.	N.A.	N.A.
	10	8.97	3.06	−1.17	32.03 ± 2.87	8.97	98.68 ± 6.63	105.49 ± 7.80	101.30 ± 4.77	111.57 ± 11.56
HA	2	4.35	15.2	0.17	33.56 ± 1.46	4.35	93.24 ± 5.68	100.36 ± 7.15	98.89 ± 3.59	97.16 ± 6.95
	5	2.9	2.02	−4.82	31.15 ± 0.90	2.9	N.A.	N.A.	N.A.	N.A.
	10	4.34	1.51	3.29	33.46 ± 1.45	4.34	104.40 ± 3.42	104.03 ± 5.74	101.13 ± 4.34	104.66 ± 8.08
MA	2	3.82	15.19	0.42	26.02 ± 0.99	3.82	94.97 ± 8.67	100.26 ± 9.16	100.46 ± 0.77	90.20 ± 10.51
	5	3.2	1.59	−13.39	23.74 ± 0.76	3.2	N.A.	N.A.	N.A.	N.A.
	10	4.65	3.26	−0.49	25.60 ± 1.19	4.65	100.93 ± 5.61	104.19 ± 6.67	101.38 ± 2.04	93.22 ± 7.50
BAC	2	3.38	18.14	−0.37	4.80 ± 0.16	3.38	109.55 ± 14.53	100.56 ± 6.29	105.67 ± 18.95	225.25 ± 1.44
	5	9.9	6.8	−8.87	4.46 ± 0.44	9.9	N.A.	N.A.	N.A.	N.A.
	10	4.08	6.48	−4.88	5.78 ± 0.24	4.08	103.91 ± 5.44	96.30 ± 2.65	90.73 ± 6.52	184.82 ± 11.92
BHA	2	3.22	7.16	3.48	5.21 ± 0.17	3.22	89.92 ± 7.02	95.97 ± 8.77	86.65 ± 10.79	166.17 ± 14.83
	5	6.26	4.47	−13.7	3.65 ± 0.23	6.26	N.A.	N.A.	N.A.	N.A.
	10	4.5	4.28	−1.36	7.42 ± 0.33	4.5	101.38 ± 1.24	100.01 ± 6.63	88.47 ± 3.88	1&7.87 ± 4.76
BMA	2	9.48	15.56	−1.38	4.15 ± 0.39	9.48	144.55 ± 14.24	94.36 ± 16.14	98.54 ± 6.93	229.74 ± 7.66
	5	3.12	5.88	−2.41	3.97 ± 0.12	3.12	N.A.	N.A.	N.A.	N.A.
	10	3.52	2.54	−3.12	4.74 ± 0.17	3.52	83.7 ± 2.74	105.89 ± 5.97	83.46 ± 5.53	136.53 ± 5.32

N.A.: No requirements for stability test on analytes' Middle of Quantification (MOQs) in U.S. FDA [28] Guidelines.

Table A5. Summary of accuracy, precision, recovery, matrix effect and stability information of AC, HA, MA, BAC, BHA and BMA in rat urine ($n = 5$).

Analytes	Concentration (ng/mL)	Intra-Day (%, RSD)	Inter-Day (%, RSD)	Accuracy (%, RE)	Absolute Recovery (%, Mean ± SD)	Matrix Effect (%, RSD)	Stability (%, Mean ± SD)			
							4 h at Room Temperature	12 h at 8 °C	3 Freeze-Thaw Cycles	30 Days at −80 °C
AC	2	3.88	4.67	5.4	84.41 ± 4.15	3.88	87.05 ± 13.58	126.77 ± 4.77	79.91 ± 12.50	73.12 ± 7.69
	10	10.24	11.5	−2.94	93.28 ± 5.62	10.24	N.A.	N.A.	N.A.	N.A.
	50	2.7	1.87	0.31	98.79 ± 3.53	2.7	117.59 ± 8.17	101.92 ± 6.69	115.14 ± 5.79	109.04 ± 3.36
HA	2	6.11	3.87	3.15	100.06 ± 1.70	6.11	76.89 ± 5.39	95.39 ± 4.69	84.95 ± 1.67	88.36 ± 2.51
	10	6.21	7.69	−10.93	94.11 ± 1.80	6.21	N.A.	N.A.	N.A.	N.A.
	50	3.98	1.62	1.63	101.24 ± 3.59	3.98	113.66 ± 2.41	98.69 ± 1.66	129.99 ± 1.85	107.34 ± 0.85
MA	2	3.5	5.74	9.99	96.95 ± 3.15	3.5	80.31 ± 1.09	97.99 ± 4.14	81.96 ± 0.65	82.86 ± 1.14
	10	4.12	6.04	−7.28	92.25 ± 2.45	4.12	N.A.	N.A.	N.A.	N.A.
	50	6.59	3.77	0.46	98.47 ± 3.07	6.59	111.56 ± 0.51	98.19 ± 2.31	119.99 ± 1.65	103.40 ± 3.53
BAC	2	3.73	6.12	9.14	106.46 ± 2.39	3.73	89.96 ± 13.19	100.22 ± 10.17	85.13 ± 11.07	81.30 ± 6.12
	10	7.11	3	0.47	99.19 ± 0.80	7.11	N.A.	N.A.	N.A.	N.A.
	50	3.3	11.7	2.59	103.41 ± 1.48	3.3	102.55 ± 2.13	122.66 ± 4.15	111.31 ± 0.85	105.31 ± 3.33
BHA	2	3.81	8.28	5.11	98.27 ± 2.68	3.81	76.31 ± 8.32	99.16 ± 6.30	94.79 ± 3.84	96.08 ± 4.23
	10	5.89	4.05	−1.71	96.69 ± 1.86	5.89	N.A.	N.A.	N.A.	N.A.
	50	7.29	6.7	−1.64	100.71 ± 2.40	7.29	96.92 ± 1.62	116.89 ± 2.81	135.02 ± 1.75	101.46 ± 2.66
BMA	10	4.99	7.41	13.21	100.05 ± 2.61	4.99	83.77 ± 4.06	103.23 ± 4.57	92.87 ± 11.00	21.21 ± 3.08
	50	8.53	8.53	0.27	104.94 ± 1.96	8.53	N.A.	N.A.	N.A.	N.A.
	100	6.52	6.52	3.07	110.31 ± 3.53	6.52	109.94 ± 1.16	126.82 ± 1.82	103.55 ± 5.31	90.84 ± 0.72

N.A.: No requirements for stability test on analytes' Middle of Quantification (MOQs) in U.S. FDA [28] Guidelines.

Table A6. Summary of accuracy, precision, recovery, matrix effect and stability information of AC, HA, MA, BAC, BHA and BMA in rat plasma (n = 5).

Analytes	Concentration (ng/mL)	Intra-Day (%, RSD)	Inter-Day (%, RSD)	Accuracy (%, RE)	Absolute Recovery (%, Mean ± SD)	Matrix Effect (%, RSD)	Stability (%, Mean ± SD)			
							4 h at Room Temperature	12 h at 8 °C	3 Freeze-Thaw Cycles	30 Days at –80 °C
AC	2	6.61	5.34	−2.34	71.28 ± 4.71	6.61	101.30 ± 7.24	100.54 ± 0.29	112.76 ± 9.08	164.14±8.87
	5	3.98	8.03	-6.10	67.46 ± 2.69	3.98	N.A.	N.A.	N.A.	N.A.
	20	5.79	3.74	5.20	82.80 ± 4.80	5.79	100.21 ± 3.55	88.89 ± 4.26	108.61 ± 3.01	108.22 ± 0.03
HA	2	3.23	5.56	−9.89	67.00 ± 2.16	3.23	125.20 ± 7.65	100.01 ± 3.38	119.58 ± 1.07	118.18±7.50
	5	2.19	8.19	−14.04	59.65 ± 0.53	0.88	N.A.	N.A.	N.A.	N.A.
	20	4.03	6.06	−1.04	69.54 ± 2.80	4.03	133.03 ± 4.53	98.81 ± 4.70	121.92 ± 5.95	114.49 ± 2.00
MA	2	5.31	5.57	−13.57	71.34 ± 3.79	5.31	118.52 ± 11.80	80.92 ± 3.75	103.79 ± 1.63	99.84 ± 5.24
	5	2.00	2.05	−6.56	88.82 ± 1.78	2.00	N.A.	N.A.	N.A.	N.A.
	20	3.74	5.04	8.29	92.04 ± 3.45	3.74	115.71 ± 4.75	87.06 ± 4.38	106.95 ± 1.82	98.85 ± 0.90
BAC	2	3.24	4.98	−3.11	82.85 ± 2.67	3.23	114.26 ± 6.75	113.52±9.71	110.02 ± 5.14	79.65 ± 3.17
	5	3.63	7.28	2.41	111.31 ± 4.04	3.63	N.A.	N.A.	N.A.	N.A.
	20	2.55	3.27	−8.37	90.32 ± 2.31	2.55	123.28 ± 5.74	95.37 ± 0.09	114.72 ± 2.84	103.64 ± 2.26
BHA	2	7.46	7.65	9.60	80.58 ± 6.01	7.46	111.19 ± 14.68	94.49 ± 3.94	108.49 ± 3.62	98.50 ± 5.31
	5	4.72	8.91	−2.83	95.82 ± 4.52	4.72	N.A.	N.A.	N.A.	N.A.
	20	4.77	5.51	−6.23	87.84 ± 4.19	4.77	122.77 ± 4.71	89.36 ± 1.46	112.25 ± 3.59	102.16 ± 2.74
BMA	2	5.25	14.89	8.22	68.49 ± 1.99	2.91	109.38 ± 4.75	132.63 ± 9.45	128.71 ± 14.56	48.11 ± 7.51
	5	9.42	12.40	5.92	100.65 ± 10.12	10.06	N.A.	N.A.	N.A.	N.A.
	20	2.52	2.15	−13.54	94.23 ± 2.39	2.91	117.83 ± 5.66	95.58 ± 1.79	106.33 ± 5.24	98.74 ± 8.44

N.A.: No requirements for stability test on analytes' Middle of Quantification (MOQs) in U.S. FDA [28] Guidelines.

Table A7. Summary of accuracy, precision, recovery, matrix effect and stability information of AC, HA, MA, BAC, BHA and BMA in rat kidney homogenate ($n = 5$).

Analytes	Concentration (ng/mL)	Intra-Day (%, RSD)	Inter-Day (%, RSD)	Accuracy (%, RE)	Absolute Recovery (%, Mean ± SD)	Matrix Effect (%, SD)	Stability (%, Mean ± SD)			
							4 h at Room Temperature	12 h at 8 °C	3 Freeze-Thaw Cycles	30 Days at −80 °C
AC	1	1.46	11.68	1.49	90.67 ± 9.23	3.20	102.31 ± 3.11	53.29 ± 5.38	106.24 ± 8.01	89.62 ± 7.12
	5	10.43	10.43	5.16	79.64 ± 2.48	0.45	N.A.	N.A.	N.A.	N.A.
	20	6.53	6.53	4.7	81.40 ± 10.34	1.85	63.75 ± 5.51	44.14 ± 5.58	75.04 ± 6.90	85.86 ± 5.49
HA	1	5.08	10.02	5.77	76.99 ± 8.35	1.60	53.22 ± 2.99	70.71 ± 0.52	47.33 ± 3.15	88.69 ± 8.19
	5	5.22	10.64	−1.03	79.36 ± 6.23	1.07	N.A.	N.A.	N.A.	N.A.
	20	0.46	1.38	0.64	92.53 ± 12.25	2.07	65.23 ± 3.45	82.41 ± 1.36	38.14 ± 1.10	87.13 ± 1.33
MA	1	5.28	10.12	8.75	87.61 ± 5.49	0.95	74.54 ± 1.05	47.25 ± 5.35	92.18 ± 5.12	98.80 ± 5.34
	5	4.78	11.87	−4.37	78.46 ± 5.63	0.87	N.A.	N.A.	N.A.	N.A.
	20	2.18	1.77	1.86	98.72 ± 12.45	1.74	68.62 ± 2.81	34.29 ± 2.14	84.91 ± 0.92	107.02 ± 2.69
BAC	1	9.8	12.28	1.64	101.51 ± 10.51	1.12	106.14 ± 5.48	49.46 ± 4.12	120.73 ± 8.11	103.41 ± 5.29
	5	7.15	7.47	−3.42	91.28 ± 14.16	1.26	N.A.	N.A.	N.A.	N.A.
	20	8.65	1.46	0.91	86.44 ± 13.89	1.21	108.06 ± 12.04	56.57 ± 3.82	127.04 ± 4.56	87.07 ± 5.42
BHA	1	6.11	6.53	1.63	96.88 ± 14.86	1.66	93.91 ± 13.38	63.66 ± 10.36	98.44 ± 13.02	108.47 ± 2.67
	5	9.47	4.83	0.91	72.70 ± 9.55	0.98	N.A.	N.A.	N.A.	N.A.
	20	6.29	1.76	1.32	87.82 ± 14.79	1.33	108.66 ± 2.18	46.05 ± 6.01	120.96 ± 2.55	92.94 ± 3.62
BMA	1	10.16	8.06	−2.43	93.99 ± 11.26	1.75	79.79 ± 1.43	77.59 ± 9.88	96.94 ± 10.13	97.18 ± 7.90
	5	7.74	3.58	3.52	78.27 ± 8.32	0.64	N.A.	N.A.	N.A.	N.A.
	20	3.43	6.06	2.84	76.90 ± 13.40	1.01	97.12 ± 7.38	49.49 ± 10.83	112.43 ± 2.61	82.85 ± 1.04

N.A.: No requirements for stability test on analytes' Middle of Quantification (MOQs) in U.S. FDA [28] Guidelines.

References

1. Chan, T.Y.K. Aconite poisoning. *Clin. Toxicol.* **2009**, *47*, 279–285. [CrossRef] [PubMed]
2. Zhou, G.; Tang, L.; Zhou, X.; Wang, T.; Kou, Z.; Wang, Z. A review on phytochemistry and pharmacological activities of the processed lateral root of Aconitum carmichaelii Debeaux. *J. Ethnopharmacol.* **2015**, *160*, 173–193. [CrossRef] [PubMed]
3. Huang, Y.J. Study on "Shang Han Lun" Dosage and Related Problems. Ph.D. Thesis, Beijing University of Chinese Medicine, Beijing, China, 2007.
4. He, F.; Wang, C.J.; Xie, Y.; Cheng, C.S.; Liu, Z.Q.; Liu, L.; Zhou, H. Simultaneous quantification of nine aconitum alkaloids in *Aconiti Lateralis Radix Praeparata* and related products using UHPLC-QQQ-MS/MS. *Sci. Rep.* **2017**, *7*, 1–12. [CrossRef] [PubMed]
5. Alert on Chinese Medicine from Department of Health. Available online: http://www.cmd.gov.hk/text/gb/important_info/alert.html (accessed on 28 May 2019).
6. Aconitine Poisoning. Poisoning Watch. Volume 4, No. 1. 2011 from Department of Health. Available online: https://www.chp.gov.hk/files/pdf/poisoning_watch_vol4_num1_20111228.pdf (accessed on 28 May 2019).
7. Tong, P.; Wu, C.; Wang, X.; Hu, H.; Jin, H.; Li, C.; Zhu, Y.; Shan, L.; Xiao, L. Development and assessment of a complete-detoxication strategy for Fuzi (lateral root of Aconitum carmichaelii) and its application in rheumatoid arthritis therapy. *J. Ethnopharmacol.* **2013**, *146*, 562–571. [CrossRef] [PubMed]
8. Li, X.J.Y.; Luan, Y.; Sun, R. Comparative Study on Acute Toxicity of Different Components of *Aconiti Lateralis Radix Praeparata* on Normal Mice. *Chin. J. Pharmacovigil.* **2013**, *10*, 583.
9. Chang, H.M.; But, P.P.; Yao, S.C. *Pharmacology and Applications of Chinese Materia Medica*; World Scientific: Singapore, 1986; Volume 1. [CrossRef]
10. Singhuber, J.; Zhu, M.; Prinz, S.; Kopp, B. Aconitum in Traditional Chinese Medicine-A valuable drug or an unpredictable risk? *J. Ethnopharmacol.* **2009**, *126*, 18–30. [CrossRef] [PubMed]
11. Ameri, A. The effects of Aconitum alkaloids on the central nervous system. *Prog. Neurobiol.* **1998**, *56*, 211–235. [CrossRef]
12. Sun, G.B.; Sun, H.; Meng, X.B.; Hu, J.; Zhang, Q.; Liu, B.; Wang, M.; Xu, H.B.; Sun, X.B. Aconitine-induced Ca^{2+} overload causes arrhythmia and triggers apoptosis through p38 MAPK signaling pathway in rats. *Toxicol. Appl. Pharmacol.* **2014**, *279*, 8–22. [CrossRef]
13. Grishchenko, I.I.; Naumov, A.P.; Zubov, A.N. Gating and selectivity of aconitine-modified sodium channels in neuroblastoma cells. *Neuroscience* **1983**, *9*, 549–554. [CrossRef]
14. Feng, Q.; Li, X.Y.; Luan, Y.F.; Sun, S.N.L.; Sun, R. Study on effect of aqueous extracts from aconite on "dose-time-toxicity" relationships in mice hearts. *Zhongguo Zhong Yao Za Zhi* **2015**, *40*, 927–932.
15. Fu, M.; Wu, M.; Qiao, Y.; Wang, Z. Toxicological mechanisms of Aconitum alkaloids. *Pharmazie* **2006**, *61*, 735–741. [PubMed]
16. Lei, H.C.; Yi, J.H.; Liu, T. Observation of hepatocyte apoptosis induced by aconite poisoning. *J. Health Toxicol.* **2004**, *18*, 199–200. [CrossRef]
17. Lei, H.C.; Yi, J.H. Observation of apoptosis in renal tubule epithelial cell after aconitine poisoning. *J. Ind. Health Occup. Dis.* **2005**, *31*, 83–85. [CrossRef]
18. Lei, H.C.; Xiang, W.C.; Shi, L. Apoptosis in brain nerve cells of aconitine poisoning rats. *J. Shandong Med.* **2006**, *46*, 21–22.
19. Mizugaki, M.; Ito, K. Aconite toxins. In *Drugs and Poisons in Humans—A Handbook of Practical Analysis*; Suzuki, O., Watanabe, K., Eds.; Springer Verlag Press: New York, NY, USA, 2005; pp. 456–467.
20. The Compile Commission of Zhonghua Bencao of the State Administration of Traditional Chinese Medicine of the People's Republic of China. *Zhonghua Bencao*; Shanghai Science and Technology Press: Shanghai, China, 1999; Volume 5, pp. 101–120.
21. Xiao, P.G. *New Chinese Materia Medica (Xin Bian Zhong Yao Zhi)*; Chemical Industry Press: Beijing, China, 2002; Volume 1, pp. 536–541, 645–660.
22. Lu, X.Q. *Zhongyao Paozhi Daquan*; Hunan Science and Technology Press: Changsha, China, 1999; pp. 271–274.
23. Hospital Authority of Hong Kong. Cluster of Aconite Poisoning after Taking Chinese Herbal Medicine. Poisoning Alert. 6 January 2017. Available online: http://www.cmd.gov.hk/html/gb/important_info/resource/Alert_on_herbal_medicine_poisoning_20170106.pdf (accessed on 26 January 2019).

24. Yang, M.B.; Ji, X.Y.; Zuo, Z. Relationships between the Toxicities of Radix Aconiti Lateralis Preparata (Fuzi) and the Toxicokinetics of Its Main Diester-Diterpenoid Alkaloids. *Toxins* **2018**, *10*, 391. [CrossRef] [PubMed]
25. Yang, B.; Xu, Y.Y.; Wu, Y.Y.; Wu, H.Y.; Wang, Y.; Yuan, L.; Xie, J.B.; Li, Y.B.; Zhang, Y.J. Simultaneous determination of ten Aconitum alkaloids in rat tissues by UHPLC–MS/MS and its application to a tissue distribution study on the compatibility of Heishunpian and Fritillariae thunbergii Bulbus. *J. Chromatogr. B* **2016**, *1033–1034*, 242–249. [CrossRef] [PubMed]
26. Ren, M.Y.; Song, S.; Liang, D.D.; Hou, W.T.; Tan, X.M.; Luo, J.B. Comparative tissue distribution and excretion study of alkaloids from Herba Ephedrae-Radix Aconiti Lateralis extracts in rats. *J. Pharm. Biomed. Anal.* **2017**, *134*, 137–142. [CrossRef] [PubMed]
27. Wang, Z.H.; Wang, Z.P.; Wen, J.; He, Y. Simultaneous determination of three aconitum alkaloids in urine by LC-MS-MS. *J. Pharm. Biomed. Anal.* **2007**, *45*, 145–148. [CrossRef]
28. Industry, G. *Bioanalytical Method Validation*; Center for Drug Evaluation and Research (CDER), Food and Drug Administration, US Department of Health and Human Services: Washington, DC, USA, 2001.
29. Song, L.; Zhang, H.; Liu, X.; Zhao, Z.L.; Chen, S.L.; Wang, Z.T.; Xu, H.X. Rapid determination of yunaconitine and related alkaloids in aconites and aconite-containing drugs by ultra high-performance liquid chromatography–tandem mass spectrometry. *Biomed. Chromatogr.* **2012**, *26*, 1567–1574. [CrossRef]
30. Wong, S.K.; Tsui, S.K.; Kwan, S.Y. Analysis of proprietary Chinese medicines for the presence of toxic ingredients by LC/MS/MS. *J. Pharm. Biomed. Anal.* **2002**, *30*, 161–170. [CrossRef]
31. The Chinese Pharmacopoeia Commission. *Chinese Pharmacopoeia*; China Medical Science Press: Beijing, China, 2015; Volume 1, pp. 191–193.
32. Ye, L.; Yang, X.S.; Lu, L.L.; Chen, W.Y.; Zeng, S.; Yan, T.M.; Dong, L.N.; Peng, X.J.; Shi, J.; Liu, Z.Q. Monoester-Diterpene Aconitum Alkaloid Metabolism in human Liver Microsomes: Predominant Role of CYP3A4 and CYP3A5. *Evid. Based Complement. Altern. Med.* **2013**, *2013*, 941093. [CrossRef] [PubMed]
33. Wu, J.J.; Lin, N.; Li, F.Y.; Zhang, G.Y.; He, S.G.; Zhu, Y.F.; Ou, R.L.; Li, N.; Liu, S.Q.; Feng, L.Z.; et al. Induction of p-glycoprotein expression and activity by aconitum alkaloids: Implication for clinical drug-drug interactions. *Sci. Rep.* **2016**, *6*, 25343. [CrossRef] [PubMed]
34. Niitsu, H.; Fujita, Y.; Fujita, S.; Kumagai, R.; Takamiya, M.; Aoki, Y.; Dewa, K. Distribution of Aconitum alkaloids in autopsy cases of aconite poisoning. *Forensic Sci. Int.* **2013**, *227*, 111–117. [CrossRef] [PubMed]
35. Zhang, P.P.; Kong, D.Z.; Du, Q.; Zhao, J.; Li, Q.; Zhang, J.H.; Li, T.; Ren, L. A conscious rat model involving bradycardia and hypotension after oral administration: A toxicokinetical study of aconitine. *Xenobiotica* **2017**, *47*, 515–525. [CrossRef] [PubMed]
36. Zhou, J. Pharmacokinetics and Cardiac Distribution of Hypaconitine in Rats. Master's Thesis, Beijing University of Chinese Medicine, Beijing, China, 2010.
37. Wang, R. Quality Evaluation of Fuzi and Pharmacokinetic Study of Aconitine. Ph.D. Thesis, Beijing University of Chinese Medicine, Beijing, China, 2007.
38. Li, W.D.; Ma, C. Determination of Aconitine in Tissues of Rats by HPLC. *J. Exp. Tradit. Med. Formulae* **2007**, *13*, 56–58. [CrossRef]
39. Huang, D.C. Study on Principle of Formation of the School of Warming Yang. Master's Thesis, Shandong University of Chinese Medicine, Shandong, China, 2009.
40. Li, W.D.; Ma, C. Tissue distribution of Aconitum alkaloids extracted from Radix aconiti preparata after oral administration to rats. *Yao Xue Xue Bao* **2005**, *40*, 539–543. [PubMed]
41. Zhou, H.; Zhang, P.; Hou, Z.; Xie, J.; Wang, Y.; Yang, B.; Xu, Y.; Li, Y. Research on the relationships between endogenous biomarkers and exogenous toxic substances of acute toxicity in Radix Aconiti. *Molecules* **2016**, *21*, 1623. [CrossRef]
42. Tang, L.; Gong, Y.; Lv, C.; Ye, L.; Liu, L.; Liu, Z.Q. Pharmacokinetics of aconitine as the targeted marker of Fuzi (Aconitum carmichaeli) following single and multiple oral administrations of Fuzi extracts in rat by UPLC/MS/MS. *J. Ethnopharmacol.* **2012**, *141*, 736–741. [CrossRef]

Article

Toxicity Studies of Chanoclavine in Mice

Sarah C. Finch [1,*], John S. Munday [2], Jan M. Sprosen [1] and Sweta Bhattarai [1]

1 AgResearch Limited, Ruakura Research Centre, Private Bag 3123, Hamilton 3240, New Zealand; jan.sprosen@agresearch.co.nz (J.M.S.); sweta.bhattarai@agresearch.co.nz (S.B.)
2 Institute of Veterinary, Animal and Biomedical Sciences, Massey University, Private Bag 11 222, Palmerston North 4442, New Zealand; J.Munday@massey.ac.nz
* Correspondence: sarah.finch@agresearch.co.nz

Received: 16 April 2019; Accepted: 30 April 2019; Published: 2 May 2019

Abstract: *Epichloë* endophytes have been used successfully in pastoral grasses providing protection against insect pests through the expression of secondary metabolites. This approach could be extended to other plant species, such as cereals, reducing reliance on pesticides. To be successful, the selected endophyte must express secondary metabolites that are active against cereal insect pests without any secondary metabolite, which is harmful to animals. Chanoclavine is of interest as it is commonly expressed by endophytes and has potential insecticidal activity. Investigation of possible mammalian toxicity is therefore required. An acute oral toxicity study showed the median lethal dose of chanoclavine to be >2000 mg/kg. This allows it to be classified as category 5 using the globally harmonized system of classification and labelling of chemicals, and category 6.1E using the New Zealand Hazardous Substances and New Organisms (HSNO) hazard classes, the lowest hazard class under both systems of classification. A three-week feeding study was also performed, which showed chanoclavine, at a dose rate of 123.9 mg/kg/day, initially reduced food consumption but was resolved by day seven. No toxicologically significant effects on gross pathology, histology, hematology, or blood chemistry were observed. These experiments showed chanoclavine to be of low toxicity and raised no food safety concerns.

Keywords: chanoclavine; toxicology; subchronic feeding study; acute toxicity; endophyte; *Epichloë*

Key Contribution: Acute and subchronic toxicity testing of chanoclavine on mice raised no food safety concerns.

---

1. Introduction

Epichloë endophytes are fungal symbionts of cool-season grasses that produce a diverse array of secondary metabolites. One such association, between *Epichloë festuca* var. *lolii* (formerly *Neotyphodium lolii*) and perennial ryegrass (*Lolium perenne*) is essential for the plants survival in New Zealand agricultural conditions [1]. This is due to the expression of secondary metabolites, such as peramine, which protect the plant from insect attack [2–4]. Similarly, the association between *E. coenophiala* (formerly *N. coenophialum*) and tall fescue (*Lolium arundinaceum* (Schreb.) Darbysh (syn *Festuca arundinacea*)) is of vital importance to farming systems in the USA because expression of the loline alkaloids confers insect resistance [5–7]. Unfortunately, in addition to these beneficial secondary metabolites, endophytes can also produce compounds that cause animal disease. In New Zealand, livestock grazing the naturalised (wild-type or common-toxic) ryegrass–endophyte association can develop the neurological disease ryegrass staggers caused by the compound lolitrem B [8,9]. In the USA, the naturalised fescue–endophyte association is responsible for heat stress of cattle and sheep, fescue-foot syndrome of cattle, as well as animal production issues, which are induced by ergot-alkaloids, including ergovaline [10–12]. To harness the beneficial effects of endophytes on insects while minimising the detrimental effects on livestock, research has focused on the screening of endophytes found in other grasses from around the world. Endophytes with a favourable chemical profile can be selected and inoculated into commercial grass cultivars. This method has been successful, and a number of perennial ryegrass and tall fescue products are now commercially available that contain selected endophytes, which provide superior animal safety and pasture persistence and production compared to the naturalised endophyte–grass associations [1].

Because of the success of selected endophyte technology in the pastoral sector, it has been hypothesised that this approach could be used in other plant species such as cereals [13]. Asexual *Epichloë* endophytes have been found to be naturally occurring in Hordeeae grasses, although they are not present in modern cereal grasses. Hordeeae grasses such as wheat (*Triticum aestivum*), barley (*Hordeum vulgare*), and rye (*Secale cereale*) are essential components of both human and domestic animal diets [14], but the current production of these cereal crops is reliant on the use of pesticides. As an alternative, development of endophytic cereals has the potential to provide crops with protection against insect pests with a reduced reliance on synthetic chemicals. However, for this approach to be successful, the selected endophyte must not only provide insect resistance but also be safe for consumption. In our research, a number of fungal endophytes have been isolated from wild relatives of modern cereals (*Elymus* and *Hordeum*) [13]. The targeted endophytes are those that do not express known animal toxins such as lolitrem B and ergovaline, but that do express compounds that are active against insects such as the lolines and peramine. Lolines are not associated with toxicity of grazing animals [15,16]. A short-term toxicity study using mice raised no toxicity issues [17], suggesting that these compounds are an appropriate target.

In addition to lolines and peramine, the clavines may be another class of compounds to target for inclusion into cereals. Clavines are the simplest compounds of the ergot-alkaloid class produced by the early steps of the ergot-alkaloid biosynthetic pathway, which then progresses to produce lysergic acid and its amides and ultimately the highly complex ergopeptines such as ergovaline [18]. Some endophytes express only clavine compounds, and others express the whole range of ergot-alkaloid derivatives. It is often found that compounds such as chanoclavine (of the clavine class) can accumulate to be at a similar concentration as the ergopeptine alkaloids. This inefficient flow of intermediates to the biosynthetic end product is unusual, and it has been hypothesised that intermediates, such as chanoclavine, may provide some benefit to the endophyte or to its host plant [19,20]. While it is known that ergovaline has an effect on some insect pests such as the African black beetle (*Heteronychus arator*) [21], research into the effects of chanoclavine (Figure 1) is limited. While initial research suggests that ergot-alkaloids of the clavine class may be less effective in reducing insect damage than the animal-toxic ergopeptine class, the clavines have been shown to deter the feeding of fall armyworm (*Spodoptera frugiperda*) [22]. Ergovaline is well documented to cause fescue toxicosis, which is associated with a dramatic lowering of serum prolactin levels in grazing animals [12]. In vitro testing of ergovaline and chanoclavine on cultured anterior pituitary cells showed that ergovaline lowered prolactin secretion at a concentration of 0.5 nM, whereas an effect of chanoclavine was observed only at a concentration of 1000 nM [23]. This suggests that chanoclavine is unlikely to have a significant effect on serum prolactin levels in grazing animals, and there are no documented reports of animal toxicity induced by chanoclavine.

Further insect testing would be required to determine whether endophytes that produce clavines, such as chanoclavine, should be actively targeted for inclusion into cereal crops. However, if the clavines are found to deter insects, it would be equally important to determine whether these compounds have any mammalian toxicity. Even if it is found that the bioactivity of the clavines is low, information on the toxicity of chanoclavine is important because this compound can be produced by endophytes, which express lolines and peramine without lolitrem B or ergovaline, and are therefore of interest to the cereal endophyte research program. The objective of this study was to provide preliminary information on the toxicity of chanoclavine by performing an oral, acute toxicity determination as well as a three-week subchronic feeding study in mice.

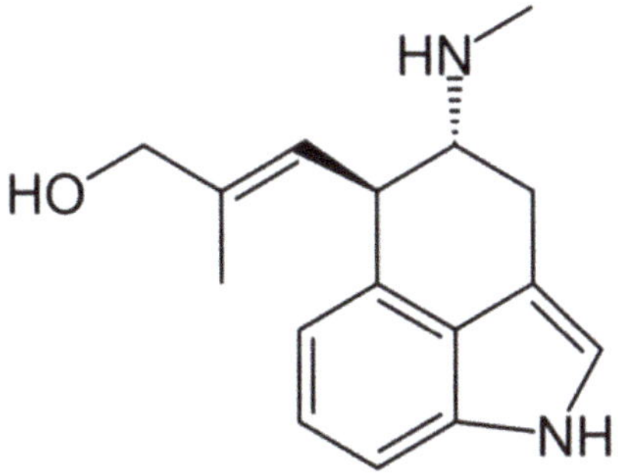

HO HN NH

Figure 1. Structure of chanoclavine.

2. Results

2.1. Acute Toxicity Testing

All five mice dosed with chanoclavine at the limit dose (2000 mg/kg) survived, although some signs of neurotoxicity were observed in three of the five mice. Tremors were detected in two mice, and movement was characterised by splaying of the back legs. Time of onset of these toxic symptoms was between 1 and 5 h post-dosing, but the mice were moving and feeding normally by 24 h. One further mouse exhibited a weak grip 4 h post-dosing, but this was resolved within 2 h. The appearance, behaviour, and growth of all mice remained normal throughout the subsequent 14-day observation period. No abnormalities were observed in any of the animals at necropsy.

Results of the acute oral toxicity testing of chanoclavine indicated that the median lethal dose (LD_{50}) was greater than the given dose rate of 2000 mg/kg. In accord with OECD (Organisation of Economic Co-operation and Development) guideline 425, this dose rate was the limit dose such that further dosing at higher rates was not required. A median lethal dose of greater than 2000 mg/kg allowed classification of the oral toxicity of chanoclavine as category 5 using the globally harmonized system of classification and labelling of chemicals (GHS) [24]. Using the New Zealand HSNO hazard classes, this equates to category 6.1E [25].

2.2. Short-Term Toxicity Study

2.2.1. Analysis of Mouse Diet

Taking into account moisture content of the prepared diet, the theoretical concentration was 563 μg/g chanoclavine. Analysis of the three diet chunks taken from the prepared discs yielded concentrations of 565 μg/g (100%), 629 μg/g (112%), and 565 μg/g (100%). This indicated that chanoclavine was stable through the diet-making process and was present in a homogenous manner throughout the prepared diet.

2.2.2. Clinical Observations and Appearance

The appearance, movement, and behaviour of all mice remained normal throughout the experimental period.

2.2.3. Bodyweight and Food Consumption

Ergovaline is well recognised to reduce food intake in a range of animals including cattle, birds, rabbits, and small mammals [26–28]. An experiment using perennial ryegrass-Neotyphodium sp. Lp1 with altered ergot alkaloid profiles (due to the knockout of endophyte genes) showed that endophyte-infected plants, devoid of ergot-alkaloids, were preferred by rabbits over endophyte-free plants. This preference was reduced in plants containing either all of the ergot-alkaloids (ergopeptines and clavines) or in those containing only clavines. However, only those plants containing ergovaline were shown to affect food consumption of rabbits [29]. In the current experiment, which used diets of far higher chanoclavine concentrations than those used by Panaccione (2006) (563 μg/g compared with 1.2 μg/g), food intake of the male and female mice fed a diet containing chanoclavine was lower compared with those fed a control diet in the initial stages of the experiment. However, by day seven, there was no difference in the food intake per gram of bodyweight for mice fed the two different treatment diets, and when expressed in this way, there was no overall treatment effect for the male ($p = 0.221$) or female ($p = 0.228$) mice (Figure 2). This initial food intake difference was reflected in the mouse bodyweights although, overall, there was no treatment effect on the bodyweights of the male ($p = 0.139$) or female ($p = 0.141$) mice (Figure 3).

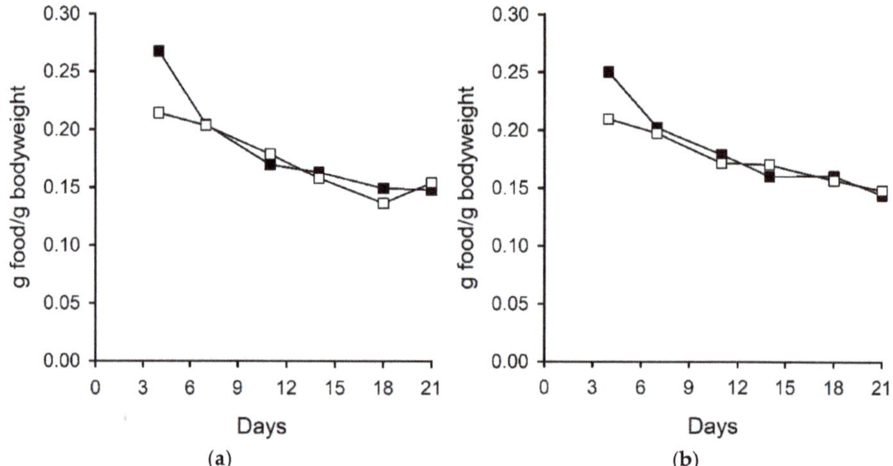

Figure 2. Food consumption of (**a**) male and (**b**) female mice fed control (■) and chanoclavine (□, 563 μg/g) diets.

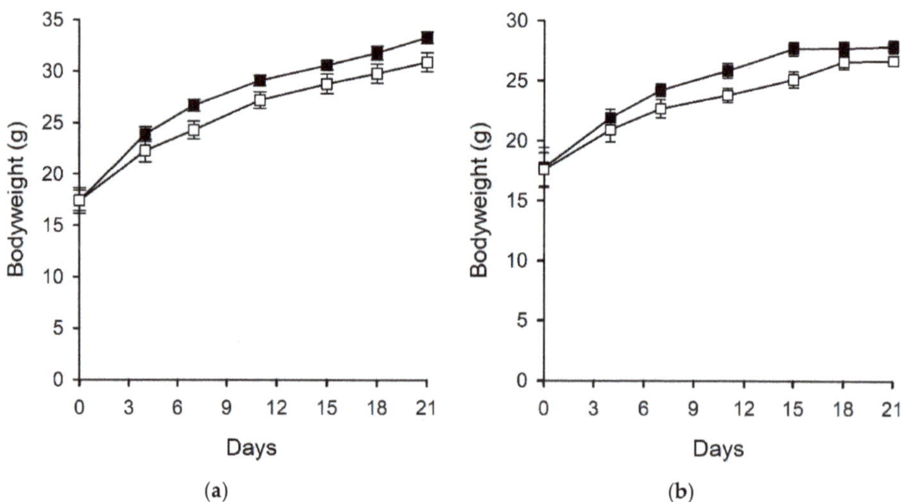

Figure 3. Bodyweights of (**a**) male and (**b**) female mice fed control (■) and chanoclavine (□, 563 μg/g) diets. Error bars represent standard errors of the mean but are too small to be visible in some cases.

2.2.4. Hematological and Serum Biochemical Data

The hematological and serum biochemical data for day 21 are presented in Tables 1 and 2. The only statistically significant difference observed in the hematological data was a higher neutrophil percentage in the female mice fed a control diet compared with those fed a diet containing chanoclavine. The only statistically significant difference observed in the serum biochemical data of male mice was that the levels of total bilirubin were higher in control mice compared with those fed a diet containing chanoclavine. For the female mice, a statistically significant difference was observed in the concentrations of total protein and albumin, which were higher in mice fed a diet containing chanoclavine compared with those fed a control diet. These differences were not considered to be toxicologically meaningful.

Table 1. Hematology data from mice fed treatment diets for 21 d.

Item	Control [1]	Chanoclavine [1]	SED	p-Value
Males				
HCT (L/L)	0.46 ± 0.01	0.47 ± 0.01	0.01	0.620
HB (g/L)	140.8 ± 3.6	142.8 ± 4.9	2.72	0.483
RBC ($\times 10^{12}$/L)	8.8 ± 0.30	8.9 ± 0.4	0.21	0.527
MCV (fL)	52.8 ± 1.5	53.2 ± 1.3	0.88	0.663
MCH (pg)	16.2 ± 0.4	16.2 ± 0.4	0.28	1.000
MCHC (g/L)	303.8 ± 2.6	303.6 ± 4.6	2.35	0.934
WBC ($\times 10^{9}$/L)	10.3 ± 1.6	10.4 ± 1.7	1.06	0.884
Neutrophil (%)	17.6 ± 1.5	19.4 ± 7.5	3.42	0.613
Lymphocyte (%)	75.8 ± 5.2	69.6 ± 8.9	4.62	0.216
Monocyte (%)	3.2 ± 1.6	4.8 ± 1.5	0.99	0.145
Eosinophil (%)	3.0 ± 2.0	3.0 ± 1.9	1.23	1.000
Basophil (%)	0.4 ± 0.6	3.2 ± 7.2	3.21	0.408
Females				
HCT (L/L)	0.47 ± 0.01	0.49 ± 0.04	0.02	0.213
HB (g/L)	145.4 ± 1.3	151.0 ± 12.5	5.61	0.347
RBC ($\times 10^{12}$/L)	8.9 ± 0.1	9.4 ± 0.8	0.35	0.149
MCV (fL)	52.8 ± 0.8	51.8 ± 1.3	0.69	0.187
MCH (pg)	16.4 ± 0.5	16.0 ± 0.0	0.25	0.141
MCHC (g/L)	311.0 ± 3.7	308.6 ± 3.0	2.14	0.294
WBC ($\times 10^{9}$/L)	7.4 ± 2.1	8.5 ± 1.9	1.27	0.405
Neutrophil (%)	13.6 ± 3.1 [a]	9.4 ± 1.7 [b]	1.59	0.029
Lymphocyte (%)	78.0 ± 3.7	81.4 ± 2.4	1.97	0.122
Monocyte (%)	3.6 ± 1.1	4.0 ± 1.6	0.87	0.659
Eosinophil (%)	4.8 ± 1.3	4.6 ± 0.9	0.71	0.784
Basophil (%)	0.0 ± 0.0	0.6 ± 1.3	0.60	0.347

[1] Values are means ± standard deviation (n = 5). Fisher's least significant difference was used to compare treatment means. Two means that have no letter in common are statistically different at the 5% level. If no letters are given, the overall p-value was not statistically significant at the 5% level. HCT, hematocrit value; HB, hemoglobin level; MCV, mean corpuscular volume; MCHC, mean corpuscular hemoglobin concentration.

Table 2. Serum biochemical data from mice fed treatment diets for 21 d.

Item	Control [1]	Chanoclavine [1]	SED	p-Value
Males				
CK (IU/L)	294 ± 69	437 ± 206	91.6	0.164
AST (IU/L)	75 ± 15	75 ± 20	11.1	0.972
ALT (IU/L)	37 ± 12	35 ± 6	6.0	0.748
T.Bil (μmol/L) [2]	1.8 ± 0.4 [a]	0.9 ± 0.2 [b]	0.2	0.004
TP (g/L)	54.0 ± 2.2	53.8 ± 2.0	1.4	0.886
ALB (g/L)	33.2 ± 0.8	33.0 ± 1.4	0.7	0.786
Globulin (g/L)	20.8 ± 2.9	20.0 ± 1.4	1.5	0.616
A/G	1.622 ± 0.230	1.660 ± 0.178	0.1	0.787
CRN (μmol/L)	14.0 ± 2.8	17.0 ± 5.8	2.9	0.328
Na (mmol/L)	146 ± 2	146 ± 1	0.9	0.829
K (mmol/L)	15.0 ± 1.4	15.3 ± 1.3	0.9	0.700
Cl (mmol/L)	103 ± 1	105 ± 1	0.6	0.101
Females				
CK (IU/L)	569 ± 210	594 ± 209	132.6	0.851
AST (IU/L)	110 ± 20	121 ± 17	11.7	0.366
ALT (IU/L)	31 ± 5	39 ± 9	4.5	0.126
T.Bil (μmol/L) [2]	1.5 ± 0.7	0.9 ± 0.2	0.3	0.108
TP (g/L)	54.8 ± 1.1 [b]	57.8 ± 2.3 [a]	1.1	0.029
ALB (g/L)	36.6 ± 1.1 [b]	38.8 ± 1.5 [a]	0.8	0.030
Globulin (g/L)	18.2 ± 0.8	19.0 ± 1.6	0.8	0.347
A/G	2.015 ± 0.140	2.053 ± 0.181	0.1	0.721
CRN (μmol/L)	17.0 ± 0.7	16.8 ± 3.7	1.7	0.908
Na (mmol/L)	148 ± 1	148 ± 4	1.8	0.913
K (mmol/L)	12.4 ± 1.0	12.8 ± 2.1	1.0	0.682
Cl (mmol/L)	106 ± 1	106 ± 2	0.9	0.838

[1] Values are means ± standard deviation (n = 5). Fisher's least significant difference was used to compare treatment means. Two means that have no letter in common are statistically different at the 5% level. If no letters are given, the overall p-value was not statistically significant at the 5% level. [2] When total bilirubin (T. Bil) was below the detection limit it was assigned a value of half the detection limit (0.5 μmol/L). CK, creatine kinase; AST, aspartate aminotransferase; ALT, alanine aminotransferase; TP, total protein; ALB, albumin; CRN, creatinine.

2.2.5. Organ Weights

Absolute and relative organ weights are presented in Tables 3 and 4. The only statistically significant difference was the absolute and relative weight of the kidneys of the male mice. This difference was not considered to be toxicologically meaningful.

Table 3. Absolute organ weight data (g) from mice fed treatment diets for 21 d.

	Brain [1]	Heart [1]	Kidneys [1]	Liver [1]	Spleen [1]
Males					
Control	0.467 ± 0.017	0.181 ± 0.011	0.512 ± 0.019 [a]	1.832 ± 0.129	0.135 ± 0.029
Chanoclavine	0.453 ± 0.021	0.176 ± 0.019	0.441 ± 0.033 [b]	1.656 ± 0.183	0.126 ± 0.017
SED	0.012	0.010	0.017	0.100	0.015
p-value	0.269	0.623	0.003	0.117	0.579
Females					
Control	0.457 ± 0.020	0.167 ± 0.012	0.355 ± 0.017	1.384 ± 0.203	0.146 ± 0.028
Chanoclavine	0.466 ± 0.010	0.155 ± 0.006	0.328 ± 0.029	1.280 ± 0.093	0.132 ± 0.020
SED	0.010	0.006	0.015	0.100	0.016
p-value	0.385	0.086	0.117	0.328	0.420

[1] Values are means ± standard deviation (n = 5). Fisher's least significant difference was used to compare treatment means. Two means that have no letter in common are statistically different at the 5% level. If no letters are given, the overall p-value was not statistically significant at the 5% level.

Table 4. Relative organ weight data (% of bodyweight) from mice fed treatment diets for 21 d.

	Brain [1]	Heart [1]	Kidneys [1]	Liver [1]	Spleen [1]
Males					
Control	1.439 ± 0.039	0.559 ± 0.041	1.578 ± 0.079 [a]	5.638 ± 0.259	0.417 ± 0.097
Chanoclavine	1.500 ± 0.110	0.586 ± 0.090	1.456 ± 0.023 [b]	5.458 ± 0.251	0.417 ± 0.046
SED	0.052	0.044	0.037	0.161	0.048
p-value	0.280	0.555	0.011	0.298	0.999
Females					
Control	1.671 ± 0.136	0.610 ± 0.037	1.296 ± 0.059	5.030 ± 0.510	0.530 ± 0.090
Chanoclavine	1.775 ± 0.049	0.592 ± 0.031	1.250 ± 0.101	4.873 ± 0.322	0.504 ± 0.074
SED	0.065	0.022	0.052	0.270	0.052
p-value	0.147	0.439	0.404	0.578	0.626

[1] Values are means ± standard deviation ($n = 5$). Fisher's least significant difference was used to compare treatment means. Two means that have no letter in common are statistically different at the 5% level. If no letters are given, the overall p-value was not statistically significant at the 5% level.

2.2.6. Histological Examination

Examination of tissues from male mice in the control group revealed that one mouse had a single small focus of inflammation within the liver. This focus was not associated with significant hepatocellular injury. Similar foci were also visible within three male mice that were fed chanoclavine. No other histological lesions were observed on examination of tissues from the male mice. Examination of tissues from the control female mice revealed all five mice had scattered foci of inflammation within the hepatic parenchyma. While inflammation was mild and not associated with significant hepatocellular damage in two mice, some of the larger foci in the other three animals contained swollen eosinophilic hepatocytes that demonstrated loss of cell features, consistent with mild hepatocellular damage. Examination of tissues from the female mice that received chanoclavine revealed small foci of inflammation present within the liver of four of the mice. This inflammation was associated with scattered, degenerate hepatocytes in three animals. In addition, one mouse had a small focus of lymphocytes and macrophages within the submucosa of the bladder. No other lesions were observed on examination of tissues from the female mice. The cause of the hepatic changes was uncertain; however, as they developed with approximately equal frequency and severity in the control and treated mice, this change was considered extremely unlikely to be due to chanoclavine toxicity.

3. Discussion

The toxicity of chanoclavine was investigated. Acute toxicity testing showed neurotoxicity in some animals at the limit dose of 2000 mg/kg. This is of no concern, as this extremely high dose rate is not biologically relevant. Since no deaths were observed at this dose rate, the median lethal dose (LD_{50}) of chanoclavine is, therefore, greater than 2000 mg/kg. This allowed it to be classified in category 5 using the GHS labelling system and in category 6.1E using the New Zealand HSNO hazard classes. This is the lowest hazard class under both systems of classification. A short-term feeding trial was also performed, whereby a group of male and female mice were fed either a diet containing pure chanoclavine or regular mouse food. Using food consumption and bodyweight data, the dose rate of chanoclavine was calculated for each of the biweekly measurement periods. This showed that the male and female mice ate on average 123.3 and 124.4 mg/kg/day chanoclavine, respectively. In the initial stages of the three-week experiment, mice fed the chanoclavine diet ate less and gained weight more slowly than the mice fed a control diet. However, by day seven, the food consumption of both groups was equivalent on a food intake per gram of bodyweight basis. No toxicologically significant effects were noted in hematological or serum biochemical parameters. Although foci of inflammation were noted in the livers of a number of mice, this was observed in both the control and chanoclavine-fed animals, so it was considered to be unrelated to the treatment diet. No other significant effects were noted on histological examination.

The concentration of chanoclavine that could be present in endophyte-infected cereals is unknown. However, in the acute oral toxicity study, the limit dose of 2000 mg/kg was reached, and in the feeding trial a very high dose rate was chosen. The amounts fed to mice in this study would be equivalent to a 70 kg human consuming 8.6 g of chanoclavine per day. These experiments, therefore, show that chanoclavine is of low toxicity and should be safe for inclusion into cereal crops. However, once a chanoclavine-producing endophyte is successfully inoculated into a cereal crop, further toxicological work will be required.

4. Materials and Methods

4.1. *Chanoclavine*

Chanoclavine (>98.69% purity) was purchased from Alfarma s.r.o (Prague, Czech Republic) and was stored at 4 °C before use.

4.2. *Animals*

For acute toxicity testing, Swiss albino mice (5 females, 4–5 weeks old, and 18–20 g) were used. For the short-term toxicity study, Swiss albino mice (10 male and 10 female, 3–4 weeks old, and 14–19 g on day −3) were again used. Animals were housed in groups of five in a temperature-controlled room (21 ± 1 °C) with a 12 h light–dark cycle. They were allowed access to food and water ad libitum. All animal manipulations were approved by the Ruakura Animal Ethics Committee established under the Animal Protection (code of ethical conduct) Regulations Act, 1987 (New Zealand). Acute toxicity: project Number 13242, approval date 1 May 2014; Subchronic toxicity: project Number 13508, approval date2 April 2015.

4.3. *Acute Toxicity Protocol*

Acute toxicity was determined according to the principles of OECD guideline 425 [30] using the limit dose of 2000 mg/kg. Pairs of mice were randomly selected and fasted overnight. Mice were weighed immediately before dosing, and the amount of chanoclavine required was calculated on a mg/kg basis. The required amount of chanoclavine was added into the glass pipette tip of a positive displacement pipette along with a small portion of ground mouse food (20 mg) and water (50 µL) to yield a paste. This paste was applied over the tongue of the mouse, with care taken to ensure that the entire dose was swallowed. Control mice were dosed in an analogous manner with a ground mouse food (20 mg) and water (50 µL) mixture. To avoid diurnal variations in response, all dosing was conducted between 9.30 and 10.00 am. The mice were monitored intensively during the day of dosing and then examined and weighed each day for a further 14 d, after which they were killed by CO_2 inhalation and necropsied. In total, five pairs of mice were dosed.

4.4. *Short-Term Toxicity Study*

Mouse diets were prepared from Teklad Global 2016 mouse food pellets (Harlan UK, Bicester, England), which were finely ground using a Udy cyclone sample mill (Udy Corporation, Fort Collins, CO, USA). Initially, a test batch of chanoclavine diet was prepared to check stability and diet homogeneity by mixing chanoclavine (29.8 mg) with ground mouse food (50 g) using a cake mixer, adding water (45 mL), and drying as described below. Three separate samples of diet were taken for analysis and ground using a mortar and pestle. The diets for the feeding study were prepared by mixing chanoclavine (141.6 mg) with ground mouse food (200 g) using a cake mixer. Water (90 mL) was added to 100 g aliquots of this mixture and combined to form a paste. This was split evenly into 14 portions, which were formed into discs and dried in a fan oven (50 °C, 24 h). This resulting diet had a moisture content of 11.3% and a chanoclavine concentration of 628 µg/g (wet weight). Control diets were prepared in an analogous manner to yield diets with a moisture content of 13.8%. Diets were prepared twice weekly during the experimental period.

Mice were randomly assigned to their cage groups 3 d before the start of the experiment. They were fed prepared control diets so that they could get used to the new form of their food. Each group was comprised of five female and five male mice (housed by gender); group 1 was the control group, and group 2 was the chanoclavine group. During the experiment, food consumption of each cage group of five mice was measured every second day. At this time, the appearance, movement, and behaviour of each mouse was observed to check for any changes. Biweekly the body weight of each mouse was measured. Food consumption of each cage group of mice was calculated from the quantity of diet eaten by the five mice divided by the total bodyweight. At the conclusion of the three-week experiment, all mice were killed by CO_2 inhalation, and a blood sample was collected by heart puncture using heparin as the anticoagulant. Whole blood was analysed to measure hematocrit values (HCT), hemoglobin levels (HB), mean corpuscular volumes (MCVs), mean corpuscular hemoglobin (MCH), mean corpuscular hemoglobin concentrations (MCHCs), and red and white cell counts, while activities of alanine aminotransferase (ALT), aspartate aminotransferase (AST), and creatine kinase (CK) as well as concentrations of bilirubin, total protein (TP), albumin (ALB), globulin, sodium, potassium, chloride, and creatinine (CRN) were determined in plasma (New Zealand Veterinary Pathology, Hamilton, New Zealand). Necropsies of all mice were performed to check for any macroscopic changes, and the weights of brain, heart, kidneys, liver, and spleen were measured and expressed as a percentage of body weight for each mouse. These organs, along with adrenals, lungs, pancreas, gastrocnemius, jejunum (3 mm section), ovary and uterus or testes, spinal cord (3 × 2 mm sections), stomach (washed), thymus, and urinary bladder were placed in 4% buffered formaldehyde to allow fixation before being routinely processed for histology. The same pathologist examined all samples and was unaware of the treatment groups.

4.5. Measurement of Chanoclavine in Mouse Diet

One of the prepared chanoclavine diets was taken after the drying process, and three samples were taken from different areas of the disc. These samples were ground using a mortar and pestle. Aliquots (50 mg) were extracted with 50% methanol (1 mL) containing an internal standard (festuclavine, 0.2 μg/mL). Filtered extracts were analysed according to the method of Rasmussen et al. [31].

4.6. Statistical Analysis

Bodyweight, food intake, hematology, serum biochemistry, and organ weight data from the short-term toxicity study were analysed by one-way ANOVA followed by Fisher's protected least significant difference test at the 5% significance level. Data from each measurement time, and for each gender, were analysed separately. Since the genders for each treatment group were housed together, there was no gender replication of the treatments in the trial, so the assumption was made that mice within a cage behaved independently of one another. Boxplots and residual plots were inspected for departures from normality and homogeneity of variance. All analyses were conducted in GenStat version 16 (VSN International, Hemel Hempstead, UK).

Author Contributions: Conceptualization, S.C.F.; Data curation, S.C.F., J.S.M., J.M.S. and S.B.; Methodology, S.C.F. and J.S.M.; Writing—original draft, S.C.F. and J.S.M.; Writing—review & editing, S.C.F., J.S.M., J.M.S. and S.B.

Funding: This research was funded by the New Zealand Ministry for Business, Innovation and Employment (grant numbers C10X0807 and C10X1403), Grasslanz Technology Ltd., the New Zealand Foundation for Arable Research and the Australian Grains Research and Development Corporation.

Acknowledgments: We thank Wade Mace (AgResearch Grasslands) for alkaloid analysis, Catherine Cameron and Vanessa Cave (AgResearch Ruakura) for statistical analysis and Ric Broadhurst and Bobby Smith (AgResearch Ruakura) for the care of animals.

Conflicts of Interest: The authors declare no conflict of interest.

References

1. Johnson, L.J.; de Bonth, A.C.M.; Briggs, L.R.; Caradus, J.R.; Finch, S.C.; Fleetwood, D.J.; Fletcher, L.R.; Hume, D.E.; Johnson, R.D.; Popay, A.J.; et al. The exploitation of epichloae endophytes for agricultural benefit. *Fungal Divers.* **2013**, *60*, 171–188. [CrossRef]
2. Rowan, D.D.; Hunt, M.B.; Gaynor, D.L. Peramine, a novel insect feeding deterrent from ryegrass infected with the endophyte *Acremonium loliae*. *J. Chem. Soc. Chem. Commun.* **1986**, 935–936. [CrossRef]
3. Rowan, D.D.; Dymock, J.J.; Brimble, M.A. Effect of the fungal metabolite peramine and analogs on feeding and development of Argentine stem weevil (*Listrontus bonariensis*). *J. Chem. Ecol.* **1990**, *16*, 1683–1695. [CrossRef]
4. Rowan, D.; Gaynor, D. Isolation of feeding deterrents against Argentine stem weevil from ryegrass infected with the endophyte *Acremonium loliae*. *J. Chem. Ecol.* **1986**, *12*, 647–658. [CrossRef] [PubMed]
5. Jensen, J.G.; Popay, A.J.; Tapper, B.A. Argentine stem weevil adults are affected by meadow fescue endophyte and its loline alkaloids. *New Zealand Plant Prot.* **2009**, *62*, 12–18.
6. Reidell, W.; Kieckhefer, R.; Petroski, R.; Powell, R. Naturally occurring and synthetic loline alkaloid derivatives: Insect feeding behaviour modification and toxicity. *J. Entomol. Sci.* **1991**, *26*, 122–129. [CrossRef]
7. Wilkinson, H.H.; Siegel, M.R.; Blankenship, J.D.; Mallory, A.C.; Bush, L.P.; Schardl, C.L. Contribution of fungal loline alkaloids to protection from aphids in a grass-endophyte mutualism. *Mol. Plant-Microbe Interact.* **2000**, *13*, 1027–1033. [CrossRef] [PubMed]
8. Di Menna, M.E.; Finch, S.C.; Popay, A.J.; Smith, B.L. A review of the *Neotyphodium lolii/Lolium perenne* symbiosis and its associated effects on animal and plant health, with particular emphasis on ryegrass staggers. *New Zealand Vet. J.* **2012**, *60*, 315–328. [CrossRef]
9. Fletcher, L.R.; Harvey, I.C. An association of a *Lolium* endophyte with ryegrass staggers. *New Zealand Vet. J.* **1981**, *29*, 185–186. [CrossRef] [PubMed]
10. Bacon, C.W.; Porter, J.K.; Robbins, J.D.; Luttrell, E.S. *Epichloë typhina* from toxic tall fescue grasses. *Appl. Environ. Micro.* **1977**, *34*, 576–581.
11. Schmidt, S.P.; Hoveland, C.S.; Clark, E.M.; Davis, N.D.; Smith, L.A.; Grimes, H.W.; Holliman, J.L. Association of an endophytic fungus with fescue toxicity in steers fed Kentucky 31 tall fescue seed or hay. *J. Anim. Sci.* **1982**, *55*, 1259–1263. [CrossRef]
12. Klotz, J. Activities and effects of ergot alkaloids on livestock physiology and production. *Toxins* **2015**, *7*, 2801–2821. [CrossRef] [PubMed]
13. Simpson, W.R.; Faville, M.J.; Moraga, R.A.; Williams, W.M.; McManus, M.T.; Johnson, R.D. Epichloë fungal endophytes and the formation of synthetic symbioses in Hordeeae (=Triticeae) grasses. *J. Syst. Evol.* **2014**, *52*, 794–806. [CrossRef]
14. Feuillet, C.; Muehlbauer, G.J. *Genetics and genomics of the Triticeae*; Springer Science & Business Media: New York, NY, USA, 2009.
15. Daccord, R.; Schmidt, D.; Arrigo, Y.; Gutzwiller, A. Endophytes in meadow fescue: a limiting factor in ruminant production? In Proceedings of the 2nd International Conference on Harmful and Beneficial Microorganisms in Pastures, Turf and Grassland, Paderborn, Germany, 22–24 November 1995; pp. 155–159.
16. Fletcher, L.R.; Popay, A.J.; Stewart, A.V.; Tapper, B.A. Herbage and sheep production from meadow fescue with and without the endophyte *Neotyphodium uncinatum*. In Proceedings of the 4th International Neotyphodium/Grass Interactions Symposium, Soest, Germany, 27–29 September 2000; pp. 447–453.
17. Finch, S.C.; Munday, J.S.; Munday, R.; Kerby, J.W.F. Short-term toxicity studies of loline alkaloids in mice. *Food Chem. Toxicol.* **2016**, *94*, 243–249. [PubMed]
18. Young, C.; Schardl, C.; Panaccione, D.; Florea, S.; Takach, J.; Charlton, N.; Moore, N.; Webb, J.; Jaromczyk, J. Genetics, Genomics and Evolution of Ergot Alkaloid Diversity. *Toxins* **2015**, *7*, 1273–1302. [CrossRef]
19. Panaccione, D.G. Origins and significance of ergot alkaloid diversity in fungi. *FEMS Microbiol. Lett.* **2005**, *251*, 9–17. [CrossRef] [PubMed]
20. Robinson, S.; Panaccione, D. Diversification of ergot alkaloids in natural and modified fungi. *Toxins* **2015**, *7*, 201–218. [CrossRef]
21. Ball, O.J.P.; Miles, C.O.; Prestidge, R.A. Ergopeptine alkaloids and *Neotyphodium lolii*-mediated resistance in perennial ryegrass against adult *Heteronychus arator* (Coleoptera: Scarabaeidae). *J. Econ. Entomol.* **1997**, *90*, 1382–1391. [CrossRef]

22. Clay, K.; Cheplick, G.P. Effect of ergot alkaloids from fungal endophyte-infected grasses on fall armyworm (*Spodoptera frugiperda*). *J. Chem. Ecol.* **1989**, *15*, 169–182. [CrossRef]
23. Piper, E.; Denard, T.; Johnson, Z.; Flieger, M. Effect of chanoclavine on in vitro prolactin release. In Proceedings of the 4th International Neotyphodium/Grass Interactions Symposium, Soest, Germany, 27–29 September 2000; pp. 531–534.
24. United Nations. *Globally harmonized system of classification and labelling of chemicals (GHS)*, 4th ed.; United Nations Publications: New York, NY, USA, 2011.
25. New Zealand Government—Environmental Protection Authority Correlation between GHS and New Zealand HSNO hazard classes and categories. 2009.
26. Finch, S.C.; Pennell, C.G.L.; Kerby, J.W.F.; Cave, V.M. Mice find endophyte-infected seed of tall fescue unpalatable—Implications for the aviation industry. *Grass Forage Sci.* **2016**, *71*, 659–666. [CrossRef]
27. Coley, A.B.; Fribourg, H.A.; Pelton, M.R.; Gwinn, K.D. Effects of tall fescue endophyte infestation on relative abundance of small mammals. *J. Environ. Qual.* **1995**, *24*, 472–475. [CrossRef]
28. Matthews, A.K.; Poore, M.H.; Huntington, G.B.; Green, J.T. Intake, digestion, and N metabolism in steers fed endophyte-free, ergot alkaloid-producing endophyte-infected, or non ergot alkaloid-producing endophyte-infected fescue hay. *J. Anim. Sci.* **2005**, *83*, 1179–1185. [CrossRef]
29. Panaccione, D.G.; Cipoletti, J.R.; Sedlock, A.B.; Blemings, K.P.; Schardl, C.L.; Machado, C.; Seidel, G.E. Effects of ergot alkaloids on food preference and satiety in rabbits, as assessed with gene-knockout endophytes in perennial ryegrass (*Lolium perenne*). *J. Agric. Food Chem.* **2006**, *54*, 4582–4587. [CrossRef] [PubMed]
30. OECD guidelines for the testing of chemicals Guideline 425. *Acute Oral Toxicity-Up-and-Down-Procedure (UDP)*; OECD: Paris, France, 2008.
31. Rasmussen, S.; Lane, G.A.; Mace, W.; Parsons, A.J.; Fraser, K.; Xue, H. The use of genomics and metabolomics methods to quantify fungal endosymbionts and alkaloids in grasses. *Methods Mol. Biol.* **2012**, *860*, 213–226. [PubMed]

Review

Impact of Ergot Alkaloids on Female Reproduction in Domestic Livestock Species

Rebecca K. Poole and Daniel H. Poole *

Department of Animal Science, North Carolina State University, Raleigh, NC 27607, USA; rkpoole@ncsu.edu
* Correspondence: dhpoole@ncsu.edu; Tel.: +1-919-515-0033; Fax: +1-919-515-6884

Received: 31 May 2019; Accepted: 14 June 2019; Published: 21 June 2019

Abstract: Fescue toxicosis is a multifaceted syndrome that elicits many negative effects on livestock consuming ergot alkaloids produced by endophyte-infected tall fescue. The economic losses associated with fescue toxicosis are primarily due to reproductive failure including altered cyclicity, suppressed hormone secretion, reduced pregnancy rates, agalactia, and reduced offspring birth weights. For decades, a multitude of research has investigated the physiological and cellular mechanisms of these reproductive failures associated with fescue toxicosis. This review will summarize the various effects of ergot alkaloids on female reproduction in grazing livestock species.

Keywords: ergot alkaloids; livestock; reproduction

Key Contribution: Exposure to ergot alkaloids negatively impacts many reproductive processes in domestic livestock species and this manuscript explores the various effects of ergot alkaloids on reproductive tissues, provides an up-to-date summary of the latest data, and emphasizes the need for more mechanistic studies to increase our understanding of ergot alkaloid toxicity.

---

1. Introduction

Ergot alkaloid mycotoxins were first identified in *Claviceps*, a parasitic fungus that infects many grasses and grains [1]. These mycotoxins are produced by a variety of fungi, including *Neotyphodium* and *Epichloë*, and classified as the tall fescue endophyte [1–3]. Specifically, *Epichloë coenophiala* is recognized as the endophyte that shares a symbiotic relationship with Kentucky (KY)-31 tall fescue (*Lolium arundinaceum* [Schreb.] Darbysh; [4]). Chronic consumption of *Epichloë coenophiala*-produced ergot alkaloids in grazing livestock results in a syndrome known as fescue toxicosis [5]. With KY-31 being estimated to be grown on 35 million acres of land in the Southeast to Midwest U.S. regions and that 90% of these pastures are infected with the *Epichloë* endophyte [6–8], it is speculated that fescue toxicosis contributes to over $2 billion in annual economic loss to the U.S. livestock industries [9]. Detrimental signs of this syndrome include reduced intake, weight gain, circulating prolactin concentrations, reproductive performance, milk production, and hyperthermia [10]. Many of these signs of fescue toxicosis can be attributed to the structural similarities between ergot alkaloids and monoamine neurotransmitters (e.g., serotonin, norepinephrine, and dopamine), which elicit agonistic effects on numerous monoamine receptors [11–15]. While many livestock species do experience ergot alkaloid-induced effects on productivity, there is a high variability between individual animal responses to exposure. However, across livestock species, there have been consistent reports indicating that ergot alkaloids cause issues with reproductive performance, including reduced pregnancy rates, circulating hormone concentrations, blood flow to reproductive organs, and offspring birth weight [16]. Therefore, this review will focus on the various effects of ergot alkaloids (e.g., physiologic, mechanistic, etc.) on female reproduction in livestock species.

2. Ergot Alkaloids and the Brain

Ergot alkaloids share many structural similarities to monoamine neurotransmitters, and thus interact on the various monoamine receptors within the brain. Specifically, the cells within the anterior pituitary contain monoamine receptors and therefore ergot alkaloids may interfere with the physiological processes regulated by the anterior pituitary. The anterior pituitary can be divided into five endocrine cell types based on morphology and functional role: Somatotrophs (growth hormone (GH)), corticotrophs (adrenocorticotropic hormone (ACTH)), thyrotrophs (thyroid-stimulating hormone (TSH)), lactotrophs (prolactin (PRL)), and gonadotrophs (luteinizing hormone (LH) and follicle stimulating hormone (FSH)) [17]. While it does not appear that ergot alkaloids affect secretion of GH and TSH, there is great evidence that ergot alkaloids alter secretion of ACTH, PRL, and LH and FSH [12,18,19].

2.1. Dopamine Receptor D2 and Prolactin Secretion

A common symptom of fescue toxicosis is a suppression in prolactin secretion (i.e., hypoprolactinemia; [20,21]). Secretion of prolactin is predominantly regulated by hypothalamic prolactin-inhibiting factors, specifically the monoamine neurotransmitter dopamine [22]. Dopamine is considered the major regulator of prolactin secretion, and secretion is inhibited when dopamine is bound to its receptor on the lactotrophs of the anterior pituitary [23,24]. There are five isoforms of dopamine receptors that are divided into two subfamilies: D1-like, which is comprised of D1 and D5, and D2-like, which is comprised of D2, D3, and D4 [25]. On the pituitary lactotrophs, the dopamine receptors are primarily the dopamine receptor D2 (DRD2) [26]. The dopamine receptor D2 is coupled to a Giα protein, which inhibits adenylyl cyclase activity, cyclic adenosine monophosphate (cAMP), and cytoplasmic calcium concentrations when dopamine or an agonist is bound [27,28], thus resulting in a suppression of prolactin secretion.

Prolactin concentrations have been shown to be reduced in animals consuming endophyte-infected tall fescue and this is often used as an indicator of fescue toxicosis [10]. Ergot alkaloids, specifically ergovaline, have a high affinity towards DRD2 in vitro and elicit agonistic effects that result in a decrease in prolactin secretion [29]. Moreover, a recent study conducted gene expression profiles of pituitaries from steers grazing either high (HE; 746 parts per billion (ug/kg)) or low (LE; 23 ug/kg) endophyte-infected tall fescue pastures [19]. Li et al. [19] demonstrated DRD2 and PRL expression decreased in HE steers by 53% and 82%, respectively. Since PRL synthesis is directly related to DRD2 signaling pathway, a single nucleotide polymorphism (SNP) in the DRD2 gene could serve as a marker for resistance to fescue toxicosis as proposed by Campbell et al. [30].

Prolactin plays a role in numerous biological functions, most notably in lactation, but also in reproduction, immune responses, and metabolism [24]. Additionally, an elevation in prolactin is associated with a rise in environmental temperature or longer day length (i.e., photoperiod; [31,32]), and has been related to hair coat shedding in numerous species [33–36]. It has been speculated that hyperprolactinemia due to consumption of ergot alkaloid prevents shedding of the winter hair coat, therefore resulting in an elevation in core body temperature and increased vulnerability to heat stress (i.e., hyperthermia; [36]). In regards to reproduction, the heat stress-like signs greatly associated with fescue toxicosis have a large impact on reproductive success in livestock species and will be described in further detail in further sections of this review. Additionally, the interaction between ergot alkaloids and lactation, in which prolactin plays a critical role, will be discussed in greater detail in subsequent sections of this review.

2.2. Gonadotropins

The gonadotropins, luteinizing hormone (LH) and follicle stimulating hormone (FSH), are produced by the gonadotrophs of the anterior pituitary in response to gonadotropin releasing hormone (GnRH) secreted from the hypothalamus and govern reproductive cyclicity [37]. Browning Jr. et al. [38]

demonstrated in steers receiving injections of ergotamine tartrate (23.8 μg/kg body weight) that LH concentrations were reduced. Therefore, it was then investigated if acute ergot alkaloid exposure (19 μg/kg body weight) would alter LH and FSH in primiparous cows during the late luteal phase (day 15 or 16 post-estrus). Luteinizing hormone concentrations were reduced 4 h post injection, however FSH concentrations did not differ between the cows receiving ergotamine tartrate and the saline control [39]. An in vitro study using ovine pituitary cells demonstrated that bromocriptine (ergocryptine derivative) inhibits LH and FSH secretion in response to GnRH [40]. Conversely, a separate study found no differences in LH secretion in postpartum beef cows nor in cycling heifers and cows [41]. Additional studies further described that ergot alkaloids do not suppress LH or equine chorionic gonadotropin (eCG) in ewes or mares, respectively [42,43]. Most recently, Li et al. [19] found minimal differences in pathways utilized for FSH and LH production, secretion, or signaling from pituitary tissue collected from steers grazing either HE or LE fescue pastures. Variations in ergot alkaloid source and concentration, route of administration, environmental conditions, and physiological status of the animal could account for these discrepancies, and thus additional research is needed to better understand the actions ergot alkaloids have on gonadotropin synthesis, secretion, and functionality.

3. Ergot Alkaloids and Ovarian Function

Reproductive failure in cattle following ergot alkaloid exposure can be attributed to altered ovarian follicle development, luteal dysfunction, and reduced circulating steroid hormone concentrations, subsequently leading to reduced pregnancy rates [44]. Conversely, ergot alkaloid exposure has minimal impact during reproductive cyclicity and a greater impact during pregnancy and post-partum in sheep and mares. Specifically, in the mare resulting in prolonged gestation, late-term foal loss, dystocia, thickened placentas, and agalactia [45], thus discussion on the effects of ergot alkaloids on ovarian function will emphasize the data collected on cattle.

3.1. Ovarian Blood Flow

Ergot alkaloids induce a vasoconstrictive response by interacting with biogenic amine receptors including serotonergic and adrenergic receptors [46]. Numerous studies have demonstrated ergot alkaloid, specifically ergovaline and ergotamine, induced vasoconstriction via the serotonin 2A (5-HT2A) receptor utilizing an in vitro bovine lateral saphenous vein bioassay [13,14]. Similarly, ergot alkaloids have a high affinity towards α_2-adrenergic receptors [47,48]. Additionally, a couple of studies have used Doppler ultrasonography to show that heifers chronically exposed to endophyte-infected tall fescue have reduced caudal artery area and blood flow to the peripheral arteries when compared to heifers consuming endophyte-free tall fescue [49,50], however literature describing the extent to which vasoconstriction occurs to the internal organs is limited [16].

Recently, Poole et al. [50] investigated if chronic exposure of ergot alkaloids would decrease the diameter of the utero-ovarian blood vessels thus reducing systemic blood flow to the ovary during various stages of the estrous cycle. Ovarian artery and vein area was measured via Doppler ultrasonography on days 0, 4, 10, and 17 to represent both the follicular and luteal phases of the estrous cycle. Ovarian artery area was not different on days 0 and 4, however ovarian artery area was reduced on days 10 and 17 in heifers consuming ergot alkaloids. Additionally, minimal changes were observed in the ovarian vein area, most likely due to the reduction of vascular smooth muscle cells surrounding veins compared to arteries. Previous studies have demonstrated estrogen-induced vasodilation in various arteries and veins [51], and increased estrogen concentration early in the estrous cycle may have prevented the ergot alkaloid induced reduction in ovarian artery area that was observed during the luteal phase [50]. Ultimately, ergot alkaloid induced vasoconstriction of the utero-ovarian vessels would limit nutrients essential to ovarian function potentially altering sex steroid synthesis, altered follicular, and/or luteal development as well as dysregulation of the estrous cycle.

3.2. Folliculogenesis

As previously mentioned, the gonadotropins (LH and FSH) are produced by the anterior pituitary in response to GnRH and govern reproductive cyclicity [37]. Follicle stimulating hormone initiates the proliferation of granulosa cells and aid in the transition from primordial to primary and secondary follicles. As these follicles grow and undergo selection, granulosa cells become more responsive to LH and theca cells will produce androgens, which are converted primarily to estradiol (E2) in the granulosa cells [52,53]. As E2 and inhibin concentrations produced by the growing selected follicles increase, FSH production by the anterior pituitary is suppressed via negative feedback [54,55]. A few selected follicles will continue to grow into dominance and produce more E2, while others will become atretic. During the final follicular wave (post-luteolysis), large concentrations of E2 produced by the preovulatory follicle will act via positive feedback to trigger the preoptic nucleus of the hypothalamus to release a surge of GnRH, which will stimulate the anterior pituitary to release the preovulatory surge of LH and ovulation will occur [56–58].

McKenzie and Erickson [59,60] observed a decrease in the diameter and number of large follicles in heifers consuming endophyte-infected tall fescue (Table 1). Likewise, Burke and Rorie [61] examined follicular development and estrogen concentrations in lactating beef cows grazing endophyte-free (EF) or endophyte-infected (EI) tall fescue. No differences were found in the number of class 1 (small; 3 to 5 mm) and class 3 (large; >10 mm) antral follicles between treatments. Conversely, the number of class 2 (medium; 6–9 mm) follicles was reduced in cows grazing EI fescue compared to cows grazing EF fescue (Table 1). Similarly, Poole et al. [50] observed that a 6 to 9 mm follicle number was reduced in heifers consuming ergot alkaloids (Table 1). The 6 to 9 mm follicle size can be classified as selected follicles, and are of critical importance to follicular development with the gonadotropin dependence switching from FSH to LH. These results suggest that exposure to ergot alkaloids may hinder follicular selection though inadequate delivery of gonadotropins and other nutrients due to insufficient blood flow to the ovary.

Table 1. A review of the effects of ergot alkaloids and/or heat stress on follicular dynamics.

Alkaloid (μg/kg/day)	Heat Stress	Average Daily Gain	Animal	Effect	Source
N/A	N/A	N/A	Beef heifer	Decrease in the diameter and number of large follicles.	[59]
N/A	N/A	N/A	Beef heifer	Decrease in the diameter and number of large follicles.	[60]
N/A	Possibly, grazing from April to Sept. in Arkansas	Reduced	Mature beef cows	Diameter of the largest follicle tended to be smaller. No difference in the number of small or large follicles. Number of medium sized follicles reduced.	[61]
10,310 total; 3910 ergovaline	Possibly, barn temperature. 27 °C and average THI [1] 72	Reduced, same intake	Beef heifer	Number of medium sized follicles reduced. No difference in the number of small or large follicles or diameter of largest follicle.	[50]
1900 ergovaline	Yes, between 25 °C and 31 °C	N/A, reduced intake	Beef heifer	Decrease in the diameter and number of large follicles.	[62]
1000 total; 421 ergovaline	Yes, average THI 88	Same	Beef heifer	Decrease in the diameter and number of small and large follicles.	[63]

[1] THI = temperature-humidity index; THI = $(1.8 \times T_{db} + 32) - ((0.55 - 0.0055 \times RH) \times (1.8 \times T_{db} - 26.8))$; T_{db} = dry bulb temperature (°C), and RH = relative humidity. Recovery (non-life threatening, 75) and emergency (high-risk, 85) thresholds as described by Hahn [64].

Furthermore, many of these signs in cattle exposed to ergot alkaloids are amplified during periods of heat stress. The inability to maintain a thermoneutral body temperature due to increased ambient temperatures has been previously shown to impair folliculogenesis [65,66]. Therefore, Burke et al. [62]

investigated the interaction between heat stress and consumption of endophyte-infected fescue on follicular dynamics. These authors controlled dietary intake for the heat stressed heifers on both diets to minimize variation in intake. Heifers consuming the EI seed in heat stress conditions resulted in a decreased number of large follicles (>9 mm) in addition to having a smaller preovulatory follicle diameter compared to control heifers (Table 1). Recently, a genetic trait has been identified in Bos taurus-influenced breeds, Senepol, and other Criollo cattle breeds, that is associated with high heat tolerance and a slick hair coat [67–70]. This slick hair coat phenotype is due to a frame-shift mutation in the prolactin receptor [70]. Therefore, Poole et al. [63] evaluated the effect of the slick trait and exposure to ergot alkaloids on follicular dynamics in beef heifers. Dietary intake and average daily gain (ADG) remained constant between groups. Heifers consuming the EI fescue with a wild-type hair coat (lacking the slick hair mutation) had an increase in the number of preselected follicles (2 to 4 mm), however, no change in the number of selected follicles (5 to 8 mm), yet a decrease in the number of preovulatory follicles (>9 mm) compared to the other heifer groups (wild-type hair coat consuming EF fescue and heifers with a slick-type hair coat consuming EI or EF fescue; Table 1). Intriguingly, this lack of follicular transition indicates a dysregulation during follicular selection during folliculogenesis that was not observed in heifers possessing the slick hair trait and consuming the EI fescue. Together, both Burke et al. [62] and Poole et al. [63] demonstrated that heat stress altered fescue toxicosis and alters the efficiency of ovarian follicular selection and dominance.

3.3. Corpus Luteum

The corpus luteum (CL) is a transient endocrine organ that forms on the ovary following ovulation. When a follicle ovulates, the granulosa and theca cells undergo dramatic changes into luteal cells, a process known as luteinization. Progesterone (P4) is the primary hormone produced by the CL and has numerous functions including suppression of ovulation and maintenance of pregnancy. It has been demonstrated that LH, not prolactin, is the predominant luteotropic hormone responsible for maintenance of the CL and production of P4 in livestock species [71–73]. Specifically, the functionality of the CL (i.e., production of P4) is dependent on the degree of vascularization or angiogenesis [74,75].

Due to the vasoconstrictive effects of ergot alkaloids, many researchers speculated that chronic exposure would result in luteal dysfunction, thus reducing pregnancy rates. Estienne et al. [76] observed a reduction in circulating P4 concentrations in heifers on endophyte-infected fescue even though no differences in CL size or presence were observed via ultrasonography. Interestingly, a separate study found that even if heifers appeared to be cycling and ovulating normally, there were cellular changes (fewer nuclei and a greater number of large luteal cells with increased diameter) of the CL in heifers grazing endophyte-infected fescue, which may contribute to altered functionality [77]. As previously mentioned, Poole et al. [63] observed a decrease in the diameter of the ovulatory follicle in EI heifers, and because of the process of luteinization it is not surprising that the luteal area (mm^2) was also reduced in heifers consuming the EI fescue.

There have been varying reports regarding the impact of fescue toxicosis on luteal formation and function with numerous findings observing no differences [50,78,79]. This variation in responses could be due to the fact that circulatory ovarian steroid concentrations, specifically progesterone, are not only dependent on the rate of secretion but also on the metabolism in the liver and on the incidence of vasodilation or vasoconstriction.

3.4. Ovarian Steroidogenesis

One theory to explain the altered follicular dynamics and luteal dysfunction in animals consuming endophyte-infected fescue is a reduction in the steroid hormone (E2 and P4) precursor, cholesterol. Following synthesis in the liver from low-density and high-density lipoproteins; cholesterol must be transported to the ovary for sex steroid synthesis in the thecal, granulosal, and luteal cells [80,81]. Few studies have shown a decrease in circulating cholesterol concentrations in cattle consuming EI tall fescue [62,82,83]. Moreover, Burke et al. [62] observed that heat stress conditions further reduced cholesterol concentrations in heifers consuming EI fescue.

3.4.1. Estradiol

Estrogen production is a critical component of a healthy developing follicle [52] and essential for reproductive success. Burke et al. [62] observed that E2 concentrations were reduced in heifers consuming EI fescue compared to the control heifers in the thermoneutral environment, whereas the additive effect of heat stress reduced E2 concentrations regardless of ergot alkaloid exposure. However, these results were not observed in postpartum cows consuming either EF or EI tall fescue [61]. Interestingly, Poole et al. [63] demonstrated that EI heifers with a wild-type hair coat (heat stressed) had decreased E2 concentrations, yet EI heifers possessing the slick hair trait had similar E2 concentrations to EF heifers. Collectively, ergot alkaloids impair follicular development and E2 secretion; however, it remains unknown if ergot alkaloids directly impact granulosal and thecal cell function or indirectly alter folliculogenesis through reduced blood flow to the ovary.

3.4.2. Progesterone

Similar to the varying reports regarding the impact of fescue toxicosis on CL formation, there are contrasting reports regarding P4 synthesis and secretion from the CL. Mahmood et al. [84] examined luteal function in heifers grazing either low (0%) or high (>75%) EI fescue pastures for 168 days. Heifers were synchronized with prostaglandin $F_{2\alpha}$ ($PGF_{2\alpha}$) on days 101 and 112 of the trial and P4 concentrations were determined on days 112, 116, 120, and 124. Heifers on the high EI fescue pastures had either low P4 concentrations (<1.5 ng/mL) after synchronization or relatively high P4 concentrations (>1.5 ng/mL) that would sharply decrease, thus indicating luteal dysfunction or shorten luteal phase [85], respectively. Both Burke et al. [62] and Poole et al. [63] demonstrated that heat stress exacerbated fescue toxicosis and observed a reduction in P4 concentrations. Conversely, numerous reports have found no differences in P4 secretion in animals consuming EI fescue [50,78,79]. Interestingly, Jones et al. [86] evaluated P4 concentrations in EF, EI, and EI heifers treated with domperidone (EID; dopamine antagonist) during the months of May and June in Southern Illinois. Heifers consuming EI fescue had reduced mid-cycle P4 concentrations when compared to EF and EID heifers. Furthermore, cultured luteal cells collected from a subset of heifers from each treatment group revealed no differences observed in P4 secretion in vitro [86]. The authors suggest that utilization of domperidone in vivo may alleviate some of the signs associated with fescue toxicosis (i.e., reduced PRL and heat stress) to improve CL function.

4. Ergot Alkaloids and Uterine Function

4.1. Uterine Blood Flow

Dyer [87] was the first to report the interaction between ergovaline and serotonin (5-HT) receptors in bovine uterine arteries. Furthermore, Poole et al. [50] measured uterine artery and vein area via Doppler ultrasonography on days 0, 4, 10, and 17 of the estrous cycle in EF and EI beef heifers. It was observed that uterine artery and vein areas were not different on days 0 and 4, however, uterine artery and vein areas did differ on days 10 and 17 with heifers consuming ergot alkaloids and this is the time of the estrous cycle when P4 concentrations are greatest. The ergot alkaloids induced vasoconstriction of the uterine vessels occurred prior to the timing of maternal recognition of pregnancy (day 14–16) in cattle, Poole et al. [50] speculated that this could reduce hormonal communication between the ovary and uterus during this time of embryonic signaling to the endometrium, thus decrease pregnancy retention [88]. Vasoconstrictive activity has also been detected in ovine uterine arteries [89]. Pregnant Suffolk ewes were subjected to either an EF or EI (1770 μg/day ergovaline) diet from day 35 to 86 of gestation (Period 1), then were fed the same diet either throughout or received a crossover diet with fescue type opposite of the original diet until termination of pregnancy on day 133 of gestation (Period

2). Ewes fed EI fescue during Period 2 had reduced uterine vessel area, however ewes fed EI fescue during Period 1 then switched to EF fescue did not display this reduced vessel area. Moreover, utilizing an in vitro bioassay the uterine arteries collected from EF ewes during Period 2 were responsive to serotonin and ergot alkaloids (ergotamine and ergovaline) [89]. In regards to the mare, uterine arteries collected from non-pregnant mares were subjected to an in vitro bioassay, however, were not responsive to serotonin, ergotamine, or ergovaline [90]. Klotz and McDowell [90] suggested that effects of ergot alkaloids on reproductive failure in the mare might not be a consequence of vasoconstriction and restricted blood flow.

4.2. Prostaglandin $F_{2\alpha}$ ($PGF_{2\alpha}$) Synthesis

It has been suggested that ergot alkaloids have an oxytocic effect (i.e., contractile response) on the uterus [91]. During late luteal phase of the estrous cycle, oxytocin acts on localized receptors on the endometrium to stimulate synthesis of $PGF_{2\alpha}$, thus triggering luteolysis [92,93]. A few studies have evaluated the effect of ergot alkaloids on $PGF_{2\alpha}$ secretion. Browning Jr. et al. [39] injected ergotamine tartrate to primiparous cows during the late luteal phase (day 15 or 16 post-estrus) and observed that $PGF_{2\alpha}$ concentrations were elevated just one hour post injection and continued to increase every hour for four hours. This response mirrored that of pulsatile $PGF_{2\alpha}$ response observed during luteolysis; unfortunately, the authors did not evaluate luteal function or regression. Vogt Engeland et al. [94] administered 105 μg ergotamine per kilogram of body weight via oral drench twice daily from day 98 to 107 of gestation in dairy goats, and observed that ergotamine treated goats had significantly greater concentrations of $PGF_{2\alpha}$ resulting in a greater incidence of induced parturition and fetal death [94].

5. Ergot Alkaloids and Pregnancy

A common symptom of fescue toxicosis is reduced pregnancy rates, specifically in cattle [9,10,95]). According to a review by Kallenbach [9], fescue toxicosis attributes to over $2 billion in annual economic loss to the U.S. livestock industries, primarily due to reproductive loss (Figure 1). In cattle, this reproductive loss is because of a failure to conceive or early embryonic loss [9]. Additionally, many livestock species experience difficulties during late-gestation and this has an impact on fetal and neonatal development [9,45,96].

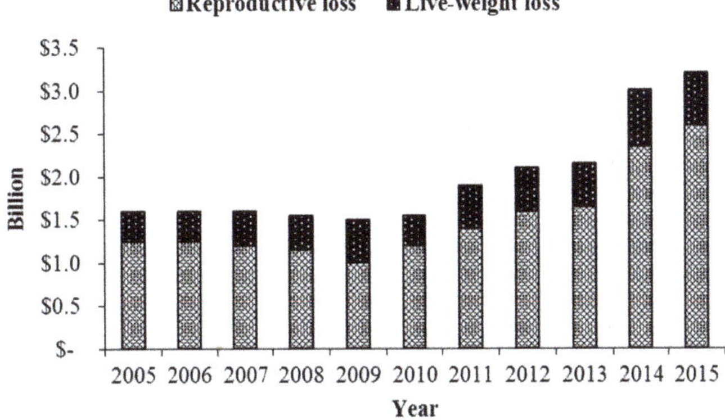

Figure 1. Fescue toxicosis severely reduces calving rates with 16% of reproductive beef cows failing to conceive or experiencing early embryonic loss. Calves born in a fescue environment have an average reduction in weaning weight by 22.6 kg (Reproduced from Kallenbach [9]. 2015, Oxford University Press).

5.1. Early Embryonic Development

A few studies have evaluated the impact that ergot alkaloids have on oocyte competency and early embryonic development. Jones et al. [97] cultured cumulus-oocyte-complexes (COCs) with either a control medium, EF-treated medium with 10% EF plasma, or EI-treated medium with 10% EI plasma supplemented. Plasma used as a treatment was previously collected when the heifers were exposed to EF or EI pastures for 24 days. There were no differences in the percent of COCs that progressed to metaphase II (MII). Additionally, ovum pick-up was performed on the heifers and the grade I oocytes (≥5 layers of compact cumulus cells and homogenous cytoplasm; [98]) were subjected to traditional in vitro procedures. Interestingly, there was a difference observed with 66% of the EF grade I oocytes progressing to MII versus 0% of the EI grade I oocytes, thus demonstrating that in vivo exposure to endophyte-infected tall fescue can directly inhibit proper oocyte maturation [97].

Schuenemann et al. [99] explored the impact of ergot alkaloids on early embryonic development (experiment 1) and uterine receptivity (experiment 2) in vivo. Cattle were allotted to receive either the control (CON) or an ergot alkaloid seed (EI) diet. In experiment 1, uterine horn ipsilateral to the CL was flushed for embryo recovery following estrous synchronization and artificial insemination. Embryo recovery tended to be more successful in CON cattle versus EI cattle. Of the embryos recovered, a greater percent of embryos from CON animals had developed to compacted morula or blastocyst, and there was a greater percent of better quality embryos from CON cattle versus ET cattle [99]. In experiment 2, two frozen-thawed good quality embryos were transferred to recipients in both treatment groups seven days following synchronized estrus. Interestingly, pregnancy rates following transfer did not differ [99]. The authors concluded that the uterine environment is suitable to maintain pregnancy after day 7 of gestation, however ergot alkaloid exposure appears to detrimentally affect either the oocyte or the early embryo prior to the blastocyst stage.

As previously mentioned, elevated body temperature and heat stress-like signs are greatly associated with cattle exposed to ergot alkaloids. It is also well established that heat stress can have a negative effect on most aspects of female reproduction including oogenesis, oocyte maturation, and early embryonic development [100]. In fact, results from numerous studies evaluating heat stress and early embryonic development mirror the results observed by Schuenemann et al. [99]. For example, Ealy et al. [101] observed that exposure to heat stress conditions at day 1 post-estrus (two-cell cleaved embryos), reduced the percent of embryos that developed to the blastocyst stage. However, heat stress exposure at days 3, 5, and 7 had no effect on the percent of embryos that were blastocysts. Likewise, Edwards and Hansen [102] found that heat stress did not impact oocytes during the first 12 h of maturation, however it greatly reduced the number of two-cell embryos that developed to the blastocyst stage in vitro. While ergot alkaloid exposure inhibits early embryonic development in vitro, in combination with the inability to maintain a thermoneutral body temperature due to heat stress severely impacts embryonic development resulting in decreased pregnancy rates in many spring calving herds throughout the Southeastern and Mid-Atlantic states.

Interestingly, a few studies have evaluated effects on embryonic development at levels greater than 300 μg/kg in mares (Table 2). At 867 μg/kg ergovaline, there was no impact on embryonic development or establishment of pregnancy in mares [43]. However, at 1171 μg/kg, there was an increased incidence of early embryonic loss and reduced pregnancy rates in mares [103]. Minimal effects were observed if mares were exposed to less than 300 μg/kg ergovaline [104]. Unlike cattle exposed to ergot alkaloids, mares do not experience elevated body temperatures, which is most likely due to evaporative cooling because of increased sweating capability [105], which may mitigate the negative effects of ergot alkaloids on early embryo loss. The effects of ergot alkaloids on pregnant mares are much greater during late gestation [106], suggesting that ergot alkaloids potentially impact placental efficiency.

Table 2. A review of the effects of ergovaline on early embryonic development and establishment of pregnancy in the mare.

Ergovaline (μg/kg)	Effect	Source
45	No negative effects	[107]
160	No negative effects	[108]
271	No negative effects	[107]
308	Suppressed PRL, no negative effects on pregnancy outcomes	[108]
867	Decreased P4, no negative effects on pregnancy outcomes	[43]
1171	Increase in early embryonic loss and reduced pregnancy rates	[104]

5.2. Placenta

It has been shown that ergot alkaloids can cross the placental barrier in rodents [109,110], but conclusive evidence in livestock species is unavailable. However, similar to the uterine artery, there have been reports that ergot alkaloids induce a vasoconstrictive response in both bovine and ovine umbilical arteries [87,89]. Britt et al. [111] examined placental characteristics in pregnant Suffolk ewes following exposure to EF or EI (1770 μg/day ergovaline) diets and found that ewes subjected to an EI diet during Period 2 experienced an overall reduction in total caruncle, cotyledon, and placentome weight. In sheep, the placenta increases in vascularity after day 80 of gestation to support rapid fetal growth [112,113]. Therefore, it is believed that this vasoconstrictive activity on the umbilical artery reduces blood flow to the placenta and subsequently the fetus, which results in reduced birth weights [89,111]. Ergot alkaloid induced effects on the placenta are evident in the mare. Early reports suggested that pregnant mares grazing endophyte-infected tall fescue experience reproductive failure [114,115]. Specifically, Monroe et al. [116] demonstrated that mares grazing EI tall fescue during late gestation experienced prolonged gestation, increased number of retained placentas, and increased placental weight and thickness when compared to mares grazing EF tall fescue. Due to these findings, it is now highly recommended that pregnant mares are removed from EI tall fescue pastures during the third trimester to avoid any serious complications.

5.3. Fetal Programming

Studies have found that ergot alkaloid exposure during gestation results in a reduction in birth weight in lambs [96,111] and calves [117,118]. There are a few theories to explain this reduction in birth weight. One being the previously described vasoconstriction to the uterine and placental arteries, however another is a decrease in daily nutrient intake by the dam, thus resulting in a reduction in dam body weight [111,118]. Either mechanism results in a reduction in placental growth and supply of nutrients to the developing fetus. Placental growth is partly regulated by the paternally imprinted gene, insulin-like growth factor-2 (IGF2), and when there are modifications or changes in expression of imprinted genes, then this is associated with developmental programming and a reduction in birth weight [119,120]. Interestingly, Britt et al. [111] found that ewes exposed to ergot alkaloids during Period 1 and 2 of gestation had increased mRNA expression of IGF2 in cotyledon tissue, however these differences were not observed in ewes only exposed to ergot alkaloids during Period 1 or 2. The authors speculated that these adverse conditions (i.e., exposure to ergot alkaloids) resulted in the increase in IGF2 expression to aid in placental adaptation to the conditions [111].

6. Lactation

The effects of ergot alkaloids on lactation vary based on the livestock species. Consumption of ergot alkaloids reduces milk yield in cattle and sheep [44]. However, while prolactin plays a critical role in mammary gland development and milk synthesis [121], it is important to note that decreased prolactin concentrations does not directly influence milk yield in these species [122] and ergot alkaloid exposure during the dry period does not impair mammary development or milk production in the following lactation [123]. More recently, Capuco et al. [124] described changes in the mammary gland transcriptome related to lipid metabolism, and molecular transport following exposure to endophyte-infected fescue seed, which could contribute to reduced milk production as previously reported by Baldwin et al. [123]. However, cofounding factors such as reduced feed intake following exposure to endophyte-infected fescue seed in these studies [123,124] limits identification of the exact mechanism of action of ergot alkaloids on the mammary gland. Therefore, other associated signs of fescue toxicosis, such as reduced feed intake or vasoconstriction, play a more critical role in the reduction of milk production in cattle and sheep. In contrast, horses exhibit complete agalactia when exposed to ergot alkaloids [45]. Unlike ruminants, which produce both placental lactogen and prolactin to initiate prepartum lactogenesis, horses rely solely on prolactin [125]. Thus, horses become agalactic when grazing endophyte-infected tall fescue due to ergot alkaloids agonistic effects of DRD2 and subsequent decrease in prolactin secretion. Many proposed strategies are available to mitigate the signs of fescue toxicosis in mares, but perhaps the one of the most effective to improve lactation is administering a dopamine antagonist, domperidone [45]. Redmond et al. [126] conducted a study to determine the minimum effective oral dose of domperidone (1.1, 1.65, 2.2 mg/kg BW) to treat fescue toxicosis in late-gestation mares. Redmond et al. [126] concluded that the minimum oral dose of 1.1 mg/kg BW was effective to alleviate signs of fescue toxicosis when provided daily for 30 days before foaling. A follow-up study determined that subcutaneous administration of 0.44 mg/kg BW domperidone 10 days before foaling was also effective to alleviate signs and improve lactation in mares [127]. While domperidone is an effective alleviator of agalactia in mares grazing EI tall fescue, most veterinarians would recommend removal from EI tall fescue pastures at least 30 days before the expected foaling date [45].

7. Conclusions

While evidence of reduced reproductive performance of animals consuming endophyte-infected tall fescue has been extensively studied in an attempt to find remedies for, or offset the negative impact of, fescue toxicosis; the complex etiology of this syndrome has hindered an exploration of specific mechanisms of action of ergovaline on specific tissues. Seasonal or annual fluctuations in ergot alkaloid concentrations in combination with the age and genetic background of the animal, elevated environmental conditions, and/or hypoxic conditions at the cellular level influence the impact ergot alkaloids have on the reproductive tissues leading to inconsistencies in reduced reproductive performance in animals consuming ergot alkaloid-contaminated diets. Further exploration into the precise mechanism of action of ergot alkaloids on the hypothalamic-pituitary-gonadal axis through innovative research combining cellular and molecular techniques with applied experimental models will lead to a better understanding of the negative impact these toxins on reproductive processes. Moreover, this knowledge will lead to inventive tools and strategies enhance best management practices to improve reproductive performance in animals consuming endophyte-infected tall fescue.

Author Contributions: R.K.P and D.H.P conceived, designed and wrote the paper.

Funding: This research received no external funding.

Conflicts of Interest: The author declares no conflict of interest.

References

1. Jakubczyk, D.; Cheng, J.Z.; O'Connor, S.E. Biosynthesis of the ergot alkaloids. *Nat. Prod. Rep.* **2014**, *31*, 1328–1338. [CrossRef] [PubMed]
2. Bacon, C.; Porter, J.; Robbins, J.; Luttrell, E. *Epichloë typhina* from toxic tall fescue grasses. *Appl. Environ. Microbiol.* **1977**, *34*, 576–581. [PubMed]
3. Glenn, A.E.; Bacon, C.W.; Price, R.; Hanlin, R.T. Molecular phylogeny of *Acremonium* and its taxonomic implications. *Mycologia* **1996**, *88*, 369–383. [CrossRef]
4. Leuchtmann, A.; Bacon, C.W.; Schardi, C.L.; White, J.F., Jr.; Tadych, M. Nomenclatural realignment of *Neotyphodium* species with genus *Epichloë*. *Mycologia* **2014**, *106*, 202–215. [CrossRef] [PubMed]

5. Lyons, P.C.; Plattner, R.D.; Bacon, C.W. Occurrence of peptide and clavinet ergot alkaloids in tall fescue grass. *Science* **1986**, *232*, 487–489. [CrossRef]
6. Burns, J.C.; Chamblee, D.S. Adaptation. In *Tall Fescue. Agronomy Monographs 20*; Buckner, R., Bush, L., Eds.; ASA, CSSA, SSSA: Madison, WI, USA, 1979; Chapter 3; pp. 9–30.
7. Strickland, J.R.; Aiken, G.E.; Spiers, D.E.; Fletcher, L.R.; Oliver, J.W. Physiological Basis of Fescue Toxicosis. In *Tall Fescue for the Twenty-First Century. Agronomy Monographs 53*; Fribourg, H.A., Hannaway, D.B., West, C.P., Eds.; ASA, CSSA, SSSA: Madison, WI, USA, 2009; Chapter 12; pp. 203–227.
8. Siegel, M.R.; Latch, G.C.M.; Johnson, M.S. *Acremonium* fungal endophytes of tall fescue and perennial ryegrass: Significance and control. *Plant Dis.* **1985**, *69*, 179–181.
9. Kallenbach, R.L. BILL E. KUNKLE INTERDISCIPLINARY BEEF SYMPOSIUM: Coping with tall fescue toxicosis: Solutions and realities. *J. Anim. Sci.* **2015**, *93*, 5487–5495. [CrossRef]
10. Strickland, J.R.; Looper, M.L.; Matthews, J.C.; Rosenkrans, C.F., Jr.; Flythe, M.D.; Brown, K.R. BOARD-INVITED REVIEW: St. Anthony's fire in livestock: Causes, mechanisms, and potential solutions. *J. Anim. Sci.* **2011**, *89*, 1603–1626. [CrossRef]
11. Berde, B. Ergot compounds: A synopsis. In *Ergot Compounds and Brain Function: Neuroendocrine and Neuropsychiatric Aspects*; Goldstein, M., Lieberman, A., Calne, D.B., Thorner, M.O., Eds.; Raven Press: New York, NY, USA, 1980; pp. 4–23.
12. Elsasser, T.H.; Bolt, D.J. Dopaminergic-like activity in toxic fescue alters prolactin but not growth hormone or thyroid stimulating hormone in ewes. *Domest. Anim. Endocrinol.* **1987**, *4*, 259–269. [CrossRef]
13. Klotz, J.L.; Brown, K.R.; Xue, Y.; Matthews, J.C.; Boling, J.A.; Burris, W.R.; Bush, L.P.; Strickland, J.R. Alterations in serotonin receptor-induced contractility of bovine lateral saphenous vein in cattle grazing endophyte-infected tall fescue. *J. Anim. Sci.* **2012**, *90*, 682–693. [CrossRef]
14. Klotz, J.L.; Aiken, G.E.; Johnson, J.E.; Brown, K.R.; Bush, L.P.; Strickland, J.R. Antagonism of lateral saphenous vein serotonin receptors from steers grazing endophyte-free, wild-type, or novel endophyte-infected tall fescue. *J. Anim. Sci.* **2013**, *91*, 4492–4500. [CrossRef] [PubMed]
15. Klotz, J.L.; Aiken, G.E.; Bussard, J.R.; Foote, A.P.; Harmon, D.L.; Goff, B.M.; Schrick, F.N.; Strickland, J.R. Vasoactivity and Vasoconstriction Changes in Cattle Related to Time off Toxic Endophyte-Infected Tall Fescue. *Toxins* **2016**, *8*, 271. [CrossRef] [PubMed]
16. Klotz, J.L. Activities and Effects of Ergot Alkaloids on Livestock Physiology and Production. *Toxins* **2015**, *7*, 2801–2821. [CrossRef] [PubMed]
17. Nakane, P.K. Classifications of anterior pituitary cell types with immunoenzyme histochemistry. *J. Histochem. Cytochem.* **1970**, *18*, 9–20. [CrossRef] [PubMed]
18. Thompson, F.N.; Stuedemann, J.A.; Sartin, J.L.; Belesky, D.P.; Devine, O.J. Selected hormonal changes with summer fescue toxicosis. *J. Anim. Sci.* **1987**, *65*, 727–733. [CrossRef]
19. Li, Q.; Hegge, R.; Bridges, P.J.; Matthews, J.C. Pituitary genomic expression profiles of steers are altered by grazing of high vs. low endophyte-infected tall fescue forages. *PLoS ONE* **2017**, *12*, e0184612. [CrossRef] [PubMed]
20. Nasr, H.; Pearson, O.H. Inhibition of prolactin secretion by ergot alkaloids. *Acta Endocrinol.* **1975**, *80*, 429–443. [CrossRef]
21. Hurley, W.L.; Convey, E.M.; Leung, K.; Edgerton, L.A.; Hemken, R.W. Bovine prolactin, TSH, T and T concentrations as affected by tall fescue summer toxicosis and temperature. *J. Anim. Sci.* **1980**, *51*, 374–379. [CrossRef]
22. Ben-Johnathan, N. Dopamine: A prolactin-inhibiting hormone. *Endocr. Rev.* **1985**, *6*, 564–589. [CrossRef]
23. Leong, D.A.; Frawley, L.S.; Neill, J.D. Neuroendocrine control of prolactin secretion. *Annu. Rev. Physiol.* **1983**, *45*, 109–127. [CrossRef]
24. Freeman, M.E.; Kanyicska, B.; Lerant, A.; Nagy, G. Prolactin: Structure, Function, and Regulation of Secretion. *Physiol. Rev.* **2000**, *80*, 1523–1631. [CrossRef] [PubMed]
25. Sibley, D.R.; Monsma, F.J., Jr. Molecular biology of dopamine receptors. *Trends Pharmacol. Sci.* **1992**, *13*, 61–69. [CrossRef]
26. Caron, M.C.; Beaulieu, M.; Raymond, V.; Gange, B.; Drouin, J.; Lefkowitz, J.; Labrie, F. Dopaminergic receptors in the anterior pituitary gland. *J. Biol. Chem.* **1978**, *53*, 2244.
27. Enjalbert, A.; Bockaert, J. Pharmacological characterization of D2 dopamine receptors negatively coupled with adenylate cyclase in the rate anterior pituitary. *Mol. Pharmacol.* **1983**, *23*, 576–584. [PubMed]

28. Lledo, P.M.; Homburger, V.; Bockaert, J.; Vincent, J.D. Differential G Protein-Mediated Coupling of D2 Dopamine Receptors to K+ and Ca2+ Currents in Rat Anterior Pituitary Cells. *Neuron* **1992**, *8*, 455–463. [CrossRef]
29. Larson, B.T.; Samford, M.D.; Camden, J.M.; Piper, E.L.; Kerley, M.S.; Paterson, J.A.; Turner, J.T. Ergovaline binding and activation of D2 dopamine receptors in GH4ZR7 cells. *J. Anim. Sci.* **1995**, *73*, 1396–1400. [CrossRef]
30. Campbell, B.T.; Kojima, C.J.; Cooper, T.A.; Bastin, B.C.; Wojakiewicz, L.; Kallenbach, R.L.; Schrick, F.N.; Waller, J.C. A single nucleotide polymorphism in the dopamine receptor D2 gene may be informative for resistance to fescue toxicosis in angus-based cattle. *Anim. Biotechnol.* **2014**, *25*, 1–12. [CrossRef]
31. Schams, D.; Reinhardt, V. Influence of the season on plasma prolactin levels in cattle from birth to maturity. *Hormone Res.* **1974**, *5*, 217. [CrossRef]
32. Wettemann, R.P.; Tucker, H.A. Relationship of ambient temperature to serum prolactin in heifers. *Proc. Soc. Exp. Biol. Med.* **1974**, *146*, 908. [CrossRef]
33. Thompson, D.L., Jr.; Hoffman, R.; DePew, C.L. Prolactin administration to seasonally anestrous mares: Reproductive, metabolic, and hair-shedding responses. *J. Anim. Sci.* **1997**, *75*, 1092–1099. [CrossRef]
34. Foitzik, K.; Krause, K.; Nixon, A.J.; Ford, C.A.; Ohnemus, U.; Pearson, A.J.; Paus, R. Prolactin and its receptor are expressed in murine hair follicle epithelium show hair cycle-dependent expression, and induce catagen. *Am. J. Pathol.* **2003**, *162*, 1611–1621. [CrossRef]
35. Nixon, A.J.; Ford, C.A.; Wildermoth, J.E.; Craven, A.J.; Ashby, M.G.; Pearson, A.J. Regulation of prolactin receptor expression in ovine skin in relation to circulating prolactin and wool follicle growth status. *J. Endocrinol.* **2002**, *172*, 605–614. [CrossRef] [PubMed]
36. Aiken, G.E.; Klotz, J.L.; Looper, M.L.; Tabler, S.F.; Schrick, F.N. Disrupted hair follicle activity in cattle grazing endophyte-infected tall fescue in the summer insulates core body temperatures. *Prof. Anim. Sci.* **2011**, *27*, 336–343. [CrossRef]
37. Hansel, W.; Convey, E.M. Physiology of the Estrous Cycle. *J. Anim. Sci.* **1983**, *57* (Suppl. 2), 404–424. [PubMed]
38. Browning, R., Jr.; Thompson, F.N.; Sartin, J.L.; Leite-Browning, M.L. Plasma concentrations of prolactin, growth hormone, and luteinizing hormone in steers administered ergotamine or ergonovine. *J. Anim. Sci.* **1997**, *75*, 796–802. [CrossRef] [PubMed]
39. Browning, R., Jr.; Schrick, F.N.; Thompson, F.N.; Wakefield, T., Jr. Reproductive hormonal repsonses to ergotamine and ergonovine in cows during the luteal phase of the estrous cycle. *J. Anim. Sci.* **1998**, *76*, 1448–1454. [CrossRef] [PubMed]
40. Hodson, D.J.; Henderson, H.L.; Townsend, J.; Tortonese, D.J. Photoperiodic modulation of the suppressive actions of prolactin and dopamine on the pituitary gonadotropin responses to gonadotropin-releasing hormone in sheep. *Biol. Reprod.* **2012**, *86*, 122. [CrossRef]
41. Mizinga, K.M.; Thompson, F.N.; Stuedemann, J.A.; Kiser, T.E. Effects of feeding diets containing endophyte-infected fescue seed on luteinizing hormone secretion in postpartum beef cows and in cyclic heifers and cows. *J. Anim. Sci.* **1992**, *70*, 3483–3489. [CrossRef]
42. Louw, B.P.; Lishman, A.W.; Botha, W.A.; Baumgartner, J.P. Failure to demonstrate a role for the acute release of prolactin at oestrus in the ewe. *J. Reprod. Fertil.* **1974**, *40*, 455–458. [CrossRef]
43. Brendemuehl, J.P.; Carson, R.L.; Wenzel, J.G.W.; Boosinger, T.R.; Shelby, R.A. Effects of grazing endophyte-infected tall fescue on eCG and progestogen concentrations from gestation days 21 to 300 in the mare. *Theriogenology* **1996**, *46*, 85–95. [CrossRef]
44. Porter, J.K.; Thompson, F.N., Jr. Effect of Fescue Toxicosis on Reproduction in Livestock. *J. Anim. Sci.* **1992**, *70*, 1594–1603. [CrossRef] [PubMed]
45. Cross, D.L.; Redmond, L.M.; Strickland, J.R. Equine Fescue Toxicosis: Signs and Solutions. *J. Anim. Sci.* **1995**, *73*, 899–908. [CrossRef] [PubMed]
46. Pertz, H.H.; Eich, E. Ergot alkaloids and their derivatives as ligands for serotoninergic, dopaminergic, and adrenergic receptors. In *Ergot*; Kren, V., Cvak, L., Eds.; Harwood Academic Publishers: Amsterdam, The Netherlands, 1999; pp. 441–450.
47. McPherson, G.A.; Beart, P.M. The selectivity of some ergot derivatives for alpha 1 and alpha 2-adrenoceptors of rat cerebral cortex. *Eur. J. Pharmacol.* **1983**, *91*, 363–369. [CrossRef]

48. Oliver, J.W.; Strickland, J.R.; Waller, J.C.; Fribourg, H.A.; Linnabary, R.D.; Abney, L.K. Endophytic fungal toxin effect on adrenergic receptors in lateral saphenous veins (cranial branch) of cattle grazing tall fescue. *J. Anim. Sci.* **1998**, *76*, 2853–2856. [CrossRef] [PubMed]
49. Aiken, G.E.; Kirch, B.H.; Strickland, J.R.; Bush, L.P.; Looper, M.L.; Schrick, F.N. Hemodynamic responses of the caudal artery to toxic tall fescue in beef heifers. *J. Anim. Sci.* **2007**, *85*, 2337–2345. [CrossRef] [PubMed]
50. Poole, D.H.; Lyons, S.E.; Poole, R.K.; Poore, M.H. Ergot alkaloids induce vasoconstriction of bovine uterine and ovarian blood vessels. *J. Anim. Sci.* **2018**, *96*, 4812–4822. [CrossRef] [PubMed]
51. Miller, V.M.; Duckles, S.P. Vascular actions of estrogens: Functional implications. *Pharmacol. Rev.* **2008**, *60*, 210–241. [CrossRef]
52. Fortune, J.E.; Quirk, S.M. Regulation of steroidogenesis in bovine preovulatory follicles. *J. Anim. Sci.* **1988**, *66* (Suppl. 2), 1–9.
53. Richards, J.S. Hormonal control of gene expression in the ovary. *Endocr. Rev.* **1994**, *15*, 725–751. [CrossRef]
54. Law, A.S.; Logue, D.N.; O'Shea, T.; Webb, T. Evidence for a novel factor in steroid free bovine follicular fluid (bFF) which acts to directly suppress follicular development. *J. Reprod. Fertil. Abstr. Ser.* **1990**, *5*, 4.
55. Turzillo, A.M.; Fortune, J.E. Suppression of the secondary FSH surge with bovine follicular fluid is associated with delayed ovarian follicular development in heifers. *J. Reprod. Fertil.* **1990**, *89*, 643–653. [CrossRef] [PubMed]
56. Wettemann, R.P.; Hafs, H.D.; Edgerton, L.A.; Swanson, L.V. Estradiol and progesterone in blood serum during the bovine estrous cycle. *J. Anim. Sci.* **1972**, *34*, 1020–1024. [CrossRef] [PubMed]
57. Kinder, J.E.; Garcia-Winder, M.; Imakawa, K.; Day, M.L.; Zalesky, D.D.; D'Occhio, M.J.; Kittok, R.J.; Schanbacher, B.D. Influence of different estrogen doses on concentrations of serum LH in acute and chronic ovariectomized cows. *J. Anim. Sci.* **1983**, *57* (Suppl. 1), 350.
58. Day, M.L.; Imakawa, K.; Garcia-Winder, M.; Kittok, R.J.; Schanbacher, B.D.; Kinder, J.E. Influence of prepubertal ovariectomy and estradiol replacement therapy on secretion of luteinizing hormone before and after pubertal age in heifers. *Dom. Anim. Endocrinol.* **1986**, *3*, 17–25. [CrossRef]
59. McKenzie, P.P.; Erickson, B.H. The effects of fungal-infected fescue on hormonal secretion and ovarian development in the beef heifer. *J. Anim. Sci.* **1989**, *67* (Suppl. 2), 58.
60. McKenzie, P.P.; Erickson, B.H. Effects of fungal-infested fescue on gonadotrophin secretion and folliculogenesis in beef heifers. *J. Anim. Sci.* **1991**, *69* (Suppl. 1), 387.
61. Burke, J.M.; Rorie, R.W. Changes in ovarian function in mature beef cows grazing endophyte infected tall fescue. *Theriogenology* **2002**, *57*, 1733–1742. [CrossRef]
62. Burke, J.M.; Spiers, D.E.; Kojima, F.N.; Perry, G.A.; Salfen, B.E.; Wood, S.L.; Patterson, D.J.; Smith, M.F.; Lucy, M.C.; Jackson, W.G.; et al. Interaction of endophyte-infected fescue and heat stress on ovarian function in the beef heifer. *Biol. Reprod.* **2001**, *65*, 260–268. [CrossRef]
63. Poole, R.K.; Devine, T.L.; Mayberry, K.J.; Eisemann, J.H.; Poore, M.H.; Long, N.M.; Poole, D.H. Impact of slick hair trait on physiological and reproductive performance in beef heifers consuming ergot alkaloids from endophyte-infected tall fescue. *J. Anim. Sci.* **2019**, *97*, 1456–1467. [CrossRef]
64. Hahn, G.L. Dynamic response of cattle to thermal heat loads. *J. Anim. Sci.* **1999**, *77*, 10–20. [CrossRef]
65. Murphy, M.G.; Enright, J.W.; Crowe, M.A.; McConnell, K.; Spicer, L.J.; Boland, M.P.; Roche, J.F. Effect of dietary intake on pattern of growth of dominant follicles during the oestrous cycle in beef heifers. *J. Reprod. Fertil.* **1991**, *92*, 333–338. [CrossRef] [PubMed]
66. Roth, Z.; Meidan, R.; Shaham-Albalancy, A.; Braw-Tal, R.; Wolfenson, D. Delayed effect of heat stress on steroid production in medium-sized and preovulatory bovine follicles. *Reproduction* **2001**, *121*, 745–751. [CrossRef] [PubMed]
67. Olson, T.A.; Lucena, C.; Chase, C.C.; Hammond, A.C. Evidence of a major gene influencing hair length and heat tolerance in *Bos taurus* cattle. *J. Anim. Sci.* **2003**, *81*, 80–90. [CrossRef] [PubMed]
68. Mariasegaram, M.; Chase, C.C., Jr.; Chaparro, J.X.; Olson, T.A.; Brenneman, R.A.; Niedz, R.P. The slick hair coat locus maps to chromosome 20 in Senepol derived cattle. *Anim. Genet.* **2007**, *38*, 54–59. [CrossRef] [PubMed]
69. Porto-Neto, L.R.; Bickhart, D.M.; Landaeta-Hernandez, A.J.; Utsunomiya, Y.T.; Pagan, M.; Jimenez, E.; Hansen, P.J.; Dikmen, S.; Schroeder, S.G.; Kim, E.; et al. Convergent Evolution of Slick Coat in Cattle through Truncation Mutations in the Prolactin Receptor. *Front. Genet.* **2018**, *9*, 57. [CrossRef] [PubMed]

70. Littlejohn, M.D.; Henty, K.M.; Tiplady, K.; Johnson, T.; Harland, C.; Lopdell, T.; Sherlock, R.G.; Li, W.; Lukefahr, S.D.; Shanks, B.C.; et al. Functionally reciprocal mutations of the prolactin signaling pathway define hairy and slick cattle. *Nat. Commun.* **2014**, *5*, 5861. [CrossRef] [PubMed]
71. Kaltenbach, C.C.; Graber, J.W.; Niswender, G.D.; Nalbandov, A.V. Luteotrophic properties for some pituitary hormones in nonpregnant or pregnant hypophysectomized ewes. *Endocrinology* **1968**, *82*, 818–824. [CrossRef] [PubMed]
72. Murdoch, W.J.; Dailey, R.A.; Inskeep, E.K. Preovulatory changes in prostaglandins E2 and F2α in ovine follicles. *J. Anim. Sci.* **1981**, *53*, 192–205. [CrossRef]
73. Farin, C.E.; Moeller, C.L.; Mayan, H.; Gamboni, F.; Sawyer, H.R.; Niswender, G.D. Effect of luteinizing hormone and human chorionic gonadotropin on cell populations in the ovine corpus luteum. *Biol. Reprod.* **1988**, *38*, 413–421. [CrossRef]
74. Reynolds, L.P.; Grazul-Bilska, A.T.; Redmer, D.A. Angiogenesis in the corpus luteum. *Endocrine* **2000**, *12*, 1–9. [CrossRef]
75. Stocco, C.; Telleria, C.; Gibori, G. The Molecular Control of Corpus Luteum Formation, Function, and Regression. *Endocr. Rev.* **2007**, *28*, 117–149. [CrossRef] [PubMed]
76. Estienne, M.J.; Schillo, K.K.; Fitzgerald, B.P.; Hielman, S.M.; Bradley, N.W.; Boling, J.A. Effects of endophyte-infected fescue on puberty onset and luteal activity in beef heifers. *J. Anim. Sci.* **1991**, *68* (Suppl. 1), 468.
77. Ahmed, N.M.; Schmidt, S.P.; Arbona, J.R.; Marple, D.N.; Bransby, D.I.; Carson, R.L.; Coleman, D.A.; Rahe, C.H. Corpus luteum function in heifers grazing endophyte-free and endophyte-infected Kentucky-31 tall fescue. *J. Anim. Sci.* **1990**, *68* (Suppl. 1), 468.
78. Fanning, M.D.; Spitzer, J.C.; Cross, D.L.; Thompson, F.N. A preliminary study of Growth, serum prolactin and reproductive performance of beef heifers grazing *Acremonium coenophialum*-infected tall fescue. *Theriogenology* **1992**, *38*, 375–384. [CrossRef]
79. Seals, R.C.; Schuenemann, G.M.; Lemaster, J.W.; Saxton, A.M.; Waller, J.C.; Schrick, F.N. Follicular dynamics in beef heifers consuming ergotamine tartrate as a model of endophyte-infected tall fescue consumption. *J. Anim. Vet. Adv.* **2005**, *4*, 97–102.
80. Pate, J.L.; Condon, W.A. Effects of serum and lipoproteins on steroidogenesis in cultured bovine luteal cells. *Mol. Cell. Endocrinol.* **1982**, *28*, 551–562. [CrossRef]
81. Pate, J.L.; Condon, W.A. Regulation of steroidogenesis and cholesterol synthesis by prostaglandin F-2 alpha and lipoproteins in bovine luteal cells. *J. Reprod. Fertil.* **1989**, *87*, 439–446. [CrossRef]
82. Rice, R.L.; Blodgett, D.J.; Schurig, G.G.; Swecker, W.S.; Fontenot, J.P.; Allen, V.G.; Akers, R.M. Evaluation of humoral immune response in cattle grazing endophyte-infected or endophyte-free fescue. *Vet. Immunol. Immunopathol.* **1997**, *59*, 285–291. [CrossRef]
83. Nihsen, M.E.; Piper, E.L.; West, C.P.; Crawford, R.J., Jr.; Denard, T.M.; Johnson, Z.B.; Roberts, C.A.; Spiers, D.A.; Rosenkrans, C.F., Jr. Growth rate and physiology of steers grazing tall fescue inoculated with novel endophytes. *J. Anim. Sci.* **2004**, *82*, 878–883. [CrossRef]
84. Mahmood, T.; Ott, R.S.; Foley, G.L.; Zinn, G.M.; Schaeffer, D.J.; Kesler, D.J. Growth and ovarian function of weanling and yearling beef heifers grazing endophyte-infected tall fescue pastures. *Theriogenology* **1994**, *42*, 1149–1158. [CrossRef]
85. Odde, K.G.; Ward, H.S.; Kiracofe, G.H.; McKee, R.M.; Kittok, R.J. Short estrous cycles and associated serum progesterone levels in beef cows. *Theriogenology* **1980**, *14*, 105–112. [CrossRef]
86. Jones, K.L.; King, S.S.; Griswold, K.E.; Cazac, D.; Cross, D.L. Domperidone can ameliorate deleterious reproductive effects and reduced weight gain associated with fescue toxicosis in heifers. *J. Anim. Sci.* **2003**, *81*, 2568–2574. [CrossRef] [PubMed]
87. Dyer, D.C. Evidence that ergovaline acts on serotonin receptors. *Life Sci.* **1993**, *53*, 223–228. [CrossRef]
88. Roberts, R.M.; Xie, S.; Mathialagan, N. Maternal recognition of pregnancy. *Biol. Reprod.* **1996**, *54*, 294–302. [CrossRef] [PubMed]
89. Klotz, J.L.; Britt, J.L.; Miller, M.F.; Snider, M.A.; Aiken, G.E.; Long, N.M.; Pratt, S.L.; Andrae, J.G.; Duckett, S.K. Ergot alkaloid exposure during gestation alters: II. Uterine and umbilical artery vasoactivity. *J. Anim. Sci.* **2019**, *97*, 1891–1902. [CrossRef] [PubMed]
90. Klotz, J.L.; McDowell, K.J. Tall fescue ergot alkaloids are vasoactive in equine vasculature. *J. Anim. Sci.* **2017**, *95*, 5151–5160. [CrossRef]

91. Saameli, K. Effects on the uterus. In *Ergot Alkaloids and Related Compounds. Handbook of Experimental Pharmacology*; Berde, B., Schild, H.O., Eds.; Springer: New York, NY, USA, 1978; Volume 49, pp. 233–319.
92. Wathers, D.C.; Lamming, G.E. The oxytocin receptor, luteolysis and the maintenance of pregnancy. *J. Reprod. Fertil.* **1995**, *49*, 53–67.
93. McCracken, J.A.; Custer, E.E.; Lamsa, J.C. Luteolysis: A neuroendocrine-mediated event. *Physiol. Rev.* **1999**, *79*, 263–323. [CrossRef]
94. Vogt Engeland, I.; Andresen, O.; Ropstad, E.; Kindahl, H.; Waldeland, H.; Daskin, A.; Olav Eik, L. Effect of fungal alkaloids on the development of pregnancy and endocrine foetal-placental function in the goat. *Anim. Reprod. Sci.* **1998**, *52*, 289–302. [CrossRef]
95. Paterson, J.; Forcherio, C.; Larson, B.; Samford, M.; Kerley, M. The effects of fescue toxicosis on beef-cattle productivity. *J. Anim. Sci.* **1995**, *73*, 889–898. [CrossRef]
96. Duckett, S.K.; Andrae, J.G.; Pratt, S.L. Exposure to ergot alkaloids during gestation reduces fetal growth in sheep. *Front. Chem.* **2014**, *2*, 68. [CrossRef] [PubMed]
97. Jones, K.L.; King, S.S. Effect of domperidone supplementation of fescue-fed heifers on plasma and follicular fluid fatty acid composition and oocyte quality. *J. Anim. Sci.* **2009**, *87*, 2227–2238. [CrossRef] [PubMed]
98. Hazeleger, N.L.; Hill, D.J.; Stubbing, R.B.; Walton, J.S. Relationship of morphology and follicular fluid environment of bovine oocytes to their developmental potential *in vitro*. *Theriogenology* **1995**, *43*, 509–522. [CrossRef]
99. Schuenemann, G.M.; Hockett, M.E.; Edwards, J.L.; Tohrbach, N.R.; Breuel, K.F.; Schrick, F.N. Embryo development and survival in beef cattle administered ergotamine tartrate to simulate fesecue toxicosis. *Reprod. Biol.* **2005**, *5*, 137–150.
100. Hansen, P.J. Effects of heat stress on mammalian reproduction. *Philos. Trans. R. Soc. Lond. B Biol. Sci.* **2009**, *364*, 3341–3350. [CrossRef]
101. Ealy, A.D.; Drost, M.; Hansen, P.J. Developmental changes in embryonic resistance to adverse effects of maternal heat stress in cows. *J. Dairy Sci.* **1993**, *76*, 2899–2905. [CrossRef]
102. Edwards, J.L.; Hansen, P.J. Differential responses of bovine oocytes and preimplantation embryos to heat shock. *Mol. Reprod. Dev.* **1997**, *46*, 138–145. [CrossRef]
103. Brendemuehl, J.P.; Boosinger, T.R.; Pugh, D.G.; Shelby, R.A. Influence of enophyte-infected tall fescue on cyclicity, pregnancy rate and early embryonic loss in the mare. *Theriogenology* **1994**, *42*, 489–500. [CrossRef]
104. Smith, S.R.; Schwer, L.; Keene, T.C. Tall Fescue Toxicity for Horses: Literature Review and Kentucky's Successful Pasture Evaluation Program. *Forage Grazinglands* **2009**. Available online: https://forages.ca.uky.edu/files/11_tall_fescue_toxicity_for_horses_lit_review_and_kys_pe_program.pdf (accessed on 2 May 2019). [CrossRef]
105. Putnam, M.R.; Bransby, D.I.; Schumacher, J.; Boosinger, T.R.; Bush, L.; Shelby, R.A.; Vaughan, J.T.; Ball, D.; Brendemuehl, J.P. Effects of the fungal endophyte *Acremonium coenophialum* in fescue on pregnant mares and foal viability. *Am. J. Vet. Res.* **1991**, *52*, 2071–2074.
106. Hovermale, J.T.; Craig, A.M. Correlation of ergovaline and lolitrem B levels in endophyte-infected perennial ryegrass (*Lolium perenne*). *J. Vet. Diagn. Investig.* **2001**, *13*, 323–327. [CrossRef] [PubMed]
107. Youngblood, R.C.; Filipov, N.M.; Rude, B.J.; Christiansen, D.L.; Hopper, R.M.; Gerard, P.D.; Hill, N.S.; Fitzgerald, B.P.; Ryan, P.L. Effects of short-term early gestational exposure to endophyte-infected tall fescue diets on plasma 3,4-dihydroxyphenyl acetic acid and fetal development in mares. *J. Anim. Sci.* **2004**, *82*, 2619–2929. [CrossRef] [PubMed]
108. Arns, M.J.; Pruitt, J.A.; Sharp, C.; Wood, C.; Northcutt, S. Influence of Endophyte-Infected Tall Fescue Seed Consumption on the Establishment and Maintenance of Pregnancy in Mares. *Prof. Anim. Sci.* **1997**, *13*, 118–123. [CrossRef]
109. Indänpään-Heikkilä, J.E.; Schoolar, J.C. LSD: Autoradiographic study of the placental transfer and tissue distribution in mice. *Science* **1969**, *164*, 1295–1297. [CrossRef] [PubMed]
110. Leist, K.H.; Grauwiler, J. Transplacental passage of 3H-ergotamine in the rat, and the determination of the intra-amniotic embryotoxicity of ergotamine. *Experientia* **1973**, *29*, 764.
111. Britt, J.L.; Greene, M.A.; Bridges, W.C.; Klotz, J.L.; Aiken, G.E.; Andrae, J.G.; Pratt, S.L.; Long, N.M.; Schrick, F.N.; Strickland, J.R.; et al. Ergot alkaloid exposure during gestation alters. I. Maternal characteristics and placental development of pregnant ewes. *J. Anim. Sci.* **2019**, *97*, 1874–1890. [CrossRef] [PubMed]

112. Ehrhardt, R.A.; Bell, A.W. Growth and metabolism of the ovine placenta during mid-gestation. *Placenta* **1995**, *16*, 727–741. [CrossRef]
113. Borowicz, P.P.; Arnold, D.R.; Johnson, M.L.; Grazul-Bilska, A.T.; Redmer, D.A.; Reynolds, L.P. Placental growth throughout the last two thirds of pregnancy in sheep: Vascular development and angiogenic factor expression. *Biol. Reprod.* **2007**, *76*, 259–267. [CrossRef]
114. Garrett, L.M.; Heimann, E.D.; Wilson, L.L.; Pfander, W.H. Reproductive problems of pregnant mares grazing fescue pastures. *J. Anim. Sci.* **1980**, *51* (Suppl. 1), 237.
115. Poppenga, R.H.; Mostrum, M.S.; Haschek, W.M.; Lock, T.F.; Buck, W.B.; Beasley, V.R. Mare agalactia, placental thickening, and high foal mortality associated with the grazing of tall fescue: A care report. In Proceedings of the Annual Meeting–American Association of Veterinary Laboratory Diagnosticians, Fort Worth, TX, USA, 21–23 October 1984; p. 325.
116. Monroe, J.L.; Cross, D.L.; Hudson, L.W.; Henricks, D.M.; Kennedy, S.W.; Bridges, W.C. Effect of selenium and endophyte-contaminated fescue on performance and reproduction in mares. *J. Equine Vet. Sci.* **1988**, *8*, 148. [CrossRef]
117. Beers, K.W.; Piper, E.L. Effect of grazing endophyte-infected fescue on heifer growth, calving rate and calf birth weight, of first calf heifers. *Ark. Farm Res.* **1987**, *36*, 7.
118. Watson, R.H.; McCann, M.A.; Parish, J.A.; Hoveland, C.S.; Thompson, F.N.; Bouton, J.H. Productivity of cow-calf pairs grazing tall fescue pastures infected with either the wild-type endophyte or a nonergot alkaloid-producing endophyte strain, AR542. *J. Anim. Sci.* **2004**, *82*, 3388–3393. [CrossRef] [PubMed]
119. Antonazzo, P.; Alvino, G.; Cozzi, V.; Grati, F.R.; Tabano, S.; Sirchia, S.; Miozzo, M.; Cetin, I. Placental IGF2 expression in normal and intrauterine growth restricted (IUGR) pregnancies. *Placenta* **2008**, *29*, 99–101. [CrossRef] [PubMed]
120. Fowden, A.L.; Sibley, C.; Reik, W.; Constanicia, M. Imprinted genes, placental development and fetal growth. *Horm. Res.* **2006**, *65* (Suppl. 3), 50–58. [CrossRef] [PubMed]
121. Akers, R.M.; Bauman, D.E.; Capuco, A.V.; Goodman, G.T.; Tucker, H.A. Prolactin regulation of milk secretion and biochemical differentiation of mammary epithelial cells in periparturient cows. *Endocrinology* **1981**, *109*, 23–30. [CrossRef] [PubMed]
122. Walner, B.M.; Booth, N.H.; Robbins, J.D.; Bacon, C.W.; Porter, J.K.; Kiser, T.E.; Wilson, R.; Johnson, B. Effect of an endophytic fungus isolated from toxic pasture grass on serum prolactin concentrations in the lactating cow. *Am. J. Vet. Res.* **1983**, *44*, 1317.
123. Baldwin, R.L., 6th; Capuco, A.V.; Evock-Clover, C.M.; Grossi, P.; Choudhary, R.K.; Vanzant, E.S.; Elsasser, T.H.; Bertoni, G.; Trevisi, E.; Aiken, G.E.; et al. Consumption of endophyte-infected fescue seed during the dry period does not decrease milk production in the following lactation. *J Dairy Sci.* **2016**, *99*, 7574–7589. [CrossRef]
124. Capuco, A.V.; Bickhart, D.; Li, C.; Evock-Clover, C.M.; Choudhary, R.K.; Grossi, P.; Bertoni, G.; Trevisi, E.; Aiken, G.E.; McLeod, K.R.; et al. Effect of consuming endophyte-infected fescue seed on transcript abundance in the mammary gland of lactating and dry cows, as assessed by RNA sequencing. *J Dairy Sci.* **2018**, *101*, 10478–10494. [CrossRef]
125. Forsyth, I.A. Variation among species in the endocrine control of mammary growth hormone and placental lactogen. *J. Dairy Sci.* **1986**, *69*, 886. [CrossRef]
126. Redmond, L.M.; Cross, D.L.; Kennedy, S.W. Effect of three levels of domperidone on gravid mares grazing endophyte (*Acremonium coenophialum*) infected tall fescue. *J. Anim. Sci.* **1993**, *71* (Suppl. 1), 16.
127. Altom, E.K.; Cross, D.L.; Roach, D.K.; Strickland, J.W.; Greene, E.M.; Clare, K.A.; Oliver, J.W. The effect of short duration domperidone therapy on gravid mares consuming endophyte infected fescue. *J. Anim. Sci.* **1995**, *73*, 20.

MDPI
St. Alban-Anlage 66
4052 Basel
Switzerland
Tel. +41 61 683 77 34
Fax +41 61 302 89 18
www.mdpi.com

Toxins Editorial Office
E-mail: toxins@mdpi.com
www.mdpi.com/journal/toxins

www.ingramcontent.com/pod-product-compliance
Lightning Source LLC
LaVergne TN
LVHW070142170726
843515LV00007B/1847